STUDY GUIDE

BIOLOGICAL SCIENCE

THIRD EDITION

SCOTT FREEMAN

EDITOR:

Warren Burggren
University of North Texas

CONTRIBUTORS:

Brian Bagatto
University of Akron

Jay Brewster
Pepperdine University

Laurel Hester
University of South Carolina

PEARSON

Benjamin
Cummings

San Francisco Boston New York
Cape Town Hong Kong London Madrid Mexico City
Montreal Munich Paris Singapore Sydney Tokyo Toronto

Editor-in-Chief: Beth Wilbur
Acquisitions Editor: Becky Ruden
Executive Director of Development: Deborah Gale
Editorial Project Manager: Carol Whitney, nSight, Inc.
Executive Marketing Manager: Lauren Harp
Managing Editor: Mike Early
Production Supervisor: Shannon Tozier
Production Service, Composition, and Illustrations: Pre-PressPMG
Text Design: Pre-PressPMG
Cover Design: Marilyn Perry and Kim Quilin
Cover Production: Seventeenth Street Studios
Manufacturing Buyer: Michael Penne
Text and Cover Printer: Courier, Stoughton

Cover Image: Octopus – he'e, Octopus sp. (juvenile): David Liittschwager/Susan Middleton

ISBN 10-digit 0-13-224984-7
13-digit 978-0-13-224984-3

Contents

Preface

Introductory textbooks are written to provide a foundation for successful learning. In addition to the textbook, students also make use of additional learning materials associated with their text, typically including CDs, websites, and study guides. In our experience, a study guide can be a wonderful asset to students, helping them assess their progress and move to the all-important stage of thinking critically about the material they are learning. Unfortunately, it is also our experience that study guides are sometimes created as a hurried afterthought—simply a repackaging of the text material along with a series of multiple-choice questions to determine if biological facts and principles have been learned.

We wanted to do something different—and something far better. So, while Scott Freeman was writing the first edition of *Biological Science*, Brian Bagatto, Jay Brewster, Laurel Hester, and I were discussing with the then Project Manager, Karen Horton, how to produce a more interactive, effective, and enjoyable study guide that would be an important component of the *Biological Science* support package. As this study guide emerged, we realized that it would be helpful to do more than just review the material found in the textbook. We also wanted to put together a guideline for how to study biology, how to write in biology, and how to understand why biological experiments underpin everything the student learns. Consequently, Part I of the book, unique among study guides, is intended to be a "survival guide" for your biology course. Part II subsequently delves into the actual material presented in *Biological Science*, but in a way that challenges students to think critically and analytically about the material they are reading in the text and hearing in their lectures.

Each chapter has a similar format to aid your learning, beginning with an outline of "Key Concepts" that summarizes the concepts discussed in the chapter. Woven through this outline are boxes entitled "Check Your Understanding", essentially a "self-test" where you can write down brief responses to questions to give you a sense of your own understanding of the material as you progress through the chapter. A section entitled "Integrating Your Knowledge" ties together concepts throughout the book, followed by a "Mastering Challenging Topics" section that walks you through the most challenging topics and concepts. Finally, a comprehensive section called "Assessing What You've Learned" tests your understanding of the material and helps you prepare for exams. Thus, each chapter provides a formula for successful learning, knowledge, and retention.

With this third edition of the Study Guide, we feel that we have fine-tuned the concept of the Study Guide as an important tool for mastering a course in biological sciences. Together with the third edition of Scott Freeman's *Biological Science* and the other support material available to the student and instructor, we hope this third edition of the Study Guide will help instill our own love of biology, and biological experimentation, into future generations of biology students.

Our special thanks go of course to our Project Managers, Mercedes Grandin, Deborah Gale, and Carol Whitney, without whose enthusiastic support this project would not have been possible. We also thank the team of editors and other specialists at Benjamin Cummings. Chris Thillen copyedited this Study Guide, bringing to bear not only her language skills and attention to detail but also her insight from working on the *Biological Science* text, instructional material, and all of the media. The team at PrePress PMG also expertly formatted the pages.

Finally, we are extremely grateful to all of those important people in our lives—family, friends, and students—who put up with our temporary neglect while we focused on the creation of this third edition of the Study Guide.

Warren Burggren
Denton, Texas
November 2007

PART **1**

Experimentation and Research in the Biological Sciences

Experimentation, or actively altering the course of an activity, process, or event to gauge the potential outcome of that alteration, seems to be as innate to humans as breathing or eating. We have always tinkered with the world around us. In the early days of *Homo sapiens*, or humans, the tinkering may have had immediate, practical purposes—does the wood from the tree with the heart-shaped leaves burn hotter and longer in a cooking fire than the wood from trees with needles? Does an arrowhead with carefully sculpted edges bring down a deer more reliably than a simple triangular arrowhead? Finding answers to these questions by trying different approaches and learning from the outcomes was often a matter of life or death, so it is hardly a wonder that the tendency for "experimentation" appears to have been naturally selected for in human genes.

Modern humans carry out experimentation in the highly organized format we call scientific research. Certainly, humans also continue to carry out some "life or death" experimentation and research, such as the ongoing search for an HIV vaccine, or better screening procedures for various types of cancers. Such research is called applied research because of the hope that there may be an immediate application of the results of such experiments. Much of the current experimentation in biomedical research is considered to be applied research. However, a great deal of experimentation and research is considered to be basic research (sometimes called curiosity-driven research) that provides basic scientific knowledge and a foundation for applied research.

Descriptions of both basic and applied research form a central, integrated theme in Scott Freeman's textbook, *Biological Science*. Rather than just asking you to memorize "facts," *Biological Science* offers important insights into the generation of the fascinating information you will be learning in this course, and it should leave you with a new appreciation for how scientific information is gathered.

Here in Part One, **Chapter 1**, we will consider the process of experimentation and research that leads to the creation of new data in an attempt to answer questions posed by scientists. In **Chapter 2** of Part One, we will consider how data are tabulated and graphed for the most efficient communication of the patterns they portray.

A. KEY CONCEPTS OF EXPERIMENTATION

Setting up and carrying out successful experiments depends on careful organization and thought. Let's discuss a few important concepts of experimentation:

- Hypothesis testing
- Independent and dependent variables
- Experimental controls
- Precision and accuracy
- Consistent treatment of experimental populations

1.1 Hypothesis Testing

In **Chapter 1** of *Biological Science*, Scott Freeman discusses the creation of a hypothesis. His example stems from a question about why giraffes have long necks (see p. 11–12 of your text). A hypothesis is a statement that is proposed to explain a certain set of events, which can be tested and proved or disproved. You might alternatively think of a hypothesis as a theory waiting to be put to the test. *Biological Science* poses the question, "Why do giraffes have long necks?" One hypothesis to explain this observation is: "The long necks of giraffes enable them to reach food that is beyond the reach of other mammals." In other words, when a scientist poses a hypothesis as a statement, he or she is initially "taking a position" when trying to answer the question. The scientist may have no idea what the answer to the question may be, but still suggests a possible explanation in the form of a hypothetical, positive statement.

Once a hypothesis is put forth, the next step is to design an experiment to test that hypothesis. The experimental design should be such that the experiment's results will either prove the hypothesis to be valid—the results support the proposed explanation—or the results will disprove, or falsify, the hypothesis. Proving a hypothesis is often a matter of gathering so much evidence in support of a hypothesis that any reasonable person would believe it to be proven. In reality, falsifying a hypothesis is often easier. The philosopher Karl Popper (1902–1994) explained the concept of falsifying hypotheses with an example about ravens. If researchers propose the hypothesis, "All ravens are black," then they might set out to collect as many black ravens as possible. After collecting hundreds of black ravens, they may believe the hypothesis to be true (that is, to be "proved"). However, the researchers can't be sure that there isn't a white raven out there somewhere! In contrast, if they set out to find a *white* raven and succeed, then the hypothesis that "all ravens are black" is completely and efficiently falsified. The researchers might then suggest the new hypothesis that "All ravens are black except for a very, very small population of white ravens."

Falsifying a hypothesis leads to the creation of a new and better hypothesis that takes new results and observations into account. When a hypothesis has undergone numerous modifications and can no longer be falsified, then scientists may permit themselves to refer to the hypothetical statement as a fact. When a fact or body of facts has stood the test of time, it may become a law.

Your Turn

In *Biological Science*, Scott Freeman describes many interesting experiments performed by biologists. Turn to page 218 in Chapter 11 (How Do Genes Work?). Here, Freman describes an experiment in the early 1940s by Geadle and Tatum, who investigated the source of hereditary material. Can you figure out what the original hypothesis proposed by the authors might have been? Try doing the same for the experiments by Hardy and Weinberg on allele frequencies that Freeman describes on pages 504 of Chapter 25 in your textbook.

1.2 Independent and Dependent Variables

Virtually all experiments, as well as the tables and graphs that present the information these experiments have produced, possess independent and dependent variables. It is crucial that you understand the difference between them. The independent variable is typically the variable that can be controlled, or fixed, by the scientist. For example, consider a plant biologist who is interested in the adaptations of a certain species of cactus to the dry conditions of the Mojave Desert. (You should pay particular attention to this example, because we will use it repeatedly in this and the next few chapters.) The biologist wonders whether the amount of water available to cactus seedlings affects their growth rate. To test this, she might take cactus seeds from native plants and grow cactus seedlings in a greenhouse (**Figure 1.1**).

Here are the experimental conditions: All the seedlings are exposed to a constant 18 hours of light. The amount of water that each pot receives is controlled as part of the experiment. Different pots of cactus seedlings are given different amounts of water, and the amount of growth stimulated by each watering regime is carefully measured and recorded each week. The amount of water given to the seedlings is the independent variable, because the amount of water the seedlings receive is fixed and independent of any physiological response the seedlings might have to the water. Their growth in response to watering is the dependent variable because the amount they grow more or less *depends* on the water they receive during the experiment.

The design of an experiment can be simple, as in this case involving cactus seedlings, or it can be very complex (for example, simultaneously varying water, nutrients, light, and genetic stock of the cactus and tracking any effects on seedling growth—an experiment with four independent variables). Nevertheless, in every experiment, you should be able to identify a dependent variable (the variable that unpredictably changes) and one or more independent variables (controlled as part of the experimental design). Once you have sorted out the meaning of dependent and independent variables, it will be

FIGURE 1.1   In this experiment carried out by a plant biologist, five groups of cactus seedlings are grown in a greenhouse and provided with different amounts of water each week.

easier to understand how tables and graphs present their information.

Find a few descriptions of experiments in *Biological Science*, and see if you can identify the independent and dependent variables in each.

1.3 Experimental Controls

Every student taking this course has carried out an experiment—both formally and informally. Your formal experience might include experiments in other college or high school science courses. Informally, think of the experiments you carry out every day. For example, have you ever tried to study less and still get the same grade, or tried to study more to get a better grade? Have you ever fiddled with a recipe to see if you can improve it? Have you ever checked to see whether you get better gasoline mileage with one brand of gas over another? These examples of real-life "tinkering" are actually experiments producing data.

While we all may be experimenters in our own right, what is often lacking in our informal experiments is a built-in control. A control is the portion of the experiment in which all possible variables are held constant, contrasting the portion of the experiment in which the variables are altered to observe the resulting effects. The presence of a control ensures that any effects observed in the experimental portion of the experiment can be attributed to the altered variable. For example, you can change a recipe for cookies by adding more chocolate chips, but you don't know if the cookies taste better due to the extra chips or because you are especially hungry when you taste them. What this experiment needs is a *control*, which would involve baking one batch of cookies using the normal recipe side by side with another batch using more chocolate chips and then tasting the two batches at the same time.

Many otherwise good experiments can suffer from a lack of adequate controls. Controls help a scientist draw valid conclusions from the observed results of an experiment, which is why trained scientists are scrupulously careful to create and include controls with every experiment. For example, a team of researchers trying to test a new drug they are developing will test the drug on one experimental population of animals or patients while simultaneously giving a second, equivalent population a "placebo," or inactive drug treatment, as the control. Any effects subsequently observed in the experimental population can then be attributed to the drug being tested.

To help develop the concept of controls a little further, let's return to our example about cactus seedlings. The average amount of water that falls in the seedlings' natural habitat during the period when cactus seeds might sprout is about 5–10 millimeters per week (mm/week). This quantity translates into about 20 milliliters (ml) of water per week, allowing the experimenter to give some of the seedlings the same amount of water they would normally receive in the wild. These plants constitute the control population for this experiment.

The experimental populations are those that receive more or less water than is given to the controls. (By the way, the term *population* is frequently used to describe a group of organisms, usually of the same species.) By setting up the experiment in this way, the plant biologist can be sure to take any extraneous or unexpected factors into account. For example, perhaps the amount of nutrients or light that all of the plants receive in the greenhouse is different than the amount they would normally receive in the wild. The biologist is still able to test the single effect of different amounts of watering by comparing all data to the control population. Returning to our example of changing a cookie recipe, this is similar to baking two batches of cookies—one the control batch and one the experimental batch.

Even if you are ravenous when you taste the two different populations of cookies, you still can make a comparison between the two recipes.

All of the experiments that Scott Freeman mentions in *Biological Science* involve careful use of controls, enabling them to survive scientific scrutiny to become landmark studies.

Your Turn

Turn to a chapter in *Biological Science* that you have already read. Find two experiments introduced by the author to help you understand the material. What do you think would constitute an appropriate control in each of the experiments you located?

1.4 Precision and Accuracy

We often use the terms *precision* and *accuracy* fairly interchangeably; yet in the context of scientific experiments, they mean two quite different things. Let's consider a vegetable scale at the grocery store. Is it accurate, precise, or both? Let's say you place a bunch of broccoli on the scale and observe that it weighs 2.0 pounds. You then weigh the same bunch of broccoli five more times, and it weighs 2.0 pounds each time. In this case, we can say that the scale is "precise" because it repeatedly gives the same weight for the same object. But let's also assume that the broccoli *actually* weighs 1.7 pounds. In this case, the scale—while precise because it repeatedly gives the same weight of 2.0 pounds—is not accurate. It would be accurate only if it indicated that the broccoli weighed 1.7 pounds, and it would be accurate *and* precise only if it repeatedly gave a weight of 1.7 pounds **(Figure 1.2)**.

FIGURE 1.2 A measuring device such as a grocery scale can be (1) neither precise nor accurate, (2) accurate but not precise, (3) precise but not accurate, or (4) both precise and accurate.

Scientists carrying out experiments must be both precise and accurate with their measurements. For this reason, experiments involving measurement with instruments (and most do) also involve the process of calibration. This means that an instrument is adjusted until it reliably produces the accurate measurement of a known standard. For example, rather than simply trusting the reading on a thermometer, a scientist will compare the temperature it reads against a known temperature (or a previously calibrated thermometer). Simple mercury thermometers can't be adjusted, but electronic ones can be adjusted to read the correct temperature. Ultimately, the accuracy of data generated depends on how carefully the scientist has calibrated the measurement devices used to collect data. The data on cactus seedling growth, for example, will be useful only if the ruler used for measuring plant height is accurate, the beaker used for measuring water is accurate, and the measurements and water amounts can be carried out precisely.

1.5 Consistent Treatment of Experimental Populations

When performing experiments, it is important to ensure that all experiments are carried out in exactly the same way every time. Otherwise, results may be due to experimental error and may not accurately reflect the results of the experiment. For example, in the case of the cactus seedlings, if some plants are measured in the morning and some in the late afternoon, wilting in the late afternoon sun might cause some plants to measure smaller than they actually are. If some plants are watered on Mondays and some on Tuesdays, some measurements taken Monday afternoon would be made on recently watered, non-wilted plants, while some plants measured will be dehydrated and wilted. Any height differences will be due to the day on which the measurements were taken rather than to how much water they received weekly.

Your Turn

Let's imagine that you are carrying out an experiment to see if studying for an extra night each week will improve your grade point average. For the first two weeks, you study on Tuesday night in the quiet library. But in the following two weeks, you study on Friday night in your dorm room with the sounds of loud music and parties all around you. When you tally the scores from all of your quizzes, exams, and assignments for the month, you find that your GPA remains unchanged. Is there something wrong with your experimental design, or does studying just not matter?

B. SETTING UP AN EXPERIMENT

Scientists may spend weeks, months, or even years planning an experiment, and they may face limitations due to lack of manpower, funding, or equipment. Some experiments are relatively simple to set up and require little maintenance. Our plant biologist's experiment with cactus seedlings, for example, is fairly straightforward, requiring only that the biologist obtain the seedlings, plant them, water them regularly according to the experimental design, and record their growth. In contrast, research into the effects of a new drug on the heart rate of animals and humans may require the combined efforts of numerous graduate students, postdoctoral fellows, research scientists, and clinical staff, as well as large amounts of expensive equipment—and make take several years to complete. Yet, despite great differences in effort and expense to carry out research experiments, the processes of generating a hypothesis, testing the hypothesis by performing experiments, and analyzing the data are common to all good scientific experimentation.

Your Turn

Try setting up your own hypothetical experiment using this scenario: Major Oil Company needs to build roads through some wilderness areas in its search for fresh oil fields. These roads will cross some streams that are important migratory routes for spawning salmon as they swim upstream. Thus, Major Oil needs to install culverts under the new roads as they are built to allow the water to flow under them. The project's chief engineer wants to install culverts with a 24-inch diameter, which will produce a water velocity of about 25 centimeters per second (cm/sec). The biological consultant hired by Major Oil fears that the fish in the stream won't be able to swim upstream against such a strong current and thinks they should use more expensive culverts with a 36-inch diameter, producing a water velocity of only 10 cm/sec. Can you design an experiment that would resolve this conflict? Construct a formal hypothesis. Think of the materials and supplies you would need, assuming that all costs will be covered by Major Oil. What would be the experimental procedure? What type of data would you collect? How would you present it to the engineer to make your case?

C. HOW MUCH DATA SHOULD BE COLLECTED?

You are probably starting to realize that biological research can be an elaborate and time-consuming enterprise. Poorly planned experiments can result in either an insufficient amount of data for accurate analysis and robust conclusions or the collection of more data than necessary, which is a waste of both time and money.

Your Turn

Let's return to the experiment with cactus seedlings. How many seedlings do you think the plant biologist should use? Should she set up just five flower pots with one seed each, giving each pot a different amount of water? This would be the easiest design, as well as simplest experiment to care for. Should she set up 20 pots? How about 100 or 500 pots? What criteria should she use to decide the size of her experiment?

All scientific data—especially biological data—have inherent variation. The most important sources of this variation are

- **Variation from inherent differences**—natural variation in the samples being examined. For example, some cactus seedlings will grow more rapidly than others simply because of their genetic disposition.
- **Variation from measurement error**—every scientific measurement has some inherent error. Different people measuring one seedling's height will produce different values, and the same person measuring one seedling's height at different times will produce different values. (Remember our earlier discussion of precision and accuracy.)

Your Turn

Carry out a simple experiment to illustrate measurement error. Using a ruler, tape measure, or even the tiles on a floor, measure the length of your normal walking stride. Now estimate the length of a corridor, the width of a small field, or some other distance by counting the number of paces you take as you walk from one end to the other. Do this five times. Ask five of your friends, relatives, or roommates each to estimate the length in question using the same method. Compare the results of your measurements. What would be the most *precise* way to measure this distance? What would be the most *accurate* way to measure this distance? If you owned a land surveying company and your reputation depended on accurate measurements, would you have a team of surveyors carry out the measurements, or would you ask one surveyor to repeat his or her measurements more than once?

Biologists typically spend a lot of time thinking about exactly how many replications—repetitions of an experiment—to carry out. Scientists want to ensure that the measurements they make are as accurate as possible, and one way to ensure accuracy is to take the average of several replicates. Of course, some experiments are relatively easy to replicate; for example, repeatedly measuring the height of a cactus seedling in a flower pot or growing several cactus seedlings rather than one. Consider, however, measuring the compression strength of the bite of a great white shark—and trying to get replicate measurements of the *same* great white shark, as well as several others in the same size range!

A variety of factors affect the design of an experiment with replicate measurements. Assuming that a researcher can manage the practical limitations of time and resources, the most important factor to consider when designing an experiment is the statistical reliability of the data produced. Statistics is a vibrant sub-branch of mathematics, and to discuss the numerous statistical methods used to test the validity of biological research would be beyond the scope of this study guide. However, two simple and useful conventions to learn are the mean and the standard deviation. The mean is the average of all data points collected, and the standard deviation of the mean describes the variation in the data about the calculated mean value in statistical terms.[1] A large variation about the mean (that is, a large standard deviation) indicates the possibility that the data is inconclusive.

To ensure that statistically sound conclusions can be made about our example experiment, the plant biologist might decide to use one seedling in each of five flower pots at every watering amount, for a total of 25 flower pots (**Figure 1.1.**)

This experimental protocol, or plan for the experiment, will allow the biologist to calculate for the five flower pots mean values and standard deviations that represent a single watering amount. Without going into statistical detail on the methods used for calculation, taking five accurate measurements (with replicate measurements taken on each seedling) often proves sufficient to determine statistical plausibility.

Your Turn

What do you think are the inherent strengths and weaknesses of using five seedlings to calculate the mean value? (*Hint:* Murphy's Law also applies to biological research—whatever might go wrong, often does.)

In our example experiment (but not in an experiment involving bite strength of great white shark!), it is a relatively small investment of time and money to compensate for error by growing additional seedlings. Thus the biologist may elect to grow 10 seedlings at each watering level to ensure against the unexpected death of a few seedlings or other possible disasters, such as a dropped flower pot.

D. ANALYZING AND INTERPRETING DATA

Analyzing data is a crucial step in the experimentation process. It's not enough to simply produce and record data; a scientist must then attempt to make sense of the gathered information to decide whether to accept or reject the hypothesis. You have already undergone similar analytic procedures in your daily life. For example, you may have had a feeling that you are spending more money than you have—a thought that is really an informal hypothesis in disguise. You may then balance your checkbook in an attempt to learn how much money you have, but you shouldn't stop there—you should also compare your monthly balance with your monthly income to see whether you are saving money or falling deeper into debt.

Once a scientific researcher has acquired enough data, he or she will begin to use the individual "trees" to see the overall "forest"; that is, to use the data to recognize broader patterns. Typically, this will be done either by putting the data in tables or by plotting it in graphs. In fact, tables and graphs are such an important part of the data analysis and presentation process that **Chapter 2** in Part One of this study guide is devoted entirely to the topic of scientific data presentation. In a curious way, analyzing scientific data is like solving a word puzzle in which the letters of words are scrambled and must be rearranged to form pieces of words before whole words begin to take shape. Scientific data alone are like the scrambled letters—a long and tedious process of interpreting the data, comparing it with previous experiments on similar topics, and fitting the outcome into the hypothesis formulated at the beginning of the experiments is required before the scrambled letters begin to resemble words. The truly great scientists of our era are able to put their data in the broadest possible context, so not only do their experiments allow them to accept or reject their own hypotheses, but their conclusions lead to advancements in the field of biology as a whole through insights gained from their experiments.

---

[1]Biologists use several conventions to describe variation—standard deviation, standard error, confidence intervals, and so on. Calculating each of these can be complex, so a rule of thumb is useful. Generally, if the standard deviations about two means substantially overlap each other, then the two means may not be statistically different.

E.   PUBLISHING DATA

The final step in experimentation is publishing the data and conclusions. Scientists receive financial and intellectual support for their work not just to indulge their own curiosity, but to encourage the results of their experiments to be distributed throughout the scientific community. Most journals subject submitted articles to rigorous peer review by other scientists in the same area of research. A manuscript is published only after intense scrutiny and often several rounds of revisions.

Scientific data has many outlets, most of them in the form of scientific journals. There are thousands of scientific journals, with titles and topics as broad as *Science* or as focused as *Pediatric Developmental Pathology*. Your institution's library probably has a large section devoted to scientific journals of all kinds. Of course, as electronic publishing on the Web has become commonplace in the last several years, most journals now publish in both traditional paper format and on the Web for subscribers.

Your Turn

Go to your library's journal section and look up a biological research journal such as *Journal of Experimental Biology* or *Physiological and Biochemical Zoology*. Find an article about experimentation with "zebra fish." Now, use your Web browser to find that same journal on the WWW, and again try to find an article involving zebra fish. Which format do you find most convenient? Which is easier for finding articles? Which format is better for viewing figures (tables, graphs, etc.)? What do you think will be the future of the electronic versus paper publishing of scientific information? Ask your biology professor to comment about his or her preference for scientific information.

From the inception of a question posed as a hypothesis through the gathering of resources, the completion of experiments, the analysis and interpretation of data, and finally the publication of results, the overall scientific endeavor can often take years and many thousands of dollars. The next time you are in your library, walk down the aisles of scientific journals and think of the enormous investment of time and resources that has produced our scientific knowledge base. Appreciate the fact that every page of Freeman's *Biological Science* is based on information gathered through careful scientific experimentation.

Your Turn

Canadian science commentator David Suzuki once stood outside a grocery store in the dead of the Canadian winter, asking shoppers what role science played in their lives as they left the store with bags containing mangoes, pineapples, bananas, and other exotic tropical fruits. Most replied, "None!" Let's consider the same question from a different angle. You have probably recently applied some sort of chemical to your body in the form of deodorant, cosmetics, soap, shampoo, and so on. Pick one, and try to imagine the scientific experimentation that was involved in the creation, production, and packaging of that product. What do you think of science's role in your own life?

All science, including biological science, depends on the clear, concise presentation of data. Let's turn to that topic in the next chapter.

2

Presenting Biological Data

*In **Chapter 1** of Part One, we considered how experiments are set up, carried out, and to some extent how they are analyzed. In this chapter, we will examine ways in which the information produced from such experimental procedures is presented and interpreted.*

A. INTRODUCTION TO DATA

Scientists thrive on data, the information gathered from scientific observation or experimentation. They think about it all the time—what type of data they need, how much data they need, how they can get it, and what it might indicate. Whether they are taking the temperature of the lava from a smoldering volcano or measuring the rate of oxygen consumption in a growing zebra-fish embryo, all scientists rely on data.

Data are the currency of science, and learning to interpret the many ways it can be presented will be an important part of your scientific training. But do not think that only scientists deal with data, or that its interpretation is a completely new skill, because you have been handling data in many everyday situations for years! Try to imagine the data you encounter regularly; you might think of the money in your checking account, or your credit card balance, or your grade point average. The data on the monthly statement from your credit card company, for example, tracks the date and amount of your purchases. Interpreting this data involves determining if you are over your limit and budgeting your future credit card purchases to reflect your income.

As our world grows more technologically advanced, it is becoming increasingly important to understand complex forms of data and to apply the knowledge gained from these data. Drawing accurate conclusions from various forms of data presentation will do more than help you correctly answer questions on your biology tests; the critical thinking and analysis skills you develop will help you excel in other aspects of your education as well as throughout your life.

This chapter introduces the typical ways in which biologists use, present, and interpret data—methods you are also likely to encounter in your classroom and laboratory. Scientists generally present their data in one of several forms:

- Writing
- Tables and graphs
- Simulations and animations

Let's discuss each form in greater detail.

B. PRESENTING DATA IN WRITING

You're probably familiar with the saying, "The pen is mightier than the sword," meaning that more can be achieved by effective written communication than by legions of armies. For thousands of years, learned people have recorded information (we might call it data) about their observations of the natural world. For example, Aristotle recorded his observations of the beating heart of a chick embryo, which he exposed by carefully making a small window in the shell of a chicken egg. Socrates left similar written observations of the chick embryo heart, as did Galileo. Since that time, other writings have left a rich legacy of knowledge about the anatomical and physiological development of the chick embryo. These contributions could not be understood without the clear, concise writing that describes the observations—the data—of each writer. Today, writing remains a mainstay of scientific communication. In *Biological Science*, Scott Freeman writes about the great wealth of biological information. While your textbook incorporates a multitude of tables and graphs to support the text, it is apparent that the written word is the primary tool for scientific communication.

You have already used writing to convey scientific information—think of the lab reports, essays, and book reports you have produced so far in your academic career. While you may already understand the components of good scientific writing, **Chapter 4** in Part One of this study guide, "Reading and Writing to Understand Biology," explores these components in considerable detail. It is important to remember that the ability to write clearly is not limited to those who want to become scientists. Those of you studying for careers in business, education, or virtually any other discipline will be required to communicate your ideas and data to others, and your study of the presentation of scientific information will help you successfully communicate all types of data.

C. PRESENTING DATA IN TABLES AND GRAPHS

In the previous section, you recalled past lab reports from science courses such as biology, chemistry, and physics. Did you ever use tables or figures to help explain the data you produced? While the pen may be mightier than the sword, you are probably also familiar with the phrase, "a picture is worth a thousand words." A table or a graph is a picture of scientific data. Though you can describe your favorite summer vacation to your friends by talking to them, you may want to take out your photo album as well, using the pictures to supplement your stories. The combination of the pictures and your verbal description provides the most complete communication of your ideas. In the same way, a scientist explaining his or her experiments or an author describing biological principles will almost always resort to a combination of words and visual tools, such as tables and graphs.

Tables and graphs are similar in that they are visual, highly structured ways of looking at data, often displaying independent and dependent variables as discussed in **Chapter 1** of Part One.

2.1 Types of Tables

Tables are such a convenient way of communicating data that they can be found all around us. As you read this book, you may have a beverage or a snack close by. Take a minute to look at the table on the packaging. It is probably a relatively simple table showing nutrition information such as caloric content and the amount of fat, cholesterol, sodium, and so on found in the product. While the packaging could describe this information with words, a simple table provides the same information more efficiently. To consider various table formats, let's return to the example of cactus seedlings introduced in **Chapter 1** of this part. We could describe the experiment in paragraph form, but consider the following table:

Cactus Group (5 pots/group)	Amount of Water Given to Seedlings (ml/week)	Average Seedling Growth (mm/week)
Group 1	10	3
Group 2	20	7
Group 3	30	13
Group 4	40	10
Group 5	50	5

A substantial amount of information about the experiment can be gathered from this table, such as:

- There are five cactus groups represented in the experiment.
- There are five pots in each group.
- A standard watering procedure is shown, in which the most-watered group of seedlings received five times more water than the least-watered group.
- The growth column indicates that all of the seedlings grew to some extent.
- Seedlings receiving a certain amount of water grew the most; seedlings receiving more or less water than this amount did not grow as well.

Tables can be relatively straightforward, as in this example, or relatively complex, with many columns and many rows. Regardless of layout, however, all tables will contain independent and dependent variables.

2.2 Types of Graphs

Although tables can be extraordinarily useful to organize data, graphs are also used extensively to communicate this information. If you pick up a newspaper (especially *USA Today*, which uses more graphics than many other national publications), you will probably find several examples of graphs (look in the "Business" section, where you may find the recent performance of the stock market presented both in a table and a graph).

There are many different types of graphs, each with advantages and disadvantages that suit it for a particular kind of data. Most graphs have another feature in common besides dependent and independent variables. They all have an x-axis, also called an abscissa, and a y-axis, or ordinate. The x-axis runs horizontally along the bottom of the graph, while the y-axis runs vertically (**Figure 2.1**). Typically, the x-axis is used to plot the independent variable (for example, watering

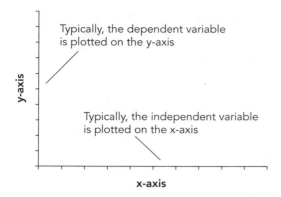

FIGURE 2.1 Graphs typically have an x-axis, showing the independent variable controlled by the researcher, and a y-axis, showing how dependent variable values change as the independent variable is altered.

amounts for our seedlings), while the dependent variable (seedling growth) is plotted on the y-axis.

Let's identify the specific types of graphs you will encounter in *Biological Science*. We'll continue to use the example of cactus seedlings and growth, because you are now familiar with these data in tabular form.

Scatter Plots

A scatter plot is aptly named; each data point is individually represented on the graph (in contrast to graphs in which only the averages of several data points are recorded) (**Figure 2.2**). It may appear that data are simply scattered on the graph. The scatter plot in Figure 2.3 presents the growth data from every pot in the cactus seedling experiment.

Looking at the scatter plot again, you should be able to see that, at each amount of water, there is variation between individual pots. Further, notice that the amount of variation differs for each amount of water, with peak growth when seedlings are given 30 ml of water per week. Scatter plots are very useful in that variation and trends in the data can be examined simultaneously. Sometimes, however, scatter plots are awkward because they can contain too much data, which makes the detection of patterns in the data more difficult. For this reason, scientists sometimes choose to communicate their data using line graphs.

Line Graphs

A line graph does not show individual data points, but instead shows the averages of several data points, connected by one or more lines that best fit the data points. In other

FIGURE 2.2 A scatterplot is so named because all individual data points are plotted, and may therefore appear to be "scattered" on the graph.

FIGURE 2.3 A line graph displays either individual or mean values of a dependent variable against the independent variables, and, as the name implies, joins these points with a line to help reveal overall trends.

words, line graphs appear to "connect the dots" of a scatter plot. The line graph in **Figure 2.3** displays the average seedling growth for each group at each watering.

This line graph clearly shows the pattern that is only hinted at by the scatter plot; namely, that seedlings do poorly with little water, but they are also adversely affected by too much water. The graph indicates that an intermediate watering amount, slightly more than the seedlings would receive in their natural habitat, is just right for optimal growth.

Your Turn

Look at some graphs in a printed version of a newspaper or on a news website. Which graph type is more common, scatter plots or line graphs? Look at examples of each—which format works better to communicate the data being presented?

While line graphs communicate trends very clearly, they often fail to provide any sense of variation in the data. This problem can be minimized by plotting one of several kinds of statistical variables to describe variation in the data. The most common of these variables is the standard deviation of the mean, which is defined in **Chapter 1** in Part One of this study guide. **Figure 2.4** is a replicate of the first line graph we looked at (**Figure 2.3**) with the standard deviation now added. The vertical line drawn through each point represents the amount of variation in the independent variable at each dependent variable value.

Bar Graphs

Scientists often illustrate data with a bar graph in which the height of the bar represents the value of the dependent

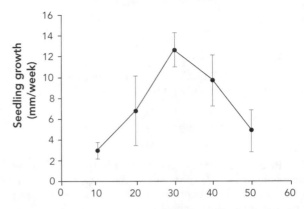

Line Graph of Seedling Growth Data, with Standard Deviations of the Mean

FIGURE 2.4 Line graphs may also show how the data varies about the plotted mean values, as in this line graph showing the standard deviation of the mean for each mean seedling growth value.

Your Turn

Go back to your newspaper or news website. Find a simple line graph with no standard deviations. Without seeing any indication of the variability in the data, can you tell if the trends depicted are "trustworthy?"

variable. Bar graphs can be plotted either horizontally or vertically, and there isn't much difference between the two—aesthetics and a sense for which communicates the data more effectively are often key factors when deciding whether to orient a bar graph horizontally or vertically. The two bar graphs presented plot the cactus seedling data in each form (**Figure 2.5**).

Bar Graph of Seedling Growth Data

FIGURE 2.5 Bar graphs are often an effective means of showing the relationship between dependent and independent variables. In this figure, the same data set is plotted as a vertical bar graph (upper) and a horizontal bar graph (lower). Which do you think more effectively presents this data?

Vertical Bar Graph of Seedling Growth Data, with Standard Deviations

FIGURE 2.6 As with line graphs, bar graphs can also show variation in the data by displaying standard deviations of the men.

Just as line graphs can show variation in data, bar graphs can also be designed to show standard deviations (**Figure 2.6**).

Bar graphs are often most effective when showing the results of an experiment that compares several different variables, or when comparing several different experiments. On a line graph, separate lines often fall on top of one another, which can make data analysis more difficult. A bar graph, however, displays data by using different bars for each dependent variable value.

Let's consider a possible follow-up cactus seedling experiment for our biologist. In this experiment, not only does the amount of water differ for each batch of seedlings, but the amount of light given to the seedlings varies. One complete set of pots (5 pots in each of 5 groups) is given 16 hours of light per day, while another set is given 18 hours of light per day. The biologist concludes that the amount of light doesn't make much difference, but she still wants to communicate these findings graphically.

Your Turn

Has the biologist added another dependent variable or another independent variable?

Now let's combine these data using both a line graph and a bar graph to highlight the differences between the two graph types (**Figure 2.7**).

Which do you think is more effective?

Pie Charts

Another common type of graph is a pie graph, or pie chart. As the name implies, a pie chart resembles a pie divided into more than one slice. A pie chart is ideally suited for presenting data expressed in the form of percentages of fractions of a whole. The size of each slice, when appropriately labeled, indicates the relative amount of the graphed variable. For example, a pie chart can effectively present data about the elemental composition of seawater (with each element shown as a percentage) or the fraction of each phylum represented in the total number of species in an ecosystem.

Your Turn

Can you think of a good way to present the cactus seedling data using a pie chart?

Seedling Growth Data from Two Experiments, Plotted as Line and Bar Graphs

FIGURE 2.7 When the results of more than one experiment are plotted on a line graph, the data can be difficult to interpret. Sometimes a bar graph (lower), with separate bars representing every dependent variable value, presents the results more clearly than a line graph (upper).

2.3 Putting Your Knowledge to Work

Now that you've learned the common types of graphs, let's assess your understanding by asking you to try a little graphing of your own. For this exercise, you will need your copy of *Biological Science* and some graph paper (or a computer program that allows graphing).

1. Try to find an example of each type of graph in *Biological Science*. For each type, identify the x- and y-axis (except for a pie chart) and the dependent and independent variables.

2. Pick five chapters from *Biological Science* at random. For each chapter, record the total number of graphs, regardless of graph type. Let's consider the chapter number to be the independent variable and the number of graphs per chapter to be the dependent variable. Now, create these graphs:
   - Scatter plot, recording each individual data point
   - Line plot, using the average number of graphs for each chapter
   - Two bar graphs, one oriented vertically, the other horizontally

   Which graph do you think is the most effective way to show the data you collected?

3. Refer to the five chapters from *Biological Science* that you used to make the graphs in step 2. Create a worksheet so that you can keep track of the number of each type of graph (line graph, bar graph, etc.) as you flip through these chapters. Using a compass (or a coffee cup or another similar object), draw a circle on a piece of graph paper. Create a pie chart showing the fractions of the total number of graphs represented by each type.

2.4 Presenting Data with Simulations and Animations

Some of your favorite sites on the World Wide Web probably contain animations, which use iconic symbols or cartoon drawings that move in a lifelike way. Animations are becoming increasingly important in the communication of scientific data. They are especially effective when describing ongoing processes occurring over a period of time; for example, an animated graph that reveals its data in a series of stages rather than all at once.

Your Turn

Your Student CD-ROM that accompanies *Biological Science* includes many animations. For example, look at Activity 52.2 on your CD-ROM, "Habitat Fragmentation." What are the advantages of using animation to communicate this information?

Related to animations are simulations, often in the form of animated graphs or tables, that allow the recreation of a real-life situation. Rather than passively watching animation unfold, you can manipulate data (usually independent variables) in a model to see how the outcome (the dependent variable) changes in response to different data sets. An increasingly important tool for data communication and interpretation, simulations are used in everything from predicting how a forest fire might spread to anticipating a drug's effects on a patient's blood pressure.

Your Turn

Can you think of any times you have used a simulation in your personal life—that is, fed data into a model to see what the outcome would be? You might recall an example involving your personal finances, or perhaps you've calculated the effects of several potential test grades on your grade point average.

As you can see, there are many ways to effectively present scientific information, each with different characteristics that may suit it for specific types of data. Experiments may yield unexpected results, however, and careful interpretation is often necessary before data can be understood. In the next chapter we will explore the challenges of data interpretation, and you'll find techniques to help you understand scientific information and excel in your study of biology.

Understanding Patterns in Biology and Improving Study Techniques

Before you can begin a comprehensive study of life, it is necessary to learn background details about the physiology of life, from large organisms to microbes that cannot be seen without the help of a microscope. This is an immense amount of information, and textbook authors are faced with the challenge of organizing it. How can details about everything from molecules and viruses to ecosystems and evolution be compiled and presented so that they will be understood by even the most inexperienced student? Fortunately, biology is not just a haphazard accumulation of facts. Life is highly ordered and involves many ongoing processes that result in recognizable patterns—some simple, others more complex—that are easily presented in textbooks.

An understanding of detail is crucial to the pursuit of biology, because though individual structures and their functions may not seem important at first, each will play an important role in your discovery of the elaborate patterns found throughout biology. When used to complement your lectures and labs, your textbook is an important tool that will help you master details and discover a rich tapestry of biological patterns.

This chapter will help you use the features of Freeman's *Biological Science* to gain full understanding of your course material and of biology as a whole.

A. DISCOVERING PATTERNS IN BIOLOGY

Researchers and students alike face the common challenge of gathering enough specific knowledge about biology to be able to recognize general patterns. Before attempting to understand complex biological concepts, you must learn about specific structures and processes that will explain biology in a broader context. In **Chapter 1** of Part One, we considered the pitfalls of "not seeing the forest for the trees," which describes the idea that broad patterns and conclusions can be overlooked when attention is focused solely on detail. Let's apply this metaphor to the study of biology.

3.1 Seeing Both the Forest and the Trees— Use Details to Understand Concepts

As a beginning biology student, you will notice that there is no shortage of trees! Hundreds of terms, many of which may be unfamiliar, are used to describe organisms, their constituent parts, and the ways they interact. Learning each tree in the forest of biology is a challenge that can be approached in several different ways. Understanding these details is necessary to do well in a biology course—for example, learning the basic structure of a plant leaf and the fundamental steps of cellular respiration are prerequisites for studying the more complex process of photosynthesis. Learning these details will be easier and more interesting if you try to understand *why* they are important. As you progress through this course, constantly question yourself about the material to assess your understanding: Why is a plant leaf so structurally complex? What are the functions of its separate components? How do a plant's leaves contribute to its survival?

Your Turn

To understand how patterns and relationships in biology are organized, go to the section in your library that contains introductory biology textbooks such as Freeman's *Biological Science*. Look at the tables of contents in several different books. Do you see much variation in how the chapters are arranged in each book? What themes are present in every book? Looking at the publication dates of these books, can you see any ways in which the organization of biological information has evolved over the years?

If you keep the *function* of an organism's structural components (the trees, to return to our metaphor) in mind, these details will be easier to learn and will help clarify a broader concept (the forest). As you read your textbook, try not to passively memorize terms; instead, continually ask yourself, "What does this structure do, and how does it relate to the terms and concepts I already know?"

This study guide emphasizes the idea that the material you study will build upon your knowledge from previous chapters and will help you understand future topics. Each chapter in Part Two contains a section entitled "Integrating Your Knowledge," an example of which can be found on page 44 of this study guide. Its two subsections, "Looking Back—Concepts from Earlier Chapters" and "Looking Forward—Concepts in Later Chapters," remind you of concepts you already know and indicate connections to concepts you will learn about in chapters to come. If you look for general patterns and broad concepts as you learn the details of biology, you will eventually discover that the shadowy outline of the forest is starting to emerge from your study of the trees.

3.2 Look for Hierarchical Relationships

At this point in your academic career, you are probably familiar with outlines. Every outline, whether it clarifies the structure of a college-level essay assignment or describes the layout of an entire book, follows the same basic format. Main ideas are identified by the major headings, and these topics are divided into more specific subheadings. While some ideas stand alone as key concepts, others need the support of more specific facts; key words or phrases can be included within subsections to indicate additional details that will be discussed in the essay or book.

In an outline, facts and ideas are presented in the form of a hierarchy. A hierarchy is a system of categorical groupings that is arranged according to relative importance. A good example of a hierarchy is a book's table of contents, which is essentially a complex outline; its hierarchical structure distinguishes it from an index, in which ideas and facts are listed alphabetically regardless of their relative importance. A good table of contents is more than a guide to help the reader locate information; it also establishes a hierarchical relationship between topics.

It is important to understand the hierarchy of information in *Biological Science* as well as in other textbooks you may encounter. To help you with this concept, a section entitled "Key Concepts" is included in each chapter in Part Two of this study guide. This section presents the main ideas of the corresponding textbook chapter in a hierarchical format, allowing you to easily

distinguish pivotal concepts from supporting details based on their relative position. For example, consider this study guide excerpt from Part Two, Chapter 40, on plant reproduction:

40.1 An Introduction to Plant Reproduction

Sexual Reproduction

- Most plants reproduce sexually. **Sexual reproduction** is based on the reduction division known as **meiosis** and on **fertilization**, the fusion of haploid cells called **gametes**.
- **Sperm** are small cells from the male that contribute genetic information in the form of DNA but few or no nutrients to the offspring. Female gametes are called **eggs**, which contribute a store of nutrients to the offspring.
- **Outcrossing** occurs when male and female gametes are exchanged between individuals of the same species. When **self-fertilization** occurs, a sperm and an egg from the same individual unite to form a progeny.
- The primary advantage of selfing is that successful **pollination** is virtually assured.

The Land Plant Life Cycle

- Plants are the only organisms that have both a multicellular form that is diploid and a multicellular form that is haploid. This life cycle is called **alternation of generations**.
- The diploid phase is called the **sporophyte** and the haploid phase is called the **gametophyte**.
- A **spore** is a cell that grows directly into an adult individual. A **gamete** is a reproductive cell that must fuse with another gamete before growing into a new individual.
- Meiosis occurs in sporophytes and results in the production of haploid spores, which are produced in structures called **sporangia**.
- Spores divide by mitosis to form multicellular, haploid gametophytes.
- Fertilization occurs when two gametes fuse to form a diploid zygote. The zygote grows by mitosis to form the sporophyte.
- Study the diagram below to make sure you understand these life-cycle intricacies.

This example introduces several details about plant reproduction, and their relative positions in the outline indicate relationships between them. Based on the structure of the outline, you can determine that there are three important points under the heading of plant reproduction—sexual reproduction, plant life cycles, and asexual reproduction. Each of these topics is supported by specific details; for example, meiosis and fertilization involving gametes are both important components of sexual reproduction. Such details are also included under the topics of plant life cycles and asexual reproduction. Although this information could be discussed in paragraph form, a hierarchical outline allows relationships between topics to be defined clearly by their appearance on the page.

3.4 Accept Variability

To understand biology, it is necessary to accept that individual organisms, structures, or behaviors may not always fit into expected patterns. Variability is a natural part of biological science; occasional variations actually emphasize that most organisms match a standard pattern. While *Biological Science* concentrates on the typical components of general patterns, it also includes realistic variation. This variation is especially apparent in your book's many graphs, which often show individual data points as well as an overall pattern. Consider this graph from Chapter 25, page 510, which illustrates the relationship between human birth weight and mortality (**Figure 3.1**).

Logically, we might expect to observe lowest mortality at one ideal birth weight, with a greater mortality rate if birth weight is either higher or lower than this optimum. In fact, the actual data do support this hypothesis. However, if you examine the data points representing mortality (in Figure 3.1, the gray points connected by a U-shaped curve) and the actual distribution of body weights at birth (the gray bars), you will see that while there are obvious trends, there is variability among individual data points. In this example, though variability is illustrated by any individual points lying off the indicated curve (drawn from the average of all data points), most points fall fairly close to this line. In some experiments, variability is far more pronounced, requiring sophisticated statistical analysis to identify a pattern in a widely distributed set of data.

Accepting variation will prevent you from being distracted by any data that do not fit the predicted pattern. Variation, though it may appear to disrupt experimental data, is not just a nuisance to be ignored. After all, it is variation among individual organisms that enables a species to change over time, and variation is therefore the foundation of evolution.

3.5 Be Comfortable with the Unknown

In addition to facing the challenges of variability, biology students must be prepared to encounter uncertainty. New facts emerge and new hypotheses are tested daily, making biology a field that is constantly evolving and reinventing itself. Though much is already known about biology, an even greater amount is waiting to be discovered and understood. *Biological Science* discusses both accepted facts and current theories, distinguishing what is proven from what is conjectured. You may pose questions to your professor and lab instructor that they may not be able to answer, because the answer is not yet known. This uncertainty simply reflects the rapid growth of the field and explains biologists' constant search for greater understanding.

B.   SURVIVING—AND THRIVING— IN THE CLASSROOM

As with any course, you will need to thoroughly understand biology in order to succeed on quizzes and tests and achieve your desired grade. Several tools are at your disposal, many of them integrated directly into your textbook, that can make the learning process easier and more rewarding. Additionally, certain study habits may improve your ability to understand the material you are studying.

3.6 Budget Your Time Carefully

The amount of time you spend on your biology course outside of class—studying, preparing for tests, and working on your laboratory assignments—will be reflected in the grade you achieve. Your other classes also require that you take the time to study, read the material, and write papers. How can you be sure that you are devoting enough time to each class?

It is important to be realistic about your time commitments. Everyone in your biology class has different schedules, course loads, and activities, so it is important

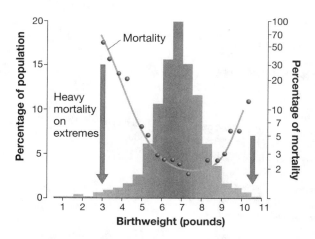

FIGURE 3.1   The histogram shows the percentages of newborns with various birth weights. The gray dots are datapoints indicating the percentage of newborns in each weight class that died, plotted on the logarithmic scale shown on the right. The gray line is a function that fits the datapoints.

to develop a study pattern that suits your life. For example, let's say that your biology course involves two lectures and a lab each week; therefore, you will probably spend at least five hours a week in class for your biology course alone. You may need at least another five hours a week to complete assignments and prepare for exams, meaning that biology alone might take up 10 hours of your time each week. Considering that there are five classes in a typical course load, you may devote 50 hours each week to academics alone—leaving very little time for family, friends, a part-time job, and relaxation! Realistically consider what you can expect to achieve in the time you have available. If you concentrate on what you hope to achieve in college and on your future goals, your courses will become a top priority.

3.7 Go to Lectures and Labs

Lectures and labs are not optional; if they were, your course syllabus might say, "Please come to class occasionally." Instead, it probably lists the exact times for all lectures and labs, implying (or specifically stating) that they are mandatory. By attending every lecture and lab, taking careful notes, and participating in class while you are there, you can take an active role in the learning process. Going to class will give you the opportunity to question your instructor and classmates about material you may not understand and allow you to gain different perspectives and practical experience. Attending class

regularly, therefore, may be the one thing you can do to most improve your performance in a biology course.

3.8 Take Good Notes in Class

Attending class is necessary for success, but it is also important to make the most of your class time by taking good notes. Though you may have personal note-taking methods that work well for you, try structuring your notes in the form of an outline to illustrate a hierarchy of ideas. A hierarchy, as introduced earlier in this chapter, arranges concepts according to relative importance. Try to mimic your textbook as you take notes in class, reflecting the hierarchy of facts outlined in the table of contents and in the chapters themselves.

Your Turn

Examine the hierarchical format used in these sample notes on Chapter 44, "Gas Exchange and Circulation," and compare it with the format of notes you've recently taken for this class. Are they similar? Can you think of ways to improve your own note-taking technique?

Even if you do not choose to follow this method precisely, consider modifying your own note-taking technique to include a hierarchy of ideas. This format

Air and Water as a Way of Breathing
Oxygen and Water in Air
- Air is primarily N_2, with 21% O_2 and traces of CO_2.
- Gases move by diffusion.
- Partial pressure is the pressure of gas in a mixture.
  o The partial pressure of air at sea level is about 160 mmHg.
  o Partial pressure on top of Everest is only 53 mmHg because of much higher altitude.

Oxygen and Water in Water
- CO_2 is 30 times more soluble in water than O_2.
- Fish and other aquatic organisms can easily get rid of CO_2, but getting O_2 is a problem.
  o Fish die-offs are almost always due to lack of O_2 rather than buildup of CO_2, while animals breathing air can have trouble getting rid of the CO_2 they produce.
- Gas solubility decreases as water temperature increases.
  o Fish in warm ponds suffer from O_2 deprivation more than in cold ponds, because of the lower solubility of O_2 in warm water.

will help to solidify your understanding of the relationships between concepts.

3.9 Integrate Lecture and Laboratory Material

Lectures and laboratories are often taught at different times by different instructors, so it may sometimes seem that they are two separate courses. Remembering that the laboratory is designed to support the lecture, and vice versa, will help you understand the material you are studying. Though they may not fall on the same day, the scheduling of laboratories and lectures often coincides so that in a given week, the material discussed in each will be related. For example, in the same week as your lecture on animal diversity, you may have a laboratory examining the morphological diversity of a wide array of animal specimens. Similarly, your lectures on the role of water and ions in an animal's physiology may be accompanied by a laboratory about the structure and function of the mammalian kidney. If you don't immediately see a connection between your lectures and labs, ask your professor or lab instructor to explain how they are related.

3.10 Take Advantage of Your Textbook

While this study guide is designed to help maximize your understanding of your biology course, it is not meant to replace your textbook. The vast amount of information contained in its chapters makes your textbook an indispensable supplement to your biology lectures and laboratories.

First, familiarize yourself with the book's features—the table of contents, glossary, index, and useful learning features such as "Key Concepts" and " Check Your Understanding"—as well as with the structure and content of the chapters. Parts of the textbook can be used like a reference book if you need to review topics you recognize from previous courses. You can use the index to locate the page on which a particular term is presented, thus avoiding a tedious search of an entire chapter. The glossary can be a useful tool to test your knowledge: Write down any key terms in your lecture notes or reading assignment, and then define these concepts in your own words and compare your definitions to those offered in the glossary.

Chapter content is carefully designed to introduce and thoroughly explain each topic, because lectures and labs may not always provide enough information to fully understand a concept. For more difficult topics, you may need to refer to the book for additional explanations. A complex subject like genetics, for example, might demand a careful reading of the entire chapter before you can fully understand the topic.

3.11 Take Advantage of the Media for Students

Be sure to take advantage of the robust media support package that accompanies your textbook. The Student CD-ROM provides resources to help you visualize difficult concepts, explore complex biological processes, and review your understanding of the material. The CD-ROM media fall broadly into two types—animations and tutorials. The simple animations will help make the one-dimensional art in *Biological Science* come to life. The more complex tutorials will walk you through difficult concepts and include opportunities to test your knowledge.

You'll know when there are animations or tutorials because you'll see "Web Animation" icons on some of your book pages—see page 924, for example. But if you are not inclined to stop reading in the middle of the chapter, look for the "Web Tutorial" icons at the end of each chapter (under "Chapter Review"). You'll also find a helpful summary there, including an estimate of how long the learning activity takes.

Another feature of the student media is an ability to test your knowledge. The quizzes on the website (www.aw-bc.com/freeman) include hints and feedback to help you assess your understanding. You'll also find research tools, including a broad collection of science and research links for the subject areas described in each chapter.

3.12 Share Your Knowledge

One of the best ways to learn is to tell others what you know. Find a study partner and meet regularly to discuss material. Take turns explaining topics to each other as thoroughly as possible while using any new terms and concepts you are learning. The partner who is listening can then critique the other's knowledge, identifying any missing information to ensure that both partners thoroughly understand the topic. In this way, you will be able to quickly identify your particular areas of weakness. You will eventually need to explain these concepts on laboratory and lecture exams (especially if these tests involve essay questions); any self-testing you do before an exam will build your confidence and ensure your mastery of the subject.

Though biology can be complex at times, understanding principles of data interpretation and improving your study technique can make learning this subject both challenging and rewarding. Reading effectively and writing to clearly communicate information will also help you in this course, throughout your education, and in the "real world" as well. In **Chapter 4** of Part One, we will discuss these skills and give you the opportunity to assess your abilities.

4

Reading and Writing to Understand Biology

In Part One, **Chapter 2** *of this study guide, we noted that while scientific information is often conveyed visually with graphs and animations, writing remains the most common and most effective means of scientific communication. This chapter will help you improve your ability to read and interpret information about biology, evaluate your note-taking skills, and learn to convey biological ideas in your own words.*

A. READING FOR UNDERSTANDING

Your success in this biology course (as well as in other classes) may depend on your ability to gather important information from what you read. Do you carefully interpret the material you read? Can you pick out the topic sentence in a paragraph? When you read a description of an experiment, can you identify the hypothesis being tested? Do you have the ability to skim your text when reviewing for a test?

Good readers easily understand written material because they are aware of the structure of the text as they read. When a text is well written, in each paragraph it is easy to identify a topic sentence that defines the main idea. A topic sentence is usually located at or near the beginning of each paragraph and is followed by further support, elaboration, or explanation. This supporting information may describe examples or analogies, explain how the topic relates to a larger context, or present alternate views to help the reader understand the information.

Certain reading methods can increase the amount of information you gather from the material. When reading something the first time, read carefully, paying attention to any figures you encounter. If you come across a topic you do not understand, refer to your study guide; it may present the same information in different words and may include other examples. Use the glossary or a dictionary to look up the definitions of any unfamiliar words. As you read, stop periodically to ask yourself what you have learned; these mini-reviews will help you

remember information later. If you have difficulty understanding a topic, ask your professor, teaching assistant, or another student to explain the subject.

You can use several techniques to improve your understanding and memory of what you read. If you plan to keep your textbook after you complete this biology course, you may want to underline topic sentences and other important points; but remember that underlining is effective only if it is done selectively. If you do not want to mark your book, making notes on sticky pads and attaching them to pages with key information will serve the same purpose.

Note-taking is another effective way to increase the amount of information you retain while reading, but certain methods may work better than others. Take notes only after you have read the material completely and thoroughly understand it. Taking notes as you read may distract you from learning the material and may tempt you to copy the text directly into your notes; you will learn and remember more from writing and reviewing your notes if you use your own words. In the next section of this chapter, "Writing for Understanding," we discuss effective note-taking methods in greater detail.

When you review for a test, look through the chapters you are studying carefully to make sure that you understand the material. Go over the figures in your textbook, testing yourself to see if you can explain each figure without referring to the text. Be sure you can accurately define any boldfaced vocabulary words. As a final review of important topics, re-read your class notes and any notes you took while reading the textbook.

Your Turn

Let's practice these reading skills. Read the following passage from Chapter 1, pages 4–5 of *Biological Science* (*"What Is Natural Selection?"*), and underline the topic sentences. (An answer key is at the end of this chapter.)

Natural selection occurs whenever two conditions are met. The first is that individuals within a population vary in characteristics that are **heritable**—*meaning, traits that can be passed on to offspring. A* **population** *is defined as a group of individuals of the same species living in the same area at the same time. Darwin and Wallace had studied natural populations long enough to realize that variation among individuals is almost universal. In wheat, for example, some individuals are taller than others. As a result of work by wheat breeders, Darwin and Wallace knew that short parents tend to have short offspring. Subsequent research has shown that heritable variation exists in most traits and populations. The second condition of natural selection is that in a particular environment, certain versions of these heritable traits help individuals survive better or reproduce more than do other versions. For example, if tall wheat plants are easily blown down by wind, then in windy environments shorter plants will tend to survive better and leave more offspring than tall plants will.*

If certain heritable traits lead to increased success in producing offspring, then those traits become more common in the population over time. In this way, the population's characteristics change as a result of natural selection acting on individuals. This is a key insight: Natural selection acts on individuals, but evolutionary change affects only populations. In this example, populations of wheat that grow in windy environments tend to become shorter from generation to generation. But in any given generation, none of the individual wheat plants get taller or shorter as a result of natural selection. This sort of change in the characteristics of a population, over time, is evolution. Evolution occurs when heritable variation leads to differential success in reproduction. If you understand this concept, you should be able to graph how average stem height will change over time in a population that occupies a windy environment versus a nonwindy environment—one where tall plants have an advantage due to better access to light. (For help with reading and making graphs, see **BioSkills** *1 in the back of this book.)*

Darwin also introduced some new terminology to identify what is happening during natural selection. For example, in everyday English the word fitness means health and well-being. But in biology, **fitness** *means the ability of an individual to produce offspring. Individuals with high fitness produce many offspring. Similarly, the word adaptation in everyday English means that an individual is adjusting and changing to function in new circumstances. But in biology, an* **adaptation** *is a trait that increases the fitness of an individual in a particular environment.*

Once again, consider wheat: In windswept habitats, wheat plants with short stalks have higher fitness than do individuals with long stalks. Short stalks are an adaptation to windy environments.

Take notes on the passage using techniques discussed in this section. (Refer to the answer key for sample notes after you've taken your own.) Look at the passage again and consider the following questions: Is the boldfaced vocabulary word familiar? (If not, look up the definition in the glossary and write it in your notes so you can easily review it later.) Do you recognize the underlined words? (If not, you might want to look them up in a dictionary.) Using only your notes and your memory of the passage, answer the following questions:

- Using the Brussels sprouts example described in the text, explain the three conditions that must be present for natural selection to take place.
- What is the difference between artificial and natural selection?

Thinking about these types of questions as you read will help you understand and remember the content.

Reading and writing are closely related. Taking notes and/or making flash cards as you read helps you understand and remember information and will make it easier for you to study later on.

B.   WRITING FOR UNDERSTANDING

4.1 Taking Notes

The ability to communicate your knowledge to your professor in writing will be essential to your success in this biology course. Don't wait until you get your first test back to assess your writing skills; you can start to practice writing clearly about biology as you take notes and answer the short-answer and essay questions in your text and in this study guide. Taking notes on the material you read will help you analyze, understand, and remember the information; answering short-answer and essay questions will allow you to consider facts from a different angle and put your thoughts into your own words.

Although it is acceptable to use informal language—incomplete sentences, abbreviations, and so on—when writing notes and flash cards, it is essential to write with clarity, brevity, and logic. Include all important facts, eliminate any extraneous details, and ensure that your notes will still make sense to you months later. (Why are "extraneous details," that you probably won't need to know for quizzes and tests, included in your textbook? These details are meant to support important points in the text. Though it may not appear on exams, this information will help you understand the more complex topics you are required to know.)

If note-taking seems insufficient and you need additional help memorizing information, try making flash cards of important terms and concepts. Write the name of the concept on one side, and use the other side for definitions and examples. Flash cards are convenient because they can be transported easily, enabling you to study even when you may not have your textbook or class notes with you. You can also tape small flash cards around your room in places where they will catch your eye, such as on your mirror or next to the light switch. Once you know a term, replace it with a new flash card. The benefits of this technique are that you can learn new vocabulary and concepts gradually throughout the semester just by walking around your room, and you can avoid last-minute cramming before tests.

Be sure your notes will allow you to distinguish key concepts from supporting details and examples when you review them later. Consider using the Cornell Note Taking System (you can learn about this method on the Internet; the "Internet References" section at the end of this chapter gives the URL of a helpful website) or another similar method. Though it is a good idea to take notes both in lecture and while reading the textbook, your lecture notes should be more detailed than your text notes. You can always refer to the textbook for further information to support your notes, but lectures are given only once.

Your Turn

Let's try reading this passage called "Evolutionary Legacy" from David S. Goodsell's article, "Biomolecules and Nanotechnology" in *American Scientist* 88(3), May–June 2000. First read for understanding, underlining topic sentences as you go, and then take notes about the passage. Be selective, distinguishing main points from supporting examples or details.

Now that you've improved your note-taking ability, let's discuss more formal writing techniques.

The process of evolution by natural selection places strong constraints on the form that biological molecules may adopt. Because genetic information is passed directly from generation to generation, cells must maintain a living line back to the earliest primordial cells. If a cell fails to generate a living descendent, all of its biological discoveries will be lost. This is far more limiting than the technology of our familiar world. If we create machines that don't function, we scrap them and go back to the drawing board. But if a cell takes a gamble and changes a critical machine, it had better get it right the first time or the result will be disastrous.

The picture is not entirely grim, however, as cells have several levels of redundancy within which to develop new machines. First, the plans for a given machine may be duplicated, which allows the duplicate to be modified and ultimately perfected to perform a function different from the original. This is very common in the evolution of life. Hemoglobin, the protein that carries oxygen in our blood, is an example. Our cells contain information for building several different types of hemoglobin. One is optimized for carrying oxygen in the blood of adults, whereas another is found in the blood of a fetus. The fetal hemoglobin has a higher affinity for oxygen, allowing it to capture oxygen from the mother's blood. About 200 million years ago, a gene duplication allowed the fetal hemoglobin to be perfected separately.

Second, biology seldom involves a single cell. A population of cells—billions, trillions—is the biologically relevant entity. Within this population there exists ample room for experimentation. Millions of modifications may be tried, even if most are ultimately lethal. The population will still survive and individuals with rare improvements may grow to dominate in later generations. Human immunodeficiency virus (HIV) shows the benefits of evolutionary change, accelerated so that we can see the effects in months instead of millennia. HIV reverse transcriptase, the enzyme that copies the virus's genetic information, is particularly error-prone. Because of this, the population of viruses within an infected individual contains viruses with all possible single-site mutations—thousands of variants on the wild-type virus. The best of these will dominate, but even the weakest are continually created and recreated in subsequent generations by the low-fidelity copying mechanism. Thus, when an infected individual is treated with anti-HIV drugs, the population has a wide range of different mutants to choose from, some of which may be resistant to the drug: The virus is made more efficient by its very inefficiency. The hallmark of biological evolution is the plasticity provided by mutation and genetic recombination. Within a population, or through genetic duplication within a single cell, a great many variants may be tested and the occasional improvement saved.

Evolution carries with it one important drawback, however: the problem of legacy. Once a key piece of machinery is perfected, it is difficult to replace it or make major modifications without killing the cell. This is particularly true for major molecular processes, such as protein synthesis, energy production and molecular machines. This leads to the remarkable uniformity of all earthly living things when observed at the molecular level. All are built of the same basic components.

Compare your notes with those shown in the answer key at the end of this section. Do your notes mention the same main points as those in the answer key? Do you prefer your notes or the sample notes? You probably find your own notes easier to understand because they are in your own words, phrased in a way that makes sense to you. If you prefer the sample notes, try to figure out why. For example, were your notes too detailed? Were they too brief? Did you miss some main points or fail to make a distinction between major topics and supporting details? Use these observations to identify ways in which you can improve your note-taking skills.

4.2 Writing Essays and Other Assignments

Why Is Writing So Important?

Many students don't consider writing to be an important part of their science classes. Yet, everyone benefits from good writing skills, including people in science-related professions. A scientist's work is not limited to research conducted in a laboratory; he or she must be able to effectively communicate research goals, procedures, and results in grant applications and scientific papers. As a biology student, you will need to take notes, write laboratory reports, and take tests that may include essay or short-answer exercises. The ability to write clearly and concisely will improve your performance in all these areas.

Evaluating and improving your writing technique will benefit your study of biology in several ways:

- Taking coherent notes in your own words will make studying easier.
- Writing down information will help you recall it later on exams.
- Clearly written short-answer and essay responses will effectively communicate your knowledge.
- Well-written laboratory reports will demonstrate your understanding of the process studied.

Your Turn

Compare these two short-answer responses to the following question: How does what we know about artificial selection support the theory of evolution?

1. Artificial selection is like near to natural selection which darwin said caused evolution and they both are types of selection therefore evolution must occur. When you select something it must be good, so plants and animals change to get better buds or something quickly because their reproduction is helped by it.

2. When humans artificially determine the reproductive success of domesticated organisms with variable, heritable traits, the artificially selected plants or animals show major changes over a relatively short period of time. Thus natural selection could also cause the changes over time (evolution) that have led to the diversity of organisms we find today.

Which answer more effectively communicates the material? Why? Which student would you be more likely to hire to work in your lab?

Elements of Good Writing

We've discussed the many ways in which the ability to write well can be an advantage and explored this idea with examples of short-answer responses. But maintaining clarity and presenting ideas cohesively can be more difficult when completing an assignment requiring a longer response. What can you do to improve your essay-writing technique?

Think First

Before you start writing, be sure you understand your topic (refer to the "Reading for Understanding" section of this chapter) and the writing assignment. Taking the specific demands of the assignment into consideration, decide on a thesis statement; this sentence introduces the topic to be addressed and identifies a particular position that your paper will defend. After you locate examples or supporting arguments to support your thesis, you are ready to make an outline that will provide a layout for the organization of your essay or lab report. Though making an outline may seem unnecessary, it will ensure that your argument is thoroughly developed and can be easily understood by other readers.

Creating an outline is a worthwhile step only if you carefully consider what information it should contain. Whether you prefer to draft detailed outlines or less elaborate ones, certain elements should be included in all outlines:

- A well-defined thesis statement located near the beginning
- Supporting points introduced in a logical order
- Explanations of how supporting details relate to the thesis
- A strong conclusion

Your Turn

You are applying to work with a professor on his or her summer research project. Draft an outline for the cover letter you plan to attach to your résumé and application. See the answer key for a sample.

Write

After completing your outline, use it to write a first draft of your essay. Though it is not necessary to be a perfectionist at this stage—you will edit later—try to create a clear topic sentence for each of your paragraphs. Use active verbs while avoiding passive voice; for example, it is more effective to write, "We inadvertently contaminated our sample," than "Mistakes were made and the sample was contaminated." Be concise, but thoroughly support each of your main points. Consider using metaphors, similes, analogies, and examples to explain your topic; these creative techniques encourage readers to make their own connections, enabling them to understand your position more clearly.

Most important, do not plagiarize. Plagiarism, stealing someone else's ideas and passing them off as your own, is unethical. When you plagiarize, you also sabotage your educational experience by discouraging yourself from identifying and explaining your own ideas and perceptions. This does not mean you cannot use material from outside sources; if you find relevant material on your topic, include it in your essay with a citation acknowledging its source.

Let's take another look at the previous sample essay by David Goodsell and analyze the components of a well-written piece. It can be difficult to examine good writing because while bad writing distracts you as you read, alerting you to problems, good writing may seem almost unnoticeable because it allows you to easily understand the information being discussed.

In the first paragraph, Goodsell introduces his thesis statement—that evolution limits the forms of biological molecules—and explains it with a comparison between the evolution of life and the development of machines. He uses the next two paragraphs to explore ideas that contradict this thesis, suggesting that evolution's limitations are not strong enough to be substantial. He supports this alternate view by modifying his previous analogy between machinery and life to discuss a similarity between the two; an example involving hemoglobin is included for further clarification. Goodsell uses the words *first* and *second* to describe the topics of the second and third paragraphs, ensuring that the reader understands that these two points are meant to support this alternative argument. By presenting a position that opposes the thesis statement, Goodsell uses a popular and effective technique of essay writing. This method assures the reader that the author thoroughly understands the topic and gives the author the opportunity to directly refute arguments that contradict his or her thesis. In the final paragraph, Goodsell explains that his thesis is still relevant despite the conflicting arguments

presented in the preceding two paragraphs. He then restates his original thesis and concludes the essay by identifying a situation in which his position is particularly relevant.

Throughout his essay, Goodsell uses clear logic, explaining and supporting his statements appropriately. He employs many techniques of strong writing, such as varying the sentence structure and using repetition—in moderation—to emphasize important points. He also avoids using passive voice. When you consider this example, however, remember that it is an excerpt from a longer article. Had he addressed the same topic in an essay of this length, Goodsell would probably have developed and supported his thesis while spending less time discussing opposing arguments.

How can you, a student writer, be sure that you are applying enough of these writing techniques to your own assignments? While drafting an outline will help you organize information before you begin writing, you may not always remember to vary sentence structure, use active verbs, or support your thesis with examples as you write. This is why editing is such an important step in the writing process.

Edit

The best way to improve a paper you have written is to re-read and edit your work a day after you finish writing. Allowing time to pass before editing will give you a fresh perspective; you will often detect problems that you may not have noticed otherwise. This method may not be an option for procrastinators, however; another, more immediate editing technique involves reading your paper aloud to yourself or a friend. Speaking your words out loud will allow you to catch and eliminate any grammatical errors and awkward phrasing, and you will notice if certain words are used frequently. Avoid using the same word repeatedly by looking up alternative words in a thesaurus. Finally, if your essay is on a computer, use the spell- and grammar-check programs. Misspelled words and poorly constructed sentences will interrupt the flow of your argument and distract your reader.

Your Turn

Practice identifying effective writing techniques as you read this passage from Carl Zimmer's article entitled "Do Parasites Rule the World?" (*Discover* 21(8), August 2000). You might first read the article for understanding and then take notes on the passage. Finally, look at the details of sentence structure, observing strengths of the author's writing style.

As scientists discover more and more parasites and uncover the extent and complexity of their machinations, they are fast coming to an unsettling conclusion: Far from simply being along for the ride, parasites may be one of nature's most powerful driving forces. At the Carpinteria salt marsh, Kevin Lafferty has been exploring how parasites may shape an entire regions' ecology. In a series of exacting experiments, he has found that a single species of fluke—Euhaplorchis californiensis—journeys through three hosts and plays a critical role in orchestrating the marsh's balance of nature.

Birds release the fluke's eggs in their droppings, which are eaten by horn snails. The eggs hatch, and the resulting flukes castrate the snail and produce offspring, which come swimming out of their host and begin exploring the marsh for their next host, the California killifish. Latching onto the fish's gills, the flukes work their way through fine blood vessels to a nerve, which they crawl along to the brain. They don't actually penetrate the killifish's brain but form a thin carpet on top of it, looking like a layer of caviar. There the parasites wait for the fish to be eaten by a shorebird. When the fish reaches the bird's stomach, the flukes break out of the fish's head and move into the bird's gut, stealing its food from within and sowing eggs in its droppings to be spread into marshes and ponds.

In his research, Lafferty set out to answer one main question: Would Carpinteria look the same if there were no flukes? He began by examining the snail stage of the cycle. The relationship between fluke and snail is not like the one between predator and prey. In a genetic sense, infected snails are dead, because they can no longer reproduce. But they live on, grazing on algae to feed the flukes inside them. That puts them in direct competition with the marsh's uninfected snails.

To see how the contest plays out, Lafferty put healthy and fluke-infested snails in separate mesh cages at sites around the marsh. "The tops were open so the sun could shine through and algae could grow on the bottom," says Lafferty. What he found was that the uninfected snails grew faster, released far more eggs, and could thrive in far more crowded conditions. The implication: In nature, the parasites were competing so intensely that the healthy snails couldn't reproduce fast enough to take full advantage of the salt marsh. In fact, if flukes were absent from the marsh, the snail population would nearly double. That explosion would ripple out through much of the salt marsh ecosystem, thinning out the carpet of algae and making it easier for the snails' predators, such as crabs, to thrive.

After you read and analyze Zimmer's article, answer the following questions:

1. What is Zimmer's thesis?
2. What hypothesis was Lafferty testing?
3. What is the larger context for Lafferty's study?
4. Note the verbs that Zimmer uses. Are they active or passive?
5. Can you identify examples of a simile and a metaphor in this passage?
6. Does sentence structure vary?
7. What other strengths and/or weaknesses can you identify in Zimmer's writing?

Your Turn

Now that you can identify the characteristics of strong writing, take out an essay or lab report you have written recently. Reread it, looking for the elements of good writing described in this chapter. Is your thesis well supported? Which are more common in your paper, active or passive verbs? Does the sentence structure in the essay show enough variation? How might you further edit the piece?

As you can see, there is more to good writing than simply identifying a topic and carelessly listing information about it. The ability to convey an idea or support a position thoroughly in writing is a skill that will improve your grade in this biology course, help you as your education continues, and will be useful throughout your professional career.

REFERENCES

The following references contain further information about ways to improve writing technique and the fundamentals of language and grammar.

Text References

Goddin, N., and E. Palma, eds. 1993. *Grammar Smart: A Guide to Perfect Usage*. Villard Books.

Kirscht, J., and M. Schlenz. 2002. *Engaging Inquiry: Research and Writing in the Disciplines*. Prentice-Hall. Learning Express. 1998. *Writing Skills for College Students*. Prentice-Hall.

Pechenik, J. 2000. *A Short Guide to Writing about Biology*, 4th ed. Longman.

Mark, Peter (Roget), and B. Kipfer, eds. *Roget's 21st Century Thesaurus*. Dell Publishing Company.

Strunk, W. Jr., and E. B. White. 2000. *The Elements of Style*, 4th ed. Allyn & Bacon.

Internet References

The Cornell Note Taking System

http://www. dartmouth. edu/ ~acskills/ docs/ cornell_ note_taking. doc

College Tutor Study Guide

http://www. amelox. com/ study. htm

INTRODUCTION 4—ANSWER KEY

Reading for Understanding

Topic sentences are highlighted below:

Natural selection occurs whenever two conditions are met.

If certain heritable traits lead to increased success in producing offsping, then thos traits become more common in the population over time

To clarify how the process works, consider the origin of the vegetables called the cabbage family plants.

Darwin's great insight was that natural selection changes the characteristics of a wild population over time, just as the deliberate manipulation of "artificial selection" changes the characteristics of a domesticated population over time.

Your notes might look like this:

Natural selection = process that causes evolution. Occurs when...

1. Individuals in population vary.
2. Variable traits are heritable.
3. Some heritable traits help individuals survive better, reproduce more.

Above 3 conditions lead to change in a population's traits over time = evolution examples: wheat, artificial selection in mustard family.

Writing for Understanding

Taking Notes

These notes are modeled after the Cornell Note Taking System (**Figure 4.1**). To learn more about this method, refer to the URL given in the "Internet References" section of this chapter.

Writing Essays and Other Assignments

Why Is Writing So Important?

The second answer communicates the information more clearly. Although the first response contains relevant information, the poor spelling, grammar errors, and misplaced punctuation in this answer are distracting. Its sentences are not structured well, forcing the reader to search for key information and piece it together. To what do the "its" in the last sentence refer? If this was a student answer from an exam, would you be convinced that the student knows what artificial selection is? Careful consideration of this response reveals that the student understands similarities between artificial and natural selection and that evolution is change over time, but a tired or distracted test grader might not take the time to draw these conclusions, deducting points because the response is hard to read.

Elements of Good Writing

Think First

I. Thesis: Hire me this summer

    A. Introduce myself—how did I hear about professor?

    B. Professor's research sounds interesting.

    C. Want to get research experience.

II. I like biology

    A. Have done well in biology classes

    B. Interested in nature: hiking, gardening, animals

    C. Want to major in biology, am especially interested in professor's research area

III. I work hard and am responsible

    A. Held jobs for the past two summers (references available)

    B. Volunteer work at local animal shelter

IV. Conclusion

    A. Hire me this summer

    B. Contact information

    C. Thanks for your consideration

Write and Edit

1. Zimmer's thesis is that "parasites may be one of nature's most powerful driving forces."
2. Lafferty tested the hypothesis that flukes affect snail population dynamics.

evolution and biological diversity	1. Evolution by natural selection limits the diversity of biological molecules. • All cell lines must descend from progenitor cells. • Machines designs can be scrappped and restarted from scratch. • But cell innovations must succeed the first time or the cell won't leave any descendants.
cell redundancy	2. However, cells do have some redundancy. • The plans (DNA) for some "functions" can be copied, and the copies can be adjusted without fatal results. - e.g. adult vs. fetal hemoglobin
cell populations and modifications	3. Within the whole population of cells, many modifications, even lethal ones, may occur. • Those few cells with changes that are advantageous will increase in frequency in the population over time. - e.g. HIV (high mutation rate causes lots of variants— difficult to kill whole population with drugs.) • Through evolution variants are tested and the rare improvements saved.
evolution limited by legacy	4. Nevertheless, evolution is still limited by history. • Major modifications to important cell equipment almost always kill a cell. • This especially applies to major molecular processes (e.g., protein synthesis). • So, all life is very similar at molecular level.

In Biomolecules and Nanotechnology, David Goodsell argues that despite (1) the occasional functional redundancy within a cell and (2) the large population of cells within which modifications can appear and be tested, evolution is limited by legacy — especially at the molecular level.

Figure 4.1   An example of notes modeled after the Cornell Note Taking System.

3. Lafferty's study was conducted within the context of a marsh ecosystem; it also observed how snail populations affected algae, crab, and bird populations.

4. Zimmer mostly uses active verbs in the present or past tense, for example: "scientists *discover*," "Lafferty *found*," "birds *release*," "eggs *hatch*," and "flukes *castrate*."

5. An example of a simile is "looking like a layer of caviar," and an example of a metaphor is "that explosion would ripple out."

6. Sentence structure varies well; for example, Zimmer begins some sentences with prepositional phrases and others with the subject. Zimmer also uses sentences of many different lengths.

PART **2**

Biology and the Tree of Life

1

A. KEY CONCEPTS

1.1 The Cell Theory

- Scientists Robert Hooke and Anton van Leeuwenhoek were the first to observe cells in the late 1660s (**Figure 1.1**).

- Additional observations made by many researchers confirmed that all organisms are made of cells.

Are All Organisms Made of Cells?

- Yes! Cells are the fundamental building blocks of life, from the smallest single-celled bacteria (200 nm wide) to the largest multicellular organisms (e.g., sequoia trees).

- Cells from different organisms or from various tissues of a multicellular organism may look quite diverse, but they all consist of membranes that surround compartments with concentrated chemicals dissolved in water.

Where Do Cells Come From?

- Scientific theories usually have two parts: the description of a pattern observed in nature and an explanation of the process that causes the observed pattern.

- The pattern component of the cell theory states that all organisms are made of cells. Virchow's 1858 hypothesis that all cells come from preexisting cells provides the process component of the cell theory.

- An alternative hypothesis that organisms can just appear from nonliving materials (**spontaneous generation**) was disproved by **Louis Pasteur**, who showed that cells do not appear in nutrient broth when colonizing cells cannot enter the broth (**Figure 1.2**).

- Bacteria only seem to magically appear in spoiled foods, because bacterial cells blowing in on air dust are too small to see with the naked eye.

- Pasteur's experiment was well designed because there was only one difference between his two treatments: the neck shape of the flask. Cells in the air could not get to the broth in the swan-necked flask.

- All cells come from preexisting cells, so single-celled organisms in a population are related to a single common ancestor, and all cells in a multicellular organism also descend from a single ancestral cell (among humans, this ancestral cell is the fertilized egg).

- Charles Darwin and Alfred Russel Wallace realized that all species are also related.

CHECK YOUR UNDERSTANDING

(a) Why didn't biologists discover earlier that all life is made of cells?

_____

_____

_____

_____

1.2 The Theory of Evolution by Natural Selection

- Scientists Darwin and Wallace proposed natural selection as a mechanism for how evolution occurs in 1858.

What Is Evolution?

- The theory of evolution states that species change through time ("descent with modification") and are connected by shared ancestry (the pattern part of the theory).

- Natural selection explains *how* evolution occurs (the process part of the theory).

What Is Natural Selection?

- Individuals of the same species living in the same area at the same time are called a **population**. Populations evolve by natural selection when two conditions occur.

- First, individuals must differ from each other (show variation) for one or more heritable trait(s).

- Second, in a specific environment, individuals with certain heritable traits must survive and/or reproduce better than individuals with other traits.

- Consequently, individuals with the favorable traits increase in frequency and the population's characteristics change over time; this is evolution through natural selection.

- Note that which traits are favorable depends on the environment. For example, shorter wheat plants may survive and reproduce better in windy environments, but height may not be important in sheltered environments.

- Natural selection affects individuals, but it is the population that evolves.

 CHECK YOUR UNDERSTANDING

(b) If you were a hunter-gatherer without any modern technology, what heritable traits might help you survive and/or reproduce?

_____

_____

_____

_____

- **Artificial selection** of cabbage-family plants is an example of many varieties descending from one common ancestor (**Figure 1.3**). In this case, humans determined which plants grew in succeeding generations (thus, how the population's traits changed over time).

- The cell theory and the theory of evolution are two central, unifying ideas of biology.

CHECK YOUR UNDERSTANDING

(c) How does what we know about artificial selection support the theory of evolution by natural selection?

_____

_____

_____

_____

_____

1.3 The Tree of Life

- Natural selection can cause populations of one species to diverge and form new species.

- Evolution by natural selection implies that all species are descended from a single common ancestor such that a family tree of all organisms—the **tree of life**—could be drawn.

Linnaean Taxonomy

- In studying species and their relationships, it is useful to categorize organisms. In 1735, Carolus Linnaeus created the classification system (or **taxonomy**) we still use today.

- Each organism is given a two-part name. The first name is the **genus**, which includes one to several closely related species. The second name is the **species**. Individuals that breed together are included in the same species.

- Scientific names are italicized, and genus names are capitalized. Each two-part name is unique (e.g., *Homo sapiens* for humans). This naming system is called **binomial nomenclature** and is still used today.

Taxonomic Levels

- The **family** is a higher taxonomic level that includes one to several related **genera** (the plural of *genus*). Other still higher taxonomic levels are **order**, **class**, **phylum**, and **kingdom** (**Figure 1.5**). Each **taxon** combines related groups of lower (less specific) **taxa**.

 CHECK YOUR UNDERSTANDING

(d) Make up a mnemonic phrase of words starting with K, P, C, O, F, G, and S to help you remember the taxonomic groups ordered from largest to smallest.

_____

_____

_____

How Many Kingdoms Are There?

- Linnaeus proposed that all organisms be placed in the plant or animal kingdom. But the diversity of organisms cannot be accurately separated into these two groups.

- A more fundamental division among organisms is that of **eukaryotes** (single- or multicelled organisms having complex cells with a membrane-bound nucleus) versus **prokaryotes** (unicellular and without a nucleus). See **Figure 1.6**.

- Another alternative to Linnaeus' two-kingdom system is a five-kingdom system (**Figure 1.7**). Other systems have also been proposed.

Using Molecules to Understand the Tree of Life

- Carl Woese and colleagues attempted to discover the evolutionary relationships, or **phylogeny**, of all organisms by studying small subunit rRNA.

- Small subunit rRNA is found in all organisms because it is required for cell growth and reproduction. RNA is made of four ribonucleotides (A, U, G, and C) connected in a linear sequence (**Figure 1.8**).

- Our knowledge of evolution suggests that small subunit rRNA has changed over time. Because they share a more recent common ancestor, closely related species should have more similar small subunit rRNA than distantly related species do.

- Therefore, small subunit rRNA sequences can be analyzed to produce a **phylogenetic tree** showing the probable evolutionary relationships of the organisms studied.

The Tree of Life Estimated from an Array of Genes

- An analysis of small subunit rRNA and other genes from many different species can be used to produce a tree of life (**Figure 1.9**).

- This tree shows three major groups of organisms: the eukaryotes (Eukarya) and two groups of prokaryotes—the Bacteria and the Archaea.

- To accommodate this diversity, Woese proposed a new taxonomic level called the **domain**. Each of the three domains (Bacteria, Archaea, and Eukarya) includes several related kingdoms.

- Surprisingly, the Bacteria and the Archaea are much more diverse (at the molecular level) than anyone had expected, and fungi are more closely related to animals than to plants.

The Tree of Life Is a Work in Progress

- Currently, other molecules are being analyzed as well as other types of data. The placement of specific branches is debated, but many major findings from small subunit rRNA analysis have been confirmed.

CHECK YOUR UNDERSTANDING

(e) Dogs are more closely related to bears than to cats. Draw a phylogenetic tree depicting this relationship. What hypothesis might you make about RNA sequence similarity among these groups?

_____

_____

_____

_____

1.4 Doing Biology

- Biologists test ideas about how the natural world works by evaluating the predictions made by alternative hypotheses. Carefully designed experiments are an important tool for hypothesis testing.

Why Do Giraffes Have Long Necks? An Introduction to Hypothesis Testing

- Data collected by Robert Simmons and Lue Scheepers suggest that long necks in giraffes may be favored because they are useful in fights rather than because they allow giraffes to reach food that is not available to other mammals.

- Alternative hypotheses must be evaluated by testing their predictions.

The Food Competition Hypothesis: Predictions and Tests

- The hypothesis that giraffes evolved long necks in order to reach high food sources leads to three predictions, each of which can be separately evaluated.

- First, neck length is variable. Studies in zoos and in nature confirm this prediction.

- Second, neck length is heritable. This prediction is still untested, because giraffe breeding experiments are difficult to conduct.

- Third, giraffes typically feed high in trees, especially when food is scarce. Data do not support this prediction; giraffes feed with bent necks (**Figure 1.10**). So, there may be another, better explanation regarding why giraffes have long necks.

The Sexual Competition Hypothesis: Predictions and Tests

- An alternative hypothesis proposes that giraffes evolved long necks because longer-necked males win more fights than shorter-necked males do. Winners gain access to estrous females and presumably father more offspring.

- Long necks may improve the fitness of male giraffes competing for matings with females more than they affect giraffe ability to compete for food.

CHECK YOUR UNDERSTANDING

(f) Write out the predictions you would make based on the sexual competition hypothesis, and compare your predictions to the data given to support this hypothesis.

_____

_____

_____

_____

_____

1.5 Why Are Chili Peppers Hot? An Introduction to Experimental Design

- Chili peppers are hot to mammals because they contain capsaicin, a molecule that binds to mammalian heat receptors. Birds, however, do not appear to have the same response to capsaicin.

- Chili seeds eaten by an animal may either be destroyed in the digestive tract (seed predation) or voided unharmed with natural fertilizer (seed dispersal).

- The most important fruit- and seed-eating animals in the natural habitat of chili are the cactus mouse and a bird—the curve-billed thrasher. The mouse appears to destroy chili seeds that it eats; thrashers do not destroy the seeds, they disperse them (**Figure 1.11**).

- John Tewksbury and Gary Nabham tested the hypothesis that capsaicin in chili peppers is an adaptation that discourages seed predation while allowing seed dispersal. This hypothesis predicts that only thrashers will eat hot chilies, but both mice and thrashers will eat fruit that does not contain capsaicin.

- A **null hypothesis** expresses the alternative possibility that the explanation offered by the hypothesis does not apply. In this case the null hypothesis predicts that mice and thrashers do not differ in their response to capsaicin and therefore should not differ in their fruit preferences.

- Several cactus mice and curve-billed thrashers were captured and offered fruits from (1) a variety of chili that does not make capsaicin, (2) a hot chili, and (3) hackberries (no capsaicin).

- These three fruits were offered to the mice and birds in the same manner, and the amounts of each eaten during a set time interval were recorded (**Figure 1.12**).

- Thrashers and mice ate similar amounts of hackberries; but the birds ate more chili peppers than hackberries (similar amounts of the hot and the mild kind), and the mice ate only a few mild chili peppers and no hot chilies.

- Researchers concluded that capsaicin is an adaptation that prevents cactus mice, but not thrashers, from eating chili fruit; the directed dispersal hypothesis is supported.

- The experiment is well designed: (1) It included a control group (the hackberries) to check for other factors that might influence the results; (2) experimental conditions were controlled to eliminate extraneous variables; and (3) the test was repeated on several individuals to reduce the effects of random variation (sample size).

- Researchers even performed a follow-up experiment to demonstrate that chili seeds are, in fact, destroyed in the mouse digestive system but pass through the thrasher unharmed.

CHECK YOUR UNDERSTANDING

(g) What results would have supported the null hypothesis?

_____

_____

_____

_____

B. INTEGRATING YOUR KNOWLEDGE

Looking Back—Key Concepts from Earlier Chapters

Chapter 1 of Part 1 of this study guide provides additional information on how biologists "do" biology.

Looking Forward—Key Concepts in Later Chapters

Cells—Chapter 6 and Unit 2, especially Chapter 7

Evolution of the first cells is discussed in **Chapter 6**, followed by Unit 2 (Chapters 7–11), which covers cell structure and function, cell–cell interactions, some chemical pathways found in many cells, and cell replication (how cells make more cells).

Evolution and Natural Selection—Chapter 4, Unit 5 (Chapters 24–27)

Chapter 4 investigates hypotheses on how first life might have evolved from nonliving macromolecules. Evolutionary processes in general, including natural selection, are more thoroughly examined in Unit 5.

Phylogeny and Taxonomy—Unit 6 (Chapters 28–35)

Chapter 1 introduces Linnean taxonomy (**Figure 1.5**) and the major groups (the five kingdoms) of organisms. The five kingdoms (**Figure 1.7**), and how these groups are related, are described further in Unit 6.

C.   MASTERING CHALLENGING TOPICS

This chapter presents several key unifying concepts in biology. Many topics introduced in this chapter are covered in greater depth later on in this text. At this point you do not need to be too concerned about the details of evolutionary processes or what rRNA does. Do, however, spend some time thinking about how biologists use hypotheses to generate predictions that can be tested in order to discover information about the natural world. You might find it useful to practice generating your own hypotheses and testable predictions. For example, if plants seem to do better in one window than in another, what hypothesis might explain this difference? What testable predictions follow from this hypothesis, and how could you test them? Can you design an experiment that controls all variables except the one of interest? What kind of sample size would you use? Throughout the following chapters, description of important experiments will often be used to explain key biological concepts. If you master the ability to learn about biology through evaluating experiments, you will not have to resort to rote memory because you will *understand* the topic.

D.   ASSESSING WHAT YOU'VE LEARNED

1. Which statement is *not* part of the cell theory?
    a. All living organisms are made of cells.
    b. All living organisms are multicellular.
    c. Cells arise from preexisting cells.
    d. All cells in a multicellular organism are connected by common ancestry.

2. In Pasteur's experiment, what different results would have supported the spontaneous generation hypothesis?
    a. no cell growth in either flask
    b. cell growth only in the swan-necked flask
    c. cell growth in the straight-necked flask
    d. cell growth in both flasks

3. Pasteur's experiment with a swan-necked flask rejected the spontaneous generation hypothesis by demonstrating:
    a. Cells can grow in broth after it has been boiled.
    b. Cells grow in broth unless the broth is sealed off from the air.
    c. Cells grow only in broth exposed to a source of preexisting cells.
    d. Bacteria, but not fungi, can grow in broth.

4. A group of individuals of the same species living in an area at the same time constitutes a:
    a. community
    b. population
    c. genus
    d. family

5. The _____ is the fundamental structural unit of all organisms.
    a. DNA
    b. tissue
    c. species
    d. cell

6. The theory of evolution by natural selection states that:
    a. All species are descended from a common ancestor.
    b. No traits are heritable.
    c. Natural selection makes individuals more fit.
    d. Species remain unchanged over time.

7. Traits that are _____ are passed from one generation to the next.
    a. variable
    b. heritable
    c. selectable
    d. prokaryotic

8. A moth learns to visit white flowers, and the good food source lets it lay lots of eggs. What do you conclude?
    a. Evolution has occurred.
    b. The moth will definitely pass this adaptation on to its offspring.
    c. Artificial selection has occurred.
    d. The moth has a high fitness.

9. _____ was/were domesticated from *Brassica oleracea*, a wild mustard species.
   a. Broccoli
   b. Cabbage
   c. Kale
   d. all of the above

10. Linnaeus' taxonomic system specifies a name unique to each organism. The first part of the name is the organism's _____ and the second part is its _____.
    a. species; family
    b. family; genus
    c. genus; species
    d. genus; order

11. Modern taxonomy attempts to describe the _____, or historical relationships, among organisms.
    a. variation
    b. natural selection
    c. phylogeny
    d. physical similarity

12. Which of the following kingdoms contain prokaryotes?
    a. Monera
    b. Fungi and Monera
    c. Monera and Protista
    d. Archaea and Eukarya

13. Use the tree of life (**Figure 1.9**) to determine which of the following is most closely related to animals.
    a. Archaea
    b. Slime Molds
    c. Ciliates
    d. Land Plants

14. Where would you be more likely to find a population of mild chilies?
    a. in an area with few birds but many mice
    b. in an area with few mice but many birds
    c. in an area with both mice and birds
    d. in the all areas where you find hot chilies

15. Which of the following steps is *not* vital to a well-designed experiment?
    a. A control group is included.
    b. All variables other than those being tested are controlled or kept the same between treatments.
    c. Experiments are repeated on many individuals (large sample size).
    d. The steps described in a, b, and c are all important.

CHAPTER 1—ANSWER KEY

Check Your Understanding

(a) Prior to the mid-1600s, microscopes powerful enough to let researchers see cells were not available.

(b) A keen sense of smell and taste might have helped a hunter-gatherer survive better by improving his or her plant identification skills.

(c) Artificial selection shows that differential survival and reproduction cause change in a population over time because those heritable traits that are selected for increase in frequency in each succeeding generation.

(d) One possible mnemonic is Kings Play Checkers or Other Fun Games on Sunday. (Kingdom, Phylum, Class, Order, Family, Genus, Species)

(e)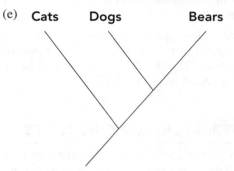

Hypothesis: Dog RNA is more similar to bear than to cat RNA sequence.

(f) Prediction 1: Males with longer necks win more fights—supported by data.

Prediction 2: Males that win fights have more offspring; this is indirectly supported by data showing that winners of fights gain access to estrous females.

Prediction 3: Neck length is heritable.

Predictions 2 and 3 require further testing to confirm the sexual competition hypothesis as proposed by Simmons and Scheepers.

(g) Observation of similar eating patterns in cactus mice and curve-billed thrashers would have supported the null hypothesis.

Assessing What You've Learned

1. b; 2. d; 3. c; 4. b; 5. d; 6. a; 7. b; 8. d; 9. d; 10. c; 11. c; 12. a; 13. b; 14. b; 15. d

2

Water and Carbon: The Chemical Basis of Life

A. KEY CONCEPTS

- **Chemical evolution** proposes that complex carbon-containing compounds (and eventually life) could have formed from simpler molecules as kinetic energy from sunlight and heat was converted to chemical energy of molecular bonds.

- First let's discuss how atoms bond together to form molecules during chemical reactions.

2.1 The Building Blocks of Chemical Evolution

- Molecules in cells are primarily made of hydrogen, carbon, nitrogen, and oxygen. Throughout this text you will see that structure determines function; so examining the structure of these four atoms will improve our understanding of how they bond together.

What Atoms Are Found in Organisms?

- An atom's nucleus contains protons and neutrons. Negatively charged electrons orbit the nucleus and are attracted to the positively charged protons (neutrons are neutral). An atom is electrically neutral when the number of protons equals the number of electrons (**Figure 2.1**).

- **Figure 2.2** shows the periodic table. The subscript indicates the **atomic number** (number of protons) and defines an **element**. Neutrons and proton masses are similar and equal about 1 **atomic mass unit (AMU)**.

- Atoms of an element with different neutron numbers are called **isotopes**. The superscript indicates the sum of the protons and neutrons or **mass number** for the most common isotope.

- Electrons are the key to understanding how atoms interact. Electron **orbitals** hold up to two electrons, and **electron shells** are groups of orbitals numbered by distance from the nucleus. Electrons fill the inner shells first (**Figure 2.3**).

- Elements commonly found in organisms have at least one unpaired electron in their outer (**valence**) shell. The number of unpaired electrons present is known as an atom's valence. The presence of unfilled electron orbitals allows the formation of **chemical bonds** to attach atoms together.

How Does Covalent Bonding Hold Molecules Together?

- Atoms are more stable when each orbital has two electrons. Orbitals with one electron each can overlap so that the nuclei share the two electrons, forming a **covalent bond** (**Figure 2.4**). A **molecule** is formed when two or more atoms are joined by covalent bonds.

- A **nonpolar covalent bond** forms when electrons are evenly shared between two atoms. If one atom holds onto the shared electrons more tightly than the other does, a **polar covalent bond** forms (**Figure 2.5**).

- An atom's affinity for electrons is called its **electronegativity**. Oxygen has a particularly high electronegativity, thus water has two polar covalent bonds; electrons stay closer to the oxygen atom, giving it a partial negative charge ($\delta-$) and the hydrogen atoms partial positive charges ($\delta+$).

- Carbon and hydrogen have approximately equal electronegativities. Nitrogen's electronegativity is greater than C and H but less than oxygen's.

How Does Ionic Bonding Hold Ionic Compounds Together?

- **Ionic bonds** form when electrons are completely transferred from one atom to another (**Figure 2.6**) to give the resulting **ions** (charged atoms) a full outermost electron shell.

- An atom that loses an electron becomes positively charged (a **cation**), and an atom that gains an electron becomes negatively charged (an **anion**). Opposite charges attract, and the ionic bonds between cations and anions form salts (e.g., table salt, NaCl).

- Electron sharing in chemical bonds exists in a continuum from equal sharing (nonpolar covalent bonds) to partial sharing (polar covalent bonds) to complete transfer or no sharing (ionic bonds). See **Figure 2.7**.

- Molecules found in organisms mostly have the stronger covalent bonds.

CHECK YOUR UNDERSTANDING

(a) Examine Figures 2.2 and 2.3.
   What type of bond is found in lithium fluoride and why?

_____

_____

_____

   In glucose ($C_6H_{12}O_6$), what type of bonds are found between: C and H?

_____

   C and O?

_____

   Why?

_____

_____

_____

Some Simple Molecules Formed from H, C, N, and O

- The number of unpaired electrons in the valence shell determines the number of bonds an atom can make.

- Thus, C bonds with 4 H to form methane (CH_4), N bonds with 3 H to form ammonia (NH_3), and O bonds with 2 H to form water (H_2O). See **Figure 2.8**.

Double and Triple Bonds

- When there are two unpaired electrons in the valence shell, two nuclei can share four electrons in a double bond (e.g., carbon dioxide, CO_2). A triple bond can form if there are three unpaired electrons (e.g., molecular nitrogen, N_2).

The Geometry of Some Simple Molecules

- CH_4, NH_3, H_2O, CO_2, and N_2 were prevalent in early Earth's atmosphere and ocean. Molecular function is influenced by shape, so let's examine the shape of these molecules.

- Molecular shape depends on bond angles, which depend on the orbitals participating in the bond. An electron orbital is the area of space an electron inhabits around a nucleus.

- The four C orbitals with unpaired electrons have a tetrahedral shape, the three unpaired N orbitals are pyramid shaped, and the two unpaired O orbitals are bent (**Figure 2.9**). Consequently, methane (CH_4) is shaped like a tetrahedron, ammonia (NH_3) is shaped like a pyramid, and water (H_2O) is planar and bent.

Representing Molecules (**Figure 2.10**)

- The **molecular formula** states the numbers and types of atoms in a molecule (e.g., H_2O, CH_4).

- **Structural formulas** show which atoms are bonded together, and they indicate single, double, or triple bonds. Other models show three-dimensional geometry.

CHECK YOUR UNDERSTANDING

(b) Draw H_2CO and HCN, showing the electrons and orbitals involved in the bonds as in **Figure 2.8** (*Hint:* C forms a double bond with O and a triple bond with N in these two molecules).

Quantifying the Concentration of Key Molecules

- The **molecular weight** of a molecule equals the sum of the mass numbers of all the atoms in the molecule. One **mole**, 6.022×10^{23} molecules, has a mass equal to the molecular weight expressed in grams.

- Molecules bond to each other in whole-number combinations; and by measuring in moles, you know how many molecules you have.

- The concentration of a substance dissolved in a liquid (a **solution**) is typically expressed as molarity (M)—the number of moles present per liter of solution.

(c) A nitrogen atom has 7 protons and 7 neutrons. What is its atomic number?

_____

What is its mass number?

_____

What is the molecular weight of molecular nitrogen (nitrogen gas, N_2)?

_____

What type of bond holds N_2 together (be specific)?

_____

Is this bond weaker or stronger than the bond holding NaCl together?

_____

How much would a mole of N_2 weigh?

_____

2.2 The Early Oceans and the Properties of Water

- Life originated in and is based on water because water is a great **solvent** (substances dissolve easily in it).

Why Is Water Such an Efficient Solvent?

- The bonds between H and O in water are polar covalent because hydrogen attracts electrons less strongly than oxygen does. The whole molecule is **polar** because the O side of the molecule has a partial negative charge ($\delta-$) and the H atoms carry partial positive charges ($\delta+$).

- Partial charges on the H and O in different H_2O molecules attract each other and form weak **hydrogen bonds**. Hydrogen bonds are weaker than covalent or ionic bonds, but many together have important biological effects.

- Hydrogen bonds also form between H_2O and other polar molecules or charged substances dissolved in water. These bonds help these **hydrophilic** substances stay in solution. Uncharged and nonpolar or **hydrophobic** substances interact with each other and avoid water (**Figure 2.13**).

(d) Compare and contrast hydrogen bonds and ionic bonds.

_____

_____

_____

_____

How Does Water's Structure Correlate with Its Properties?

Cohesion, Adhesion, and Surface Tension

- Water molecules stick together due to hydrogen bonding. This binding between molecules of the same type is called **cohesion**. When a water molecule hydrogen-bonds with a polar or charged surface, the process is called **adhesion**.

- Cohesion and adhesion explain many properties of water, including the formation of a concave **meniscus** when water adheres to the sides of a glass container and **surface tension** when water sticks together at the surface (**Figure 2.14**).

Water Is Denser as a Liquid than as a Solid (**Figure 2.15**)

- Hydrogen bonds in ice connect molecules in a crystal-like pattern, whereas there are fewer hydrogen bonds in water, causing water molecules to pack more tightly. Thus, ice is less dense than liquid water, an extremely unusual property. Large bodies of water rarely freeze through because ice floats.

Water Has a High Capacity for Absorbing Energy

- The amount of energy it takes to raise the temperature of 1 gram of a substance 1°C is called its **specific heat**. Water has a very high specific heat because hydrogen bonds must be broken in order for water molecules to move faster (**Table 2.1**).

- Water also has a high **heat of vaporization** (the amount of energy required to change 1 gram of liquid to a gas).

- Compounds important in chemical evolution might have rained down from the atmosphere and dissolved in ocean water, where water's temperature-buffering capacity would have protected them from energy sources that could have broken them apart.

(e) Ocean temperatures have increased about 1/10th of a degree in the past 10 years. This does not sound like much—so why is it viewed as strong evidence for global warming?

_____

_____

_____

_____

Acid–Base Reactions and pH

- A water molecule can lose a proton (also called a **hydrogen ion** H^+) which then tends to associate with another water molecule to make H_3O^+. This dissociation reaction can be written: $H_2O + H_2O \leftrightarrow; H_3O^+ + OH^-$ (a **hydroxide ion**).

- **Acid–base reactions** are chemical reactions in which one molecule or ion gives up protons (an **acid**), and another accepts protons (a **base**). Thus, water can act as both an acid and a base. Water is a very weak acid because only a few molecules dissociate.

- Strong acids and bases can greatly increase or decrease (respectively) a solution's proton concentration. This in turn may affect the amount of hydrogen bonding and other interactions for all compounds in the solution.

- Pure water at 25°C has a proton concentration, symbolized as $[H^+]$, of 1.0×10^{-7} M. The **pH scale** indicates $[H^+]$: $pH = -\log[H^+]$ (**Figure 2.16**).

- Pure water has a pH of 7 and is considered neutral. Solutions with a pH lower than 7 are acidic (more likely than water to give up protons); those with a pH greater than 7 are basic and tend to accept protons from solution.

- Most living cells have a pH of about 7. See **Box 2.1** for a discussion of how **buffers** cause solutions to resist pH change.

2.3 Chemical Reactions, Chemical Evolution, and Chemical Energy

- A **chemical reaction** occurs when one molecule combines with or is broken into others. The chemical evolution theory proposes that the simple molecules present on ancient Earth reacted with one another to create larger, more complex molecules. Under what conditions could this have occurred?

How Do Chemical Reactions Happen?

- A common chemical reaction in volcanic gases is $CO_2(g) + H_2O(l) \leftrightarrow H_2CO_3(aq)$, meaning that carbon dioxide gas (g) and liquid (l) water (**reactants**) can form an aqueous solution (aq) of carbonic acid (**product**). The double arrow means the reaction is reversible, and the equation is balanced with the same number of each atom per side.

- **Chemical equilibrium** occurs when forward and reverse reactions proceed at the same rate. Changes in temperature or concentration alter the chemical equilibrium. For example, adding more reactants will cause more product to be made until a new equilibrium is achieved.

- **Endothermic** reactions such as the conversion of water from liquid to gas form must absorb heat to proceed; **exothermic** reactions release heat.

What Is Energy?

- **Energy** is the ability to do work or supply heat, and **potential energy** is stored energy. For example, electrons are attracted to the protons in the nucleus so electrons in outer shells have more potential energy than those in inner shells and will fall into an inner shell if possible (**Figure 2.17**).

- **Kinetic energy** is movement energy. Molecules are always moving; this type of kinetic energy is called **thermal energy**. **Temperature** measures thermal energy. If an object is "cold," its molecules are moving slowly.

- When a cold object touches a hot object (with faster-moving molecules), thermal energy (**heat**) passes from the hot to the cold object. The cold object's molecules heat up and move faster, while the hot object's molecules cool down and move slower.

- One type of energy can change into another type of energy, but it is never created or destroyed. The **first law of thermodynamics** expresses this concept by stating that energy is conserved.

- An electron in an outer shell has high potential energy, similar to that of a water molecule at the top of a waterfall (**Figure 2.18**).

- As an electron falls to a lower energy shell, potential energy is converted to kinetic energy that, in turn, is changed into light or heat as the electron reaches the lower shell. The potential energy lost equals the light or heat energy released.

- Molecules on ancient Earth were exposed to a lot of thermal energy from volcanic eruptions and ultraviolet radiation from the Sun. This energy input affected the reactions that could occur.

Chemical Evolution: A Model System

- Computer models can simulate chemical reactions that might have occurred on ancient Earth. Researchers investigated whether formaldehyde (H_2CO) and hydrogen cyanide (HCN) could have been produced because formation of these molecules is the first step of chemical evolution.

- Formation of these molecules is not spontaneous: Energy input is necessary.

What Makes a Chemical Reaction Spontaneous?

- Reactions occur automatically or spontaneously without any energy input when products have less potential energy than reactants (exothermic reactions) and/or when products are less ordered than reactants.

- Specifically, a reaction is spontaneous when the **Gibbs free-energy change** (ΔG) is negative (Δ means change).

- $\Delta G = \Delta H - T\Delta S$, where H is potential energy, T is temperature in degrees Kelvin, and S is **entropy** (a measure of disorder). See **Figures 2.19 and 2.20**. Changes in both potential energy and entropy determine a reaction's ΔG, with changes in entropy becoming more important at high temperatures.

- The **second law of thermodynamics** states that in an isolated system, entropy increases over time.

- Reactions that proceed spontaneously ($\Delta G < 0$) are **exergonic**. Reactions that require energy input to occur ($\Delta G > 0$) are **endergonic**. When $\Delta G = 0$, a reaction is at equilibrium and no net change in reactants or products occurs.

Energy Inputs and the Start of Chemical Evolution

- A computer model of chemical reactions among CH_4, NH_3, H_2O, CO_2, H_2, and N_2 molecules included both spontaneous reactions and reactions that occur in the presence of sunlight energy (photons).

- High-energy photons reached ancient Earth because there was little ozone (O_3) in the atmosphere. These photons can knock electrons away from valence shells, breaking apart molecules and forming **free radicals**—highly reactive atoms with unpaired electrons (**Figure 2.21**).

- Reactions involving these free radicals were included in the computer simulation of reactions that could occur on ancient Earth.

The Roles of Temperature and Concentration in Chemical Reactions

- Spontaneous reactions ($-\Delta G$) may nonetheless occur extremely slowly. However, high temperatures and high concentrations cause more reactant collisions and faster reaction rates (**Figure 2.22**).

- As computer models have shown, significant amounts of H_2CO and HCN form under temperature and concentration conditions likely found in early Earth's atmosphere.

CHECK YOUR UNDERSTANDING

(g) In the experiment shown in **Figure 2.22**, why did the students test temperature and concentration separately instead of together (for example, by increasing temperature and concentration at the same time)?

_____

_____

What would happen to the reaction rate if temperature and reactant concentration were increased at the same time?

_____

_____

What might happen to the reaction rate when the temperature was increased but the reactant was decreased?

_____

_____

CHECK YOUR UNDERSTANDING

(f) Consider the following reaction: $CO_2(g) + 2\,H_2O(g) \rightarrow H_2CO(g) + H_2O(g)$.

Is ΔH positive or negative?

_____

Is ΔS positive or negative?

_____

How about ΔG?

_____

Will this reaction proceed spontaneously?

_____

How Did Chemical Energy Change during Chemical Evolution?

- Electrons in H_2CO bonds are farther from the atom nuclei than in the reactant bonds, and thus have more potential energy. $CO_2(g) + 2\ H_2(g) + \text{sunlight} \rightarrow H_2CO(g) + H_2O(g)$

- The formation of H_2CO and HCN is a critical step in chemical evolution because energy from sunlight has been converted to **chemical energy** (potential energy in chemical bonds). This new source of chemical energy makes possible the formation of larger, more complex molecules.

2.4 The Importance of Carbon

- Carbon has four unpaired electrons, allowing it to make four covalent bonds. This flexibility in bonding makes possible an incredible diversity of carbon-based molecules. Almost all molecules made by organisms contain carbon, so molecules with a C–C bond are called **organic molecules** (**Figure 2.23**).

Linking Carbon Atoms Together (Figure 2.24)

- Larger organic molecules with additional carbon–carbon bonds easily form when CH_2O and HCN are heated.

- Chemical evolution was possible because energy was available to power endergonic reactions. Some of the molecules produced by chemical evolution are still used in cells today.

Functional Groups

- Groups of H, N, or O atoms (**functional groups**) bonded to C determine the behavior of organic compounds (**Table 2.3**). Carbon provides the structural framework, but the functional groups determine a molecule's chemistry.

- For example, amino groups attract protons but carboxyl groups tend to donate them, and carbonyl groups often participate in reactions that link molecules together. Molecules with hydroxyl groups are weak acids whereas phosphate groups give molecules a negative charge and sulfhydryl groups link different molecules together.

- During chemical evolution, further reactions produced the common large molecules still found in cells today.

B. INTEGRATING YOUR KNOWLEDGE

Looking Back—Key Concepts from Earlier Chapters

Evolution—Chapter 1

Having learned that species change over time and that all organisms descend from a single common ancestor,

(h–m) Label the following functional groups:

h) $O^- \!\!-\!\! \underset{\underset{O^-}{|}}{\overset{\overset{O}{||}}{P}} \!\!-\!\! O^-$ i) $\text{HO} \!\!-\!\! \underset{R}{\overset{\overset{O}{||}}{C}}$ j) $R\!-\!SH$

k) $\underset{R\quad H}{\overset{\overset{O}{||}}{C}}$ l) $R\!-\!\underset{H}{\overset{H}{N}}$ m) $R\!-\!OH$

in Chapter 2 you explored the conditions under which this common ancestor (the first life) evolved.

Looking Forward—Key Concepts in Later Chapters

Chemical Evolution—Chapters 3–6

Chapter 2 sets the stage for and introduces the major players in chemical evolution. **Chapters 3 through 6** continue this story and describe several hypotheses about how life evolved from nonlife.

Chemical Reactions—Chapters 9 and 10

Respiration, photosynthesis, and other important metabolic pathways involve a series of chemical reactions.

Water Is Special—Chapters 37, 42, 54, and others

Water has some unusual properties that make it essential to life. **Chapters 37** and **42** discuss how plants and animals deal with their water requirements, and **Chapter 54** describes how water cycles through the ecosystem.

Functional Groups—Chapters 3, 4, 5, 9, 38, 43, and others

Functional groups introduced in this chapter are discussed in greater detail later with reference to specific chemical pathways and nutrition.

C. MASTERING CHALLENGING TOPICS

Chapter 2 is a whirlwind tour of basic chemistry. You need to know these basics in order to understand (1) molecules and chemical reactions that are essential to life, and (2) how life could have evolved from nonlife. You may want to look at an introductory chemistry text to ensure you understand concepts like **electron orbitals**, **potential** and **kinetic energy**, different **bond** types, **pH**, and how **acid–base reactions** work. You may also want to go back and carefully examine the figures in this chapter. Make a list for yourself, itemizing how the concepts introduced in this chapter relate to our best hypotheses about the chemical reactions that could have led to life.

D. ASSESSING WHAT YOU'VE LEARNED

1. Two isotopes of an element vary in the number of _____.
   a. electrons
   b. protons
   c. neutrons
   d. atoms

2. The number of _____ within an atom never varies among atoms of a specific element.
   a. neutrons
   b. protons
   c. electrons

3. An organic molecule always contains one or more atoms of:
   a. nitrogen
   b. sulfur
   c. carbon
   d. oxygen

4. Which group of atoms determines a molecule's behavior?
   a. functional group
   b. carbon skeleton
   c. hydrogen atoms

5. 96% of matter found in living organisms is composed of these four elements:
   a. hydrogen, carbon, oxygen, sulfur
   b. carbon, sulfur, nitrogen, iron
   c. hydrogen, carbon, oxygen, nitrogen
   d. sulfur, hydrogen, oxygen, nitrogen

6. A large boulder is balanced on top of a hill. You give the boulder a push, and it rolls down the hill. This is an example of transferring _____ energy into _____ energy.
   a. kinetic; potential
   b. potential; kinetic
   c. kinetic; thermal

7. Which of these factor(s) determine(s) whether a chemical reaction will occur spontaneously?
   a. amount of entropy (disorder in a group of molecules)
   b. temperature
   c. amount of potential energy differences between products and reactants
   d. all of the above

8. Which phase of matter has the highest entropy?
   a. solid
   b. liquid
   c. gas

9. Which statement regarding the properties of water is *not* correct?
   a. Water is a good solvent.
   b. The oxygen atom in water has a high electro-negativity.
   c. Water is a polar molecule.
   d. Solid water (ice) is denser than liquid water.

10. Substances that give up protons H^+ during acid–base reactions are:
    a. isotopes
    b. acids
    c. bases
    d. ions

11. How many protons are present in 1 liter of pure water? (Hint: See definition of molarity.)
    a. 1.0×10^{-7}
    b. 6.022×10^{23}
    c. 6.022×10^{16}
    d. No protons are present.

12. A cation has a _____ charge.
    a. positive
    b. negative
    c. neutral

13. The bonds between the oxygen and hydrogen atoms within a water molecule are _____ bonds.
    a. hydrogen
    b. ionic
    c. polar covalent
    d. nonpolar covalent

14. Which of the following is part of the theory of chemical evolution?
    a. Evolution causes nonpolar covalent bonds to become polar covalent bonds.
    b. Elements evolve through time as their bonds change.
    c. Natural selection led to formation of complex molecules found in living cells today.
    d. On ancient Earth, simple chemical compounds combined to form larger complex molecules.

15. Which of the following functional groups can act as a base?
    a. carboxyl
    b. amino
    c. hydroxyl
    d. carbonyl

CHAPTER 2—ANSWER KEY

Check Your Understanding

(a) Lithium fluoride has ionic bonds; lithium, like sodium, tends to transfer one electron to fluoride which, like chlorine, accepts one electron to fill its outer electron shell. In glucose C and H form nonpolar covalent bonds because their electronegativities are similar, whereas C and O form polar covalent bonds because O has a higher electronegativity than C.

(b)

(c) The atomic number for nitrogen is 7, and its mass number is 14.

The molecular weight of N_2 is 28. A triple covalent bond holds N_2 together, and a mole of N_2 would weigh 28 g.

Covalent bonds are stronger than ionic bonds. For example, ionic bonds easily dissolve in water.

(d) Both hydrogen bonds and ionic bonds form due to attraction of opposite charges for each other. However, only partial charges (from unequally shared electrons in covalent bonds) are involved in hydrogen bonds, while ions have gained or lost whole electrons. Hydrogen bonds are much weaker than ionic bonds.

(e) A rise in ocean temperatures is strong evidence for global warming because water has such high specific heat. It takes a *lot* of energy to warm all the ocean water even a tenth of a degree.

(f) ΔH is positive because more bonds are created than broken; the products have more potential energy. ΔS is negative because there are fewer molecules and thus less entropy in the products than in the reactants. ΔG must be positive, and this reaction cannot proceed spontaneously—an energy source (e.g., sunlight) must be present in order for it to occur.

(g) Good experimental design varies just one variable at a time. If the students had changed temperature and concentration at the same time, they would not have been able to tell which variable caused the reaction rate change. However, if temperature and reactant concentration were increased at the same time, we would expect the reaction rate to increase even more than if only one of these variables were changed. It is hard to predict what would happen to the reaction rate if the temperature were increased, but the reactant concentration decreased since these effects oppose each other. The end reaction rate would depend on the amount of temperature increase and reactant concentration decrease.

(h) phosphate

(i) carboxyl

(j) sulfhydryl

(k) carbonyl

(l) amino

(m) hydroxyl

Assessing What You've Learned
1. c; 2. b; 3. c; 4. a; 5. c; 6. b; 7. d; 8. c; 9. d; 10. b; 11. c; 12. a; 13. c; 14. d; 15. b

3

Protein Structure and Function

A. KEY CONCEPTS

- Alexander I. Oparin and J.B.S. Haldane proposed the theory of chemical evolution in the 1920s.

- This theory proposes that energy from sunlight and other sources was converted into chemical energy (potential energy in molecular bonds) as more and more complex carbon compounds formed due to re-actions among molecules present in ancient Earth's atmosphere and ocean.

- The four steps of chemical evolution (each requiring energy input) are (1) production of small molecules such as H_2CO and HCN; (2) creation of a **prebiotic soup** in shallow ocean waters as larger organic subunit molecules formed; (3) linkage of these organic sub-units to make the large organic molecules important in cells today; and (4) evolution of a self-replicating molecule.

- This chapter focuses on proteins: one of the four major classes of large organic molecules.

3.1 What Do Proteins Do?

- Proteins do most of the work in a cell (**Table 3.1**). Protein structure largely determines protein func-tion.

- Antibodies and complement proteins defend cells against disease-causing viruses and bacteria.

- Motor and contractile proteins move cells and cell cargo.

- Proteins called enzymes speed up cell reactions (**catalysis**).

- Proteins and peptide hormones carry and receive signals between cells.

- Structural proteins provide mechanical support and form accessory structures like hair.

- Transport proteins move molecules in and out of cells as well as throughout the body.

- These protein functions will be further examined in later chapters.

3.2 Early Origin-of-Life Experiments

- Stanley Miller (1953) was first to simulate chemical evolution in the laboratory in order to experimen-tally test whether the first stages of chemical evolu-tion would have occurred on ancient Earth.

- Methane (CH_4), ammonia (NH_3), and hydrogen (H_2) were placed in a large flask representing the atmos-phere. This flask was connected to another flask with a model ocean.

- Miller boiled the "ocean" water to add water vapor to the atmosphere gases. As the vapor cooled, it con-densed and rained down, carrying molecules with it through tubing that connected back to the water flask so that the water and molecules dissolved in it circulated through the system (**Figure 3.1**).

- Miller added sparks (simulated lightning) to the "atmosphere" as an energy source.

- Presto! HCN and H_2CO, important precursors for more complex organic molecules, were made along with other more complex organic compounds, including amino acids, the subunits that connect to form proteins.

3.3 Amino Acids and Polymerization

- Recent experiments similar to Miller's show that amino acids and other organic molecules easily form under conditions that more accurately simulate con-ditions on ancient Earth.

- Most biologists agree that amino acids and other im-portant organic compounds would have easily been made on early Earth, forming the prebiotic soup in this second stage of chemical evolution.

The Structure of Amino Acids

- The huge diversity of proteins in living organisms is made from just 21 amino acid subunits. All **amino acids** have a central carbon atom bonded to H, an amino functional group (NH_2), a variable side chain, and a carboxyl functional group, COOH (**Figure 3.2**).

- In water (pH 7), the amino group attracts a proton to form NH_3^+ and the carboxyl group loses a proton to form COO⁻. This polarity helps amino acids stay in solution and makes them more reactive. Amino acids are so named because the carboxyl group acts as a weak acid.

The Nature of Side Chains

- The 21 major amino acids differ only in the variable side chain (R-group) attached to the central carbon (**Figure 3.2**). R-groups differ in their size, shape, reactivity, and interactions with water. Many include functional groups discussed in Chapter 2.

- Amino acids can be grouped as nonpolar, polar, or electrically charged (**Figure 3.3**).

- Nonpolar **hydrophobic** R-groups cannot form hydrogen bonds with water and tend to group together in solution, whereas polar or charged **hydrophilic** R-groups interact readily with water (**Table 3.2**).

- R-group structure largely determines an amino acid's properties and function. For example, amino acid side chains containing solely carbon and hydrogen are hydrophobic and are not very reactive.

CHECK YOUR UNDERSTANDING

(a) Look at the structures of valine and serine in **Figure 3.3**. Why is serine more likely to interact with water than valine?

_____

_____

_____

What Are Optical Isomers?

- **Isomers** are molecules with the same molecular formula but different structures (**Figure 3.4**).

- **Structural isomers** have the same atoms, but the atoms are attached in different orders.

- **Geometric isomers** differ in the way atoms or groups attach to the two sides of a double bond or ring.

- **Optical isomers** differ in the arrangement of atoms or groups around a carbon atom that has four different groups attached to it.

- All amino acids except glycine have optical isomers—two mirror-image forms.

- Surprisingly, only "left-handed" amino acids are found in cells. The theory of chemical evolution does not yet include a good explanation for this observation.

CHECK YOUR UNDERSTANDING

(b) Explain why only glycine does not have optical isomers.

_____

_____

_____

How Do Amino Acids Link to Form Proteins?

- **Polymerization** is the process of linking molecular subunits (**monomers**) to make **polymers** (macromolecules of attached monomers; **Figure 3.5**).

- Proteins are macromolecules (large polymers) made by connecting amino acid monomers. These polymers presumably formed in the prebiotic soup.

- Monomers do not form polymers spontaneously, because polymers have less entropy ($-\Delta S$) and higher potential energy ($+\Delta H$) than do unlinked monomer subunits, making ΔG positive at all temperatures.

Could Polymerization Occur in the Energy-Rich Environment of Early Earth?

- Monomers polymerize through **condensation reactions** (also known as **dehydration reactions**) that release a water molecule as the polymer bond forms. In the reverse reaction, **hydrolysis**, water reacts with the polymer to release a monomer subunit (**Figure 3.6**).

- Condensation reactions require energy input since ΔG is positive for these reactions (hydrolysis is energetically favorable and increases entropy); but even with energy inputs, few polymers are produced in an aqueous environment.

- Experiments show that polymers can form on tiny mineral particles like those found in clay or mud. Adsorption to these tiny particles protects a growing macromolecule from hydrolysis. Thus, chemical evolution of polymers could have occurred in coastal environments.

The Peptide Bond

- Amino acid condensation reactions bond the carboxyl group of one molecule to the amino group on another to form a **peptide bond** (**Figure 3.7**). A chain of linked amino acids (or **residues**) is a **polypeptide**.

- Peptide bonds are especially strong covalent bonds, having some double-bond qualities.

- Side chains stick out from linked amino acid residues (**Figure 3.8**).

- Polypeptides are flexible and have directionality (the N-terminus has an unlinked amino group, and the C-terminus has a carboxyl group; **Figure 3.9**).

- By convention, polypeptides are shown with the N-terminus on the left and the free carboxyl on the right. Amino acids are numbered starting from the N-terminus.

- Peptide bonds cannot rotate, but the bonds to each side of them can, so polypeptides are flexible molecules. **Oligopeptides** have less than 50 amino acids, whereas proteins have 50 or more linked amino acids.

3.4 What Do Proteins Look Like?

- Proteins have incredibly diverse shapes and functions; let us examine the four levels of protein structure (**Figure 3.10**).

Primary Structure

- The **primary structure** of a protein is its unique sequence of amino acids. With 21 amino acids, just one protein that is 10 amino acids long has billions of possible amino acid sequences! More specifically, a polypeptide of length n has 21^n possible sequences.

- Because the amino acid R-groups affect a polypeptide's size, shape, chemical reactivity, and interactions with water, just one amino acid change can radically alter protein function (**Figure 3.11**).

Secondary Structure

- Hydrogen bonding between the carboxyl O of one amino acid residue and the amino H of another creates a protein's **secondary structure**. The partial charges on these atoms attract each other only when a polypeptide is bent such that these functional groups lie close to each other (**Figure 3.12**).

- Usually these hydrogen bonds cause the polypeptide either to coil into an α-helix or to fold into a β-pleated sheet (like the folds in a paper fan). Ribbon diagrams show α-helices as coils and β-pleated sheets as arrows lying next to each other with alternating directionality.

- Primary structure affects the secondary structures that form; different amino acids facilitate formation of different secondary structures.

- A protein's secondary structure increases its stability; although one hydrogen bond is fairly weak, many hydrogen bonds contribute to each of these secondary structures.

Tertiary Structure

- Amino acid R-groups are largely responsible for a polypeptide's three-dimensional shape or **tertiary**

structure. R-groups interact with other side chains or with the peptide-bonded backbone, making the polypeptide bend and fold into a precise shape (**Figure 3.13**).

- Hydrogen bonds form between H and the COO– group found in all amino acids and between H and R-groups with partial negative charges.

- Hydrophobic side chains group together in water and are stabilized by weak electrical attractions called **van der Waals interactions**. These are weaker than hydrogen bonds, but a large number of these attractions can significantly affect the folding and stability of proteins, often causing a globular shape.

- Covalent **disulfide bonds** or bridges can form between sulfur atoms such as those in cysteine.

- Ionic bonds can form between side chains with opposite charges.

- A protein's primary and secondary structure determine its tertiary structure.

Quaternary Structure

- Some proteins have several polypeptide subunits; bonds or other interactions between subunit R-groups or peptide backbones produces **quaternary structure** (**Figure 3.14**). This quaternary structure depends, in turn, on the primary, secondary, and tertiary structure of the protein's subunits.

- The combined effects of primary, secondary, tertiary, and sometimes quaternary structure (**Table 3.3**) allow for amazing diversity in protein form and function.

CHECK YOUR UNDERSTANDING

(c) Your text explains that a change from glutamate to valine in one of hemoglobin's amino acids causes sickle-cell disease. Which structural levels were likely affected and why?

_____

_____

_____

Would you expect the effects to be more or less serious if glutamate were changed to aspartate? Why?

_____

_____

Folding and Function

- Polypeptides in solution automatically fold so that hydrophobic side chains group together away from water and hydrophilic side chains face water. This folding is often spontaneous and exergonic because the bonds and van der Waals interactions stabilize the molecule.

- The folded three-dimensional shape is basic to protein function. If chemical agents or high temperatures disrupt the hydrogen and disulfide bonds that stabilize the folded shape, a protein is **denatured** and unable to function normally (**Figure 3.15**, also see **Box 3.1**).

- Proteins called **molecular chaperones** help proteins fold correctly in cells. Many of these molecular chaperones are heat-shock proteins, produced after a cell experiences high temperature or other conditions that disrupt protein folding.

3.5 Enzymes: An Introduction to Catalysis

- Just because a chemical reaction is spontaneous $(-\Delta G)$ does not mean it happens quickly.

- Molecules must collide with the correct orientation and have enough kinetic energy to break and/or form chemical bonds despite the repulsion of the electrons involved in the bond.

- Enzymes bring reactant molecules, or **substrates**, together in the right position and decrease the amount of kinetic energy necessary by facilitating the formation of a **transition state** in which a combination of old and new bonds is present.

- **Activation energy** is the free energy necessary to achieve this transition state. Less-stable transition states require more activation energy; the higher the activation energy, the slower the reaction rate.

- Chemical reactions occur faster at higher temperatures because more molecules have enough energy to reach the transition state (**Figure 3.17**).

- **Figure 3.18** graphically represents free-energy changes during a chemical reaction. ΔG is negative, so this reaction is exothermic and spontaneous. But the activation energy (E_a) needed to form the transition state is high, so this reaction would occur very slowly.

- Less-stable transition states have higher activation energies and proceed slowly. Reactions can be sped up by stabilizing the transition state and lowering the activation energy and thus the amount of kinetic energy necessary for the reaction to proceed.

(d) Under identical conditions, which of the reactions shown below will proceed faster and why?

_____

_____

_____

- Electrons in the transition state can be stabilized by interactions with other ions, atoms, or molecules. A **catalyst** is a molecule that lowers the activation energy of a reaction by stabilizing the transition state (**Figure 3.19**).

- Catalysts are not used up in chemical reactions and do not change ΔG or the energy of the reactants or products. Only E_a, the amount of energy required to achieve the transition state, is affected.

- **Enzymes** are protein catalysts that typically catalyze only one reaction. They speed up reactions by bringing reactants together in the correct orientation and stabilizing the transition state.

- Most biological chemical reactions occur only at meaningful rates in the presence of an enzyme; many enzymes allow reactions to proceed a million times faster than they would without the enzyme.

How Do Enzymes Work?

- An early model of enzyme action, the **lock-and-key model** (Fischer, 1894), described enzymes as being like locks into which reactant "keys" fit. Inside the enzyme "lock," Fischer said, the **substrates** participated in the chemical reaction catalyzed by that enzyme.

- This early model correctly identified (1) that enzymes bring substrates together in specific positions that facilitate reactions, and (2) that enzymes are very specific as to which reactions they catalyze because substrates bind only to sites with the appropriate shape and chemical properties.

- The substrate-binding site on an enzyme is called the **active site** (**Figure 3.20**) which is usually a fold or hole within the globular protein.

- The lock-and-key model has been replaced by the **induced fit** model because enzymes are not rigid like locks; instead, they are flexible and often change shape as substrates bind to the active site.

- Interactions between the substrate and active-site R-groups include hydrogen bonding, formation of temporary covalent bonds involved in atom transfer, and gain or loss of protons aided by acidic or basic R-groups (**Figure 3.21**).

- Reaction products have a low affinity for the active site and are therefore released from the enzyme.

- Substrate interaction with the enzyme active site causes stabilization of the transition state and a decrease in the reaction's activation energy.

- Enzyme action has three steps: **initiation**, when reactants are correctly oriented as they bind to the active site; **transition state facilitation**, as interactions between the substrate and active site R-groups lower the activation energy; and **termination**, when reaction products are released from the enzyme (**Figure 3.22**).

Do Enzymes Act Alone?

- Enzymes often require **enzyme cofactors**—atoms or molecules not included in an enzyme's primary structure. These cofactors are either metal ions or small organic molecules called **coenzymes**.

- Enzyme cofactors often bind to the active site and are involved in transition state stabilization.

- Many vitamins are necessary because they are precursors for enzyme cofactors. Vitamin deficiency diseases occur when there are not enough enzyme cofactors to allow normal enzyme function.

- Most enzymes are regulated by other molecules that bind to the enzyme, causing a shape change.

- **Competitive inhibition** occurs when a regulatory molecule similar in size and shape to the substrate competes with the substrate for active site binding (**Figure 3.23a**).

- **Allosteric regulation** occurs when a molecule causes a change in enzyme shape by binding to the enzyme at a location other than the active site. Depending on the way in which the enzyme's shape changes, allosteric regulation can either increase or decrease enzyme activity (**Figure 3.23b**).

What Limits the Rate of Catalysis?

- Enzyme kinetics describes how various factors affect enzyme-catalyzed reaction rates.

- **Figure 3.24** shows how the rate of product formation increases linearly for a given increase in substrate concentration at low substrate concentrations, but levels out at high substrate concentrations. Uncatalyzed reactions show linear increases in reaction rate across all ranges of substrate concentrations.

- Initially, each increase in substrate concentration allows the available enzymes to facilitate that many more reactions. But, as the substrate concentration increases, enzyme active sites eventually become filled as fast as they are emptied, and no further increase in reaction rate is possible.

- All enzymes show this type of kinetics, but enzymes with high substrate affinities rapidly achieve their maximum reaction rate as substrate concentration increases (left-shifted curve), while enzymes with low affinities for their substrate require fairly high substrate concentrations before reaching their maximum rate of reaction (right-shifted curve).

- Enzyme-catalyzed reaction rates are influenced by both substrate concentration and an enzyme's affinity for the substrate.

How Do Physical Conditions Affect Enzyme Function?

- Enzymatic activity is strongly influenced by temperature (effect on enzyme movement and substrate kinetic energy) as well as pH (effect on R-group charge and proton- and electron-transfer ability of the active site).

- Most enzymes function best at some particular temperature and pH. Temperature and pH optima usually reflect the environment in which the enzyme functions because natural selection favors the survival and reproduction of organisms with enzymes that function efficiently (**Figure 3.25**).

(e) Examine **Figure 3.21:** Explain how a decrease in pH might affect this enzyme's activity, including specific reference to its active site R-groups.

_____

_____

_____

- Primary structure variations that increase enzyme activity at the temperatures and pH the enzyme typically experiences help organisms adapt to their environment.

Was the First Living Entity a Protein?

- A self-replicator would need to catalyze the polymerization of its copy, and enzymes are the most efficient catalysts known.

- Proteins also could have formed in the prebiotic soup, but the first self-replicator likely also needed a copy-making template—something not found in proteins. Consequently, the first living entity was probably not a protein.

B. INTEGRATING YOUR KNOWLEDGE

Looking Back—Key Concepts from Earlier Chapters

Chemical Evolution—Chapter 2

The start of chemical evolution and basic chemistry needed to understand the chemical reactions discussed in this chapter are described in **Chapter 2**.

Chemical Bonds, Chemical Reactions, and Chemical Energy—Chapter 2

Chapter 2 also provides background information on how chemical bonds form and how Gibbs free energy (ΔG) determines whether reactions can occur spontaneously.

Looking Forward—Key Concepts in Later Chapters

Proteins are the workhorse molecules of cells, so these molecules will appear in many later chapters. Here, a few protein functions are highlighted.

Structural Proteins and Transport Proteins— Chapters 6–8, 37, 42, 43, and others

Proteins form most of the major structures in cells and are responsible for transporting molecules into and out of cells.

Enzymes—Chapters 9, 10, 16, and others

Chapters 9 and 10 focus on some of the most important cell reactions. These reactions are, of course, catalyzed by enzymes. You will learn more about enzyme regulation in these chapters. In Chapter 16 you will learn about how enzymes are involved in transcription and translation—the processes by which cells make enzymes and other proteins.

Other Protein Functions—Chapters 39, 47, and 48 (hormones) and Chapter 49 (defense)

C. MASTERING CHALLENGING TOPICS

This chapter gives you lots of information about proteins and their amino acid subunits—how they fit together and the chemical reactions involved. Be sure to carefully study the text figures; you might even want to make your own model of alanine using toothpicks and gumballs to get a better understanding of chirality, or practice drawing a generic amino acid so that you can reliably reproduce it on a test if necessary.

Sometimes analogies help us understand processes that we cannot directly observe. For example, the end-products of spontaneous chemical reactions typically have less **potential energy** (ΔH) than the starting reactants do. Similarly, an object on the floor has less potential energy than an object on the table does, because it is closer to Earth's center of gravity. But the object on the table needs **kinetic energy** to reach the **transition state** of passing the table edge before it will fall to the floor. If you have many marbles on the table and a fan blowing in the room, air currents may give some of the marbles enough kinetic energy to reach the table edge, at which point they fall to the floor. But if you put the marbles in a box, you have just increased the **activation energy** needed to reach the transition state of reaching the table edge.

As in the example of marbles on a table, even spontaneous chemical reactions may proceed so slowly as to be hardly detectable if the activation energy needed to reach the transition state is high and/or the kinetic energy of the molecules is low. Recall that enzymes increase reaction rates by stabilizing transition states and decreasing activation energy. Reaction rates can also be speeded up by heating the molecules to give them more kinetic energy (this is why sugar dissolves in hot water much faster than in cold water).

D. ASSESSING WHAT YOU'VE LEARNED

1. Stanley Miller showed experimentally that complex, carbon-containing compounds could form from simple molecules present in early Earth

atmosphere and oceans. In Miller's experiment, _____ was the energy source that permitted these complex molecules to form.
   a. heat
   b. light
   c. electricity
   d. wind

2. Amino acids are the building blocks of:
   a. nucleic acids
   b. proteins
   c. carbohydrates
   d. lipids

3. All 21 amino acids have a(n) _____ functional group and a(n) _____ functional group, but vary in their _____ group.
   a. carboxyl; amino; R-
   b. R-; carboxyl; amino
   c. amino; R-; carboxyl

4. Monomers join to form polymers via:
   a. acid–base reactions
   b. hydrolysis
   c. condensation reactions

5. The _____ structure of a protein is the sequence of amino acids.
   a. primary
   b. secondary
   c. tertiary
   d. quaternary

6. Which of the following statements regarding proteins is correct?
   a. Alpha helices and beta sheets are part of a polypeptide's tertiary structure.
   b. Nucleotides join via condensation reactions to form polypeptides.
   c. The words *polypeptide* and *protein* have identical meaning.
   d. Protein structure is correlated with protein function.

7. Which of the following side-chain interactions are important in determining tertiary structure?
   a. covalent bonds
   b. ionic bonds
   c. hydrogen bonds and van der Waals interactions
   e. a and c
   f. b and c
   g. a, b, and c

8. Why do proteins have so many different cell functions?
   a. They are the simplest cell molecules.
   b. They are structurally diverse.
   c. They were the first life-form.
   d. They have strong peptide bonds.

9. Which condition(s) will speed the rate of a chemical reaction?
   a. high kinetic energy of reactants
   b. presence of a catalyst
   c. both a and b

10. How do catalysts increase reaction rates?
    a. by stabilizing the transition state
    b. by providing chemical energy
    c. by decreasing ΔG
    d. by increasing the kinetic energy of reactants

11. Further research led to modifications in the lock-and-key model of enzyme action. Which assumption is *not* correct?
    a. Enzymes are rigid structures analogous to a lock.
    b. Enzymes bring substrates together so that reactions are more likely to occur.
    c. Enzymes are very specific as to the reaction they catalyze.

12. Why is lysine more likely to interact with water than valine?
    a. Lysine is more hydrophobic than valine.
    b. Lysine has more hydrogen than valine does.
    c. Lysine is charged at cell pH.
    d. Lysine's central carbon bonds to an amino group and a carboxyl group.

13. A molecule is a competitive inhibitor of an enzyme if:
    a. It changes enzyme shape by binding to a site other than the active site.
    b. It can bind in an enzyme's active site because it is chemically similar to the substrate.
    c. It competes with the enzyme for substrate binding.

14. At low substrate concentrations, enzyme-catalyzed reaction rates _____ as substrate concentration increases, and at high substrate concentrations, enzyme-catalyzed reaction rates _____.
    a. show an exponential increase; increase linearly
    b. increase linearly; increase linearly
    c. decrease linearly; plateau at a maximum speed
    d. increase linearly; plateau at a maximum speed

15. Which of the following statements is correct?
    a. All enzymes function best in the neutral pH range.
    b. Sometimes enzymes have different versions with different temperature and pH optima.
    c. Most enzymes are not affected by temperature or pH.
    d. Cofactors help enzymes by providing energy for chemical reactions.

CHAPTER 3—ANSWER KEY

Check Your Understanding

(a) Serine has an OH group, which is polar due to oxygen's high electronegativity; valine just has C and H in its R group, and these atoms form non-polar covalent bonds. Thus serine is more likely to interact with water.

(b) Glycine's R-group is a hydrogen, which means that the central carbon has two hydrogen atoms attached to it, making its mirror image the same as the molecule itself (like if your hand had two thumbs instead of a thumb and a pinky).

(c) All structural levels were likely affected because all are influenced by primary structure. However, the effects would have been less serious if gluta-mate were changed to aspartate because these two R-groups are similar (both are charged and soluble in water).

(d) Reaction 1 will proceed faster than reaction 2 because it has a lower activation energy. Activation energy, not ΔG, determines the speed at which a reaction takes place.

(e) If the pH were decreased, the nitrogen in R1, which accepts a proton from the substrate, might pick up a proton from the surrounding solution. R1, then, could no longer accept the substrate's proton and the enzyme would not catalyze the reaction.

Assessing What You've Learned

1. c; 2. b; 3. a; 4. c; 5. a; 6. d; 7. f; 8. b; 9. c; 10. a; 11. a; 12. c; 13. b; 14. d; 15. b

4

Nucleic Acids and the RNA World

A. KEY CONCEPTS

- Key features distinguishing life from nonlife are (1) ability to reproduce and (2) metabolism, the ability to control chemical reactions. All modern living organisms have a cell membrane, which creates a microenvironment favorable to cell metabolism.

- The theory of chemical evolution suggests that the first step in the evolution of life was the formation of a self-replicating molecule. When a descendent of this self-replicator became enclosed in a membrane, it created the first cell.

- Most biologists think the first self-replicator was a molecule of ribonucleic acid (RNA). This chapter's discussion of this RNA world hypothesis will also introduce you to nucleic acid structure and function.

4.1 What Is a Nucleic Acid?

- A **nucleic acid** is a polymer of nucleotides in which each **nucleotide** is made up of a phosphate group bonded to a sugar, which is in turn bonded to a nitrogenous base (**Figure 4.1**).

- A **sugar** is an organic compound with a carbonyl group and several hydroxyl groups; the sugar is *ribose* in **ribonucleotides** and *deoxyribose* in **deoxyribonucleotides**—these two sugars differ only in that deoxyribose has an H instead of an –OH on the second ring carbon.

- Nitrogenous bases include adenine (A) and guanine (G) from the purine structural group and cytosine (C), uracil (U), and thymine (T) from the pyrimidine group. The first three are found in both types of nucleotides, but ribonucleotides contain uracil whereas deoxyribonucleotides contain thymine.

Could Chemical Evolution Result in the Production of Nucleotides?

- Chemical evolution simulations have not yet produced nucleotides; sugars and purines are easily made, but researchers think ribose would have had to dominate for nucleic acids to form ("the ribose problem"), and pyrimidines are not easily synthesized under conditions thought to exist on ancient Earth.

How Do Nucleotides Polymerize to Form Nucleic Acids?

- Nucleotides polymerize to form nucleic acids when a condensation reaction forms a **phosphodiester linkage** between the phosphate group of one nucleotide and the –OH group on the 3' carbon of another (**Figure 4.2**).

- **Ribonucleic acid** (RNA) is made from nucleotides containing ribose, and **deoxyribonucleic acid** (DNA) is made from nucleotides containing deoxyribose. Sugar-phosphate linkages form the 'backbone' of a nucleic acid; this chain is polar because one end has an unlinked 5' carbon and the other end has an unlinked 3' carbon (**Figure 4.3**).

- The nucleotide base sequence creates a nucleic acid's primary structure, which is written out from the 5' to the 3' direction using the first letter of each nucleotide.

- In cells, enzyme-catalyzed polymerization occurs after nucleotide free energy has been raised by adding two extra phosphate groups per nucleotide (**Figure 4.4**).

- "Activated" or **phosphorylated** nucleotides can form RNA when incubated for several days with tiny mineral particles (**Box 4.1, Figure 4.6**). Raising potential energy through phosphate addition is vital to cell metabolism and makes otherwise endergonic reactions possible.

- If all the nucleotides were able to form during chemical evolution, they could have polymerized on clay particles to form RNA and DNA.

CHECK YOUR UNDERSTANDING

(a) Why can't the condensation reaction illustrated in Figure 4.2 happen as shown?

_____

_____

_____

4.2 DNA Structure and Function

- Both nucleic acids and proteins have a backbone-like primary structure formed via condensation reactions (phosphodiester bonds connecting sugars and phosphates in DNA and RNA and peptide bonds in proteins).

- Nitrogenous bases extend outward from the nucleic acid backbone, whereas amino acid R-groups extend outward from the protein backbone.

- DNA and RNA have different secondary structures. These secondary structures also differ from those found in proteins.

What Is the Nature of DNA's Secondary Structure?

- James Watson and Francis Crick discovered DNA's secondary structure in 1953. Their hypothesis built on the following results from other laboratories.

- Chemists had determined the structure of nucleotides and the nature of phosphodiester linkages, so Watson and Crick knew that DNA had a sugar-phosphate backbone.

- Chargaff's rules stated that in DNA the number of purines equals the number of pyrimidines, the numbers of T's and A's are equal, and the number of C's and G's are equal.

- Rosalind Franklin and Maurice Wilkins used **X-ray crystallography** to calculate distances between atom groups that regularly repeated; they concluded that the structure of DNA was either helical or spiral.

- Analysis of the size and geometry of deoxyribose, nitrogenous bases, and phosphate groups led Watson and Crick to conclude that Franklin and Wilkins' 2.0-nm measurement represented the width of the helix, and 0.34 nm was the distance between bases in the spiral.

- Using models (**Figure 4.5**), Watson and Crick concluded that two antiparallel DNA strands form a **double helix** with the hydrophilic sugar-phosphate backbone facing the exterior and with purine-pyrimidine pairs of nitrogenous bases on the interior, packed closely together.

- *Antiparallel* means that if one DNA strand starts with the 5′ end and ends with a 3′ end, the other strand is oriented such that its 3′ end matches up with the first strand's 5′ end.

- Within the helix, adenine hydrogen bonds with thymine and guanine hydrogen bonds with cytosine: this is known as **complementary base pairing** or **Watson-Crick pairing** (**Figure 4.6**). The guanine-cytosine pair is stronger than the adenine-thymine pair because this pair forms three, instead of two, hydrogen bonds.

(b) Why did Watson and Crick conclude that adenine bonds to thymine and guanine bonds to cytosine within the double helix?

_____

_____

_____

_____

- DNA's double helix is like a twisted ladder where the complementary base pairs form the rungs. Twisting occurs to maximize hydrogen bonding between the pairs (**Figure 4.7**).

- Each turn of the helix (34 nm) has 10 complementary base pairings, and the double helix has a major and minor groove (**Figure 4.8**).

- DNA is water soluble due to its hydrophilic sugar-phosphate backbone, but it is extremely stable due to its tightly packed hydrophobic interior.

DNA Is an Information-Containing Molecule

- The nitrogenous base sequence of DNA stores information required for the growth and **reproduction** of all living cells.

- Complementary base pairing provides a simple mechanism for DNA replication since each strand can serve as a mold or template for formation of a new complementary strand (**Figure 4.9**).

- DNA's double helix can be pulled apart by enzymes or heating. The exposed bases on the original **template strand** can then base-pair with free deoxyribonucleotides, which are joined together by phosphodiester linkages to create a new **complementary strand** with an antiparallel directionality.

- DNA replication requires many enzymes in today's cells—DNA cannot self-replicate.

Is DNA a Catalytic Molecule?

- DNA's stability makes it a reliable store for genetic information; DNA has few exposed chemical groups that could participate in chemical reactions, and the presence of an H instead of OH on the sugar's 2′ carbon makes DNA less reactive than RNA and more resistant to degradation.

- Stable molecules like DNA are poor catalysts because stabilization of the transition state requires interaction between the substrate and the catalyst. Enzymes, for example, have exposed functional groups that interact with substrates.

- Proteins have incredible structural diversity, leading to a similar diversity of specific binding sites. DNA, on the other hand, has an extremely simple primary and secondary structure and does not appear to be able to catalyze any chemical reactions.

- Thus, biologists think that the first life-form was RNA, not DNA.

4.3 RNA Structure and Function

- RNA's primary structure is similar to DNA's except RNA contains uracil instead of thymine, and ribose instead of deoxyribose. The hydroxyl group on ribose's 2′ carbon can participate in reactions that degrade RNA, making RNA more reactive and less stable than DNA.

- RNA's secondary structure, like DNA's, is due to complementary base pairing between purine and pyrimidine nitrogenous bases. In RNA, two hydrogen bonds form between adenine and uracil, and three form between guanine and cytosine.

- However, the bases of RNA typically form hydrogen bonds with complementary bases on the *same* strand: the RNA strand folds over, forming a **hairpin** structure where the bases on one side of the fold align with an antiparallel RNA segment on the other side of the fold (**Figure 4.10**).

CHECK YOUR UNDERSTANDING

(c) An RNA strand folds into a hairpin structure. One side of the base of the hairpin has the sequence (listed as always in 5′ to 3′ order): U-U-A-G-U-C-A. What is the nucleotide order on the complementary strand listed in 5′ to 3′ order?

_____

_____

A second RNA strand has the sequence C-A-G-G-G-U-C on one side of its hairpin stem. Which hairpin structure will be more stable? Why?

_____

_____

_____

- Other RNA secondary structures also occur spontaneously, and all are stabilized by hydrogen bonds. Hydrogen bond formation is exothermic, so that hairpin formation is exergonic even though entropy is decreased.

- Sometimes hairpins and other RNA secondary structures have additional folds or are attached to other RNA strands, giving some RNA molecules tertiary and quaternary structure; this structural complexity produces RNA molecules that vary greatly in their shapes and chemical properties.

- Overall, RNA has less structural complexity than proteins, but more structural complexity than DNA. RNA's function is also intermediate between that of proteins and DNA. For example, RNA participates in information processing and catalyzes some important reactions.

CHECK YOUR UNDERSTANDING

(d) List four structural differences between DNA and RNA.

_____

_____

_____

_____

RNA as an Information-Containing Molecule

- Due to the predictability of complementary pair bonding, a RNA molecule could theoretically copy of itself. An initial **template strand** could be used to make a **complementary strand** that, once separated, could serve as a mold for formation of a RNA strand identical to the original template.

- RNA would be a rather unstable information-containing molecule. But, it might have survived long enough in the prebiotic soup to replicate itself.

Is RNA a Catalytic Molecule?

- RNA's structure is less variable than that of proteins, but it does have some secondary and tertiary structure as well as some chemical reactivity. These qualities allow RNA to stabilize some transition states.

- RNA catalysts are called **ribozymes**, and in at least one single-celled organism (*Tetrahymena*), ribozymes catalyze the hydrolysis and condensation of phosphodiester linkages. Perhaps an RNA molecule could catalyze the reactions necessary for its own replication!

4.4 The First Life-Form

- RNA can both provide a template for copying itself and catalyze the polymerization reaction that links monomers to make the (complementary) copy. These are essential qualities for a self-replicator.

- Researchers propose that the first life-form was RNA. This is known as the RNA world hypothesis. Because no self-replicating molecules exist today (all were outcompeted by cellular life), researchers test the RNA world hypothesis using laboratory simulations.

- Wendy Johnston and others in David Bartel's lab used artificial selection to create an RNA replicase that could catalyze the addition of ribonucleotides to a growing RNA strand.

- The researchers began by creating many large variable RNA molecules, which they then incubated with a small RNA template or "primer." A few of the large RNAs used themselves as a template and catalyzed, albeit inefficiently, the addition of a few ribonucleotides to the starting template.

- Researchers isolated the round-1 ribozymes and created the next round of large RNA molecules by copying these ribozymes in a way that caused a few random changes to the primary sequence.

CHECK YOUR UNDERSTANDING

(e) You are told that "Bartle's team copied the selected molecules in a way that introduced a few random changes to their sequence of bases" and that "most of the modified ribozymes worked worse than the original ones." Why, then, did the research team introduce these random changes?

_____

_____

_____

_____

Do you think most natural mutations are also detrimental?

_____

_____

_____

- This step is similar to natural selection in that only the "fittest" molecules produced offspring, but these offspring had a few random changes (mutations). Most of the changes decreased the ribozyme efficiency, but a few improved the ribozyme's catalyst function.

- The above process was repeated 18 times, each time selecting the best ribozymes as templates for the next round. At each step, some ribozymes showed improved catalyst function, and the final population included ribozymes that were pretty good catalysts for the addition of ribonucleotides to a growing complementary RNA strand (**Figure 4.11**).

- These results support the RNA world hypothesis; researchers may soon be able to produce a self-replicating molecule similar to the one that might have been our world's first life-form.

CHECK YOUR UNDERSTANDING

(f) Give three arguments in support of the hypothesis that the first self-replicator was an RNA molecule.

_____

_____

_____

_____

_____

B. INTEGRATING YOUR KNOWLEDGE

*Looking Back—Key Concepts
from Earlier Chapters*

Artificial Selection—Chapter 1

The ribozyme experiment described in Section 4.4 is similar in many ways to other types of artificial selection. Chapter 1 first introduced and defined both artificial and natural selection.

Catalysis and Correlation between Structure and Function—Chapter 3

RNA is able to catalyze some chemical reactions because, like proteins, it has variability in structure and chemical reactivity, permitting it to stabilize the transition state for certain chemical reactions, including reactions key to its potential as a self-replicator.

Looking Forward—Key Concepts in Later Chapters

DNA—Chapters 11–20, 25, and others

DNA encodes the genetic information of all cells. Any discussion of genetics therefore necessitates reference to DNA. DNA is equally essential to descriptions of cell reproduction and natural selection (because traits must be heritable in order to be selected).

RNA—Chapters 16 and 35

RNA's main function in extant cells is to use the information contained in DNA to produce proteins. This process is described in **Chapter 16**. **Chapter 35** describes how viruses use RNA.

C. MASTERING CHALLENGING TOPICS

The ribozyme selection experiment is cutting-edge research; it is hard to understand at first. You might want to reexamine **Figure 4.11**. Think about why researchers included each of the five steps listed in this figure and about how this process is similar to the artificial selection of cabbage-family plants described in Chapter 1. In studying nucleotide structure and function, you may want to make a table comparing the structure and functions of DNA, RNA, and proteins.

D. ASSESSING WHAT YOU'VE LEARNED

1. Which of these pairs of attributes distinguishes life from nonlife?
   a. presence of a plasma membrane and use of DNA to code genetic information
   b. ability to reproduce and presence of a plasma membrane
   c. metabolism and ability to make protein catalysts
   d. ability to control chemical reactions and ability to reproduce

2. The theory of chemical evolution predicts:
   a. Reproduction of a life-form is not possible without a cell membrane.
   b. A self-replicating molecule evolved before the cell membrane.
   c. The first self-replicator was a molecule of DNA.
   d. Self-replicating molecules exist in nature today.

3. Two major challenges to the theory of chemical evolution are:
   a. that living organisms use DNA for their genetic code and proteins as their primary catalysts
   b. that no self-replicating molecules exist today, and all organisms are cellular
   c. the ribose problem and the origin of pyrimidine bases
   d. that RNA cannot catalyze reactions and proteins cannot serve as templates for replication

4. What type of bond is a phosphodiester linkage?
   a. covalent
   b. polar covalent
   c. ionic
   d. hydrogen

5. Examine Figure 4.5: What would happen if the power supply were hooked up with the negative electrode at the top near the sample wells and the positive electrode at the bottom?
   a. Nothing; the samples would stay in the wells.
   b. No difference; the samples would still run toward the bottom of the gel.
   c. The large molecules would move further than the small molecules.
   d. The samples would run off the gel above the wells.

6. The _____ structure of a nucleic acid is the sequence of nitrogenous bases.
   a. primary
   b. secondary
   c. tertiary
   d. quaternary

7. Chargaff's rules state that in DNA, the _____.
   a. number of purines = number of pyrimidines, the number of A's = number of G's and the number of T's = number of U's.
   b. number of purines = number of pyrimidines, the number of T's = number of A's and the number of C's = number of G's.
   c. sugar-phosphate backbone is formed by phosphodiester linkages and is helical in structure.
   d. two strands are antiparallel and have complementary base pairing.

8. Which statement about DNA is *not* correct?
   a. DNA contains deoxyribose sugars.
   b. DNA has a double-helix structure.
   c. DNA contains uracil.
   d. DNA has a sugar-phosphate backbone.

9. Each nucleotide has one nitrogenous base: either a purine with a double-ring structure or a pyrimidine with a single ring. Which statement is true regarding complementary base pairing?
   a. Purines always pair with other purines, and pyrimidines always pair with other pyrimidines.
   b. The pyrimidine base cytosine (C) always pairs with the purine base guanine (A).
   c. In DNA, the purine base adenine (A) always pairs with the purine base guanine (G).
   d. Watson and Crick discovered complementary base pairing.

10. DNA is water soluble because:
    a. The nitrogenous bases form hydrogen bonds with water.
    b. The exterior-facing sugar-phosphate backbones are negatively charged and hydrophilic.
    c. The molecule has a hydrophobic interior.
    d. Cells are about 70% water.

11. Could DNA catalyze its own replication?
    a. No, its primary and secondary structures are too simple and unreactive to catalyze reactions.
    b. Yes, all living organisms store their genetic information in their DNA sequences.
    c. Yes, DNA and RNA are structurally and functionally similar.
    d. No, DNA cannot serve as a template for its own replication.

12. _____ can catalyze chemical reactions as well as carry information to copy itself (themselves).
    a. DNA
    b. Proteins
    c. RNA
    d. Nitrogenous bases

13. Which of the following is *not* true about RNA?
    a. RNA primarily functions as a catalyst in modern cells.
    b. RNA is more reactive than DNA.
    c. RNA could theoretically serve as a template for its own synthesis.
    d. RNA is less stable than DNA.

14. Due to complementary base pairing, _____ form(s) a secondary structure called the hairpin loop.
    a. protein
    b. DNA
    c. RNA
    d. sugars

15. Wendy Johnston and coworkers created RNA molecules that catalyzed:
    a. the formation of hairpin loops
    b. the formation of nucleotides from sugar, phosphate, and nitrogenous base subunits
    c. DNA replication
    d. the addition of ribonucleotides to a growing RNA strand

CHAPTER 4—ANSWER KEY

Check Your Understanding

(a) Figure 4.2 cannot happen as shown because it is an endergonic reaction; the product has both less entropy and more potential energy than the reactants.

Enzymes only lower activation energy—they cannot provide energy. The energy for this reaction is provided by "activating" a nucleotide through the addition of two more phosphate groups. This increase in reactant potential energy makes the formation of the phosphodiester linkage exergonic, allowing this reaction to occur.

(b) Watson and Crick concluded that adenine bonds to thymine and cytosine to guanine because (1) only a purine-pyrimidine pair would fit inside the double helix based on Franklin and Wilkins' X-ray diffraction measurements, and (2) this pairing would explain Chargaff's observation that the number of T's always equals the number of A's in an organism's DNA.

(c) The complementary strand's sequence is U-G-A-C-U-A-A (remember that the strands are antiparallel and that you list nucleic acid sequences in 5′ to 3′ order). The second hairpin structure will be more stable because G and C form three hydrogen bonds, but U and A form only two. Thus, the second hairpin stem will have more hydrogen bonds stabilizing it.

(d) RNA differs from DNA in that it (1) has ribose, not deoxyribose, as its nucleotide sugar; (2) uses uracil instead of thymine; (3) is single stranded, not double stranded; and (4) can have tertiary structure due to the folding of secondary structures into complex shapes (more complex/less stable structure).

(e) Most changes impaired ribozyme function, because it is much easier to break something than to fix it. For example, if you *randomly* made changes to your car engine, most of the changes would cause problems; only a very few would improve the engine's function. The researchers introduced random changes to get the very few ribozymes with improved catalyst function. In nature, most mutations are also detrimental.

(f) Arguments supporting the hypothesis that the first self-replicator was an RNA molecule: (1) RNA can serve as a template for its own synthesis (although it takes two rounds of replication to get back to the original template molecule); (2) RNA can catalyze some chemical reactions, including phosphodiester bond formation; (3) none of the other macromolecules can do both 1 and 2; and (4) researchers have already successfully created RNA replicases that catalyze the addition of ribonucleotides to a growing RNA strand.

Assessing What You've Learned

1. d; 2. b; 3. c; 4. b; 5. d; 6. a; 7. b; 8. c; 9. d; 10. b; 11. a; 12. c; 13. a; 14. c; 15. d

An Introduction to Carbohydrates

A. KEY CONCEPTS

- Single-sugar monomers (**monosaccharides**), small polymers (**oligosaccharides**), and large polymers (**polysaccharides**) with the generalized chemical formula $(CH_2O)_n$ are called **carbohydrates**.

- Carbohydrate monomers contain a carbonyl group, several hydroxyl functional groups, and many carbon–hydrogen bonds.

5.1 Sugars as Monomers

- Sugars provide chemical energy to cells and serve as molecular building blocks; for example, during chemical evolution, ribose had to be present in order for nucleotides to form.

- Monosaccharides, including ribose, are easily made under conditions simulating the prebiotic soup.

- Sugar carbonyl and hydroxyl functional groups can participate in many different chemical reactions.

- Monosaccharides vary in hydroxyl and carbonyl functional group placement; when the carbonyl is at the end of the molecule, the monosaccharide is an aldehyde sugar or *aldose*, but when the carbonyl is in the middle of the sugar's carbon chain, it is a ketone sugar or *ketose* (**Figure 5.1**).

- The number of carbons in monosaccharides varies. Three-carbon sugars are **trioses**; five-carbon sugars, like ribose, are **pentoses**; and six-carbon sugars, like glucose, are **hexoses**. Carbons are numbered starting at the end closest to the carbonyl group.

- Many sugars have the same chemical formula but different structures. For example, glucose and galactose $(C_6H_{12}O_6)$ are optical isomers (**Figure 5.2**). As always, structure affects function, and only glucose can be directly used by the cell in ATP-producing reactions.

- Sugars in aqueous solution switch back and forth between chain and ring forms, with ring forms predominating. When glucose forms a ring, carbon 1 (the carbonyl carbon) can bond with the fifth carbon's

CHECK YOUR UNDERSTANDING

(a) Draw two ketose pentose sugars that are optical isomers of each other.

hydroxyl group in two different orientations, forming either α-glucose or β-glucose (**Figure 5.3**).

- Many different monosaccharide structures are possible due to the variety of ring forms, isomers, carbon numbers, and functional group placements. Large structural variation permits a diversity of functions.

- Most monosaccharides are easily formed under conditions similar to those in ancient Earth's prebiotic soup. Additionally, a 3-C ketose and other sugar-like molecules were found on a meteorite, suggesting that simple sugars can be synthesized on space debris and might have rained down on ancient Earth.

- It is, however, unlikely that polymerization of monosaccharides occurred during chemical evolution. Researchers still do not understand how ribose could have become the most common monosaccharide in the prebiotic soup, making possible the synthesis of nucleotides.

5.2 The Structure of Polysaccharides

- **Polysaccharides** form when a condensation reaction between hydroxyl groups of two monosaccharides creates a **glycosidic linkage** (**Figure 5.4**). The simplest polysaccharide is a **disaccharide** of two linked monomers.

- Monomers joined by glycosidic linkages can be identical or different. In contrast to peptide and phosphodiester bonds, glycosidic linkages can form between any two hydroxyl groups, so that the location and geometry of these bonds varies widely.

- Polysaccharides can be hundreds to thousands of monomers long (**Figure 5.5**).

Starch: A Storage Polysaccharide in Plants

- Plants store sugars as **starch** made of many α-glucose monomers joined by glycosidic linkages between the first and fourth ring carbons (α-1,4-glycosidic bonds). The angle of these bonds causes the monomer chain to coil into a helix.

- Starch is a mix of unbranched and branched helices (amylose and amylopectin, respectively; **Table 5.1**). Starch branches when a glycosidic bond forms between carbon 1 on one strand and carbon 6 on another.

Glycogen: A Highly Branched Storage Polysaccharide in Animals

- Animals store sugars as **glycogen**, which is similar to the branched form of starch. Branches (due to α-1, 6-glycosidic linkages) occur in glycogen about once every 10 glucose monomers, whereas amylopectin branches occur about once every 30 monomers (**Table 5.1**).

- Humans store glycogen in the liver and in muscles. During exercise, glycogen is broken down, releasing glucose for use by working muscles.

CHECK YOUR UNDERSTANDING

(b) Identify two similarities and two differences between starch and glycogen.

_____

_____

_____

_____

Cellulose: A Structural Polysaccharide in Plants

- Most modern organisms have a protective barrier called a **cell wall** outside their cell membranes. This cell wall is made of polysaccharides in algae, plants, bacteria, fungi, and other groups.

- Plant cell walls are mostly made of **cellulose**, a β-glucose polymer with β-1,4-glycosidic linkages (**Table 5.1**). This linkage flips every other glucose monomer upside down, leading to a geometry that allows extensive hydrogen bonding between side-by-side strands of cellulose (**Table 5.1**).

- Each hydrogen bond is relatively weak, but many together provide great strength in cellulose structures.

Chitin: A Structural Polysaccharide in Fungi and Animals

- **Chitin** forms the cell walls in fungi. It is also found in some algae and forms the primary component of insect and crustacean exoskeletons. Although structurally similar to cellulose, chitin is composed of *N*-acetylglucosamine monomers.

- Chitin monomers, like those of cellulose, are joined by β-1,4-glycosidic linkages such that every other monomer is flipped over, allowing hydrogen bonding between adjacent strands and creating a tough, stiff, protective sheet (**Table 5.1**).

Peptidoglycan: A Structural Polysaccharide in Bacteria

- Bacterial cell walls are primarily composed of **peptidoglycan**, which is made of two alternating monosaccharides joined by β-1,4 glycosidic linkages. One of the sugar monomers is linked to a chain of amino acids, and peptide bonds link the amino acid chains of adjacent strands (**Table 5.1**).

- Some antibiotics kill bacteria by interfering with enzymes that catalyze the formation of the peptide-bond cross-links in peptidoglycan (**Box 5.2**).

- Cellulose, chitin, and peptidoglycan all consist of long parallel strands linked to each other. The resulting structure easily withstands forces that push or pull it, and it functions well as a protective sheet.

Polysaccharides and Chemical Evolution

- Polysaccharides are important to organisms today, but they probably were not involved in the origin of life.

- Glycosidic linkages appear to form only in the presence of enzymes, so polysaccharides probably did not form in the prebiotic soup.

(c) Explain how structural polysaccharides differ from storage polysaccharides and how these differences in structure relate to their different functions.

_____

_____

_____

_____

_____

- The first life-forms are thought to have been RNA based, but no known ribozyme catalyzes the formation of glycosidic linkages.

- Monosaccharides have few functional groups, so that polysaccharides have fairly simple secondary structures. They lack the structural and chemical complexity required for catalysis.

5.3 What Do Carbohydrates Do?

- Carbohydrates are important building blocks in the synthesis of other molecules. Recall that ribose and deoxyribose are nucleotide subunits; and other sugars furnish "carbon skeletons" for the synthesis of many other important molecules, including amino acids.

- Carbohydrates also form fibrous structures, indicate cell identity, and store chemical energy.

Carbohydrates as Structural Molecules

- The structural polysaccharides—cellulose, chitin, and peptidoglycan (a modified polysaccharide)—all form long strands with bonds between adjacent strands. These strands may then be organized into fibers or layered in sheets to give cells and organisms great strength and elasticity (**Figure 5.6**).

- The β-1,4-glycosidic linkages of these structural molecules are difficult to hydrolyze. Very few enzymes have active sites that can accommodate their geometry. Molecules with these bonds are thus poor energy sources for most organisms, but great structural molecules because they are so difficult to break down. Yet again, structure explains function.

- Although humans cannot digest cellulose, it aids digestion (as dietary fiber) by absorbing water and adding bulk and moisture to feces.

The Role of Carbohydrates in Cell Identity

- Polysaccharides are unable to *store* information in a way that can be replicated, but they do *display* information on the outer surface of cells.

- **Glycoproteins**—proteins with covalent bonds to oligosaccharides (or occasionally to longer polysaccharides)—project outward from cell membranes and are essential in cell–cell recognition and cell–cell signaling (**Figure 5.7**).

- In multicellular organisms, each cell type has different glycoproteins, each of which is recognized by the immune system as "self" and distinguished from foreign cells from other species (such as infecting bacteria) or individuals.

- Monosaccharide structural diversity makes possible the incredible diversity of glycoproteins necessary to indicate cell identity.

(d) Why are polysaccharides good at cell identification but unable to effectively store information or catalyze chemical reactions?

_____

_____

_____

_____

The Role of Carbohydrates in Energy Storage

- Carbohydrates are good sources of chemical energy for cells.

Carbohydrates and Free Energy

- Today, most sugars are produced via **photosynthesis**: $CO_2 + H_2O +$ sunlight $\rightarrow (CH_2O)_n + O_2$, a key process that transforms sunlight's kinetic energy into carbohydrate bond chemical energy.

- Carbohydrates have more free energy than CO_2 does, because C−H and C−C bond electrons are shared more equally and held less tightly than those in C−O bonds (**Figure 5.8**).

- Because carbohydrate electrons are held less tightly, they are farther from the nucleus and have higher potential energy than tightly held electrons like those in CO_2.

- The free energy in carbohydrate C−H and C−C bonds can be used to produce **adenosine triphosphate (ATP)** via the exergonic reaction: $CH_2O + O_2 + ADP + P_i \rightarrow CO_2 + H_2O + ATP$. The easily accessed free energy stored in ATP drives endergonic reactions and performs all kinds of cell work.

- Fatty acids have even more C–H bonds than carbohydrates do and thus store even more free energy than do carbohydrates (**Figure 5.9**). Both molecules are important sources of chemical energy for cells.

How Do Carbohydrates Store Energy?

- Starch and glycogen are convenient glucose stores because enzymes easily catalyze reactions that release glucose subunits from α–glycosidic linkages.

- **Phosphorylase** catalyzes the hydrolysis of α-glycosidic linkages in glycogen, and **amylase** catalyzes the hydrolysis of these linkages in starch.

- Very few enzymes can hydrolyze β-glycosidic linkages like those found in the structural polysaccharides.

CHECK YOUR UNDERSTANDING

(e) Termites have bacteria in their guts that enable them to eat wood. What about these bacteria might be responsible for enabling termites to eat wood? (Be specific.)

_____

_____

_____

B. INTEGRATING YOUR KNOWLEDGE

Looking Back—Key Concepts from Earlier Chapters

Chemical Evolution—Chapters 2–4

The start of evolution, environment of ancient Earth, and basic chemistry needed to understand the chemical reactions that cause chemical evolution are described in **Chapter 2**. **Chapters 3 and 4** discuss chemical evolution with respect to proteins and nucleic acids.

Structure–Function Correlation—Chapters 3 and 4

You should be familiar with the ubiquitous correlation between structure and function from the discussion in **Chapters 3** and **4** of this relationship for proteins and

nucleic acids. In **Chapter 5** you learn about how the molecular structure of carbohydrates relates to their function.

Looking Forward—Key Concepts in Later Chapters

Sugars—Chapters 9, 10, 36, and 43

Chapter 9 describes how organisms break down sugars to release and use the energy contained in their C–H bonds; **Chapter 10** describes how some organisms can use the energy in sunlight to make sugars (photosynthesis). **Chapters 37** and **43** describe the importance of sugars to plants and animals at the organismal level.

Glycoproteins—Chapter 49

Chapter 49 describes the important role that glycoproteins and proteoglycans have in forming a physical shield for a membrane, helping to prevent the invasion of microbial pathogens.

C. MASTERING CHALLENGING TOPICS

Much of this chapter relates to important structural and functional differences between α- and β-1,4-glycosidic bonds. But the *best* way to get a feel for the structural differences is to make models of each bond type that

	α-1,4-glycosidic bonds	β-1,4-glycosidic bonds
Bond structure		
Polysaccharides containing this bond and the monomers thus linked		
Cell function		
Relationship between cell function and bond structure		

joins two sets of glucose molecules. If you do not have the time and/or materials, draw your own pictures of these molecules (starch, glycogen, cellulose, chitin, peptidoglycan) to improve your understanding of their structural differences. Once you understand the structural differences of the two bond types, make a table summarizing the major structural and functional differences of polysaccharides made from monomers that are joined by these two types of bonds.

D.   ASSESSING WHAT YOU'VE LEARNED

1. Which of the following is *not* true of glucose and galactose?
   a. They are optical isomers.
   b. They are pentoses.
   c. Their formula is $C_6H_{12}O_6$.
   d. They are monosaccharides.

2. In α-glucose, the C1 hydroxyl group is: (Refer to Figure 5.3 if necessary.)
   a. gone
   b. turned into a carbonyl group
   c. on the same side of the ring as C_6
   d. on the opposite side of the ring from C_6

3. Which of the following statements regarding sugars is correct?
   a. Amino acids undergo condensation reactions to form polysaccharides.
   b. Monosaccharides undergo condensation reactions to form polysaccharides.
   c. Monosaccharides undergo hydrolysis to form polysaccharides.
   d. Glucose is the monomer for all polysaccharides.

4. The linear form of deoxyribose is:

   H–C=O
   |
   H–C–H
   |
   H–C–OH
   |
   H–C–OH
   |
   H–C–OH
   |
   H

   What type of sugar is this?
   a. a hexose aldose
   b. a hexose ketose
   c. a pentose aldose
   d. a pentose ketose

For questions 5 through 7, Identify the cell-wall polysaccharide for each type of organism using the following lettered choices:
   a. amylopectin
   b. cellulose
   c. chitin
   d. glycogen
   e. peptidoglycan

5. Bacteria _____

6. Fungi   _____

7. Plants   _____

8. The monomer in chitin is:
   a. α-glucose
   b. β-glucose
   c. amylose
   d. *N*-acetylglucosamine

9. Researchers conclude that polysaccharides played little or no role in the origin of life because:
   a. Ribozymes cannot catalyze the formation of glycosidic linkages.
   b. Polysaccharide monomers are unable to form complementary base pairs.
   c. Polysaccharides cannot catalyze reactions.
   d. All of the above are correct.

10. Carbohydrate cell functions include all of the following *except*:
   a. cell protection
   b. storage of chemical energy
   c. transport of molecules
   d. cell–cell recognition

11. Proteins covalently bonded to carbohydrates are called:
   a. glycoproteins
   b. amyloproteins
   c. cephalosporins
   d. saccharoamides

12. In carbohydrates, electrons are:
   a. closer to the carbon atom than they were in CO_2
   b. farther from the carbon atom than they were in CO_2
   c. shared less evenly in $C-H$ bonds than in $C-O$ bonds
   d. more stable in $C-H$ bonds than they are in the $C-O$ bonds of CO_2

13. Fats have more free energy than carbohydrates because:
   a. They are made during photosynthesis.
   b. They have more kinetic energy.
   c. They have more $C-H$ bonds.
   d. They have more $C-O$ bonds.

14. One thing that cellulose, chitin, and peptidoglycan all have in common is:
    a. hydrogen bonds between adjacent strands
    b. covalent bonds between adjacent strands
    c. α-1,4-glycosidic linkages
    d. β-1,4-glycosidic linkages

15. Based on this chapter's discussion of why carbohydrates have high free energy, which of the following is likely to be true of the electrons in the bond between ATP's second and third phosphates?
    a. The electrons are shared fairly equally, giving the bond a high free energy.
    b. The electrons are not shared equally, giving the bond a high free energy.
    c. The electrons are shared fairly equally, giving the bond a low free energy.
    d. The electrons are not shared equally, giving the bond a low free energy.

CHAPTER 5—ANSWER KEY

Check Your Understanding

(a)
```
      H                H
      |                |
   H–C–OH           H–C–OH
      |                |
   H–C–OH            C=O
      |                |
    C=O             H–C–OH
      |                |
   H–C–OH           H–C–OH
      |                |
   H–C–OH           H–C–OH
      |                |
      H                H
```

(b) Similarities: Both made of glucose monomers; both have α-1,4-glycosidic linkages; both are used to store chemical energy. Differences: Starch is made by plants, glycogen by animals; glycogen is more branched than starch (more α-1,6-glycosidic bonds).

(c) The β-glycosidic bonds in structural polysaccharides are not easily broken down (making these carbohydrates hard to digest) and cause every other monomer to be flipped upside down. This produces a more linear structure and allows hydrogen bonding between adjacent strands. The many hydrogen bonds between strands make these polysaccharides strong yet flexible. In contrast, α-glycosidic bonds are more easily broken down, making these polysaccharides a good source of energy. However, their helical structure is not as strong as the parallel sheets of the structural polysaccharides.

(d) Polysaccharides do not form a template that can be copied and passed on to the daughter cells, so they are poor molecules for information storage. Polysaccharides also lack the complexity of secondary and tertiary structure necessary to catalyze reactions, yet are structurally diverse enough to allow each cell type and species to display different polysaccharides.

(e) The bacteria in termite guts have an enzyme that can hydrolyze the β-1,4-glycosidic linkage found in cellulose. Very few organisms have such enzymes, which is why so few organisms can eat wood.

Assessing What You've Learned
1. b; 2. d; 3. b; 4. c; 5. e; 6. c; 7. b; 8. d; 9. d; 10. c; 11. a; 12. a; 13. c; 14. d; 15. a

6

Lipids, Membranes, and the First Cells

A. KEY CONCEPTS

- The mostly lipid **cell membrane** or **plasma membrane** surrounds a cell, separating the life inside the cell from the external environment.

- Development of a plasma membrane around a self-replicator created the first cell. Cells could now begin to control their internal environment by keeping harmful compounds out and allowing useful compounds in, greatly increasing the efficiency of cell chemical reactions.

6.1 Lipids

- Experiments simulating the chemical and energetic conditions of ancient Earth produce several types of lipid. Thus, lipids must have been present in the prebiotic soup.

- High-magnification pictures (using an **electron microscope**) of lipid-water mixtures show that some lipids spontaneously form cell-like compartments called vesicles (**Figure 6.1**).

What Is a Lipid?

- Molecules that contain only carbon and hydrogen atoms are called **hydrocarbons**; these molecules are nonpolar and consequently hydrophobic, because electrons are evenly shared in C–H bonds.

- **Lipids** have a significant hydrocarbon component; they are nonpolar, hydrophobic organic compounds that do not dissolve in water.

- **Fatty acids** (hydrocarbon chains bonded to a carboxyl COOH group) and isoprene are the key building blocks for the synthesis of lipids in organisms (**Figure 6.2**).

A Look at Three Types of Lipids Found in Cells

- Amino acids, nucleotides, and sugars are defined by their chemical structure; lipids are defined by their solubility. Thus, lipid structure varies widely.

- **Fats**, also known as *triacylglycerols* or *triglycerides*, are made when three fatty acids become linked to a three-carbon **glycerol**.

- Condensation reactions between the glycerol hydroxyl groups and fatty-acid carboxyl groups form **ester linkages** joining these subunits. Although this linkage is similar to those formed between other macromolecule subunits, phospholipids and fats are not polymers of repeating units (**Figure 6.3**).

- **Steroids** have a four-ring structure made from isoprene subunits (**Figure 6.4a**). Various functional groups or side groups attach to this ring structure.

CHECK YOUR UNDERSTANDING

(a) Examine the structure of fats in **Figure 6.3**. Why don't these fats form membranes?

_____

_____

_____

- One steroid, cholesterol, is an important part of plasma membranes. It has a long hydrocarbon tail attached to the general steroid ring structure and is a key precursor for several mammalian hormones.

- **Phospholipids** consist of a three-carbon glycerol linked to a phosphate group PO_4^{2-} and two isoprene chains (in Archaea) or two fatty acids (in Bacteria and Eukarya). The phosphate group may also bind to another small organic molecule (**Figure 6.4b**).

- Lipids have many functions in organisms, the most important being the formation of plasma membranes.

The Structures of Membrane Lipids

- Phospholipids and cholesterol are **amphipathic**—they contain both a polar, hydrophilic region and a nonpolar, hydrophobic region. The polar region interacts with water, but the nonpolar end does not, and this "dual sympathy" allows these lipids to form membranes.

- The polar "head region" can interact with water whereas the long nonpolar isoprene or fatty-acid "tails" avoid water.

6.2 Phospholipid Bilayers

- Phospholipids with short tails tend to form **micelles** (tiny droplets), and those with longer tails tend to form **lipid bilayers** (two layers) in water so that the polar phosphate "heads" face outward toward water and the hydrophobic fatty-acid "tails" face inward toward other tails (**Figure 6.5**).

- Micelles and lipid bilayers form spontaneously without any energy input. Although these structures have less entropy than free phospholipids have, they are much more stable because free phospholipids disrupt water's hydrogen bonds (**Figure 6.6**).

- Micelles and lipid bilayers form spontaneously because the decrease in potential energy more than makes up for the decreased entropy of these structures. The overall free energy of the system decreases when micelles and lipid bilayers form, making this an exergonic process.

Artificial Membranes as an Experimental System

- Lipid bilayers break and form small spherical compartments (vesicles) when shaken. Water is located both outside and inside these spherical bilayers.

- Because vesicles form spontaneously, any accumulation of phospholipids on ancient Earth would have led to the creation of water-filled vesicles similar to the lipid bilayers that form the basic structure of cell membranes.

- **Permeability**, the ease with which substances cross a given barrier, is crucial to plasma membrane function because differentiation between an internal and external environment occurs when certain molecules or ions cross a selectively permeable membrane more easily than others.

- Hypotheses about membrane permeability are tested using artificially formed vesicles called **liposomes** and **planar bilayers**—artificial membranes across a hole in a wall between two aqueous solutions (**Figure 6.7**).

- Researchers investigate membrane permeability by varying either the solute added to one side of the bilayer or the bilayer composition. They then measure the molecule or ion concentrations on both sides of the artificial membrane to determine how fast various solutes cross bilayers under different conditions.

Selective Permeability of Lipid Bilayers

- Experiments with liposomes and planar bilayers show that small, nonpolar molecules move across phospholipid bilayers quickly, but charged substances cross slowly if at all. Thus phospholipids bilayers have **selective permeability**; some molecules cross these bilayers more easily than others (**Figure 6.8**).

- Large or charged compounds cannot pass through lipid bilayers, because they are more stable in an aqueous solution than among the nonpolar, hydrophobic phospholipid tails.

- Water crosses membranes fairly rapidly, even though it is polar, because it is such a small molecule.

CHECK YOUR UNDERSTANDING

(b) Compare the permeability scale in **Figure 6.8** with the pH scale shown in **Figure 2.16**. Why are powers of 10 used for these scales?

_____

_____

Does the Type of Lipid in a Membrane Affect Its Permeability?

- Membrane permeability is affected by hydrocarbon chain length and the number of double bonds.

- A hydrocarbon chain with at least one double bond is **unsaturated** because it does not contain the maximum number of hydrogen atoms. A hydrocarbon chain without double bonds is **saturated**; these fats contain more C–H bonds and have more chemical energy than do unsaturated fats.

- Double bonds prevent rotation between the bonded carbons, thereby causing a bend or "kink" in the hydrocarbon chain that prevents the close packing of hydrocarbon tails and reduces hydrophobic interactions (**Figure 6.9**). Longer tails, on the other hand, have more hydrophobic interactions than do shorter tails.

- Experiments on liposomes show that bilayers with unsaturated hydrocarbons and/or shorter hydrocarbon tails are more fluid and permeable than those with saturated hydrocarbons and/or longer tails (**Figure 6.10**).

- The structure of phospholipids (number of double bonds and length of hydrocarbon tail) determines the amount of hydrophobic interactions, which in turn affect the phospholipid's properties and function.

• Differences in hydrocarbon fluidity are easily observed in any kitchen. Highly saturated fats are solid at room temperature (e.g., butter), and those with extremely long hydrocarbon tails form stiff waxes; unsaturated fats are liquid oils at room temperature (**Figure 6.11**).

• Cholesterol also affects membrane permeability (**Figure 6.12**); it fills spaces between phospholipids, increasing hydrophobic interactions and reducing bilayer permeability. Temperature also affects membrane permeability.

CHECK YOUR UNDERSTANDING

(c) Based on the preceding discussion, what kind of lipid composition would you expect to find in a shortening like Crisco?

_____

Examine the labels on shortening and cooking oil, and use the information you find there to explain why oil is fluid at room temperature but shortening is not.

_____

_____

_____

Why is shortening softer at room temperature than in the refrigerator?

_____

_____

Why Does Temperature Affect the Fluidity and Permeability of Membranes?

• Experiments with tagged phospholipids show their lateral movement through membranes (**Figure 6.13**). Both temperature and structure (length and saturation) affect how fast phospholipids travel within a membrane (fluidity).

• Cell membranes are fluid at room temperature because the bilayer molecules move relatively fast (in one second, phospholipids can travel the length of a bacterial cell). At lower temperatures the molecules move more slowly and the hydrophobic tails pack more tightly, making the membrane less fluid.

• Decreased fluidity causes decreased permeability (**Figure 6.12**).

6.3 Why Molecules Move Across Lipid Bilayers: Diffusion and Osmosis

• Dissolved molecules and ions (**solutes**) are in constant random motion due to their thermal or kinetic energy. Movement of molecules caused by this random kinetic motion is called **diffusion**.

• Random molecular motion causes molecules and ions to spontaneously move down their **concentration gradient**; they tend to bump into each other more often in areas of high concentration, which on average moves molecules toward areas of low concentration over time.

• Molecules or ions that are placed on one side of a phospholipid bilayer tend to diffuse toward the other side over time (**Figure 6.14**). This process is spontaneous because it increases the entropy (+ΔS; see **Chapter 2**) of the system.

• Equilibrium (no *net* movement) is achieved when molecules or ions are randomly distributed throughout the solution. Molecules still move, but they are equally likely to move in any direction.

• Diffusion of water is called **osmosis**, which occurs only when solutions of different concentrations are separated by a membrane that is permeable to water but not to the solutes (**Figure 6.15**). Water moves spontaneously across the membrane toward the solution with the higher solute concentration (i.e., lower water concentration).

• Osmosis can shrink or swell (and even burst) membrane-bound vesicles and cells (**Figure 6.16**).

• If a solution outside a membrane has a higher concentration of solutes than the interior does, and if the solutes are unable to diffuse through the membrane, water will diffuse out of the vesicle (osmosis) and make the vesicle shrivel. Solutions that shrink cells are **hypertonic**.

• **Hypotonic** solutions are less concentrated than the vesicle or cell interior. In this case, if the membrane is not permeable to the solute, water will flow into the vesicle or cell, which will then swell and perhaps burst.

• A solution with the same solute concentration as the inside of a vesicle or cell is **isotonic** and does not affect membrane shape.

• Osmosis and diffusion always reduce the differences between solutions on two sides of a phospholipid bilayer.

• Liposomes in the prebiotic soup might have contained self-replicators, but they probably did not provide a significantly different internal environment. But these simple cell-like vesicles could have grown due to the addition of new lipids, and they could have divided when shaken by wave action.

CHECK YOUR UNDERSTANDING

(d) Using your knowledge of how the concentration of solutions outside cells affects cell shape, explain why salting meat is an effective way to preserve it (and reduce bacteria spoilage).

_____

_____

_____

6.4 Membrane Proteins

- Amphipathic proteins can be added into lipid bilayers. Recall that the R-groups of amino acids vary in polarity and thus in their affinity for water. Nonpolar amino acids are stable in membranes due to hydrophobic interactions with the hydrocarbon tails, and polar amino acids are stable next to the polar heads and in the surrounding water (**Figure 6.17**).

- In fact, although phospholipids provide the basic membrane structure, cell membranes contain as much protein as phospholipids do, as measured by mass.

- Davson and J. Danielli, in an early model of membrane structure, proposed that proteins coated a pure lipid bilayer. But the more recent **fluid-mosaic model** (S. J. Singer and G. Nicolson) suggests many proteins are inserted within the lipid bilayer, making the membrane a fluid, dynamic mosaic of phospholipids and proteins (**Figure 6.18**).

- Development of **freeze-fracture electron microscopy** with a scanning electron microscope provided key evidence for the fluid-mosaic model. The technique involves freezing and fracturing membranes, splitting them to view the middle and the surface of the bilayer (**Figure 6.19**).

- Membrane proteins look like mounds sticking out of half of the bilayer, leaving pits in the other half. Some proteins form tunnels crossing a lipid bilayer, creating a channel or pore. Membrane-spanning proteins are called **integral membrane proteins** or **transmembrane proteins**.

- Other **peripheral membrane proteins** are located on the membrane surface, and they are often attached to the integral membrane proteins (**Figure 6.20**).

- Different peripheral proteins are found on the interior versus the exterior of the plasma membrane. These proteins are always moving along or through the bilayer, but they hardly ever flip from one side to the other.

- Membrane proteins on both the interior and the exterior of the plasma membrane participate in many cell functions, but this chapter focuses on how integral membrane proteins transport ions and molecules across the plasma membrane.

Systems for Studying Membrane Proteins

- To separate integral membrane proteins from cell membranes, researchers treat the membranes with small, amphipathic molecules called **detergents** (**Figure 6.21**).

- Detergents insert themselves into membranes, and their hydrophobic tails interact with the phospholipid tails, breaking apart the bilayer. Detergents also surround hydrophobic parts of membrane proteins, forming water-soluble detergent-protein complexes.

- The now-soluble membrane proteins can be isolated and purified using **gel electrophoresis**, which separates the proteins moving through a gel based on their size and charge (**Box 4.1**). Finally, the purified proteins can be inserted into artificial lipid bilayers or liposomes for experimental analysis.

CHECK YOUR UNDERSTANDING

(e) Pour some oil into a pan or bowl of water. What happens? Why?

_____

_____

Now add some dishwasher or laundry detergent and stir. What happens? Why?

_____

_____

How Do Membrane Proteins Affect Ions and Molecules?

- Channels, transporters, and pumps all affect membrane permeability.

Facilitated Diffusion via Channel Proteins

- Gramicidin is a membrane peptide that is produced and released by one type of bacteria to kill other bacteria (this makes it an **antibiotic**). Cells treated with gramicidin lose lots of ions.

(f) You learned that cholesterol is a steroid and that some steroids function as hormones. Based on what you know about the selective permeability of lipid bilayers, predict whether steroid hormones can cross membranes on their own, or whether they require a transporter. Explain your reasoning.

_____

_____

_____

_____

- Ions move from areas of like charge to areas of unlike charge as well as from areas of high concentration to areas of low concentration—they move in response to an **electrochemical gradient** (**Figure 6.22**).

- Researchers measured charge flow (electrical current) across membranes with and without gramicidin to test the effect of this peptide on ion flow. In the presence of an electrochemical gradient, current flowed across planar bilayers with gramicidin but not across those lacking gramicidin (**Figure 6.23**).

- Gramicidin makes lipid bilayers permeable to small, positively charged ions (cations) by forming a tunnel—an **ion channel**—through a membrane. Protons (H^+) pass through gramicidin more easily than other larger cations do.

- Gramicidin's amino acid sequence and tertiary structure reveal that the molecule forms a hole lined with hydrophilic R-groups. Hydrophobic R-groups on the exterior face of the tunnel interact with the membrane hydrocarbon tails (**Figure 6.24**). Once again, structure correlates with function.

- Cells have a wide variety of membrane channels. **Aquaporins**, for example, increase the rate at which water crosses membranes tenfold and are highly selective for water, and **gated channels** open only when bound to a specific molecule or in the presence of a certain electrical charge (**Figure 6.25**).

- The passage of ions through membrane channels can be controlled. For example, one type of potassium channel lets potassium ions through only when the outside of the membrane is positive relative to the inside. The positive charge causes a shape change that opens the channel.

- Although channels allow membranes to exert some control over transmembrane flow of molecules and ions, these substances always diffuse through channels down their electrochemical gradients. This process is also called **facilitated diffusion** or **passive transport**, since the cell expends no energy.

- Passive transport always occurs down an electrochemical gradient, thus decreasing charge and concentration differences between the cell's exterior and interior.

Facilitated Diffusion via Transport Proteins

- Proteins that assist in facilitated diffusion may be either channels or transport proteins that change shape to allow the ion or molecule to cross the membrane. No energy is required, since molecules only move down their electrochemical gradients in facilitated diffusion.

- Glucose is a subunit for important macromolecules and a great energy source, but lipid bilayers are only moderately permeable to glucose. How does glucose cross the membrane?

- Researchers obtained intact cell membranes by bursting red blood cells in a hypotonic solution to form red blood cell "ghosts" (**Figure 6.26**). Ghosts are more permeable to glucose than pure phospholipid bilayers are.

- Researchers isolated a glucose carrier—or **transport protein**—named GLUT-1, which increases membrane permeability to glucose. Glucose enters liposomes with GLUT-1 proteins at the same high rate observed in cell membranes.

- Only right-handed glucose, the form used by cells, is transported by GLUT-1. Researchers think glucose binds to GLUT-1 on the cell's exterior surface, causing a conformational change (just like enzymes change shape during substrate binding) that transports glucose to the cell interior (see **Figure 6.27**).

- Like channels, transport proteins or **carriers** *reduce* differences between the inside and the outside of a cell or vesicle. Facilitated diffusion by these proteins occurs only down an electrochemical gradient, so when glucose concentrations are equal, no net movement of glucose occurs by GLUT-1.

Active Transport by Pumps

- Cells extant today can also transport molecules or ions *against* their electrochemical gradient; this process requires energy and is called **active transport**.

- Adenosine triphosphate (**ATP**) provides energy in a form that proteins can easily use. When a phosphate group leaves ATP and binds to a protein, its negative charges interact with the protein's amino acids, causing a shape change. Another shape change occurs when the phosphate group leaves.

(g) Cells often break down glucose as a source of energy. How does this process alter the concentration gradient for glucose?

_____

_____

How does this process explain why many cells can rely on transporters like GLUT-1 to bring glucose into the cell, when the direction of transport is determined by the electrochemical gradient?

_____

_____

_____

- The **sodium-potassium pump**, Na^+/K^+-ATPase, was the first pump discovered. This pump uses energy from ATP to transport Na^+ and K^+ and acts like an enzyme in many respects. Ion movement against an electrochemical gradient occurs when the pump's shape changes as a phosphate group from ATP first binds and then leaves.

- Three Na^+ ions bind to Na^+/K^+-ATPase on the inside of a cell. When ATP donates a phosphate to this Na^+/K^+-ATPase, it changes shape, releasing the Na^+ ions outside the cell. The ion pump can now bind two K^+, which are in turn released inside the cell when the phosphate group drops off (**Figure 6.28**).

- The sodium-potassium pump creates a chemical and electrical gradient across the membrane because it returns only two positive K^+ charges for every three Na^+ charges removed from the cell. Other pumps are specialized for transport of other ions or molecules.

- Cells use active transport to create an internal environment that significantly differs from the environment outside the cell. Electrochemical gradients established by active transport have several critical roles in organisms—ranging from nutrient uptake to nerve impulse conduction.

B. INTEGRATING YOUR KNOWLEDGE

Looking Back—Key Concepts from Earlier Chapters

The Cell Theory—Chapter 1

Chapter 1 introduced the cell theory as one of the unifying themes of biology: All life is made of cells. In Chapter 6, you learn about the membranes that create a barrier between an organism and the external world and define the cell.

Entropy—Chapter 2

Chemical reactions can proceed spontaneously only when entropy, a measure of disorder, increases and/or potential energy decreases. Now you learn that entropy also determines the movement of molecules and ions across membranes in diffusion and osmosis.

Proteins—Chapter 3

Membrane channels, transporters, and pumps are all proteins. You may want to refer to **Chapter 3** to review protein structure so you will better understand how proteins insert into membranes and bind ions or molecules. Yet again, the complex structure of proteins is vital to their function.

Looking Forward—Key Concepts in Later Chapters

Cells—Unit 2 (Chapters 7–11), Chapter 21, 22, and other chapters

Because cells are the fundamental units of life, they reappear throughout your text; the next unit discusses the basics of cell structure, metabolism, and replication (division). Later on, you will learn how cells differentiate to form different tissue types in multicellular organisms.

Membranes and Membrane Proteins—Chapters 7–10, 37, 38, 42, 45–47, and 49

Membranes and their proteins determine what substances can cross the membrane to enter the cell. They thus have a critical role in maintaining homeostasis and transmitting signals to the cell. You'll learn more about these processes in the units on how plants and animals work. In **Chapters 7–10**, you will learn about internal cell membranes and their functions.

C. MASTERING CHALLENGING TOPICS

One good way to review and test your comprehension of **diffusion, osmosis, fluidity, concentration gradients**, and **electrochemical gradients** is to draw figures like that in **Figure 6.7c** for different experiments you could do. Then explain your expected results based on information given to you in this chapter.

If you put hydrophobic molecules like CO_2 on one side of a lipid bilayer, what would happen? How about if you put glucose on one side? What could you add to the membrane to increase glucose's permeability? In general, how could you modify the composition of the lipid bilayer to make it more or less permeable? What happens when the membrane is impermeable to a solute on one side (but permeable to water)? What happens if you add more water to one side? Ions carry a charge, and opposite charges attract; this is why ions respond to electrical gradients as well as to chemical gradients. What variation on this technique can researchers use to measure membrane permeability to ions?

Answers to all these questions are in the text, but doing your own thought experiments will help you master the information in this chapter and better understand how researchers test hypotheses about membranes.

D.   ASSESSING WHAT YOU'VE LEARNED

1. What type of molecule is represented in this figure?

Steroids

CH$_2$OH

C=O

CH$_3$ OH

CH$_3$

O

Cortisol

   a. a fat
   b. a steroid
   c. a phospholipid
   d. a fatty acid

2. Why are fatty acids called *acids*?
   a. because they have lots of hydrogens
   b. because they form ester bonds with glycerol
   c. because the carboxyl group tends to lose a proton in aqueous solution
   d. because they like solutions that have a low pH

3. _____ are artificial membrane-bound vesicles.
   a. Phospholipids
   b. Micelles
   c. Lipids
   d. Liposomes

4. Which of the following molecules will most easily cross a lipid bilayer?
   a. N$_2$
   b. ribose
   c. Ca^{2+}
   d. water

5. Why is the cell membrane less permeable to glucose than to glycerol?
   a. Glucose is larger than glycerol.
   b. Glucose contains oxygen; glycerol does not.
   c. Glycerol is nonpolar; glucose is polar.
   d. Glucose is a charged molecule; glycerol is not.

6. Lipid bilayers are most permeable when they are made of _____ fatty acids and _____ hydrocarbon chains.
   a. saturated; long
   b. saturated; short
   c. unsaturated; long
   d. unsaturated; short

7. At room temperature, a cell membrane has the consistency of olive oil. Cooling the membrane _____ its fluidity and _____ its permeability.
   a. increases; decreases
   b. decreases; increases
   c. decreases; decreases
   d. increases; increases

8. Spontaneous movement of any molecule from an area of high concentration to an area of low concentration is:
   a. osmosis
   b. diffusion
   c. equilibrium
   d. active transport

9. Distilled water is _____ relative to a solution containing dissolved salts.
   a. hypertonic
   b. hypotonic
   c. isotonic
   d. none of the above

10. Which statement best describes the fluid-mosaic model of membrane structure?
   a. Two layers of proteins are interspersed with phospholipids.
   b. A phospholipid bilayer has a few peripheral proteins associated with it.
   c. A phospholipid bilayer is coated on both sides with hydrophilic integral proteins.
   d. A phospholipid bilayer contains diverse proteins, including embedded amphipathic proteins.

11. Detergents are:
   a. amphipathic
   b. hydrophilic
   c. hydrophobic
   d. osmotic

12. Gramicidin is more permeable to H$^+$ than to K$^+$ since H$^+$ is much smaller. If the graph in **Figure 6.23** shows the size of the electric current when the concentration of H$^+$ varies, the same graph for the size of the electric current when the concentration of K$^+$ varies should be _____ the line for H$^+$.
   a. above
   b. below
   c. the same as
   d. the mirror image of

13. The glucose transporter protein GLUT-1 transports:
    a. all sugars
    b. glucose and other six-carbon sugars
    c. ribose
    d. the right-handed optical isomer of glucose

14. A protein binds to an ion, diffuses across a membrane, and releases it on the other side. No ATP or other chemical energy is used. This protein is an:
    a. aquaporin
    b. gated channel
    c. transport protein
    d. ion pump

15. Which of the following is *not* true of the sodium-potassium pump?
    a. It transports sodium into the cell and potassium out of the cell.
    b. It becomes phosphorylated by ATP.
    c. It creates an electrochemical gradient across the cell membrane.
    d. It is similar to an enzyme in many ways.

CHAPTER 6—ANSWER KEY

Check Your Understanding

(a) The fats don't form membranes because fats are not amphipathic.

(b) It is useful to show these scales as powers of 10, because a wide range of permeabilities or concentrations can easily be shown in the same figure.

(c) Oil has more unsaturated fats than shortening does. Oil is fluid at room temperature since double bonds in its unsaturated fats make kinks in the hydrocarbon chains and disrupt hydrophobic interactions. Shortening is softer at room temperature than in the refrigerator because molecules move slower at lower temperatures, causing hydrocarbon tails to pack more tightly.

(d) Bacteria reaching the salted meat shrivel up and die as water leave their cells due to osmosis. The salted meat is hypertonic to bacterial cells.

(e) The oil floats on top of the water and does not mix with the water. When the detergent is mixed in, the oil splits up into smaller and smaller droplets.

(f) The four-ring hydrocarbon structure of steroids is hydrophobic. Thus lipid bilayers are permeable to steroids, and these hormones can cross membranes on their own.

(g) As cells use glucose, the concentration of glucose in a cell decreases. This maintains a lower concentration of glucose in the cell than in, for example, the blood. Glucose moves into the cell down its concentration gradient via transporters like GLUT-1.

Assessing What You've Learned

1. b; 2. c; 3. d; 4. a; 5. a; 6. d; 7. c; 8. b; 9. b; 10. d; 11. a; 12. b; 13. d; 14. c; 15. a

7

Inside the Cell

A. KEY CONCEPTS

- The cell is the fundamental unit of life; most organisms are unicellular and, even in large multicellular organisms, complex processes originate in the cell.

- In the last chapter you learned that the **plasma membrane** is selectively permeable, so molecules and ions inside the cell are different from those found outside.

- Let's now explore cell structures and their functions, with special emphasis on the nucleus, the endomembrane system, and the cytoskeleton.

7.1 What's Inside the Cell?

- According to phylogeny, there are three main groups of organisms: Bacteria, Archaea, and Eukarya. Because Bacteria and Archaea are morphologically similar, they are called prokaryotes. We first discuss these simpler prokaryotic cells, then the more structurally complex, nucleus-containing eukaryotes.

Prokaryotic Cells

- The prokaryotic **plasma membrane** surrounds one compartment containing the **cytoplasm**, a highly concentrated solution containing all the molecules that make up the inside of the cell (**Figure 7.1**).

- The cytoplasm is typically hypertonic to the outside environment, so water tends to enter the cell by osmosis, causing it to expand. This osmotic pressure is resisted by a tough, fibrous **cell wall** that, in many prokaryotic species, consists of peptidoglycan (**Figure 7.2**).

- Cell walls protect cells and give them shape and structure. Outside the cell wall, many bacteria have yet another protective layer consisting of **glycolipids**—lipids with polysaccharides attached.

- The **chromosome**, a large DNA molecule associated with a few proteins, is the biggest cytoplasmic structure. Most prokaryotic species have one circular chromosome in the **nucleoid** region of the cell.

- DNA's nitrogenous base sequence carries the organism's genetic information. A **gene** is a section of DNA containing directions to build one polypeptide or RNA molecule.

- Because the chromosome may be hundreds of times longer than the cell, it is "supercoiled" (**Figure 7.3**).

- Prokaryotes often also contain small supercoiled DNA pieces called **plasmids**. Plasmids usually carry genes needed only under rare environmental conditions.

- All prokaryotic cells also contain ribosomes for protein synthesis. Ribosomes have a large and a small subunit and are made of 3 RNA molecules and over 50 different proteins (**Figure 7.4**).

- Some prokaryotes have tail-like **flagella** on their cell surface; these flagella are made of proteins and spin around to move the cell as fast as 60 times the cell's length per second.

- Some prokaryotes have extensive internal membranes for photosynthesis (**Figure 7.5**). The plasma membrane folds in, sometimes pinching off to form a vesicle. These membranes contain the pigments and enzymes necessary for the photosynthetic reactions.

- Certain prokaryotes have membrane-bound storage containers. Many of these qualify as **organelles** because they contain enzymes or structures specialized for a specific function. For example, one type of bacterial organelle stores calcium ions, and another type stores magnetite crystals.

- Prokaryotic cells also have a **cytoskeleton** of protein filaments. These proteins help cell division occur, and in some species they help maintain cell shape.

- Although prokaryotes have a minimal cytoskeleton and may have internal membranes, eukaryotes have a much more complex and dynamic cytoskeleton and extensive internal membranes. Eukaryotes also have a membrane-bound nucleus and tend to be much larger than prokaryotes (**Table 7.1**).

Eukaryotic Cells

- The relatively large size of eukaryotes (5 to 100 μm diameter) may have evolved as an adaptation for cell predation. However, large size makes it difficult for molecules to diffuse across the entire cell. Eukaryotes bypass this problem by breaking up the large cell volume into smaller membrane-bound compartments (**Figure 7.6**).

- The compartmentalization of eukaryotic cells increases chemical reaction efficiency by separating incompatible chemical reactions and grouping together enzymes and substrates. Groups of enzymes that work together may also be located close to each other on or within internal membranes.

The Nucleus

- All eukaryotes have a large nucleus surrounded by a double-membrane **nuclear envelope**.

- **Nuclear pores** provide passage through the envelope, and the nucleus is supported by **nuclear lamina** proteins that help organize the linear chromosomes.

- Ribosome synthesis and assembly occurs in the **nucleolus** (**Figure 7.7**).

Ribosomes

- Eukaryotic cytoplasm includes everything inside the plasma membrane except for the nucleus. Many ribosomes are found free in the **cytosol**—the fluid part of the cytoplasm.

- As in prokaryotes, eukaryotic ribosomes are made of RNA and protein and have large and small subunits that join to form a molecular machine for protein synthesis (**Figure 7.8**).

Rough Endoplasmic Reticulum

- The **rough endoplasmic reticulum** (**rough ER**) is a network of membrane-bound tubes and sacs. It connects to the outer membrane of the nuclear envelope and is studded with ribosomes. These ribosomes make proteins destined for the plasma membrane, the lysosome organelle, or cell secretion.

- Proteins made by rough ER ribosomes move inside the rough ER as they are made, and they are then folded and otherwise modified in the rough ER **lumen** (**Figure 7.9**).

- Proteins that are made by ribosomes on the rough ER enter the ER **lumen** or interior, where they are folded and modified.

Golgi Apparatus

- The **Golgi apparatus** is formed by a series of flat membrane sacs (**cisternae**) stacked together. The Golgi's *cis* side ("this side") faces the rough ER and receives products from it; the *trans* side ("across side") faces the plasma membrane and sends finished, packaged products to the cell surface in small vesicles (**Figure 7.10**).

Smooth Endoplasmic Reticulum

- Endoplasmic reticulum without ribosomes is called the **smooth endoplasmic reticulum** or **smooth ER** (**Figure 7.11**). Fatty-acid and phospholipid synthesis, as well as breakdown of hydrophobic toxins, occurs here. Enzymes that catalyze reactions with lipids facilitate these reactions.

- The smooth ER may also store calcium ions (Ca^2) that act as cell signals.

- The smooth and rough ER, together with the Golgi apparatus and the lysosomes, make up the **endomembrane system**. In each case, differences in organelle structure relate to organelle function.

CHECK YOUR UNDERSTANDING

(a) What is the difference between smooth and rough ER? Why does it make sense that rough ER is more important than smooth ER is in the synthesis and distribution of proteins?

_____

_____

_____

Peroxisomes

- **Peroxisomes** are single-membrane, globular organelles that independently grow and divide (**Figure 7.12**).

- Inside peroxisomes, oxidation reactions produce hydrogen peroxide (H_2O_2) as a by-product. The enzyme catalase quickly degrades the dangerously reactive H_2O_2 to harmless water and oxygen, so that the cell is not harmed.

- Different types of peroxisomes specialize in different types of oxidation reactions. For example, in plant leaves **glyoxisomes** convert a photosynthetic product into a sugar used to provide energy for the cell.

Lysosomes

- **Lysosomes** vary in size and shape, but they are always single-membrane-bound centers for cell storage and/or waste processing (**Figure 7.13**).

- Animal lysosomes have an acid interior (pH 5.0) and contain digestive enzymes called acid hydrolases that work best at a low pH. These enzymes break macromolecules into their monomer subunits.

- Membrane-surrounded packages sent to lysosomes originate when the animal's cell membrane engulfs a smaller cell or food particle (**phagocytosis, Figure 7.14a**) or when a cell surrounds damaged organelles with a membrane so that they can be digested and recycled in the lysosome (**autophagy, Figure 7.14b**).

- Lysosomes also receive materials via **receptor-mediated endocytosis**; macromolecules outside the cell bind to protein receptors, causing the membrane to fold inward and pinch off to form an **early endosome**. This membrane-bound vesicle receives digestive enzymes from the Golgi and has its pH lowered as it matures into a **late endosome**, which may turn into a lysosome (**Figure 7.15**).

- After macromolecules are hydrolyzed, their subunits are transported out of the lysosome by proteins in the organelle's membrane. These amino acids, sugars, and so on are then used to make new macromolecules.

- Phagocytosis and receptor-mediated endocytosis are two methods of **endocytosis,** the process whereby a cell membrane pinches off a vesicle to bring outside material into the cell. A third type of endocytosis, **pinocytosis,** brings fluid into the cell, but these vesicles do not go to lysosomes.

- Plants and Fungi have very large lysosomes called **vacuoles** that function primarily in water and/or ion storage and help the cell maintain its normal volume. Sometimes these vacuoles store other molecules (proteins, toxins, pigments) or contain digestive enzymes (**Figure 7.16**).

Mitochondria

- **Mitochondria** have two membranes; the inner one is folded into sac-like **cristae** that contain the **mitochondrial matrix (Figure 7.17)**. Most enzymes involved in making ATP (the cell's primary source of chemical energy) are found either in the inner mitochondrial membrane or in the matrix solution.

- ATP production is a mitochondrion's core function.

- Mitochondria grow and divide independently of cell division; they make their own ribosomes and contain their own supercoiled DNA in a small, circular chromosome similar to those found in bacteria.

Chloroplasts

- Most plant and algal cells have **chloroplasts** that, like mitochondria, grow and divide independently, have a double membrane, and contain **chloroplast DNA** in a circular chromosome. Via photosynthesis, chloroplasts convert light energy to chemical energy.

- Inside the chloroplast, hundreds of flat vesicles called **thylakoids** are stacked to form **grana**. Everything required for photosynthesis is located either in the thylakoid membranes or out in the chloroplast **stroma (Figure 7.18)**.

Cytoskeleton

- Eukaryotes have a complex, interconnected system of protein fibers that form the **cytoskeleton**. The cytoskeleton gives the cell shape; it also aids cell movement and the transport of materials within the cell.

The Cell Wall

- Fungi, algae, and plants have an outer **cell wall** typically made of carbohydrate fibers in a stiff matrix of polysaccharides and proteins (**Figure 7.19**). This cell wall protects the cell. Some plants produce yet another protective barrier of **lignin**, a tough molecule that, when combined with cellulose, forms wood.

CHECK YOUR UNDERSTANDING

(b) Label the organelles in the diagrams below. Are these plant, animal, or bacterial cells?

_____

_____

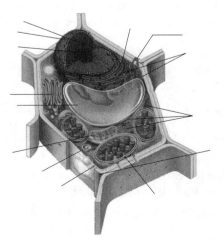

How Does Cell Structure Correlate with Function?

- An organelle's membrane and enzymes correlate closely with its function (**Table 7.2**).

- Similarly, cell structure (e.g., the type, size, and number of organelles) correlates with cell function; cells that need a lot of ATP have many mitochondria, but cells that export proteins have an extensive rough ER and Golgi apparatus (**Figure 7.20**).

The Dynamic Cell

- Microscopes reveal the basic size, shape, and structure of cells; and differential centrifugation (**Box 7.1, Figure 7.21**) allows the isolation and chemical analysis of cell components.

- High-magnification microscopes show only "snapshots" of fixed cells, and chemical analyses break up cell parts; but live cells are dynamic, living things with interacting parts and constantly moving molecules.

- Cells take in substances, synthesize molecules, get rid of wastes, and reproduce. Every second, each of your cells hydrolyzes about 10 million ATP, and your enzymes catalyze up to 25,000 or more reactions.

- Attaching **green fluorescent protein** (**GFP**) to other proteins in order to track protein movement is one modern research technique used to study living cells in action (**Box 7.2**).

- This rest of this chapter focuses on the dynamic cell by discussing (1) how molecules move in and out of the nucleus and around the cell in an ordered way, (2) how proteins move from ribosomes to their end destination, and (3) how cytoskeletal proteins facilitate movement of cell cargo and the cell itself.

CHECK YOUR UNDERSTANDING

(c) What advantage does green fluorescent protein have over other cell-staining techniques?

_____

_____

_____

_____

7.2 The Nuclear Envelope: Transport into and out of the Nucleus

- The nucleus is a highly organized information center for the cell. The nuclear lamina anchors each chromosome and maintains the overall shape and structure of the nucleus.

- Each chromosome has its own specific region within the nucleus where its genetic information is decoded and processed. Ribosomes are produced in the easily visible nucleolus.

- A double membrane called the nuclear envelope encloses the nucleus. The nuclear envelope connects to the endoplasmic reticulum, such that the space between the envelope's two phospholipid bilayers is continuous with the space inside the ER (**Figure 7.22**).

- Molecules shuttle to and from the nucleus through thousands of **nuclear pores** that connect the nucleus with the cytoplasm (**Figure 7.23**). More than 50 proteins form the **nuclear pore complex**, which extends through both nuclear membranes (**Figure 7.24**).

- Initial research on the nuclear pore complex involved injecting gold particles into cells and determining the location of the gold after various time intervals using electron microscopy. After 2 minutes, some gold was associated with nuclear pores; after 10 minutes, many particles had entered the nucleus.

- Ribonucleic acids are exported from the nucleus into the cell, whereas nucleotide triphosphates and the proteins involved in RNA synthesis must enter the nucleus; but DNA itself remains inside. The nuclear pore complex selects which particles can enter or exit the nucleus.

How Are Molecules Imported into the Nucleus?

- **Virus** proteins must enter their host cell's nucleus to reproduce themselves; but certain amino acid changes in virus proteins prevent the proteins from passing the nuclear pore.

- The preceding observation led to the hypothesis that proteins destined for the nucleus have a molecular address tag or zip code allowing them to enter the nucleus. Viral proteins have access to the nucleus if they contain this **nuclear localization signal (NLS)**.

- The NLS was studied using a protein called *nucleoplasmin*, which helps assemble chromosomes. This protein has a globular core surrounded by several long "tails." Radioactively labeled nucleoplasmin quickly enters the nucleus after being injected into a cell's cytoplasm.

- Researchers separated the nucleoplasmin core from its tails using proteases, and they radioactively labeled first one and then the other part. When injected into cells, isolated tail fragments were transported to the nucleus; but core fragments remained in the cytoplasm. Thus, the NLS must be part of the tail region (**Figure 7.25**).

- Further analysis demonstrated that the NLS, the molecular zip code, is a 17-amino-acid sequence. Other proteins destined for the nucleus have similar localization signals.

- Nuclear localization signals cause proteins to bind to **importins** in the cytoplasm. The protein/importin complex passes through the nuclear pore complex. Other proteins called **exportins** help certain large molecules exit the nucleus through the nuclear pores.

- Large molecules enter the nucleus only if they have a NLS, and their passage requires chemical energy.

CHECK YOUR UNDERSTANDING

(d) Why did researchers investigating nucleoplasmin's nuclear localization signal **(Figure 7.25)** inject radiolabeled tails and cores into different cells?

_____

_____

_____

_____

7.3 The Endomembrane System: Manufacturing and Shipping Proteins

- Some proteins made on ribosomes in the cytosol have signal sequences that cause transport to either mitochondria or chloroplasts. Other proteins are made in the rough ER and travel through the endomembrane system before reaching their final destination.

- Although small molecules like ions, ATP, and amino acids can travel throughout the cell via diffusion, the movement of large molecules is highly organized and regulated and consequently uses energy.

- Researchers formed a "secretory pathway" hypothesis after noting that extensive rough ER and Golgi are found in cells that secrete proteins. This hypothesis proposes that secreted proteins are synthesized and processed in a multistep pathway involving these organelles.

- Palade and co-workers did **pulse-chase experiments** to reveal the movements of newly synthesized proteins within secretory cells. A cell is first given a "pulse"—a high concentration of a labeled molecule for a short time—followed by the "chase"—lots of unlabeled molecules for a longer time period.

- Palade's group used pancreatic cells grown in culture (in vitro) because these cells secrete enzymes

and other molecules. First they gave the cells a 3-minute pulse of radiolabeled leucine. This amino acid became part of proteins produced during that 3-minute period, thus labeling those proteins.

- When cells were killed and examined by electron microscopy immediately after the pulse, all labeled protein was inside the rough ER. After a 7-minute chase, the protein was on the Golgi's *cis* side facing the rough ER. After 17 minutes, the protein was inside the Golgi apparatus; and after 80 minutes, the proteins were either in secretory granules on the far side of the Golgi or outside the cell (**Figure 7.27**).

CHECK YOUR UNDERSTANDING

(e) Why is the chase a necessary step in pulse-chase experiments?

_____

_____

_____

_____

- These experiments showed that secreted proteins are made in the rough ER, travel to the Golgi apparatus, and are transported out of the cell in secretory granules. The rough ER and Golgi apparatus function as an integrated endomembrane system.

Entering the Endomembrane System: The Signal Hypothesis

- Günter Blobel and colleagues proposed the **signal hypothesis**: that as ribosomes make secretory proteins, the first few amino acids act like an address tag that directs the growing polypeptide to enter the ER.

- This hypothesis is supported by the observation that secreted proteins made in a test tube are 20 amino acids longer than are the proteins secreted by cells. Blobel proposed that a 20-amino-acid **ER signal sequence** must direct the proteins to the ER and then be removed before the protein is secreted.

- Blobel's group identified the ER signal sequence. This sequence binds to an RNA–protein complex in the cytosol called the **signal recognition particle (SRP)**. The SRP then binds to a receptor in the ER membrane, the ribosome completes protein synthesis, and the signal sequence is removed (**Figure 7.28**).

- The new protein enters either the rough ER lumen (if it will be shipped to an organelle or secreted) or the rough ER membrane (if it is a membrane protein) and is folded with the help of chaperone proteins.

- Within the ER lumen, carbohydrate side chains are added to proteins (**glycosylation**), producing **glycoproteins** (**Figure 7.29**).

Getting from the ER to the Golgi

- Palade and colleagues proposed that proteins are transported from the ER to the Golgi apparatus in vesicles because the area between these two organelles contains many small membrane-bound vesicles. Differential centrifugation confirmed that these vesicles contained labeled proteins.

- Vesicles containing protein products must bud off the ER, travel to the Golgi apparatus, and deposit their contents inside as they fuse with the Golgi's *cis* face. A different type of vesicle carries proteins from one layer to the next within the Golgi apparatus.

What Happens Inside the Golgi Apparatus?

- Within the Golgi apparatus, cisternae (Golgi compartments) form at the *cis* face and then move toward the *trans* face, where they eventually break apart.

- Golgi cisternae mature as they move from *cis* to *trans* face, containing different glycosylation enzymes at different stages. Proteins within the cisternae thus undergo continual modification as they pass through the various sacs.

How Are Products Shipped from the Golgi?

- A few proteins remain within the endomembrane system, but most proteins that pass through the rough ER and Golgi are sent to lysosomes, the plasma membrane, or outside the cell.

- Molecular zip codes serve as protein shipping tags so that each protein is transported to its correct destination. For example, proteins bound for lysosomes have mannose-6-phosphate on their surface. If this sugar is removed, the proteins are not correctly shipped.

- Mannose-6-phosphate binds to a protein in vesicles with surface proteins that interact with other proteins in lysosome membranes. In this way, lysosome proteins are directed to lysosomes. Similar processes direct other proteins to vesicles destined for other locations.

- Some proteins are sent to the cell surface in vesicles that fuse with the plasma membrane, releasing their contents to the exterior of the cell (**exocytosis**).

- Protein sorting and transport from the Golgi apparatus is clearly highly organized (**Figure 7.30**). Proteins made in the cytoplasm have their own molecular zip codes directing their transport.

CHECK YOUR UNDERSTANDING

(f) What are the possible outcomes if a cell was unable to make mannose-6-phosphate?

_____

_____

_____

_____

7.4 The Dynamic Cytoskeleton

- Cytoplasm contains a complex network of fibers called the cytoskeleton, which provides structural support for cells. The cytoskeleton is dynamic; it changes in order to alter the cell's shape, transport materials in the cell, or move the cell itself.

- The three types of cytoskeletal elements are actin filaments, intermediate filaments, and microtubules (**Table 7.3**).

Actin Filaments

- The smallest-diameter cytoskeletal elements are **actin filaments**, also known as **microfilaments**. Actin, a globular protein, polymerizes to form two long, fibrous strands twisted around each other.

- Actin monomers and the filaments made from them are asymmetrical, or polar. Thus, actin filaments grow faster at the plus end than at the minus end because polymerization occurs fastest there. Microfilaments constantly grow and shrink as actin subunits are added or removed.

- Actin filaments are grouped into long fibers or dense networks, which are most often found just inside the plasma membrane. These filaments, linked by other proteins, help define the cell's shape (**Figure 7.31a**).

- Actin is also involved in cell movement through its interactions with **myosin**—a **motor protein** that can convert ATP's chemical energy into mechanical work. A conformational change of myosin's head region uses the chemical energy released by ATP hydrolysis to slide bound actin filaments (**Figure 7.32a**).

- Actin–myosin filament movement causes (1) **cell crawling** when filaments push the plasma membrane into **pseudopodia** and then pull the cell as filaments contract, (2) **cytokinesis** when a ring of filaments contracts to pinch an animal cell in two during cell division, and (3) **cytoplasmic streaming**, which moves cytosol and organelles through plant and fungal cells (**Figure 7.32b**).

Intermediate Filaments

- **Intermediate filaments** are defined by size instead of by composition (**Figure 7.31b**); all are nonpolar cytoskeletal elements that function as structural support for the cell. For example, keratins in our skin help these cells resist pressure and abrasion, and secreted keratins form hair and nails.

- **Nuclear lamins** of the nuclear lamina are another type of intermediate filament. These filaments anchor chromosomes and give shape to the nucleus.

- Some intermediate filaments reach from the nucleus to the cell membrane and form a flexible skeleton, which contributes to cell shape and holds the nucleus in place.

Microtubules

- Microtubules are large, hollow tubes made of α- and β-tubulin **dimers** (two joined monomers).

- Like actin, microtubules grow or shrink as subunits are added or removed. Microtubules are also polar; the two ends differ, and the molecule is more likely to grow from one end than from the other.

- Microtubules grow outward from a **microtubule organizing center**. Animal and fungal cells usually have only one microtubule organizing center, called the **centrosome**. Animal centrosomes have two microtubule bundles, of unknown function, called **centrioles** (**Figure 7.33**).

- Microtubules help the cell resist compression and create a structural framework for organelles.

- Microtubules also help move the cell and materials within the cell. For example, microtubules radiating from the organizing center move chromosomes to each daughter cell during cell division.

Studying Vesicle Transport

- The giant axon is an extremely large neuron that runs the length of a squid's body. If the squid is disturbed, the axon carries an electrical signal to a muscle band where the neuron releases neurotransmitters that cause the muscle to contract, propelling the animal away from danger.

- Because the giant axon is so large, it is easy to work with. Ronald Vale and co-workers were able to squeeze the cytoplasm out of the cell and observe vesicle transport of the neurotransmitters in this cell-free system.

Microtubules Act as "Railroad Tracks"

- Researchers used video-enhanced microscopy techniques to view cytoskeletal filaments. Neurotransmitter vesicles move along a track (**Figure 7.34**). The diameter and chemical composition of the track indicated that it is made of microtubules. Vesicle transport required energy because movement stopped if ATP ran out.

- The movement of vesicles from the ER to the Golgi apparatus also depends on microtubules because movement was abnormal following treatment with a drug that disrupts microtubules. How do these vesicles move through the cell along these microtubule tracks?

A Motor Protein Generates Motile Forces

- Vale and co-workers produced microtubules from purified α- and β-tubulin. They added ATP and transport vesicles isolated by differential centrifugation, but no transport occurred.

- Eventually they found a cell fraction that allowed vesicle movement. Further purification led to isolation of the **motor protein**, **kinesin**, which generates vesicle movement by converting the chemical energy of ATP into mechanical work.

- X-ray diffraction of kinesin shows three distinct regions: (1) a head with two globular pieces, each of which can bind to ATP and to a microtubule; (2) a tail that binds to a transport vesicle; and (3) a stalk that connects the head and tail regions (**Figure 7.35**). Different kinesins carry different types of vesicles.

- Researchers propose that the kinesin head region moves a step along the microtubule as it binds and hydrolyzes ATP.

- Dynein, another motor protein, transports certain organelles and contributes to cell swimming.

Cilia and Flagella: Moving the Entire Cell

- Flagella are long projections that move cells. Bacterial flagella are made of flagellin and rotate like a propeller, but eukaryotic flagella are made of microtubules and wave back and forth. Eukaryotic flagella are encased by the plasma membrane, but bacterial flagella are not.

- Structurally, eukaryotic flagella and cilia are very similar (flagella are longer). Eukaryotes with flagella typically have only one or two, whereas those with cilia have many shorter structures (**Figure 7.36**). Many eukaryotes do not have either structure.

How Are Cilia and Flagella Constructed?

- Cilia and flagella have two central microtubules surrounded by nine microtubule doublets (of one complete and one incomplete microtubule each). This 9 + 2 structure is called the **axoneme**. Spokes connect doublets to the central microtubules, molecular bridges connect doublets to one another, and each doublet has two "arms" projecting toward an adjacent doublet (**Figure 7.37**).

- The axoneme attaches to the cell at a **basal body**, which is structurally identical to a centriole but important in axoneme growth.

A Motor Protein in the Axoneme

- Ian Gibbons, in the 1960s, isolated *Tetrahymena* axonemes by using a detergent to remove the plasma membrane and then performing differential centrifugation. These axonemes beat only in the presence of ATP, showing that cilia movement requires energy.

- Further experiments with this cell-free system showed that axonemes could not beat or use ATP when protein binding was disrupted. Electron microscopy showed that the "arms" between doublets were missing. These arms are a motor protein called **dynein**, which changes shape when ATP is hydrolyzed so as to walk along microtubules.

- Cilia and flagella bend instead of elongating because the spokes and bridges constrain movement; this makes the microtubules slide past one another when the dynein arms on just one side of the axoneme walk (**Figure 7.38**).

- Key point: Cells are dynamic!

CHECK YOUR UNDERSTANDING

(g) Name three research techniques described in this chapter, and briefly state their usefulness in cell research.

_____

_____

_____

_____

B.   INTEGRATING YOUR KNOWLEDGE

Looking Back—Key Concepts from Earlier Chapters

Prokaryotes and Eukaryotes—Chapter 1

Chapter 1 first introduced you to the differences between prokaryotes and eukaryotes and showed how these groups are related on the tree of life. In this first chapter you also learned about cell theory—that cells

are the basic unit of all life, and all cells come from preexisting cells.

The Cell Membrane—Chapter 6

In the last chapter you learned about the structure and function of the plasma membrane. In this chapter you learn more about internal cell membranes—the endomembrane system. Remember that the structure (phospholipid bilayer with proteins) and function of these membranes is similar to that of the cell membrane.

Looking Forward—Key Concepts in Later Chapters

Mitochondria—Chapter 9

You will learn much more about mitochondria structure and function when you read about cell respiration.

Chloroplasts—Chapter 10

You will learn much more about chloroplast structure and function when you read about photosynthesis.

Cell Division—Chapters 11 and 12

During cell division, whole chromosomes need to be moved around the cell. Much of what you learned about intracellular transport in this chapter is thus relevant to your understanding of mitosis and meiosis.

DNA and Genes—Chapters 13–18

Because DNA contains all the information for protein synthesis, these chapters on DNA and genes will aid your understanding of cell function. For example, this chapter mentions that ribosomes are the site of protein synthesis, but the process itself (translation) is described in **Chapter 16**.

Prokaryotes—Chapter 28

Most of this chapter describes eukaryotic cell structure and function, but much of life is prokaryotic. You will learn more about bacteria and archaea in **Chapter 28**.

Microfilaments—Chapters 11, 12, 29, and 43

In Chapter 7, you learned how microfilaments help give shape to the cell. These actin filaments also play active roles in cell division (**Chapters 11** and **12**), pseudopodia motility (**Chapter 29**), and muscle contraction (**Chapter 46**).

C.   MASTERING CHALLENGING TOPICS

Chapter 7 has introduced you to many new terms and research techniques. Consider making note cards for yourself for all these new terms and techniques. If you have trouble memorizing, be sure to give yourself plenty of time before the test. It will be easier if you can study a few note cards at a time. You might even try taping up note cards around your room—above the light switch, on your computer, on your sock drawer, on your mirror. This way you can think about these new terms a little bit each day as you go through your regular daily routine.

Once you know some terms, put up new cards to learn. You can also try explaining vesicle transport to family or friends not familiar with biology. If you can explain it clearly (and correctly) to them, you know you have mastered the material.

D.   ASSESSING WHAT YOU'VE LEARNED

1. What is the fundamental difference between prokaryotic and eukaryotic cells?
   a. Prokaryotic cells do not have a plasma membrane.
   b. Prokaryotic cells do not have a nucleus.
   c. Prokaryotic cells do not have ribosomes.
   d. Eukaryotic cells never have a cell wall.

2. Which of the following is true of DNA in bacterial cells?
   a. All bacterial DNA is part of a chromosome in the cell's nucleoid region.
   b. Bacterial DNA usually forms linear chromosomes.
   c. Bacterial DNA does not contain genes.
   d. Bacterial DNA is supercoiled.

3. The nuclear envelope is:
   a. a single membrane surrounding the nucleus
   b. continuous with the Golgi apparatus
   c. supported by the nuclear lamina
   d. made of chromatin

For questions 4–9, match the given functions with the correct organelle:

For extra review, note how the structures of these organelles correlate with their functions.

a. nucleus	e. peroxisomes
b. rough ER	f. lysosomes
c. smooth ER	g. mitochondria
d. Golgi apparatus	h. chloroplasts

4. _____ processing and packaging proteins for secretion

5. _____ isolation of dangerous oxidation reactions

6. _____ lipid-processing and phospholipid manufacture

7. _____ synthesis of plasma membrane proteins

8. _____ hydrolysis of macromolecules

9. After the activation of proton pumps, this vesicle receives digestive enzymes from the Golgi apparatus.
   a. peroxisome
   b. lysosome
   c. endosome
   d. chromosome

10. Which of these proteins is not likely to be transported into the nucleus of living cells?
    a. mRNA for protein synthesis
    b. DNA polymerase, the enzyme that catalyzes DNA synthesis
    c. RNA polymerase, the enzyme that catalyzes RNA synthesis
    d. histones

11. The glucocorticoid receptor is a protein found only in the cytoplasm until the hormone glucocorticoid enters the cell and binds to this receptor. The receptor then translocates to the nucleus, where it regulates the transcription of certain genes. If a specific sequence of the receptor is deleted from the protein, the receptor still binds glucocorticoid but does not enter the nucleus. Consider the following additional hypothetical experiments. If the mutant receptor is fused with the tail fragment of nucleoplasmin, the receptor enters the nucleus regardless of whether glucocorticoid is bound. Fusion of the core fragment of nucleoplasmin with the sequence deleted from the glucocorticoid receptor results in accumulation of the core fragment in the nucleus. Which conclusion might be drawn from these observations?
    a. The binding of glucocorticoid to the receptor must expose a nuclear localization signal.
    b. The glucocorticoid receptor does not have a nuclear localization sequence.
    c. Glucocorticoid is the nuclear localization signal for the receptor.
    d. The glucocorticoid receptor enters the nucleus by a mechanism other than nucleoplasmin.

12. The signal hypothesis predicts that:
    a. Nuclear localization signal binds to signal recognition particles in cytosol.
    b. Proteins bound for the endomembrane system have a molecular zip code.
    c. Only proteins destined for the nucleus need a molecular zip code.
    d. Secreted proteins will be 20 amino acids shorter when synthesized in a test tube.

13. Microtubules, intermediate filaments, and microfilaments are made of _____, respectively.
    a. actin, keratins and other nonpolar filaments, α- and β-tubulin
    b. α- and β-tubulin, lamins, myosin
    c. actin, myosin, α- and β-tubulin
    d. α- and β-tubulin, keratins and other nonpolar filaments, actin

14. The mixing of purified microtubules with transport vesicles and ATP does not result in movement of the vesicles. Why?
    a. Actin is missing.
    b. Kinesin is missing.
    c. Glucose is missing.
    d. Intermediate filaments are missing.

15. Suppose the plasma membrane around a flagellum opens to reveal the axoneme inside. The radial spokes connecting peripheral microtubule doublets to the central pair are then broken by chemical treatment. ATP is then added. What is the expected observation?
    a. The flagellum will undergo normal bending.
    b. No movement will occur, because the plasma membrane is not present.
    c. The axoneme will elongate.
    d. No movement will occur, because the ability to use ATP is lost.

CHAPTER 7—ANSWER KEY

Check Your Understanding

(a) Rough ER has ribosomes attached, and smooth ER does not. Because ribosomes are the site of protein synthesis, it makes sense that rough ER would be responsible for the synthesis and distribution of many cell proteins.

(b) Animal cell: (a) centrioles, (b) lysosomes, (c) nuclear envelope, (d) nucleolus, (e) chromatin, (f) rough ER, (g) ribosomes, (h) peroxisome, (i) smooth ER, (j) Golgi apparatus, (k) mitochondrion, (l) cytoskeletal element, and (m) plasma membrane. Plant cell: (n) cell wall, (o) chloroplast, (p) nuclear envelope, (q) nucleolus, (r) chromatin, (s) rough ER, (t) ribosomes, (u) smooth ER, (v) Golgi apparatus, (w) vacuole, (x) peroxisome, (y) mitochondrion, (z) plasma membrane, and (aa) cytoskeletal element

(c) It can be difficult to appreciate how dynamic cells are because many techniques used to view cell details (for example, electron microscopy and many staining procedures) kill a cell. Green fluorescent protein tags allow researchers to observe movements of specific proteins in living cells.

(d) If radiolabeled tails and cores had been injected into the same cell, the researchers would not have been able to determine which had entered the nucleus by monitoring radiation—because the cores and tails themselves cannot be detected, only the radiation.

(e) Without the chase, radioactive proteins would have been found throughout the endomembrane system during the later time periods, and it would have been impossible to determine whether the radioactivity observed in the ER and Golgi was due only to new proteins or whether some previously made protein remained in the ER and Golgi.

(f) Inability to make mannose 6-phosphate (known as a congenital disorder of glycosylation) causes lysosomal problems because many lysosomal enzymes are not correctly packaged and delivered. This can sometimes be detected due to abnormally high plasma and serum concentrations of these enzymes. Organism-level effects can be quite severe.

(g) Radiolabeling (helps localize proteins within cells), pulse-chase (tracks movement of proteins within cells), green florescent protein (allows tracing of protein movement within a living cell), subcellular fractionation/differential centrifugation (helps isolate cell components for chemical testing and identification), and cell-free extracts (e.g., the squid giant axon, allow easier experimental manipulation of cell components)

Assessing What You've Learned

1. b; 2. d; 3. c; 4. d; 5. e; 6. c; 7. b; 8. f; 9. c; 10. a; 11. a; 12. b; 13. d; 14. b; 15. c

8

Cell–Cell Interactions

A. KEY CONCEPTS

- Cells constantly interact with and respond to other cells and the environment. In multicellular organisms, the cells' external environment consists primarily of other cells. This chapter focuses on how cells communicate and cooperate with each other in multicellular organisms.

8.1 The Cell Surface

- The plasma membrane's phospholipid bilayer has *peripheral* proteins attached to its surface and *integral* proteins within it. Proteins crossing the entire membrane are called transmembrane proteins. Some of them attach to the cytoskeleton inside the cell and/or to materials outside the cell (**Figure 8.1**).

The Structure and Function of an Extracellular Layer

- Most cells secrete extracellular material that glues them to other cells, defines cell shape, and/or defends the cell from external threats.

- Almost all extracellular material has a "fiber composite" structure consisting of long cross-linked filaments surrounded by a stiff ground substance. The rods or filaments protect against stretching forces, while the ground substance protects against compression (**Figure 8.2**).

- Different groups of organisms use different molecules to form this extracellular fiber composite, and many extracellular layers are flexible as well as sturdy.

The Plant Cell Wall

- The extracellular material secreted by plant cells forms a **primary cell wall** consisting of long strands of cellulose.

- Cellulose is made by enzymes in the plasma membrane. This cross-linked polysaccharide is bundled into filaments called microfibrils. The microfibrils form a crisscrossed network that becomes filled with a gelatinous polysaccharide ground substance (**Figure 8.3**).

- **Pectins** are common gelatinous polysaccharides in the ground substance of plant cell walls. Pectins are hydrophilic and keep the cell wall moist by attracting water. These carbohydrates are made in the rough endoplasmic reticulum (ER) and Golgi apparatus and secreted to the extracellular space.

- The primary cell wall determines cell shape. Normally, the solute concentration is higher inside the cell than outside, so that water enters the cell by osmosis, filling up the cell volume and pushing the plasma membrane against the cell wall.

- **Turgor pressure** is the force with which the filled cell pushes against the cell wall. In young cells, enzymes called expansins break cross-links, allowing microfibrils to slide past one another. Turgor pressure then forces the cell wall to elongate, allowing cell growth.

- Mature plant cells may secrete a **secondary cell wall** inside the primary cell wall. The structure of this secondary cell wall varies widely depending on the cell's function. For example, cells that form wood have a secondary cell wall with lots of **lignin**, a complex substance that is especially tough.

- Plant cell walls are dynamic and may be reinforced or broken down according to the needs of the plant.

The Extracellular Matrix in Animals

- Animal cells secrete a fiber composite called the **extracellular matrix** (**ECM**). Like plants, animals use gelatinous polysaccharides as a ground substance (made either by membrane proteins or in the rough ER). Animals, however, use protein fibers instead of polysaccharide filaments.

- **Collagen** is the most common ECM protein fiber (**Figure 8.4**). It and other protein fibers are synthesized in the rough ER, processed in the Golgi apparatus, and secreted via exocytosis. These proteins are more elastic than cellulose or lignin and form a flexible extracellular layer.

- In animals, the composition and amount of ECM correlates with cell function, varying widely among cell types.

- Regardless of its exact composition, the ECM provides structural support for the cell.

- The ECM typically also helps cells stick together. Cytoskeletal actin filaments connect to transmembrane **integrins**, which bind to ECM **fibronectins**, which bind to collagen. These protein–protein attachments directly link the cytoskeletons within cells to the ECM (**Figure 8.5**).

- ECM breakdown can cause major problems (e.g., cancer metastasis and other diseases; see **Box 8.1**).

CHECK YOUR UNDERSTANDING

(a) List similarities and differences between plant and animal extracellular material. How do the differences relate to differences in the function of extracellular material in animals and plants?

_____

_____

_____

8.2 How Do Adjacent Cells Connect and Communicate?

- Cells, even unicellular organisms, constantly interact with other cells. For example, some bacteria secrete a hard, polysaccharide-rich substance that accumulates outside their cell walls. This **biofilm** attaches cells to the surface on which they live.

- **Dental plaque** is a biofilm encasing a multi-species group of oral bacteria. Other biofilms are secreted as protection by a single-species group of bacteria (**Figure 8.6**).

- Unicellular organisms live together and communicate with each other, but cell–cell physical connections are the basis of **multicellularity** (**Figure 8.7**).

- In multicellular organisms, similar cells with similar functions form **tissues**, which may combine to form an **organ** specialized for one biological function.

- Multicellularity has evolved several times (**Figure 8.8**); and each multicellular group has its own type of cell–cell connections and communication. Many of these cell-connecting structures were revealed by transmission electron microscopy.

Cell–Cell Attachments

- Plant cell walls are glued together by gelatinous pectins and other molecules that form the middle lamella (**Figure 8.9**). When enzymes break down this layer, the surrounding cells separate.

- A layer of gelatinous polysaccharides also exists between cells in many animal tissues.

- Transmembrane proteins called integrins connect cell cytoskeletons to the ECM, and some tissues have additional proteins that more strongly connect neighboring cells.

- These stronger protein attachments are especially found in **epithelia**, the tissues covering external and internal body surfaces.

Tight Junctions

- **Tight junctions**, common in cells lining your stomach and intestines, are a type of cell–cell attachment whereby proteins in adjacent cell membranes line up and bind to one another, stitching the two cells together to form a watertight seal between the two plasma membranes (**Figure 8.10**).

- The exact structure and function of tight junctions depends on the type of epithelia in which they are found. For example, tight junctions in the small intestine allow small ions to pass whereas those in the bladder do not.

- Tight junctions are dynamic: Some tissues can loosen or tighten these junctions.

Desmosomes

- Epithelial cells and some muscles are held together by **desmosomes** made of proteins that link the cytoskeletons of adjacent cells. These proteins bind to each other and to the proteins that anchor cytoskeletal intermediate filaments (**Figure 8.11**).

Selective Adhesion

- When adult sponges are treated with chemicals that cause cells to separate from each other, the resulting mixed-up cell mass reconnects when the chemicals are removed.

- Reconnecting sponge cells sort themselves by species and tissue type. This is known as **selective adhesion** (**Figure 8.12**). This result suggested that cell–cell connections are species- and tissue- specific.

The Discovery of Cadherins

- Each major cell type has its own cell adhesion proteins in animals. The **cadherins**, which include desmosome proteins, are cell adhesion proteins found in plasma membranes that bind only to other cadherins of the same type, causing cells of the same tissue type to bind together.

- Identification of these cell-adhesion proteins started with the isolation of membrane proteins from a certain cell type. Pure preparations of each protein were then injected into rabbits, one at a time, causing each rabbit to produce **antibodies**—proteins that bind specifically to a part of the injected protein.

- Researchers added one type of antibody at a time to mixtures of dissociated cells and observed whether the cells could reconnect normally. Antibodies that prevented cell–cell attachment must bind to adhesion proteins, blocking their normal binding function (**Figure 8.13**).

- Cadherins are one of several major classes of cell adhesion proteins. The cell-type specificity of cadherins explains how animal cells selectively attach to other animal cells within the same individual.

CHECK YOUR UNDERSTANDING

(b) In the cadherin research, why was it necessary to isolate the membrane proteins before injecting them into the rabbits for antibody production?

_____

_____

_____

_____

_____

Cell–Cell Gaps

- Direct connections between tissue cells allow cells to work together.
- Plant cells communicate via cell-wall gaps called **plasmodesmata**, where the plasma membranes and cytoplasms of two cells connect such that smooth ER actually runs between one cell and the next (**Figure 8.14a**).
- Like the nuclear pore complex, plasmodesmata also contain proteins that regulate protein traffic. Some proteins passing through plasmodesmata help coordinate the activity of neighboring cells.
- In most animal tissues, **gap junctions** connect adjacent cells (**Figure 8.14b**). Specialized proteins that match up to form a channel between two cells line these membrane holes. Small molecules flow through these channels and can help coordinate activities of the two connected cells.
- Plant and animal cells usually retain their own organelles and large macromolecules but share most or all of the small molecules found in their cytoplasms. Cell–cell gaps allow efficient communication among tissue cells so that the tissue can act as an integrated whole.

8.3 How Do Distant Cells Communicate?

- Chemical signals travel throughout animals and plants to target cells, carrying information from one tissue or

organ to another. This intercellular signaling allows multicellular organisms to coordinate the activities of all tissues so that they work together as a unified whole.

Hormones Are Long-Distance Messengers

- Several cell–cell signals exist, but this discussion focuses on **hormones**—intercellular signaling molecules that are secreted from cells and travel through the organism to act on distant target cells.
- The functions and chemical structures of plant and animal hormones vary widely (**Table 8.1**). However, all play a role in coordinating cell activity in response to information from outside or inside the body.
- Hormone chemical structure affects the manner in which its signal is received. Steroids and other lipid-soluble hormones can diffuse across a cell's plasma membrane, but hydrophilic hormones must be recognized at the cell surface.

Signal Reception

- Hormones and other cell–cell signals must bind to a receptor molecule in order to affect a target cell. Only cells with appropriate receptors respond to a given hormone. When several tissues and organs respond to the same signal, it is because they all have appropriate receptors.
- Proteins that change their conformation or activity when a hormone binds to them are **signal receptors**.
- Steroid hormone receptors are located inside cells because steroids are lipid soluble and can easily diffuse through the plasma membrane, but most signal receptors are found in the plasma membrane.
- Receptors are dynamic and may increase or decrease in number over time. For example, receptor numbers tend to decline following prolonged hormonal stimulation. Changes in receptor number affect cell sensitivity to particular hormones.
- Receptors can be blocked. For example, beta-blockers and other drugs act by binding to and blocking certain receptors.
- Hormone binding causes a change in receptor structure. This change in receptor conformation means a signal has been received.

Signal Processing

- Steroid hormones directly initiate cell response; they enter a cell, bind to a receptor protein, and either directly affect membrane pump activity or, more commonly, are transported to the nucleus as a hormone–receptor complex that directly alters gene expression (**Figure 8.16**).
- Hormones that cannot diffuse across the plasma membrane rely on a complex series of events called a **signal transduction pathway**. **Signal transduction**

refers to conversion of the hormone signal from its original extracellular chemical form to a new intracellular form (**Figure 8.17**).

- Hormone messages are also amplified during signal transduction. Signal amplification allows small hormone concentrations to cause large cell responses.

- Signal transduction via membrane channels that alter membrane electrical properties is examined in Chapter 45. This chapter describes two other major signal transduction systems: G-proteins and enzyme-linked receptors.

- G-proteins typically trigger the production of intracellular "second" messengers, whereas enzyme-linked receptors trigger protein activation via phosphorylation cascades.

G Proteins

- **G proteins** are peripheral membrane proteins located inside the cell. They are closely associated with transmembrane signal receptors and are named for their ability to bind guanosine triphosphate (GTP) and guanosine diphosphate (GDP).

CHECK YOUR UNDERSTANDING

(c) Why are steroid receptors located inside cells, whereas most other hormone receptors are located on the plasma membrane?

_____

_____

_____

- GTP, like ATP, has high potential energy due to the close proximity of negative charges on the phosphate groups.

- G proteins are turned on, or activated, when they bind GTP; they automatically turn themselves off, or are inactivated, when they hydrolyze GTP, forming GDP. Activated G proteins trigger second messenger production within the cell.

- First, a hormone binds to a plasma membrane receptor, causing the transmembrane receptor protein to change shape. The hormone receptor's conformational change activates a G protein on the membrane's inner surface. The G protein releases GDP, binds GTP, and then splits into two parts, one of which activates a nearby enzyme in the plasma membrane (**Figure 8.18**).

- G-protein-activated enzymes catalyze the production of small nonprotein signaling molecules called **second messengers**. These intracellular signals spread the hormone message throughout the cell.

- Second messengers are small, so they are quickly produced in large quantities and rapidly diffuse throughout the cell. One hormone molecule stimulates the production of many second messenger molecules, greatly amplifying the original signal.

- Many small molecules act as second messengers (**Table 8.2**). The same second messenger may cause different effects in different cell types, and more than one type of second messenger may be involved in a cell's response to any given signal.

Enzyme-Linked Receptors

- **Enzyme-linked receptors** directly catalyze a cell reaction when activated.

- **Receptor tyrosine kinases (RTKs)** are an important type of enzyme-linked receptor. These transmembrane proteins form dimers after binding to a hormone signal and then become activated by phosphorylating each other.

- This phosphorylated RTK form activates yet other enzymes by phosphorylating them (**Figure 8.19**). These enzymes phosphorylate yet more enzymes and so on, creating a **phosphorylation cascade** that activates many enzymes and amplifies the original signal.

- Sometimes G proteins trigger phosphorylation cascades; and some phosphorylation cascades, including those initiated by RTKs, produce second messengers. Thus, signal transduction pathways form a complex network, allowing cells great flexibility in responding to extracellular signals.

- Both G-protein and receptor tyrosine kinase signal transduction converts an easily transmitted extracellular message into a greatly amplified intracellular message that carries information throughout the cell, inducing the cell response.

CHECK YOUR UNDERSTANDING

(d) Compare signal reception by a G-protein receptor versus a receptor tyrosine kinase.

_____

_____

_____

_____

Signal Response

- Steroid hormones bound to intracellular receptors directly affect gene expression. Hydrophilic hormones can also affect gene expression if the cell response includes activation or deactivation of a protein that affects gene expression.

- Second messengers or phosphorylation cascades also commonly affect cell activity via activation or deactivation of enzymes, membrane channels, or other cell proteins.

Signal Deactivation

- Cells have automatic mechanisms for signal deactivation. Rapid response, followed by equally rapid response elimination, ensures that cells remain sensitive to future hormone signals and to any change in the number or activity of signal receptors.

- Activated G proteins can hydrolyze GTP, producing GDP and a free phosphate. This reaction changes the G protein's conformation, deactivating it so that it no longer activates the enzyme producing the second messenger.

- Second messengers are either degraded (for example, phosphodiesterase deactivation of cAMP and cGMP) or pumped back into storage (Ca^{2+}).

- Phosphorylation cascades are turned off by *phosphatases*, enzymes that remove phosphate groups from proteins. These phosphatases are always present in cells, ready to quickly shut off phosphorylation cascades as soon as hormone stimulation stops.

- Problems occur when signaling pathways are not properly shut down following stimulation. For example, defective Ras proteins that do not properly deactivate can contribute to the development of cancer.

- From reception to processing to response to deactivation, cell–cell signaling allows organisms to produce an integrated whole-organism response to changing conditions both within and outside of the multicellular organism.

CHECK YOUR UNDERSTANDING

(e) Why is signal deactivation important? Generally speaking, what might happen if a signal was not deactivated?

_____

_____

_____

B.   INTEGRATING YOUR KNOWLEDGE

Looking Back—Key Concepts from Earlier Chapters

Enzyme Function—Chapter 3

Hormone–receptor interactions are similar to the enzyme–substrate interactions described in **Chapter 3**; hormones bind at a specific site and cause a conformational change to the receptor, similar to the way in which substrates often cause a conformational change in an enzyme upon binding to the active site. **Chapter 8** also introduces you to some important groups of enzymes: kinases, phosphatases, and phosphodiesterases.

The Cell Membrane and Its Associated Proteins—Chapter 6

Cadherins, gap junction proteins, hormone receptors, and G proteins are just some of many membrane proteins mentioned in **Chapter 8**. Remember that the cell membrane is dynamic—these proteins are always moving around, and their motion ensures that activated G proteins interact with the enzymes they activate.

Looking Forward—Key Concepts in Later Chapters

Tissues and Organs—Chapters 35 and 41

Chapter 8 describes how cells physically connect to form tissues and organs. You will learn more about tissue and organ structure and function in **Chapters 36** and **41** (plant and animal form and function, respectively).

Chemical Signals in Plants and Animals—Chapters 39, 47, and others

You may want to come back to this chapter later when you read about specific hormones used as chemical signals in plants and animals. **Chapter 45** looks at electrical signals.

C.   MASTERING CHALLENGING TOPICS

Antibodies are a key research tool for the cell biologist. The experiments shown in **Figure 8.13** are quite complex. If you did not completely understand these experiments after one reading, consider the following analogy.

You are told that antibodies are proteins that specifically bind to a section of another protein (animals make antibodies that bind to foreign proteins as part of their immune response). Pretend that the proteins isolated from a specific cell type and injected into the rabbit are like various-shaped objects that you place on a table for a five-year-old to use in making play-dough molds. After awhile, you have many molds of an AA battery, a Lego, some children's scissors, and various other objects. Let's pretend that these molds, when left out, eventually find and bind to the object they fit with.

For your first experiment (Experiment 1), you leave out the battery molds and discover that they bind to all the batteries in the house, causing your mini flashlights to go out, etc., but having no effect on Lego attachment. For Experiment 2, you put away the battery molds and take out the Lego molds. Now your mini-flashlights work, but the Legos won't connect anymore, because they are all covered with the molds. If you put the Lego molds away and take out the scissors molds (Experiment 3), the Legos and flashlights work, but kids in the house are no longer able to cut paper.

In Experiments 1 and 3, the molds/antibodies did not affect Lego attachment because the objects used to make the molds are not involved in that process. In Experiment 2, Lego attachment no longer works, because the mold/antibody blocks the attachment/binding site. Taking this analogy one step further, you are told that cadherin proteins are cell-type specific. In our analogy, this is comparable to discovering that the molds that block regular Legos in some houses have no effect on Duplo Legos found in other houses. Each house/cell type has its own attachment proteins that are blocked only by their specific molds/antibodies.

D. ASSESSING WHAT YOU'VE LEARNED

1. Extracellular material can function to: (Choose *all* that apply.)
   a. glue cells together
   b. actively transport nutrients into cells
   c. protect cells
   d. define cell shape
   e. catalyze chemical reactions

2. A fiber composite structure contains:
   a. cross-linked rods or filaments and contractile elements
   b. a ground substance and cross-linked rods or filaments
   c. a frame structure and contractile elements
   d. a frame structure and a ground substance

3. The primary cell wall of plants is made of:
   a. pectins and expansins
   b. cytoskeletal elements and peripheral proteins
   c. cellulose and gelatinous polysaccharides
   d. lignin and collagen

4. Turgor pressure occurs when:
   a. solute concentration is the same inside the cell and outside, so that water pushes on the cell wall
   b. solute concentration is higher inside the cell than outside, so that water leaves the cell by osmosis
   c. solute concentration is lower inside the cell than outside, so that water enters the cell by osmosis
   d. solute concentration is higher inside the cell than outside, so that water enters the cell by osmosis

5. Which of the following statements are true about the animal cell extracellular matrix (ECM)? Choose *all* that apply.
   a. Composition and amount of ECM varies widely among different cell types.
   b. Protein–protein attachments link the cell cytoskeleton to the ECM.
   c. The ECM helps maintain cell turgor pressure in animal cells.
   d. Collagen is the most common ECM protein fiber.

6. Only multicellular organisms:
   a. live together and communicate with each other
   b. have organelles
   c. have extracellular layers
   d. have long-term cell–cell physical connections

   For questions 7 to 9, use the following options to identify each structure correctly.
   a. found between plant cell walls
   b. found between inside the primary cell wall
   c. found between epithelial cells in animals, forms watertight seals
   d. found between epithelial cells, links cytoskeletons of adjacent cells

7. _____ tight junctions
8. _____ middle lamella
9. _____ desmosomes

10. What evidence suggests that cells have selective adhesion?
    a. Electron micrographs show cell–cell attachments.
    b. Randomly mixed sponge cells sorted themselves by tissue and species.
    c. Chemical treatments are unable to separate sponge cells of the same tissue type.
    d. Rabbits produce antibodies to injected membrane proteins.

11. Plant cells communicate via cell-wall gaps called
   _____.
   a. desmosomes
   b. tight junctions
   c. gap junctions
   d. plasmodesmata

12. Which of the following are shared by cells connected by gap junctions? Choose *all* that apply.
   a. ions
   b. RNA
   c. amino acids
   d. smooth ER
   e. nucleotides

13. G proteins are activated when _____.
   a. they bind to a hormone
   b they bind to ATP
   c. they bind to GTP
   d. they bind to a second messenger

14. In some cells, activation of a hormone receptor causes an increase in Ca^{2+} levels, which causes activation of an enzyme that produces nitric oxide. Nitric oxide then activates another enzyme that makes cGMP. Based on this description, nitric oxide could be considered a:
   a. steroid hormone
   b. G protein
   c. signal receptor
   d. second messenger

15. How are receptor tyrosine kinase signals deactivated?
   a. Phosphodiesterases remove phosphate groups from ATP.
   b. Phosphatases remove phosphate groups from activated proteins.
   c. Membrane pumps return calcium ions to storage.
   d. cAMP and cGMP are converted to inactive AMP and GMP.

CHAPTER 8—ANSWER KEY

Check Your Understanding

(a) *Similarities*: Both use gelatinous polysaccharides as a ground substance, and both function as structural support. *Differences*: They have different cross-linked fibers (polysaccharides vs. collagen). The fiber differences make animal cell connections more flexible. This flexibility is critical for organisms that move around more than plants do.

(b) If the proteins had not been isolated before injection into the rabbit, the rabbit would have produced antibodies to all the proteins. These antibodies would have been all mixed together. Subsequent addition of these mixed antibodies into the cell culture might have allowed researchers to determine that a cell adhesion protein was present, but researchers would not have been able to determine which of the proteins injected into the rabbit was the one involved in cell adhesion.

(c) Steroid hormones are lipid soluble and can diffuse through the phospholipid bilayer, but most other hormones cannot easily cross the plasma membrane.

(d) In general, signal reception occurs when a hormone binds to a receptor, causing a conformational change. In the case of G-protein receptors, the receptor's conformational change causes an associated G protein to release GDP, bind GTP, and split into two parts (one of which will in turn activate the next enzyme in this signal transduction pathway). In the case of receptor tyrosine kinases, hormone binding causes the receptor itself to form a dimer, bind ATP, and become phosphorylated (and able to activate the next part of this signal transduction pathway).

(e) Signal deactivation is important because if signals were not turned off, cells would be unable to respond the next time a signal was sent. Automatic signal deactivation allows cells to remain responsive. Thus, cells keep responding only if the signal continues to be sent and will stop responding if the signal dissipates.

Assessing What You've Learned

1. a, c, & d; 2. b; 3. c; 4. d; 5. a, b, & d; 6. d; 7. c; 8. a; 9. d; 10. b; 11. d; 12. a, c, & e; 13. c; 14. d; 15. b

9

Cellular Respiration and Fermentation

A. KEY CONCEPTS

- Hydrolysis of **adenosine triphosphate (ATP)** provides the chemical energy that powers most cell work. ATP is the major energy currency of the cell.

- Energy to make ATP comes from oxidation of sugars and other reduced compounds. This energy is used to add a third phosphate to **adenine diphosphate (ADP)**, making ATP.

- This chapter introduces you to **metabolism**—the chemical reactions that occur in cells.

9.1 The Nature of Chemical Energy and Redox Reactions

- Energy associated with position is called potential energy, and chemical energy is a type of potential energy. Electrons are the most important form of chemical energy in cells.

- Electrons have high potential energy when they are close to other negative charges and/or far from the positive charges of protons in atom nuclei.

- ATP has high potential energy because its four negative charges are close together. During ATP hydrolysis, the bond holding ATP's outermost phosphate breaks, releasing an inorganic phosphate ($H_2PO_4^-$, abbreviated as P_i) and ADP (**Figure 9.1**).

- ATP hydrolysis is highly exergonic and releases 7.3 kilocalories (kcal) of energy per mole of ATP under standard temperature and pressure conditions. This reaction is exergonic because the products (ADP and P_i) have more entropy and less potential energy than the reactant (ATP).

- Cells use the energy released during ATP hydrolysis to do cell work. Often the P_i released from ATP is added to a protein (**phosphorylation**). Protein phosphorylation reactions are also exergonic.

- Typically a protein's shape changes when it is phosphorylated because the protein's original electrons respond to the addition of another negative charge,

causing part of the protein to move (**Figure 9.2**). This protein movement in turn powers cell work.

How Does ATP Drive Endergonic Reactions?

- Endergonic reactions cannot occur on their own in cells. These reactions must be coupled to an exergonic reaction like ATP hydrolysis in order to proceed (**Figure 9.3**).

- An endergonic reaction between A and B to form product AB might be coupled to ATP hydrolysis as follows. First, ATP hydrolysis phosphorylates B (an exergonic reaction), increasing B's free energy. Phosphorylated B (an "activated" substrate) now reacts with A to form AB. This second reaction can occur because B's release of its phosphate makes this reaction exergonic.

What Is a Redox Reaction?

- In **reduction–oxidation** or redox **reactions**, one molecule loses electrons (is **oxidized**), and another gains electrons (is **reduced**). One reactant is an **electron donor** and another is an **electron acceptor**.

- The "loss" of an electron may just mean that it moved farther away from its atom's nucleus. Electron position around an atom's nucleus changes as the atom forms new bonds since covalent bond electrons are shared more or less depending on the electronegativities of the atoms involved.

- For example, during photosynthesis carbon's electrons move closer to this atom's nucleus, but oxygen's electrons move further away from their nucleus. Thus, carbon is reduced and oxygen is oxidized (**Figure 9.4**).

- When electrons get closer to or further from an atom's nucleus, their potential energy changes, affecting the chemical energy of reactants and products. Photosynthesis is endergonic because electrons are held more loosely (further from carbon's nucleus) in glucose than in CO_2. Consequently, this reaction proceeds only with energy input from sunlight.

(a) Compare the energy transformations in the endergonic reaction A + B → AB discussed above and in the
overall reaction for photosynthesis (**Figure 9.4**).

_____

_____

_____

_____

_____

Some Other Approaches to Understanding Redox

- Electrons that are equally shared between two atoms
have higher potential energy and are less stable than
electrons that are unequally shared and closer to a
highly electronegative atom.

- During cell redox reactions, electrons (e^-) may be
fully transferred from an atom in one molecule to an
atom in a different molecule. During these electron
transfers, a proton (H^+) may move with the e^- such
that the molecule with the reduced atom gains a hydrogen atom (since H^+ plus $e^- = H$).

- Reduced molecules with C—H bonds have high
potential energy because electrons are shared
equally in C—H bonds and are relatively far from
the protons in the C and H atom nuclei. This is why
molecules with many C—H bonds, such as carbohydrates and fats, store a lot of potential energy.

- When molecules are oxidized in cells, they often
lose a proton along with their electron. Oxidized
molecules have low potential energy and tend to
have many C—O bonds and few, if any, C—H
bonds. C—O bond electrons have low potential
energy because they are close to the highly electronegative O nucleus.

What Happens When Glucose Is Oxidized?

- When glucose burns (**Figure 9.5**), its potential energy
(686 kcal per mole!) is released as kinetic energy in
the form of heat. When cells oxidize glucose, this energy is released slowly during many controlled redox
reactions so that it can be used to make ATP.

9.2 An Overview of Cellular Respiration

- ATP is continually made and hydrolyzed, since at
any one time a cell contains only enough ATP for a
few minutes of work.

- Cellular respiration and fermentation are the metabolic pathways primarily responsible for ATP production. Glucose fuels these catabolic pathways and is a
key intermediary in cell metabolism.

- Glucose is primarily made by photosynthesis. Non-
photosynthetic organisms obtain glucose by decomposing or eating photosynthetic species. All organisms
use glucose to build fats, carbohydrates, and other
compounds. Glucose can be recovered by breaking
down these molecules when more fuel is needed
(**Figure 9.6**).

- The chemical reactions that break down glucose to
produce ATP are among the most fundamental and
universal of all cell processes; some of the enzymes
involved originated 3.5 billion years ago.

- Cellular respiration is a four-step process in many
organisms: (1) glucose is broken down to pyruvate;
(2) pyruvate is processed to form acetyl-CoA;
(3) acetyl-CoA is oxidized to CO_2; and (4) reduced
compounds from steps 1–3 are oxidized in reactions
that allow ATP production.

Processing Glucose: Glycolysis

- **Glycolysis**, a series of 10 chemical reactions that
occur in the cytoplasm, is the first step in glucose
oxidation (**Figure 9.8a**). Glycolysis is an ancient
pathway; almost all species use the same 10 chemical reactions.

- Glycolysis breaks down glucose into two 3-carbon
molecules of pyruvate. Cells use the released potential energy to phosphorylate ADP. Nicotinamide
adenine dinucleotide (NAD^+) is also reduced to
make NADH (**Figure 9.7**), an **electron carrier** that
readily donates electrons to more oxidized
molecules.

Pyruvate Processing

- Pyruvate is used to produce acetyl-CoA when an
appropriate electron acceptor (usually oxygen) is
available. During pyruvate processing, one carbon is
oxidized to CO_2 and a molecule of NADH is made
(**Figure 9.8b**). Eukaryotes process pyruvate in the
mitochondrial matrix.

The Krebs Cycle

- Acetyl-CoA from pyruvate processing enters a
sequence of reactions known as the **Krebs cycle**
(**Figure 9.8c**). In eukaryotes, Krebs cycle enzymes
are found inside mitochondria.

- Acetyl-CoA is completely oxidized in the Krebs
cycle, producing two molecules of CO_2. Some of the
released energy is used (1) to reduce NAD^+ to
NADH; (2) to reduce flavin adenine dinucleotide

(FAD) to FADH$_2$ (another electron carrier); and (3) to phosphorylate ADP to make ATP.

Electron Transport

- The high-potential-energy electrons carried by NADH and FADH$_2$ participate in a series of redox reactions. Their electrons are passed down a series of molecules in an **electron transport chain (ETC)** found in the inner mitochondrial membrane of eukaryotes (and in the cell membrane of some prokaryotes).

- At each ETC step, electrons lose potential energy; the released energy is used to pump protons to the outside of the membrane, forming a strong electro-chemical gradient. When they receive an electron, some ETC molecules change shape in a way that moves a proton across the membrane.

- Located in the same membrane, ATP synthase uses the proton gradient energy to phosphorylate ADP. Part of ATP synthase spins as protons pass through this enzyme on their way back to the matrix side of the membrane. This spinning powers ADP phosphorylation to produce ATP (see **Figure 9.9** for an analogy).

- The final electron acceptor at the end of the electron transport chain is usually O$_2$, which gains protons along with electrons, forming water. This completes the oxidation of glucose (**Figure 9.8d**).

- The combination of glycolysis, Krebs cycle, and electron transport chain is called **cellular respiration**. This term can refer to any process that produces ATP using an electron donor, an electron transport chain, and an electron acceptor.

CHECK YOUR UNDERSTANDING

(b) On the figure below, label the locations of glycolysis, pyruvate processing, the Krebs cycle, and the electron transport chain. Also label the mitochondrial matrix and indicate which part of the cell shown would have the lowest pH during cellular respiration.

Methods of Producing ATP

- ATP synthase uses the proton gradient set up by the electron transport chain to power ATP production. This method of ATP production is linked to oxidation of NADH and FADH$_2$, so it's called **oxidative phosphorylation**.

- **Substrate-level phosphorylation** occurs when a phosphate group is transferred from an intermediate substrate to ADP, forming ATP, in an enzyme-catalyzed reaction (**Figure 9.10**). In these reactions, energy comes from the phosphorylated substrate.

- ATP production during glycolysis and the Krebs cycle occurs via substrate-level phosphorylation.

- Throughout glycolysis and the Krebs cycle, the free energy of the compounds involved gradually decreases. Larger drops in free energy are associated with ATP synthesis; smaller drops are associated with the production of NADH and FADH$_2$ (**Figure 9.11**).

- In most organisms, if no oxygen is available to act as the electron acceptor, cell respiration cannot occur (a few rare organisms use other electron acceptors). Instead, **fermentation** reactions occur so that glycolysis can continue (**Figure 9.12**). In eukaryotes, fermentation produces ethanol or lactic acid.

9.3 Glycolysis

- Hans and Edward Buchner discovered glycolysis by chance in the late 1890s. They added sucrose to yeast extracts and found that the sugar was broken down and fermented, producing alcohol. This discovery showed that cell metabolism could be studied in vitro (outside the cell).

- The Buchners and others soon found that the fermentation reactions lasted much longer when inorganic phosphate was added, that fructose bisphosphate (with two phosphate groups) was one intermediate, and that all intermediate compounds are phosphorylated; only glucose and pyruvate lack phosphate.

- Glycolytic reactions stopped if the yeast extract was boiled. This suggested the involvement of enzymes, because enzymes are inactivated by heat. In fact, a different enzyme catalyzed each step of glycolysis.

(c) Why did the fermentation reactions observed by the Buchners last longer when inorganic phosphate was added?

_____

_____

A Closer Look at the Glycolytic Reactions

- Differential centrifugation techniques were used to verify that all glycolytic reactions occur in the cytosol. **Figure 9.13** shows all 10 reactions of glycolysis.

- In the first reaction, ATP is hydrolyzed to phosphorylate glucose, forming glucose-6-phosphate. Glucose-6-phosphate is then rearranged, creating fructose-6-phosphate, and a second ATP is used up to make fructose-1,6-bisphosphate. These steps are the energy-investment phase of glycolysis.

- In the second half of glycolysis, the energy investment is more than paid off. The sixth reaction reduces 2 NAD^+, the seventh reaction produces 2 ATP, and the last reaction produces another 2 ATP. The net yield is 2 NADH, 2 ATP, and 2 pyruvates per glucose.

- In glycolysis, ATP is produced by substrate-level phosphorylation, because enzymes catalyze the direct transfer of phosphates from a phosphorylated intermediate to ADP.

How Is Glycolysis Regulated?

- High levels of ATP inhibit the enzyme **phosphofructokinase**, which catalyzes step 3 of glycolysis. Whereas the products of steps 1 and 2 can be used in other metabolic pathways, the product of step 3 (fructose-1,6-bisphosphate) can be used only in glycolysis, making step 3 a good point for regulation.

- Regulation of phosphofructokinase by ATP is unusual in that ATP is a substrate for this enzyme. Addition of a substrate typically increases the reaction rate. However, because ATP is also one of the end-products of glycolysis, it makes sense that cells with lots of ATP do not need to make more.

- Inhibition of an enzyme by the product of a reaction sequence is called **feedback inhibition** (**Figure 9.14**). ATP inhibition of phosphofructokinase allows cells to stop glycolysis when ATP is plentiful and conserve glucose stores for another time.

- Phosphofructokinase has two ATP binding sites: an active site, where ATP is hydrolyzed, and a **regulatory site** (**Figure 9.15**), where ATP acts as an allosteric regulator. When ATP binds to the regulatory site, the enzyme's conformation changes in a way that significantly decreases enzyme activity.

(d) What would happen to glycolysis reactions if a cell ran completely out of ATP?

_____

_____

_____

9.4 Pyruvate Processing

- Eukaryotes transport the pyruvate made by glycolysis to the mitochondria.

- Mitochondria have two membranes. The inner membrane is folded into sac-like **cristae**. Most Krebs cycle enzymes are in the **mitochondrial matrix**—the aqueous area inside the cristae (**Figure 9.16**).

- Pyruvate crosses the outer membrane through small pores and is actively transported (an energy-using process) into the mitochondrial matrix by an inner membrane protein called the pyruvate carrier.

- Inside the mitochondria, pyruvate reacts with **Coenzyme A (CoA)** to produce **acetyl CoA**. Coenzyme A is a common enzyme cofactor that transfers acetyl ($—COCH_3$) groups to substrates.

- The reaction between pyruvate and CoA is catalyzed by an enzyme complex called **pyruvate dehydrogenase**, which is found in the mitochondrial matrix in eukaryotes (and in the cytosol of prokaryotes).

- The reaction between pyruvate and CoA results in the reduction of NAD^+ to NADH and the oxidation of one of pyruvate's carbons to CO_2. The remaining two carbons form the acetyl group transferred to CoA (**Figure 9.17**).

- Pyruvate dehydrogenase, like phosphofructokinase, is sensitive to feedback inhibition. When ATP is plentiful, pyruvate dehydrogenase becomes phosphorylated. This phosphorylation changes the enzyme's shape and inhibits catalysis. High concentrations of acetyl CoA and NADH also promote pyruvate dehydrogenase phosphorylation.

- In contrast, high concentrations of NAD^+, CoA, or AMP (an indicator for low ATP concentration) increase pyruvate dehydrogenase's enzymatic activity.

- Because pyruvate processing is under both positive and negative control, it is another key regulatory point in glucose oxidation.

9.5 The Krebs Cycle

- Biologists discovered eight small **carboxylic acids** (R—COOH) involved in redox reactions that produced CO_2. Because CO_2 is the endpoint of glucose metabolism, researchers proposed that these carboxylic acids were associated with glucose metabolism.

- Researchers observed that adding any of these carboxylic acids to cells caused an increase in cell respiration, but the added molecules were not used up.

- Biochemists determined the redox reaction order for these eight carboxylic acids (**Figure 9.18a**).

- Hans Krebs realized that the carboxylic acid reactions could occur in a cycle tied to the breakdown of pyruvate. This hypothesis explained why carboxylic acids were not used up in the reactions and was further supported by the observation that citrate (the cycle's start point) formed when oxaloacetate (the cycle's endpoint) and pyruvate were added to cells.

- This redox reaction cycle is now called the Krebs cycle (**Figure 9.18b**). Experiments with radioactive carbon isotopes confirmed that the cycle occurred as Krebs proposed.

- Later, pyruvate dehydrogenase's discovery led to the realization that acetyl CoA, instead of pyruvate, combines with oxaloacetate to form citrate (**Figure 9.19**).

- To summarize: The reaction between pyruvate and CoA oxidizes one carbon (releasing CO_2) and links glycolysis and the Krebs cycle. Acetyl CoA's two carbons are oxidized in the Krebs cycle and the energy released is used to make 3 NADH, 1 $FADH_2$, and 1 **guanosine triphosphate (GTP)**.

- GTP is made through substrate-level phosphorylation and can be converted to ATP.

- Krebs cycle reactions occur in the cytoplasm of bacteria and archaea and primarily in the mitochondrial matrix of eukaryotes.

How Is the Krebs Cycle Regulated?

- The Krebs cycle is regulated such that low ATP concentrations cause high reaction rates and high ATP concentrations inhibit Krebs cycle reactions (**Figure 9.20**). For example, ATP inhibits the enzyme that converts acetyl CoA to citrate.

- Feedback inhibition also occurs at two other points in the Krebs cycle. In one case, NADH binds to the enzyme's active site as a competitive inhibitor; in the other case, ATP binds to an allosteric site.

- Natural selection has favored organisms that regulate enzyme activity, thereby conserving glucose, pyruvate, and acetyl CoA when no new ATP or NADH is needed.

What Happens to the NADH and FADH₂?

- Two pyruvates are formed for each glucose molecule that enters glycolysis. After each pyruvate has been fully oxidized in the Krebs cycle, the cell has produced 10 NADH, 2 $FADH_2$, and 4 ATP (**Figure 9.21**):

$$C_6H_{12}O_6 + 10\ NAD^+ + 2\ FAD + 4\ ADP + 4P_i \rightarrow$$
$$6\ CO_2 + 10\ NADH + 2\ FADH_2 + 4\ ATP$$

- The four ATP were made by substrate-level phosphorylation and can be used for cell work, the CO_2 is a waste gas that you exhale when you breathe, and NADH and $FADH_2$ are used to make more ATP in a reaction that could be written as:

$$NADH + FADH_2 + O_2 + ADP + P_i \rightarrow$$
$$NAD^+ + FAD + H_2O + more\ ATP$$

- In the 1960s, researchers finally discovered how the high-potential-energy electrons carried by NADH and $FADH_2$ were eventually donated to O_2, producing H_2O in a process that allows ATP production.

9.6 Electron Transport and Chemiosmosis

- Eukaryotes oxidize NADH in their inner mitochondrial membranes. These membranes, isolated by differential centrifugation, can oxidize NADH; but the mitochondrial matrix and intermembrane fluid cannot. In prokaryotes, oxidation of NADH occurs in the cell membrane.

 CHECK YOUR UNDERSTANDING

(e) Explain why it is advantageous for the inner mitochondrial membrane to be highly folded.

_____

_____

Components of the Electron Transport Chain

- Molecules in the inner mitochondrial membrane that are involved in the oxidation of NADH and $FADH_2$ make up the **electron transport chain (ETC)**. When oxygen accepts electrons at the end of the ETC, water forms, and glucose oxidation is complete.

- The energy released by the sequential redox reactions is used to pump protons across the inner mitochondrial membrane, creating a proton gradient. Protons flow back into the mitochondrial matrix through ATP synthase, driving ATP production.

- Most ETC molecules are proteins with distinctive chemical groups (flavins, iron–sulfur complexes, heme groups, or copper atoms) that are easily reduced or oxidized in redox reactions.

- **Ubiquinone**, also called **coenzyme Q** or **Q**, is a carbon-containing ring attached to a long hydrophobic tail of isoprene subunits. Q is lipid soluble and easily moves through the mitochondrial membrane, whereas all but one of the ETC proteins are anchored in the inner mitochondrial membrane.

- ETC molecules differ in their electronegativity—their tendency to become reduced or oxidized. Electrons proceeding down an ETC are held more and more tightly (have less and less potential energy) as they are passed from molecule to molecule (**Figure 9.22**). Each reaction releases a small amount of energy.

- Researchers used poisons to confirm the existence of ETCs. A poison that inhibits a specific ETC protein causes ETC molecules that receive electrons before the poisoned protein to stay in a reduced state, whereas those "downstream" of the poisoned protein remain oxidized.

- NADH donates electrons to a flavin-containing protein at the top of the chain, but FADH$_2$ donates electrons to an iron–sulfur protein that passes electrons directly to Q.

- Oxygen is the final electron acceptor in eukaryotes, and the total difference in potential energy from NADH to oxygen is 53 kcal/mol.

The Chemiosmotic Hypothesis

- No enzymes could be found that directly used the energy released by ETC redox reactions to make ATP. Peter Mitchell proposed an indirect connection between the redox reactions and ATP production.

- Mitchell suggested that the electron transport chain pumps protons from the matrix to the intermembrane space. This would create an electrochemical gradient favoring the movement of protons back into the matrix. A protein in the inner membrane could use this **proton-motive force** to make ATP.

- Mitchell's idea is known as the **chemiosmotic hypothesis** because he used the term **chemiosmosis** to describe the production of ATP via a proton gradient. The following experiment supported his hypothesis.

- Researchers made vesicles from artificial membranes containing bacteriorhodopsin (a protein that uses light energy to pump protons) and an ATP-synthesizing enzyme found in mitochondria. When lit, bacteriorhodopsin pumped protons out of the vesicles, creating a proton gradient, and ATP was produced inside the vesicles (**Figure 9.23**).

- This experiment showed that as long as there was a proton gradient, ATP could be produced even in the absence of an electron transport chain.

How Is the Electron Transport Chain Organized?

- Electron transport chain proteins are organized into four protein–cofactor complexes. Lipid-soluble Q and the protein **cytochrome c** transfer electrons between complexes. In complexes I and IV, protons pass through the electron carriers as they are pumped out of the matrix (**Figure 9.24**).

- Q also carries a proton across the membrane; when Q accepts electrons from complex I or II, it also gains protons. Q then diffuses through to the outer side of the membrane, where its electrons are transferred to complex III and its protons are released into the intermembrane space.

- Thus, the ETC uses the energy released by a series of redox reactions to pump protons from the matrix into the intermembrane space. ATP synthase uses this proton-motive force to make ATP.

CHECK YOUR UNDERSTANDING

(f) A compound called dinitrophenol (DNP) can be used to poke holes in the inner mitochondrial membrane. What would this do to ATP synthesis by oxidative phosphorylation?

_____

_____

_____

The Discovery of ATP Synthase

- Efraim Racker observed that inside-out mitochondrial membrane vesicles had large stalk- and knob-shaped proteins on their surfaces. The knobs fell off when shaken or treated with urea.

- Isolated knobs hydrolyzed ATP, and vesicles with stalks and no knobs could not make ATP even though proton transport occurred normally. Thus, the base appeared to be a proton channel.

- Racker hypothesized that the knobs were enzymes that, depending on the conditions, can either hydrolyze or synthesize ATP. This hypothesis was supported by the discovery that when the knobs were added back to knobless vesicles, the vesicles regained their ability to synthesize ATP.

- The knob-and-stalk protein is now called **ATP synthase**. Protons flow through the F_o unit (the base), causing the stalk connecting the two units to spin. As the F_1 unit (the knob) spins with the rod, its subunits are deformed in a way that catalyzes the phosphorylation of ADP to ATP (**Figure 9.25**).

CHECK YOUR UNDERSTANDING

(g) Why wasn't it enough for Racker to show that ATP synthesis stopped when the knobs were removed? That is, why was his explanation more convincing after he added back the isolated knobs, causing the vesicles to regain the ability to synthesize ATP?

_____

_____

Oxidative Phosphorylation

- ATP production via the combination of proton pumping by electron transport chains (creating the proton-motive force) and ATP synthase action is called **oxidative phosphorylation**. ATP synthase produces about 26 of the 30 ATP molecules produced per glucose molecule during cell respiration (**Figure 9.26**).

- All eukaryotes and many prokaryotes use oxygen as the final electron acceptor of electron transport chains. These **aerobic organisms** use **aerobic respiration**. Many other prokaryotes, especially those in oxygen-poor environments, use other electron acceptors in **anaerobic respiration**.

- Oxygen, with its high electronegativity, is the best electron acceptor because the potential energy of its electrons is very low, creating a large energy difference between NADH and O_2 electrons. This large potential energy difference allows the generation of a large proton-motive force for ATP production.

- Cells that use other electron acceptors cannot pump as many protons per NADH; thus, they cannot make as much ATP per glucose. These cells grow more slowly than aerobic respirators do. Aerobic respirators

typically outcompete anaerobic respirators in environments with oxygen.

- If oxygen or other electron acceptors are temporarily unavailable, NADH electrons have no place to go. The ETC stops, yet cells must have NAD^+ to produce ATP from glycolysis. What do cells do?

9.7 Fermentation

- **Fermentation** pathways convert NADH back to NAD^+ and allow glycolysis to continue producing some ATP via substrate-level phosphorylation when the ETC electron acceptor is not available (**Figure 9.27**).

- During fermentation, pyruvate—or a molecule derived from pyruvate—accepts electrons from NADH. The molecule formed from the reduction of pyruvate is often excreted from the cell as waste.

- In humans, pyruvate accepts electrons from NADH in **lactic acid fermentation** when muscles metabolize glucose faster than oxygen can be supplied. Lactate and NAD^+ are produced.

- Yeast deprived of oxygen use **alcohol fermentation**, in which pyruvate loses CO_2 and is converted to acetylaldehyde, which then accepts an electron from NADH to produce ethanol and NAD^+.

- Other fermentation pathways also exist, and some bacteria and archaea exclusively use fermentation to break down glucose (e.g., in your small intestine; in the cow's rumen).

- Because the electron acceptors in fermentation are much more reduced than oxygen, only 2 ATP are produced per glucose—compared to about 30 ATP per glucose in aerobic respiration. Consequently, organisms never use fermentation if an appropriate electron acceptor is available for cell respiration.

- Organisms that can switch between fermentation and aerobic cell respiration are called **facultative aerobes**.

CHECK YOUR UNDERSTANDING

(h) What would happen if a winemaker followed the industrial process for making wine, but then bubbled oxygen through the vats of grape juice instead of letting it sit undisturbed?

_____

_____

_____

9.8 How Does Cellular Respiration Interact with Other Metabolic Pathways?

- Cell metabolism involves thousands of different chemical reactions. Types and amounts of molecules inside cells are constantly changing (**Figure 9.28**).

- **Catabolic pathways** involve the breakdown of molecules and the production of ATP, whereas **anabolic pathways** result in the synthesis of larger molecules from smaller components.

- Together, these pathways allow cells to survive, grow, and reproduce, supplying chemical energy (ATP) and carbon-containing building blocks for synthesis of the molecules necessary for life.

Processing Proteins and Fats as Fuel

- Cells use enzyme-catalyzed reactions to break down glycogen, starch, and most simple sugars into glucose or fructose, which can enter the glycolytic pathway.

- Fats can be broken down into glycerol, which can be phosphorylated to form an intermediate of glycolysis, and acetyl CoA, which enters the Krebs cycle.

- Carbon compounds remain after proteins are broken down into amino acids and amino groups are removed and excreted. These carbon compounds can be converted into pyruvate, acetyl CoA, and/or other intermediates of glycolysis or the Krebs cycle.

- Macromolecules are broken down in different ways, but glycolysis and the Krebs cycle are central to ATP production in these catabolic pathways (**Figure 9.29**). For ATP production, cells first use carbohydrates, then fats, and finally proteins.

Anabolic Pathways Synthesize Key Molecules

- Molecules found in carbohydrate metabolism are used in anabolic metabolic pathways too (**Figure 9.30**).

- About half of our amino acids can be synthesized from Krebs cycle compounds.

- Acetyl CoA is the carbon building block for fatty-acid synthesis.

- The second compound in glycolysis can be used to synthesize a key intermediate in nucleotide production for RNA and DNA synthesis.

- Pyruvate and lactate can be used to synthesize glucose when sufficient ATP is present to power these reactions. Glucose can, in turn, be converted to glycogen (in animals) or starch (in plants) for storage.

- Molecules may contribute to many cell functions, and metabolic pathways are both interconnected and tightly regulated.

B. INTEGRATING YOUR KNOWLEDGE

Looking Back—Key Concepts from Earlier Chapters

Natural Selection—Chapter 1

The pathways discussed in this chapter are all a product of natural selection. For example, enzyme regulation likely evolved due to the advantages it confers on an organism. This chapter also reminds us that selection is always relative to the environment; use of oxygen as an electron acceptor is highly advantageous in environments with oxygen, but it is not beneficial in anoxic environments.

ATP Provides Energy for Endergonic Reactions—Chapters 2, 3, and 7

Chapters 2 and **3** introduce you to the idea that cells perform chemical reactions, and that endergonic reactions are necessary to create the macromolecules essential to life. Most endergonic reactions in cells are made possible by coupling these reactions with ATP hydrolysis. In **Chapter 7** you learn that ATP provides the energy for intercellular transport; now, in **Chapter 9**, you learn how cells make ATP.

Membranes—Chapter 6

The structure and function of mitochondrial membranes are in many ways similar to the structure and function of the cell membranes described in **Chapter 6**. For example, Chapter 6 also discusses the transport of ions across membranes.

Looking Forward—Key Concepts in Later Chapters

Many later chapters discuss cell work powered by ATP hydrolysis.

Photosynthesis—Chapter 10

In the next chapter, you learn about another ATP-producing process that uses an electron transport chain—photosynthesis.

Bacteria and Archaea—Chapter 28

Bacteria and archaea are groups that show incredible metabolic diversity. **Chapter 9** just touches on this diversity by mentioning that many of these species use anaerobic respiration.

C. MASTERING CHALLENGING TOPICS

The chemiosmotic hypothesis can be difficult to understand. This is partly why some researchers opposed this theory for so long. But it really just proposes a slightly different type of energy transformation than those we are more familiar with. For example, in **Chapter 2** you learned how potential energy can be converted to kinetic energy. In the electron transport chain, the potential energy of electrons is simply converted to the potential energy in an electrochemical gradient. Consider an analogy. The electrochemical proton gradient can be compared to the water and gravity gradient observed in a dam. Cells use redox reactions and the potential energy of electrons to pump the protons in the same way a water pump could burn fuel to pump water up a dam. Then the protons flow back across the membrane through ATPase, and the energy in the proton gradient is converted into energy used to phosphorylate ATP. Similarly, water flowing through a dam can be used to turn turbines that then produce electricity.

D. ASSESSING WHAT YOU'VE LEARNED

1. Under standard conditions, how much energy does the hydrolysis of one mole of ATP release?
   a. 110 kcal
   b. 52 kcal
   c. 7.3 kcal
   d. No energy is released; this reaction is endothermic.

2. Why do reduced molecules tend to have a lot of C—H bonds?
   a. Reduced molecules hold electrons tightly, and C—H bonds hold electrons tightly.
   b. Reduced molecules like glucose can't have any oxygen atoms.
   c. Carbohydrates and fats have lots of C—H bonds.
   d. Reduced molecules usually gain protons along with electrons.

3. The burning of methane can be written as $CH_4 + O_2 \rightarrow CO_2 + H_2O$. Which of the following is true of this reaction?
   a. Oxygen is oxidized and carbon is reduced.
   b. Carbon is oxidized and oxygen is reduced.
   c. Both carbon and oxygen are reduced.
   d. This is not a redox reaction because no electrons are completely transferred.

4. What is the purpose in having several steps in glycolysis or the Krebs cycle rather than a single step from glucose and oxygen to carbon dioxide and water?
   a. The multistep approach is the only way to convert glucose to carbon dioxide.
   b. The multistep approach increases the amount of potential energy in the reaction.
   c. The multistep approach makes better use of potential energy in the reaction.
   d. The multistep approach increases the amount of heat produced in the reaction.

5. Which of the following is true regarding ATP synthesis in prokaryotes?
   a. They oxidize NADH on the cell membrane.
   b. They are unable to use oxygen.
   c. They do not make ATP.
   d. They use mitochondria.

6. In the Buchner experiment, why did boiling the yeast extract prevent the conversion of ethanol from glucose?
   a. Phosphate required for the reactions was vaporized.
   b. Sucrose was destroyed.
   c. Proteins were denatured.
   d. Yeast cells were killed.

7. Why is a different enyzme involved in each step of glycolysis?
   a. Each step occurs in a different subcellular location.
   b. Each step occurs in a different cell.
   c. Each step involves a different cofactor.
   d. Each step involves a different chemical reaction.

8. Nearly every living organism uses glucose as a nutrient source of energy. Why?
   a. Glucose is the only molecule capable of providing the energy to produce ATP.
   b. The ability to harvest energy from glucose appeared very early in biological evolution.
   c. Glucose contains more potential energy than any other nutrient molecule.
   d. The structure of glucose is very similar to ATP.

9. Which of the following chemical reactions does not occur during glycolysis?
   a. hydrolysis of ATP
   b. phosphorylation of ADP
   c. reduction of NAD^+
   d. oxidation of NADH

10. If glucose is labeled with ^{14}C, what molecule will become radioactive as glycolysis and the Krebs cycle are completed?
    a. water
    b. carbon dioxide
    c. ATP
    d. NADH

11. At the end of the Krebs cycle, but before the electron transport chain, the oxidation of glucose has produced a net yield of:
    a. 3 CO_2, 5 NADH, 1 $FADH_2$, and 2 ATP
    b. 6 CO_2, 10 NADH, 2 $FADH_2$, and 4 ATP
    c. 6 CO_2, 10 NADH, 2 $FADH_2$, and 6 ATP
    d. None of the above are correct.

12. What would have been the result of Krebs' experiment if the Krebs cycle were linear instead of circular?
    a. Oxaloacetate + pyruvate would have produced citric acid.
    b. Oxaloacetate + pyruvate would have produced acetyl CoA.
    c. Oxaloacetate + pyruvate would have produced no products.
    d. Oxaloacetate + pyruvate would have produced glucose.

13. If oxygen gas is labeled with ^{18}O, what molecule will become radioactive as cellular respiration occurs?
    a. water
    b. carbon dioxide
    c. ATP
    d. NADH

14. Cyanide poisons an electron transport chain protein that has less free energy than Q. Which of the following would occur during cyanide poisoning?
    a. Q would stay reduced and stop moving protons across the mitochondrial membrane.
    b. Q would stay oxidized and stop moving protons across the mitochondrial membrane.
    c. Q would not be affected, but the proton gradient would dissipate.
    d. The electron transport chain would stop functioning because the cell would run out of NADH.

15. Why does NADH donate electrons to the beginning of the electron transport chain, whereas $FADH_2$ donates electrons to the middle of the chain?
    a. $FADH_2$ is more rapidly oxidized than NADH.
    b. NADH has more potential energy than $FADH_2$.
    c. $FADH_2$ has more reducing potential than NADH.
    d. NADH has more electrons to donate than $FADH_2$.

CHAPTER 9—ANSWER KEY

Check Your Understanding

(a) Both photosynthesis (really a whole pathway, as you will see in Chapter 10) and A + B → AB are endergonic reactions. In both of these reactions, the products have less entropy than the reactants; and in photosynthesis, at least, the product has more potential energy than the reactants due to the increased distance of the electrons from their carbon nuclei. The key difference in energy transformations between these two reactions is that while photosynthesis is powered by light energy, A + B → AB is powered by chemical energy. As the text discusses, this reaction occurs only when it is coupled to ATP hydrolysis.

(b) See figure at bottom of page.

(c) Inorganic phosphate is one of the substrates for glycolysis. Thus glycolysis and fermentation stop when no more phosphate is available.

(d) If a cell ran completely out of ATP, it could not perform glycolysis because the first and third steps require ATP input before ATP is made in the final steps.

(e) The folds provide more surface area—more area for electron transport chain components and ATPases.

(f) When DNP puts holes in the inner mitochondrial membrane, no ATP can be produced by ATPase because the proton gradient disappears.

(g) Something else that was responsible for ATP synthesis might have been knocked off along with the knobs.

(h) You would produce carbon dioxide, water, and lots more yeast, but little if any alcohol because the yeast would use aerobic respiration instead of fermentation.

Assessing What You've Learned

1. c; 2. d; 3. b; 4. c; 5. a; 6. c; 7. d; 8. b; 9. d; 10. b; 11. b; 12. c; 13. a; 14. a; 15. b

SUMMARY OF CELLULAR RESPIRATION

10

Photosynthesis

A. KEY CONCEPTS

- **Photosynthesis** is the ability to convert energy from sunlight into chemical energy. Photosynthetic organisms are called **autotrophs**, organisms that make all their own food. Organisms that derive their food from other organisms are called **heterotrophs**.

- Photosynthesis is the ultimate food source for most organisms, and the evolution of this metabolic pathway is one the great events in the history of life.

10.1 An Overview of Photosynthesis

- Photosynthesis occurs in the green parts of plants. It requires sunlight, carbon dioxide, and water and produces oxygen. The overall reaction is CO_2 + $2H_2O$ + light energy $\rightarrow (CH_2O)_n$ + H_2O + O_2, where $(CH_2O)_n$ stands for carbohydrate. Usually glucose $(C_6H_{12}O_6)$ is the carbohydrate made:

$$6CO_2 + 12H_2O + \text{light energy} \rightarrow C_6H_{12}O_6 + 6O_2 + 6H_2O$$

- Photosynthesis reduces carbon dioxide to glucose (or other sugars). This endergonic reaction is approximately the reverse of cellular respiration, the exergonic reactions that oxidize glucose to carbon dioxide in order to make ATP.

Photosynthesis: Two Linked Sets of Reactions

- Cornelius van Niel studied photosynthesis in purple sulfur bacteria. These grow even when starved of sugars if exposed to sunlight and hydrogen sulfide (H_2S). Sulfur (S), not O_2, is the by-product of carbohydrate production for these cells: CO_2 + $2H_2S$ + light energy $\rightarrow (CH_2O)_n$ + 2S.

- Because CO_2 is used in the reaction just described but no O_2 was made, van Niel's result suggested that the O_2 released in most photosynthesis must come from H_2O, not CO_2.

- The hypothesis that photosynthetic O_2 comes from H_2O was supported by the observation that isolated chloroplasts make O_2 when exposed to sunlight, even if no CO_2 is present. Researchers also showed that algae and plants exposed to water containing the heavy isotope of oxygen, ^{18}O, emitted $^{18}O_2$ gas.

- Melvin Calvin fed $^{14}CO_2$ to algae to track the fate of carbon by identifying the order in which products became ^{14}C-labeled. He found that $^{14}CO_2$ was used to make carbohydrates even in the dark. This light-independent part of photosynthesis, the **Calvin cycle**, reduces CO_2 and produces sugar.

- In photosynthesis the **light-dependent reactions** produce O_2 from H_2O; the **light-independent reactions** produce sugar from CO_2. While the light-independent Calvin cycle does not require light itself, it depends on products from the light-dependent reactions.

- As H_2O splits and forms O_2, electrons are released and transferred to NADP$^+$, forming NADPH, an electron carrier similar to NADH. These electrons are later used in the Calvin cycle reduction of CO_2. ATP made by the light-dependent reactions is also used in the Calvin cycle (**Figure 10.1**).

The Structure of the Chloroplast

- Photosynthesis occurs in green plant parts—specifically, in small green organelles called chloroplasts. Chloroplast membranes release O_2 when exposed to sunlight.

- Leaf cells contain from 40 to 50 **chloroplasts** (**Figure 10.2**). Each chloroplast has an outer and an inner membrane and is filled with vesicle-like structures called **thylakoids**, which are often found in stacks called **grana**. The surrounding fluid is called the **stroma**, and the space inside the thylakoids is its **lumen**.

- Thylakoid membranes contain lots of **chlorophyll** pigment. **Pigments** absorb certain wavelengths of light and transmit others. Chlorophyll absorbs blue and red light and transmits green light, causing the green color of plants, algae, and many bacteria.

- Chloroplasts derive from colorless **proplastids** found in embryonic plant cells and rapidly dividing mature plant tissue. Proplastids mature into chloroplasts, leucoplasts, or chromoplasts depending on the function of the cell they are in (**Box 10.1**).

(a) Why are plants green?

_____

_____

_____

10.2 How Does Chlorophyll Capture Light Energy?

- Different types of electromagnetic radiation correspond to different **wavelengths** (**Figure 10.4**); the part of the **electromagnetic spectrum** that humans see is called **visible light** and encompasses wavelengths from 400 to 710 nanometers (nm). Shorter wavelengths (e.g., blue, UV) have more energy than longer wavelengths do (e.g., red, infrared).

- Light has particle-like as well as wavelike properties. Packets of light are called **photons**, and each photon or wavelength has a specific amount of energy. Pigments can absorb this energy.

Photosynthetic Pigments Absorb Light

- Photons can be absorbed, transmitted, or reflected by molecules. White light is a mix of all visible wavelengths; when pigments absorb some, but not all wavelengths, they appear to us to be the color of the reflected or transmitted wavelengths.

- Pigments can be extracted and separated via a technique called **thin layer chromatography** (**Figure 10.5**). Raw leaf extract, blotted on a stiff support coated with a thin layer of porous material, separates into components as a solvent travels up the coating, carrying along the extract molecules at different rates based on their size and/or solubility.

- Leaves contain several pigments. By using a spectrophotometer, researchers can isolate individual pigments and determine the wavelengths each absorbs (**Box 10.2, Figure 10.7**). A graph of light absorbed versus wavelength gives the **absorption spectrum** for a given pigment (**Figure 10.6**).

- The major leaf pigments are the **chlorophylls** (chlorophyll _a_ and chlorophyll _b_), which absorb red and blue light and transmit green light; and the **carotenoids**, which absorb blue and green light and thus appear yellow, orange, or red.

- Algae on a slide lit with a spectrum of colors produce oxygen (measured by bacterial growth) in areas exposed to blue and red light. These wavelengths make up the **action spectrum** for photosynthesis. Chlorophylls absorb red and blue wavelengths, so they must be the primary photosynthetic pigments.

What Is the Role of Carotenoids and Other Accessory Pigments?

- Carotenoids are **accessory pigments** that absorb light and pass the energy on to chlorophyll.

- The two classes of carotenoids found in plants are the **carotenes** (e.g., β-carotene in carrots, **Figure 10.8a**) and the **xanthophylls** (e.g., zeaxanthin in corn). These molecules are similar and are both found in chloroplasts. The accessory pigments give deciduous trees their fall colors when chlorophyll degrades.

- Carotenoids absorb wavelengths of light not absorbed by chlorophyll, thus extending the range of wavelengths that can be used in photosynthesis. Carotenoids also stabilize free radicals produced by electromagnetic radiation, protecting chlorophyll from degradation.

- **Flavonoids** also protect chlorophyll and other plant molecules from destructive radiation by absorbing high-energy ultraviolet (UV) light. Both carotenoids and flavonoids act like a plant's own internal sunscreen.

The Structure of Chlorophyll

- Chlorophyll _a_ and _b_ have similar structures but slightly different absorption spectra. Both chlorophylls have a long tail of isoprene subunits which stays embedded in the thylakoid membrane, and a large ring structure (the "head") containing a magnesium atom where light is absorbed (**Fig. 10.8b**).

When Light Is Absorbed, Electrons Enter an Excited State

- If the energy of a photon striking chlorophyll is similar to the energy difference between two electron shells in the chlorophyll head, the photon's energy can be absorbed and an electron is "excited"—it is raised to a higher energy shell where it has greater potential energy (**Figure 10.9**).

- The energy difference between electron state 0 and state 1 in chlorophyll equals the energy of a red photon; the energy difference between state 0 and state 2 equals the energy of a (higher energy) blue photon. Intermediate energy levels (e.g., those of green light) cannot easily be absorbed by chlorophyll.

- Ultraviolet wavelengths have so much energy that they can eject electrons from pigment molecules, while infrared wavelengths carry so little energy that they merely heat atoms a bit (increase their movement).

- Electrons in high-energy states may fall back to their ground state, causing **fluorescence** as they release heat and photons. Isolated chlorophyll fluoresces when exposed to UV light (**Figure 10.10**), but in chloroplasts chlorophyll shows little fluorescence.

- In the thylakoid membrane, 200–300 chlorophyll molecules and accessory pigments group together in complexes called **photosystems**. Each photosystem has an antenna complex and a reaction center.

The Antenna Complex

- When **antenna complex** electrons absorb photon energy and become excited, they pass the energy, but not the electron, to a nearby chlorophyll molecule. The original electron falls back to its ground state, but the energy transfer excites the nearby chlorophyll molecule's electron. This phenomenon is called resonance.

- Resonance energy is transmitted from chlorophyll to chlorophyll until it reaches the reaction center.

CHECK YOUR UNDERSTANDING

(b) Chlorophyll can fluoresce; so why aren't plants noticeably fluorescent?

_____

_____

_____

_____

The Reaction Center

- Excited electrons are finally passed to an electron acceptor in the **reaction center**. The reduction of the electron acceptor completes the transformation of electromagnetic energy into chemical energy.

- The reaction center couples a light-induced exergonic reaction to an endergonic redox reaction.

- Excited electrons in photosynthetic pigments can (1) fluoresce as they drop back down to a low energy level; or (2) excite adjacent pigment electrons, causing resonance in the antenna complex; or (3) be transferred to an electron acceptor in the reaction center (**Figure 10.11**).

10.3 The Discovery of Photosystems I and II

- Researchers found that green algae had similar photosynthetic responses to far-red (700 nm) and red (680 nm) light. But the rate of photosynthesis was **much** higher (more than double) when algae were exposed to both wavelengths at once (**Figure 10.12**). This result was called the enhancement effect.

- Robin Hill and Faye Bendall proposed that green algae and plants have two types of reaction centers: **photosystem II** and **photosystem I**. Photosynthesis is more efficient when both photosystems work together, causing the enhancement effect.

- Many bacteria have photosystems similar to those found in algae and plants, and several bacteria have only one of the two photosystems. This finding allowed researchers to study each photosystem in isolation before trying to understand how they work together.

How Does Photosystem II Work?

- Photosystem II is similar to the single photosystem found in purple nonsulfur and purple sulfur bacteria. When the antenna complex transmits energy to this photosystem's reaction center, a high-energy electron is donated to **pheophytin**—a molecule similar to chlorophyll that lacks magnesium in its head region.

Electrons from Pheophytin Enter an Electron Transport Chain

- When an excited electron from chlorophyll reduces pheophytin in a reaction center, pheophytin passes the electron to an electron transport chain similar to that found in mitochondria.

- Photosystem II's electron transport chain, like that in the mitochondria, contains cytochromes and quinones. Electrons participate in redox reactions and are gradually stepped down in potential energy. These redox reactions lead to proton pumping, triggering ATP synthesis via chemiosmosis.

- **Plastoquinone (PQ)**, a small hydrophobic molecule, receives electrons from pheophytin and transports them across the thylakoid membrane to a cytochrome complex where electrons are passed down a series of iron- and copper-containing proteins. The released potential energy adds protons to other plastoquinones, which transport the protons to the thylakoid's lumen side (**Figure 10.13**).

- This proton transport increases the proton concentration inside the thylakoid, causing a thousandfold difference in proton concentration (thylakoid interior at around pH 5; exterior at pH 8). This proton motive force then drives ATP production by ATP synthase.

ATP Synthase Uses the Proton-Motive Force to Phosphorylate ADP

- As in the mitochondria, protons diffuse down their electrochemical gradient through ATP synthase, causing a conformational change that drives the phosphorylation of ADP. The capture of light energy by photosystem II to produce ATP chemical energy is called **photophosphorylation**.

- In purple sulfur bacteria, cytochrome donates its electron back to the reaction center; but in cyanobacteria, green algae, and plants, the cytochrome complex of photosystem II donates its electrons to photosystem I.

Photosystem II Obtains Electrons by Oxidizing Water

- Cyanobacteria, algae, and plants produce oxygen because part of photosystem II "splits" water to replace its lost electrons: $2 H_2O \rightarrow 4 H^+ + 4 e^- + O_2$. Because oxygen is produced as a by-product, organisms that split water during photosynthesis are said to perform **oxygenic photosynthesis**.

- Purple sulfur and purple nonsulfur bacteria cannot oxidize water. They perform **anoxygenic photosynthesis**.

- Photosystem II becomes extremely electronegative when it donates its excited electrons to the ETC. This allows it to strip electrons away from water, leaving protons and oxygen. This endergonic reaction is possible because sunlight energy drives the initial removal of electrons from photosystem II. Photosystem II is the only known protein complex capable of splitting water.

- All oxygen we breathe originates from splitting of water in photosystem II. Addition of O_2 to the atmosphere by oxygenic photosynthesis had a huge impact on the history of life.

- Oxygen is toxic to anaerobic organisms. In addition, organisms that evolved the ability to use oxygen as an electron acceptor in cell respiration were able to make much more ATP than were organisms using other electron acceptors. Consequently, aerobic organisms became dominant on our planet.

CHECK YOUR UNDERSTANDING

(c) Purple sulfur bacteria were commonly used to research photosystem II. Why were purple sulfur bacteria a better model system than trees for studying these processes?

_____

_____

_____

How Does Photosystem I Work?

- Heliobacteria, which produce NADH from NAD^+ in the presence of sunlight, were used to study photosystem I. In cyanobacteria, algae, and land plants, this photosystem reduces $NADP^+$, a phosphorylated version of NAD^+, forming the electron-carrier NADPH.

- Excited electrons from the reaction center of photosystem I are passed down an electron transport chain of iron- and sulfur-containing molecules to **ferredoxin** (**Figure 10.14**).

- The enzyme ferredoxin/$NADP^+$ oxidoreductase (also called $NADP^+$ reductase) transfers a proton and two electrons from ferredoxin to $NADP^+$, forming NADPH. The entire photosystem is based in the thylakoid membrane.

- Like NADH from the Krebs cycle, the electrons from NADPH can be used to reduce other compounds. Thus photosystem I produces chemical energy stored as NADPH's reducing power, and photosystem II produces chemical energy stored in ATP.

CHECK YOUR UNDERSTANDING

(d) Explain how photosystems I and II are similar and how they differ.

_____

_____

_____

_____

_____

The Z Scheme: Photosystems I and II Work Together

- Robert Hill and Fay Bendall proposed that photosystems I and II interact as shown in **Figure 10.15**. This model is called the **Z scheme** because an electron's path and potential energy changes can be graphed in a Z shape.

- The Z scheme starts when photons excite electrons in photosystem II's chlorophyll and the energy is transmitted to the reaction center, where chlorophyll P680 (which most easily loses electrons when absorbing wavelengths of 680 nm) passes an excited electron to pheophytin.

- The electron's potential energy is gradually lowered through redox reactions with quinones and cytochromes, and the released energy is used by plastoquinone to transport protons across the thylakoid membrane into the stroma. ATP synthase then uses the proton-motive force to phosphorylate ADP, creating ATP.

- At the end of photosystem II's electron transport chain, a small protein called **plastocyanin** (PC) gains an electron from the cytochrome complex and carries it back across the thylakoid membrane to photosystem I. In this way, plastocyanin physically links photosystem II and photosystem I.

- A chlorophyll called P700 in the photosystem I reaction center loses electrons most readily upon absorbing wavelengths of 700 nm. The electrons enter an electron transport chain, are eventually passed to ferredoxin, and are then used to reduce $NADP^+$ to NADPH.

- Photosystem II's electrons are replaced by the splitting of water, and photosystem I's electrons are replaced with the electrons that plastocyanin shuttles from photosystem II.

- The Z scheme explains the enhancement effect; photosynthesis is more efficient when both 680-nm and 700-nm wavelengths are available, allowing both photosystems to run at maximal rates.

- In green plants and algae, photosystem I sometimes transfers electrons to the electron transport chain of photosystem II to increase ATP production instead of using them to reduce $NADP^+$; this is called **cyclic photophosphorylation** (Fig. 10.16).

- Surprisingly, photosystem I and ATP synthase are most commonly found in the exterior, unstacked thylakoid membranes, while photosystem II is most common in the interior stacked membranes of grana (**Figure 10.17**).

CHECK YOUR UNDERSTANDING

(e) Which photosystem's electron transport chain is more similar to that in the mitochondria, and why?

_____

_____

10.4   How Is Carbon Dioxide Reduced to Produce Glucose?

- Photosystems I and II can produce ATP and NADPH only in the presence of light. But the reactions that produce sugar from carbon dioxide are light independent. However, they require the ATP and NADPH produced by the light-dependent reactions.

- Melvin Calvin led the research team that documented the intermediate compounds produced as carbon dioxide is reduced to sugar.

The Calvin Cycle

- Melvin Calvin led the research team that used the pulse-chase technique to document the intermediate compounds produced as carbon dioxide is reduced to sugar.

- Calvin's group fed radioactively labeled carbon dioxide ($^{14}CO_2$) to green algae and then isolated and identified product molecules containing ^{14}C. By waiting different amounts of time between adding the $^{14}CO_2$ and killing the algae, they determined the order in which compounds are produced during the reduction of CO_2 to sugar.

- Product molecules from algal extract were isolated using chromatography and identified by laying X-ray film over the chromatography surface. Radioactively labeled products created a black spot on the X-ray film (**Figure 10.18**).

- Cells killed right after exposure to $^{14}CO_2$ contained a radiolabeled three-carbon compound called 3-phosphoglycerate (3PG), which is an intermediate in glycolysis. Because Calvin's pathway produces sugar, and glycolysis breaks down glucose, it makes sense that the pathways share some intermediates.

- A five-carbon compound called ribulose bisphosphate (RuBP) reacts with CO_2 to produce a six-carbon compound that immediately splits in half to form two molecules of 3PG.

- The cycle that reduces CO_2 to produce 3PG (and eventually glucose) is now known as the Calvin cycle. This cycle of reactions occurs in the chloroplast stroma (**Figure 10.19**).

- The Calvin cycle has three phases: (1) _Fixation phase_: CO_2 reacts with RuBP, producing two 3PG molecules. The attachment of CO_2 to an organic compound is called **carbon fixation**. (2) _Reduction phase:_ 3PG molecules are phosphorylated by ATP and reduced by NADPH to produce **glyceraldehyde 3-phosphate (G3P)**. Some G3P is drawn off to make glucose. (3) _Reduction phase:_ The remaining G3P is used in reactions that regenerate RuBP.

- Each turn of the Calvin cycle fixes one molecule of CO_2, so it takes three turns to produce one molecule of G3P.

- The Calvin cycle uses the substantial chemical energy stored in ATP and NADPH from the light-dependent reactions to power the reduction of CO_2 to sugar.

- Most Calvin cycle reactions also occur as part of glycolysis and/or another metabolic pathway, but the reaction between CO_2 and RuBP is unique to the Calvin cycle. This is the key reaction responsible for the fixation of atmospheric CO_2.

CHECK YOUR UNDERSTANDING

(f) Compare and contrast the Krebs cycle of cell respiration (**Chapter 9**) with the Calvin cycle. How are they similar and how are they different?

_____

_____

_____

The Discovery of Rubisco

- Arthur Weissbach and colleagues purified proteins from spinach extracts and then tested whether they could catalyze the incorporation of $^{14}CO_2$ into 3PG. Eventually they isolated a carbon-fixing enzyme called ribulose 1,5-bisphosphate carboxylase/oxygenase (**rubisco**), which is very common in leaf tissue.
- Rubisco is cubical and has eight active sites for CO_2 fixation. Despite its importance, rubisco is slow and inefficient. Oxygen competes with CO_2 at the active site—this apparently **maladaptive** trait may date back to the enzyme's evolution at a time when little O_2 was present on Earth.
- When O_2 and RuBP react in rubisco's active site, one product undergoes reactions that use up ATP and produce CO_2 (**Figure 10.20**). These reactions are called **photorespiration** because they consume O_2 and produce CO_2.
- Photosynthetic rate declines during photorespiration. However, as long as cell CO_2 concentration is high and O_2 concentration is low, carbon fixation is favored over photorespiration.

How Is Carbon Dioxide Delivered to Rubisco?

- Plants are covered with a waxy cuticle that prevents water loss via evaporation from the plant. This cuticle also prevents CO_2 from directly entering surface leaf cells.
- **Stomata** are the leaf structures where gas exchange occurs. They consist of two **guard cells** that change shape when leaf CO_2 concentration is low during photosynthesis to create an opening called a **pore**. Stomata are normally open during the day and closed at night.
- When leaf pores open, atmospheric CO_2 diffuses into the extracellular fluid and on into cells down its concentration gradient (**Figure 10.21**). The Calvin cycle continually uses CO_2, maintaining a steep concentration gradient favoring CO_2 diffusion into cells.
- Open stomata also cause leaf water loss because the atmosphere is usually much drier than the leaf interior. Thus, when conditions are hot and dry, many plants must keep their stomata closed to prevent water loss, causing photosynthesis to stop.

CHECK YOUR UNDERSTANDING

(g) Why is rubisco considered an inefficient enzyme?

_____

_____

- Research groups using Calvin's pulse-chase approach found that in sugar cane (Kortschak's group) and corn (Karpilov's group), $^{14}CO_2$ was initially incorporated into four-carbon organic acids instead of into 3PG (**Figure 10.22**). This is now known as C_4 **photosynthesis**.
- In C_4 plants, CO_2 is added to three-carbon compounds by **PEP carboxylase** in **mesophyll cells** near the leaf surface. The four-carbon organic acids produced by this carbon fixation travel to **bundle-sheath cells** surrounding the plant's **vascular tissue**, where water and nutrient transport occurs.
- Rubisco is found in the bundle-sheath cells of C_4 plants where the four-carbon organic acids release CO_2, initiating the Calvin cycle (**Figure 10.23**; model initially proposed by Hatch and Slack).
- The C_4 pathway is mostly found in plants that live in hot, dry habitats. This pathway uses energy but increases the concentration of CO_2 in the cells, where rubisco catalyzes the start of the Calvin cycle. In this way, C_4 plants are able to limit photorespiration.
- PEP carboxylase has a higher affinity for CO_2 than rubisco does and is not affected by high oxygen levels. Thus, in hot, dry habitats, C_4 photosynthesis is more efficient than regular C_3 photosynthesis. Use of this alternate carbon fixation pathway has evolved independently several times.

CAM Plants

- Flowering plants called the Crassulaceae have evolved a different solution to the problem of photorespiration in hot, dry environments. These plants keep their stomata closed all day to minimize water loss. They open their stomata at night, when they temporarily fix CO_2 to organic acids using a pathway called **crassulacean acid metabolism (CAM)**.

- CAM plants store the organic acids they make at night in the central vacuoles of photosynthesizing cells. During the day, CO_2 is released from the stored organic acids and used by the Calvin cycle.

- Thus, C_4 plants stockpile CO_2 in cells where rubisco is not active, and CAM plants store CO_2 at a time when rubisco is not active. Both C_4 and CAM plants use energy to increase the CO_2 concentration when and where rubisco is active, minimizing photorespiration when they must keep their stomata closed (**Figure 10.24**).

CHECK YOUR UNDERSTANDING

(h) Name the two pathways that allow some plants to stockpile CO_2 in order to minimize photorespiration. How do the pathways differ?

_____

_____

_____

What Happens to the Sugar That Is Produced by Photosynthesis?

- G3P molecules produced by the Calvin cycle are often used to make glucose and fructose, which can combine to form sucrose in the cytosol (**Figure 10.25**). Water-soluble sucrose is easily transported throughout the plant; it can be broken down to fuel cell respiration or be converted to starch for storage.

- In rapidly photosynthesizing cells, where sucrose is abundant, glucose is temporarily stored as starch in the chloroplast. Because starch is not water soluble, it is broken down at night and used to make more sucrose for transport throughout the plant.

- Intermediates in the production of glucose from G3P also occur in glycolysis and can be used to fuel ATP production via glycolysis and cellular respiration if necessary.

- Sugars and starch from photosynthesis provide food for herbivore growth and reproduction when herbivores eat plants and for carnivore growth and reproduction when carnivores eat herbivores. Photosynthesis is thus the ultimate source of energy for almost all cell growth and reproduction on Earth.

B.   INTEGRATING YOUR KNOWLEDGE

Looking Back—Key Concepts from Earlier Chapters

Use of Radioactive Isotopes—Chapters 7 and 9

Chapter 7 described "pulse-chase" experiments involving the use of radioactive isotopes to discover how molecules are transported throughout the cell. **Chapter 9** mentions that radioactive isotopes were used to discover intermediate compounds in the Krebs cycle. Here in **Chapter 10,** you learn how isotopes were used in research identifying the metabolic pathways involved in carbon fixation.

Redox Reactions and Electron Transport Chains—Chapter 9

Chapter 9 introduced you to redox reactions, electron transport chains and chemiosmosis—all key components of photosynthesis. As detailed in that chapter, mitochondria also use an electron transport chain to convert the energy of high potential energy electrons to more easily usable chemical bond energy (e.g., ATP).

Looking Forward—Key Concepts in Later Chapters

Oxygen Produced by Photosynthesis—Chapter 28

Photosynthetic bacteria (some similar to those used in experiments described in **Chapter 10**) were responsible for producing the oxygen that shapes our current world. Find out more about these organisms in **Chapter 28**.

C_4 and CAM Pathways for Carbon Fixation—Chapter 30

Chapter 30 gives you more information about green plants that use the C_4 and CAM strategies for carbon fixation.

Water and Sugar Transport in Plants—Chapter 37

Plants must take up water to replace water lost via stomata and transport fructose and other nutrients throughout the organism. These transport processes are detailed in **Chapter 37**.

Global Warming and CO_2 Levels—Chapter 54

You'll learn more about global warming and rising CO_2 levels in **Chapter 54**, on ecosystem ecology.

C. MASTERING CHALLENGING TOPICS

Three new major metabolic pathways are described in this chapter: photosystem I, photosystem II, and the Calvin cycle. Connections between these and other metabolic pathways are also discussed. One good strategy for learning this information is to make your own lists of the order in which events occur, drawing flowcharts depicting the connections between pathways.

For example, you might start a list describing photosynthesis as follows: (1) Photon hits chlorophyll and excites electron; (2) excited electron is passed to other chlorophyll molecules until it reaches a reaction center; (3) at the reaction center, the electron reduces an electron acceptor. Another study strategy is to look at the text figures and describe out loud to yourself the names of structures and what happens where. In assessing the importance of each step along a pathway, think about what would happen if that part of the pathway failed.

D. ASSESSING WHAT YOU'VE LEARNED

1. How would the experiments of van Niel with purple sulfur bacteria have been different if oxygen came from carbon dioxide during photosynthesis?
   a. Elemental sulfur would be produced from carbon dioxide and hydrogen sulfide.
   b. Oxygen would be produced from carbon dioxide and hydrogen sulfide.
   c. Oxygen would be produced from carbon dioxide and water.
   d. Elemental sulfur would be produced from carbon dioxide and water.

2. If green plant cells are incubated with ^{18}O-labeled carbon dioxide, what molecule will become radioactive as the cells are exposed to light?
   a. water
   b. oxygen
   c. ATP
   d. sugar

3. What two high-energy molecules produced during the light-dependent reactions of photosynthesis are used to drive the Calvin cycle?
   a. ATP and oxygen
   b. NADPH and carbon dioxide
   c. ATP and NADPH
   d. ATP and carbon dioxide

4. What is/are the function(s) of accessory pigments in plants?
   a. They extend the range of wavelengths a plant can use to drive photosynthesis.
   b. They protect plants from the damaging effects of electromagnetic radiation.
   c. Both a and b are functions of accessory pigments in plants.
   d. Accessory pigments have no function in plants.

5. Which of the following factors is *not* involved in determining whether light is absorbed by a molecule?
   a. the intensity of the light
   b. the wavelength of the light
   c. the potential energy levels of electron shells in the molecule
   d. the chemical structure of the molecule

6. What is the purpose of chemical reduction of an electron acceptor in the photosynthetic reaction center?
   a. It allows the energy of absorbed light to be trapped and converted to chemical energy.
   b. It adjusts the energy level of chlorophyll electrons to match the energy of light illuminating the leaf.
   c. It changes the wavelength of chlorophyll fluorescence so that it will not interfere with absorption of light.
   d. It provides electrons to be excited by the absorbed light.

7. What is the evidence for there being two photosystems?
   a. Microscopy reveals two colors of chloroplast within the same leaf cells.
   b. The combination of light at 680 nm and 700 nm is much more effective at stimulating photosynthesis than at either wavelength alone.
   c. Chlorophyll absorbs light of two different wavelengths: blue and red.
   d. Two different high-energy molecules are produced during the light-dependent reactions of photosynthesis: ATP and NADPH.

8. What happens to the ATP synthesis rate if the pH of the thylakoid lumen decreases?
   a. The rate of ATP synthesis will not change.
   b. The rate of ATP synthesis will decrease.
   c. The rate of ATP synthesis will increase.

9. In an experiment conducted by Jagendorf, isolated thylakoids were incubated in an acidic solution at pH 4.0 until the pH was equilibrated across the thylakoid membrane. The thylakoids were then transferred to a buffer at pH 8.0 with ADP and inorganic phosphate. ATP was synthesized. Did this experiment require light to generate the ATP?
   a. Yes; ATP synthase requires high-energy electrons from photosystem II.
   b. No; ATP synthesis is dependent only on the presence of a hydrogen ion gradient and does not require light directly.
   c. Yes; the hydrogen ion gradient is generated via transfer of electrons in photosystem II.
   d. No; ATP synthesis is part of the light-independent reactions of photosynthesis.

10. Electrons excited by absorption of light in photosystem II are transferred to plastoquinone and so must be replaced. The replacements come from:
    a. photosystem I
    b. water
    c. oxygen
    d. cytochrome

11. Electrons excited by absorption of light in photosystem I are transferred to iron–sulfur electron acceptors and must be replaced. The replacements come directly from:
    a. photosystem II
    b. ATP
    c. NADPH
    d. water

12. The biochemical function of photosystem I is to:
    a. phosphorylate ADP
    b. hydrolyze ATP
    c. oxidize NADPH
    d. reduce $NADPH^+$

13. Rubisco is an important enzyme because:
    a. It catalyzes the splitting water, which provides electrons for photosynthesis.
    b. It catalyzes carbon fixation.
    c. It catalyzes the reaction that allows CAM plants to store CO_2.
    d. It is one of the fastest enzymes known.

14. Why is it critical for plants to maintain a high concentration of carbon dioxide in the leaves?
    a. It helps to prevent photorespiration.
    b. It is the only substrate for rubisco.
    c. Oxygen cannot be produced without it.
    d. It is necessary for regeneration of RuBP from glyceraldehyde-3-phosphate.

15. How does the sequestering of carbon dioxide in CAM plants help them to survive?
    a. It keeps carbon dioxide away from places in the leaves where it would be toxic.
    b. It allows the plants to produce sugars in the winter when leaves are dead.
    c. It allows the light-dependent and light-independent reactions to occur at different times of the day.
    d. It allows carbon dioxide to be gathered and used at different times of the day.

CHAPTER 10—ANSWER KEY

Check Your Understanding

(a) Plants are green because they contain lots of chlorophyll; this pigment absorbs red and blue light but transmits and reflects green light.

(b) Plant chlorophyll molecules are part of a photosystem. They pass the energy of excited electrons to other chlorophyll molecules or to the electron acceptor in the reaction center. Therefore, this energy is not normally given off as electromagnetic radiation and does not cause fluorescence.

(c) Some advantages to working with bacteria: They are simpler, faster-growing, and easy to keep lots of in the lab. Trees are more complex, take up a lot of space, and take a long time to grow and reproduce—making them, in many ways, more difficult to work with. Purple sulfur bacteria were especially useful because they have only one of the two photosystems (II).

(d) Photosystems I and II are similar in that they both use electrons excited by photons to reduce molecules at the beginning of electron transport chains. Both are located in the thylakoid membrane. On the other hand, photosystems I and II most easily use light of different wavelengths, and they use the potential energy of the excited electrons to produce different compounds (II makes ATP and I makes NADPH). They occur in different parts of the thylakoid membrane, and they replace electrons lost from their chlorophylls in different ways (II splits water and I gets electrons from II).

(e) Photosystem II's electron transport chain is more similar to that in the mitochondria than photosystem I's because photosystem II also creates a proton motive force that is used by ATP synthase to make ATP.

(f) Both are biochemical cycles where the starting compound is regenerated at the end of the cycle. Both take place inside eukaryotic organelles. However, whereas the Krebs cycle occurs in the mitochondria, photosynthesis occurs in the chloroplast; the Krebs cycle makes ATP and NADH (and $FADH_2$), but the Calvin cycle uses up ATP and NADPH; and the Krebs cycle gives off CO_2, but the Calvin cycle uses up and/or fixes CO_2.

(g) Rubisco is inefficient because (1) it is slow, and (2) it catalyzes the toxic photorespiration reaction when cell CO_2 concentration is low and O_2 concentration is high. It evolved when there was little O_2 in Earth's atmosphere.

(h) The C_4 pathway allows some plants to stockpile CO_2 in cells with relatively high oxygen concentrations by using an enzyme (PEP carboxylase) that is not inhibited by the presence of oxygen. This CO_2 is later released in bundle-sheath cells with high rubisco content. The CAM pathway allows some plants to stockpile CO_2 at night and release it during the day when rubisco is active.

Assessing What You've Learned

1. b; 2. d; 3. c; 4. c; 5. a; 6. a; 7. b; 8. c; 9. b; 10. b; 11. a; 12. d; 13. b; 14. a; 15. d

11

The Cell Cycle

A. KEY CONCEPTS

- The cell theory states that all organisms are made of cells and that all cells come from other cells; but how does this occur? Rudolf Virchow proposed that new cells arise through **cell division**. This hypothesis was confirmed by microscopic observations of early development in **embryos**.

- There are two types of cell division in multicellular eukaryotes (**Figure 11.1**). Sexual reproduction requires **meiosis**, cell division in which only half of the parent cell's genetic material is passed to the daughter cells (**gametes**—the sperm or eggs). This type of cell division is described in Chapter 12.

- The second type of eukaryotic cell division is **mitosis**. Mitosis produces **somatic cells** (all the nongamete cells in your body), and daughter cells resulting from mitosis are genetically identical to the parent cell.

- Specifically, mitosis refers to the division of genetic material and **cytokinesis** refers to the division of the cytoplasm. Typically, mitosis is immediately followed by cytokinesis to produce two daughter cells.

- In eukaryotes, mitosis and cytokinesis allow growth, wound repair, and **asexual reproduction**—the production of genetically identical offspring.

11.1 Mitosis and the Cell Cycle

- Certain dyes stain threadlike structures in the nuclei of dividing eukaryotic cells. In 1879 Walther Flemming published his observations of salamander embryo cell division (**Figure 11.2**). Paired threads split apart in a process he called mitosis.

- Roundworm studies revealed that the total number of threads per cell remains constant, division after division. W. Waldeyer named these threads **chromosomes** in 1888.

- We now know that each chromosome contains a long double helix of deoxyribonucleic acid (DNA) wrapped around proteins. DNA carries the cell's genetic information, and the purpose of mitosis is to evenly distribute this genetic material to daughter cells.

- Chromosomes are copied before mitosis, condense to a more compact form at the start of mitosis, and then one copy of each chromosome is moved to each of the two daughter cells (**Figure 11.3**).

M Phase and Interphase

- Growing eukaryotic cells alternate between a dividing phase, known as the mitotic or **M phase**, and **interphase**, when no division occurs. Cells spend most of their time in interphase. Chromosomes are uncoiled during interphase and, consequently, are not visible using light microscopy.

- Nineteenth-century biologists assumed that cells duplicated their chromosomes at some point because, after mitosis, each daughter cell always contained the same number of chromosomes as the parent cell.

The Discovery of the Cell Cycle

- In the 1950s, Alma Howard and Stephen Pelc used radioactive isotopes to determine when chromosome replication occurs. They supplied growing cells with radioactive phosphorus or thymidine. The cells incorporated these radiolabeled compounds into newly synthesized DNA. which could then be visualized using X-ray film.

- When radioactivity was assayed right after exposure to the radioactive isotope, newly synthesized DNA, indicated by black dots, was found only in interphase nuclei. This indicated that DNA replication occurs during interphase.

- Howard and Pelc called the DNA-replication part of interphase **synthesis** (or **S**) **phase**—for DNA synthesis. They also referred to the entire cell division sequence from formation of a new daughter cell, to DNA duplication, through cell division, as the **cell cycle**.

- Two key cell-cycle events are (1) the **replication**, or copying, of the hereditary material (DNA) during the S phase of interphase; and (2) the distribution of one complete copy to each daughter cell during M phase.

(a) Explain why researchers working in the early 20th century were unable to determine when, during the cell cycle, DNA synthesis occurred.

_____

_____

_____

_____

The Discovery of the Gap Phases

- To determine the length of S phase, researchers exposed growing cells to radioactive phosphorus or thymidine and then examined the cells after various time intervals. Researchers used cultured cells because they could be manipulated more easily than cells inside living organisms (**Box 11.1**).

- A pulse of radioactive phosphorus or thymidine could be followed by a chase of nonradioactive substrate so that cells could continue through the cell cycle without incorporating additional label. Thus, only cells synthesizing DNA during a specific time period became radioactively labeled.

- It took 4 hours for labeled dividing cells to appear, suggesting that there was a 4-hour gap in the cell cycle between the end of S phase and the beginning of M phase. This part of the cell cycle is now called G_2 **phase**, for second gap (**Figure 11.4**).

- Researchers determined that S phase lasts 6 to 8 hours because labeled mitotic nuclei can be observed for 6 to 8 hours. These cells must have been in S phase when radiolabeled thymidine was available.

- Researchers subtracted G_2-, S-, and M-phase times from total cell-cycle time (24 hours) to conclude that another cell-cycle gap must exist after M phase. This first gap, or G_1 **phase**, lasts about 7 to 9 hours (**Figure 11.5**).

- During the gap phases, cells make additional cytoplasm and organelles replicate. Cells also perform other normal cell functions.

- A segment of DNA containing information for making a particular protein or RNA molecule is called a **gene**. Bacteria, as well as eukaryotes, must ensure that daughter cells receive all necessary genetic information. Prokaryotes reproduce via cell division in a process called **binary fission** (**Box 11.2, Figure 11.8**).

(b) What might happen to cells that repeatedly divided without any gap phases?

_____

_____

_____

11.2 How Does Mitosis Take Place?

Events in Mitosis

- The number of chromosomes varies widely among species, but all eukaryotic chromosomes consist of a DNA–protein complex called **chromatin**. The proteins in chromatin are called **histones**. Chromosomes become visible at the beginning of mitosis, as chromatin condenses to form a more compact structure.

- Each of the two DNA strands in a replicated chromosome is called a **chromatid**. **Sister chromatids** are exact copies of the same genetic information encoded in one long DNA double helix. A constriction of the joined sister chromatids occurs at the specialized region called the **centromere**.

- The two joined chromatids separate during mitosis to form independent chromosomes (**Figure 11.6**). Each daughter cell thus receives a complete copy of the parental cell's genetic information.

- Mitosis (M phase) is a continuous process, but biologists identify several subphases (prophase, prometaphase, metaphase, anaphase, telophase) based on specific events occurring at different stages in the process (**Fig. 11.7**).

Prophase

- During **prophase**, the previously replicated chromosomes become visible in the light microscope as they condense.

- In the cytoplasm, microtubules begin to form the **mitotic spindle**. **Spindle fiber** microtubules attach to the chromosomes. These fibers pull chromosomes into the daughter cells later on in mitosis. During prophase the spindles either form on, or begin to move toward, opposite sides of the cell.

- Spindle fibers radiate from microtubule organizing centers. In animals these centers contain **centrioles** and are known as **centrosomes**.

Prometaphase

- The nucleolus disappears and the nuclear envelope breaks down during **prometaphase**.

- Spindle fibers attach to each sister chromatid at structures called **kinetochores**, located at chromosome centromeres.
- Kinetochore microtubules now start moving chromosomes toward the middle of the cell.

Metaphase

- Centrosomes reach opposite poles of the cell, in animals. Chromosomes, pulled by kinetochore microtubules, reach the middle of the cell. The imaginary plane where the chromosomes line up is called the **metaphase plate**. The mitotic spindle is now complete—stretching from each kinetochore to one of the two cell poles.

Anaphase

- Kinetochore spindles shorten, pulling apart sister chromatids to create separate chromosomes that are pulled toward microtubule organizing centers at opposite cell poles. Chromatid separation ensures that each daughter cell receives one copy of each parent chromosome.
- Motor proteins also push apart the two cell poles.

Telophase

- A new nuclear envelope begins to form around each set of chromosomes. The spindle disintegrates, and the chromosomes become less compact and less visible as they de-condense.

CHECK YOUR UNDERSTANDING

(c) Why might there be occasions when a researcher observing a mitotic cell under a microscope would have difficulty determining whether the cell is in prophase or metaphase?

_____

_____

_____

_____

Cytokinesis

- The division of the cytoplasm to form two daughter cells typically occurs immediately after mitosis.
- Cytokinesis in animals, fungi, and slime molds occurs when a ring of actin and myosin filaments inside the cell membrane contracts, causing the cell membrane to pinch inward and form a **cleavage furrow**.

- Plant cell cytokinesis occurs as vesicles are transported from the Golgi apparatus to the middle of the dividing cell. These vesicles fuse to form a **cell plate** (**Figure 11.9**).
- Refer to **Table 11.1** for definitions of key mitotic structures and to **Figures 11.7** and **11.10** for depictions of the different mitotic subphases.

How Do Chromosomes Move during Mitosis?

Mitotic Spindle Forces

- Because a microtubule's length depends on the number of α- and β-tubulin dimers it contains, researchers hypothesized that spindle microtubules shorten as tubulin subunits are lost.
- Fluorescently labeled tubulin subunits were added to prophase or metaphase cells. The whole mitotic spindle could then be seen using a fluorescent microscope.
- After the start of anaphase, biologists marked a region of the spindle with a laser light that stops fluorescence in the exposed region. This "photobleached" area remained stationary as anaphase progressed and the chromosomes moved toward the nonfluorescing region (**Figure 11.11**).
- The researchers concluded that chromosome movement occurs because tubulin subunits are lost from the kinetochore ends. The rest of the microtubules remain stationary.

CHECK YOUR UNDERSTANDING

(d) In the experiment on spindle shortening, why was labeled tubulin added during prophase or metaphase instead of during anaphase?

_____

_____

_____

A Kinetochore Motor

- Kinetochore proteins "walk" down the microtubules as they shorten, similar to the way kinesin "walks" down microtubules during vesicle transport (**Chapter 7**).
- Dyneins and other kinetochore motor proteins appear to detach near the chromosome and reattach to the kinetochore microtubule further down its length, causing the microtubule shortening responsible for pulling chromosomes to opposite cell poles (**Figure 11.12**).

- ATP hydrolysis powers this "walking" movement as tubulin subunits are lost from the kinetochore end of the microtubules.

11.3 Control of the Cell Cycle

- Cell-cycle length varies greatly from tissue to tissue within the same individual. G_1 phase is practically eliminated in rapidly dividing cells such as those found in the intestine. Other cells get permanently stuck in G_1 phase; this arrested stage is called the G_0 state.
- The rate of cell division may also respond to changes in conditions. These variations in cell-cycle length suggest that the cell cycle is regulated and that regulation varies among cells and organisms.

The Discovery of Cell-Cycle Regulatory Molecules

- Certain chemicals, viruses, or electric shocks can fuse two cells, forming a hybrid cell with two nuclei.
- When two cells in different cell-cycle stages were fused, one of the two nuclei sometimes changed phase. For example, fusion of an M-phase cell with an interphase cell caused the interphase nucleus to enter M phase (**Figure 11.13a**).
- Researchers hypothesized that the M-phase cell cytoplasm contains a regulatory molecule that causes interphase cells to start mitosis. This hypothesis was confirmed by experiments with *Xenopus* frog eggs.
- Maturing *Xenopus* eggs start as **oocytes** and develop into mature M-phase eggs. These eggs are very large, making it feasible to purify large amounts of cytoplasm. This purified cytoplasm can be injected into eggs to test for the presence of mitotic regulatory factors.
- When *Xenopus* M-phase cytoplasm was injected into the cytoplasm of frog oocytes arrested in the G_2 phase, the immature oocytes entered M phase. Cells injected with cytoplasm from interphase cells remained in G_2 phase (**Figure 11.13b**).
- Researchers concluded that some factor is present in the cytoplasm of M-phase cells (but not in interphase cells) that induces mitosis. This factor has now been purified and is called **M-phase promoting factor (MPF)**. MPF induces mitosis in all eukaryotes.

MPF Contains a Protein Kinase and a Cyclin

- MPF contains two polypeptide subunits. One, a **protein kinase**, catalyzes a protein phosphorylation reaction that transfers a phosphate group from ATP to a target protein.
- Many protein kinases are regulatory molecules because adding a phosphate group usually alters the target protein's shape and activity, either activating

or inactivating the protein. However, the MPF protein kinase concentration does not change much throughout the cell cycle.

- The other MPF subunit, a **cyclin**, does change in concentration throughout the cell cycle; MPF cyclin concentration increases during interphase, peaks in M phase, and then decreases again.
- The MPF protein kinase is active only when it is bound to the cyclin subunit; it is a **cyclin-dependent kinase (Cdk)**. Thus, when cyclin concentrations are high, more MPF is active and the target proteins are phosphorylated, causing initiation of mitosis (**Figure 11.14**).
- MPF itself is synthesized in an inactive phosphorylated form. Late in G_2 phase, enzymes remove the phosphates from cyclin, causing the activation of MPF, which then phosphorylates many different types of proteins (**Figure 11.14b**).
- One protein activated by MPF causes chromosomes to condense, initiating mitosis. Another protein activated by MPF catalyzes the degradation of the MPF cyclin. This negative feedback produces the observed oscillations in MPF cyclin concentration.

Cell-Cycle Checkpoints

- Other regulators of the cell cycle include a different cyclin and protein kinase that trigger the transition from G_1 to S phase, and several regulatory proteins that maintain the G_0 state of quiescent cells.
- Leland Hartwell and Ted Weinert analyzed mutant yeast cells with cell-cycle defects. Some grew abnormally due to mutations that affected regulatory points in the cell cycle. These researchers coined the term **cell-cycle checkpoint** to describe a critical point in cell-cycle regulation.
- In our bodies, cells without effective cell-cycle checkpoints keep growing and form a **tumor**.
- Three checkpoints have been identified (**Figure 11.15**). At each checkpoint, interactions between regulatory molecules determine whether a cell proceeds with division.
- Factors affecting whether cells pass the G_1 checkpoint are (1) *cell size*—so they can split into two functional daughter cells; (2) *nutrient conditions*—in unicellular organisms; (3) *signaling molecules from other cells*—in multicellular organisms; (4) **tumor suppressors**—regulatory proteins that can stop the cell cycle.
- For example, tumor suppressor **p53** stops the cell cycle if a cell's DNA is damaged. It can initiate **apoptosis** (programmed cell death). Defects in this protein can lead to cancer.

- In general, G_1 checkpoints function to ensure that a cell is healthy before DNA can be replicated and cell division initiated.

- Another checkpoint exists between the G_2 and M phases. Improper chromosome replication or DNA damage can cause cells to arrest at this stage. It appears that DNA damage can block the dephosphorylation and activation of MPF, keeping cells in G_2 phase.

- A checkpoint also occurs in metaphase. Cells arrest during M phase if the chromosomes are not properly attached to the mitotic spindle. This mechanism prevents incorrect chromosome separation that could give daughter cells the wrong number of chromosomes.

CHECK YOUR UNDERSTANDING

(e) Why did Hartwell and Weinert use yeast instead of bacteria to study cell-cycle regulation?

_____

_____

_____

_____

11.4 Cancer: Out-of-Control Cell Division

- **Cancer** is a common, often lethal disease (**Figure 11.16**). There are at least 200 different types of cancer. Despite the differences, all cancers originate from cell-cycle checkpoint failures. These checkpoint failures allow cells to grow uncontrollably. As these cells spread throughout the body, they use nutrients and space needed by normal cells.

- Two types of cell-cycle defects cause cancer: (1) defects that cause continuous activation of proteins required for cell growth (e.g. defects in Ras), and (2) defects that prevent tumor suppressor genes from stopping the cell cycle (e.g., p53).

Properties of Cancer Cells

- A **tumor** is formed when one or more cells in a multicellular organism begin to divide uncontrollably. **Benign tumors** are noninvasive, but **malignant tumors** are cancerous and can spread throughout the body via the blood or lymph and initiate new tumors.

- Detachment from the original tumor and invasion of other tissues is called **metastasis**. Once a cancer has metastasized, it cannot be cured by surgery alone (**Figure 11.17**).

Cancer Involves Loss of Cell-Cycle Control

- Mature cells often remain in the G_0 state; but if they pass through the G_1 checkpoint, they continue through mitosis and divide. Biologists thus hypothesized that most cancers involve defects in the G_1 checkpoint.

Social Control

- Unicellular organisms pass the G_1 checkpoint when nutrients are available and cell size is sufficient. Cells of multicellular organisms instead respond to signals from other cells, so that cells divide only when their growth benefits the whole organism. This is known as *social control*.

- Normally, cell cultures will not grow, even if provided with nutrients, unless polypeptides or small proteins called **growth factors** are present. Growth factors were first found in **serum**, the liquid that remains after blood clots and the cells are removed from the blood.

- **Platelet-derived growth factor (PDGF)** is a serum growth factor released into the blood by **platelets** at wound sites. Cells near the wound divide in response to PDGF and help heal the injury.

- Many growth factors exist, produced by many types of cells. Different cells divide in response to different combinations of growth factors, and specific growth factors must be present for a particular type of cell culture to grow.

- Cancer cells, however, do not need growth factors in order to divide—they are no longer subject to social control at the G_1 checkpoint.

CHECK YOUR UNDERSTANDING

(f) Why is there no social control in unicellular organisms?

_____

_____

_____

_____

_____

Social Controls and Cell-Cycle Checkpoints

- Growth factors initiate cell division by triggering production of cyclins and E2F, a regulatory protein.

- E2F can trigger the expression of genes required for S phase. However, newly produced E2F normally binds to a tumor suppressor protein called Rb. While E2F is bound to R, it remains inactive and the cell remains in G_0.

- If growth factors continue to arrive, additional cyclin production eventually leads to the activation of cyclin–Cdk complexes via dephosphorylation (**Figure 11.18**). Active cyclin–Cdk catalyzes Rb phosphorylation. Phosphorylated Rb releases E2F, which can then activate its target genes, which in turn initiate S phase.

- In some human cancers the G_1 checkpoint cyclin is overproduced, permanently activating Cdk, which then continuously phosphorylates Rb. E2F remains in its active free form, sending the cell into S phase. Either the presence of excessive growth factors or cyclin production in the absence of growth factors can cause cyclin overproduction.

- In other human cancers the G_1 checkpoint fails due to defective Rb. If Rb is absent or does not bind to E2F, the free E2F will trigger the gene expression that initiates S phase.

- · The tumor suppressor protein Rb is named for its discovery in children with retinoblastoma eye cancer. Most human cancer cells either overproduce cyclins or have Rb defects, causing loss of social control over the G_1 checkpoint, uncontrolled cell growth, and tumor formation.

Cancer Is a Family of Diseases

- Many different types of defects can cause the G_1 checkpoint to fail. Most cancers result from multiple defects in cell-cycle regulation. Each type of cancer is caused by a unique combination of errors.

B. INTEGRATING YOUR KNOWLEDGE

*Looking Back—Key Concepts
from Earlier Chapters*

Cell Theory—Chapter 1

See **Chapter 1** for further discussion of the cell theory, its importance, and some supporting evidence.

Use of Radioactive Isotopes—Chapters 7, 9, and 10

Radioactive isotopes are extremely useful research tools. Several pulse-chase experiments and other experiments involving the use of radioactive isotopes are described in **Chapters 7, 9,** and **10**. The use of radioactive thymidine to identify the timing of DNA synthesis is described here in Chapter 11.

Microtubules—Chapter 7

Microtubules were introduced in their role as cytoskeletal elements in **Chapter 7**. Here in Chapter 11 you discover that microtubules also make up the mitotic spindle, which pulls apart sister chromatids during mitosis. The way that spindle fibers move chromosomes is somewhat similar to the way that cytoskeletal microtubules function in vesicle transport.

Signal Transduction—Chapter 8

Chapter 8 introduces the importance of cell–cell signaling via hormonal initiation of phosphorylation cascades. In fact, the Ras protein shown in **Figure 8.19** can be involved in the development of cancer if a defective Ras prevents signal deactivation.

*Looking Forward—Key Concepts
from Later Chapters*

Meiosis—Chapter 12

Chapter 12 describes meiosis, the type of cell division used in sperm and egg cell production. In many ways, meiosis is similar to mitosis. However, the daughter cells produced by meiotic divisions contain only half of the parental cell's genetic information.

Chromosomes—Chapters 13 and 15

Chromosomes, cell structures made of DNA and protein, are responsible for carrying the cell's genetic information. Chapter 11 discusses how copies of each chromosome are provided to each daughter cell during cell division. In **Chapters 13** and **15**, you'll learn more about chromosomes, their importance, and how genes work.

DNA Synthesis—Chapter 14

Chapter 11 describes evidence that DNA synthesis occurs during S phase of the cell cycle. Details of DNA synthesis are given in **Chapter 14**.

C. MASTERING CHALLENGING TOPICS

One effective way to learn different phases of the cell cycle is to simulate the events, using thread for the microtubules, shoelaces for chromosomes, and a plastic bag for the nuclear envelope. Recite aloud to yourself names of the phases and major events as you act out the process. You may also want to chart the different cell-cycle phases, listing their major events and any cell-cycle checkpoints. Finally, use your own mnemonic device for remembering the mitotic phases. For example, you could remember interphase, prophase, prometaphase, metaphase, anaphase, telophase, cytokinesis (IPPMATC) as "I Pet Purple Moose At The Circus."

D.   ASSESSING WHAT YOU'VE LEARNED

1. What is the fundamental difference between mitosis and meiosis?
   a. Mitosis involves two cell divisions, whereas meiosis involves only one.
   b. The number of chromosomes doubles in meiosis; it stays the same in mitosis.
   c. The amount of genetic material per cell is cut in half in meiosis, but remains constant in mitosis.
   d. Meiosis occurs in prokaryotes, and mitosis occurs in eukaryotes.

2. Chromatin is:
   a. the structure that joins sister chromatids
   b. the microtubule organizing center in animals
   c. one strand of a replicated chromosome
   d. the DNA–protein complex that makes up eukaryotic chromosomes

3. Mitotic spindle fibers are composed of:
   a. microtubules
   b. centrosomes
   c. centromeres
   d. kinetochores

4. During mitosis, it is necessary for the nuclear envelope of the parent cell to dissolve. This is accomplished, at least partly, by phosphorylation of proteins associated with the nuclear envelope. If the enzyme responsible for that phosphorylation event is inhibited, at which phase of mitosis are cells likely to arrest?
   a. prophase
   b. prometaphase
   c. metaphase
   d. anaphase
   e. telophase

5. In addition to a pulse-labeling experiment, the length of different phases of the cell cycle can also be estimated by labeling cells continuously with radioactive thymidine and waiting to see how long it takes for radioactivity to appear in mitotic chromosomes. What phase of the cell cycle could be estimated by the time between the addition of radioactive thymidine and the time that radioactivity first appears in prophase chromosomes?
   a. G_1 phase
   b. G_2 phase
   c. S phase
   d. M phase

6. Although the process of chromosome partitioning during mitosis is visible through the light microscope, the process of DNA replication is not. Why?
   a. Chromosomes form only during mitosis.
   b. Chromosomes are visible only after DNA has been duplicated.
   c. Chromosomes are too extended during S phase to be seen by light microscopy.
   d. Chromosomes do not contain protein until mitosis.

7. A certain species of animal has six pairs of chromosomes. How many molecules of DNA do the nuclei of these animals have during G_2 phase?
   a. 6
   b. 12
   c. 24
   d. 48

8. In animals, cytokinesis occurs:
   a. when microtubules pull the cell membrane into a cleavage furrow
   b. when vesicles from the Golgi build up to divide the cell along the cell plate
   c. when G_2 phase is complete
   d. when a ring of actin and myosin filaments contracts

9. The experiment where biologists photobleached spindle fibers made with fluorescent tubulin showed that microtubules shortened from the kinetochore ends. What different results would indicate that microtubules shorten from the microtubule organizing center end?
   a. movement of the chromosomes toward a stationary photobleached area
   b. movement of the photobleached region toward the cell pole
   c. rapid disappearance of the photobleached microtubules
   d. none of the above

10. Some of the earliest experiments in control of the cell cycle were cell fusion experiments. Fusion of cells in M phase with cells in interphase caused the nuclei from the interphase cells to behave as if they were in mitosis. These results could have been obtained by adding just a part of the M-phase cells to the interphase cells. What portion of the M-phase cell would suffice to cause the interphase nuclei to mimic the process of mitosis?
    a. the nucleus
    b. ribosomes
    c. the cell membrane
    d. cytoplasm

11. Active Cdk phosphorylates proteins that cause:
    a. M-phase initiation
    b. nuclear envelope breakdown
    c. cyclin degradation
    d. all of the above

12. Cells that are too small to successfully produce daughter cells fail to pass:
    a. the G_1 checkpoint
    b. the G_2 checkpoint
    c. the metaphase checkpoint
    d. all checkpoints

13. Can surgery successfully cure a cancer that has metastasized?
    a. No; all body cells are dividing uncontrollably.
    b. Yes; it could remove all cells with defective cell cycle regulation.
    c. No; cancer cells are no longer localized in one spot.
    d. Yes, if the tumor is benign.

14. What would happen if a cell had defective E2F?
    a. The cell would immediately enter the S phase.
    b. The cell would be unable to enter the S phase.
    c. The cell would become cancerous.
    d. The Rb protein would become activated.

15. One cause of cancer is mutation of genes that encode for proteins in the biochemical pathway regulated by PDGF. How might such mutations cause cancer?
    a. If the mutations caused the biochemical pathway to become permanently active, the cells might become cancerous.
    b. If the mutations caused the biochemical pathway to become permanently inactive, the cells might become cancerous.
    c. If PDGF stimulated cell division, the cells might become cancerous.
    d. Mutation of any gene will probably cause cancer.

CHAPTER 11—ANSWER KEY

Check Your Understanding

(a) Before radioisotopes became available in the mid-20th century, researchers could not determine when DNA synthesis occurred because there was no way to "see" when DNA synthesis occurred (especially since interphase chromosomes are not visible via light microscopy). When radioactively labeled nucleotides became available, researchers could use them to mark the incorporation of new nucleotides into DNA during S phase.

(b) Cells that divided without any gap phases would probably get smaller and smaller and have fewer and fewer organelles with each succeeding division. Eventually they would not have enough organelles or cytoplasm to function effectively.

(c) Recall that mitosis is a continuous process. It may thus be difficult to determine whether the centrosomes have fully completed their migration to opposite sides of the cell and whether the chromosomes have fully, or only mostly, moved to the middle of the cell.

(d) Because the mitotic spindle is complete by the start of anaphase, labeled tubulin added in anaphase would not be incorporated into the mitotic spindle.

(e) Recall that bacteria replicate very differently from eukaryotes. They do not undergo mitosis, because they have circular chromosomes instead of linear chromosomes. Yeast, however, are eukaryotes that divide quickly and live well in culture—two qualities that make them useful in research.

(f) Natural selection favors unicellular organisms that leave the most offspring. Thus, unicellular organisms should divide whenever enough nutrients are available to allow sufficient growth for cell division (asexual reproduction). Only in multicellular organisms is it favorable to the organism to allow other cells (in the same organism) to influence cell division.

Assessing What You've Learned

1. c; 2. d; 3. a; 4. a; 5. b; 6. c; 7. c; 8. d; 9. b; 10. d; 11. d; 12. a; 13. c; 14. b; 15. a

12

Meiosis

A. KEY CONCEPTS

12.1 How Does Meiosis Occur?

An Overview of Meiosis

- Examine **Figure 12.1** and consider meiosis from the perspective of a scientist trying to understand the nature of chromosome function in the **lubber grasshopper**. There is a logic to this process. Define the key outcomes of a complete cycle of meiosis.
- **Table 12.2** offers a critical summation of terminology introduced in this section.
- As you begin studying the process of meiosis, it is important to realize that the germinal cell initiating this process contains **two homologs** of each chromosome (**2n or diploid**), and that the **gamete** cell receives only **one homolog** of each chromosome upon completion of meiosis (it will be 1n or **haploid**). In 1902, **Walter Sutton** published his observations of homologous chromosomes in the lubber grasshopper (**Figure 12.1**—homologs identified by size and by pairing).
- In a diploid organism, what is the origin of each homologous chromosome? How might carrying this "extra" information be valuable? How might it be detrimental? The total number of chromosomes in an organism is known as its **karyotype** (see **Box 12.1**).
- Review **Figure 12.2** as an overview of the meiotic process—use this digram to reveal changes in ploidy during meiosis.
- Just before meiosis begins, each of the chromosomes in the 2n parental cells are **replicated**. The newly synthesized chromosomal strands remain attached to the original chromosome from which they were copied until a later stage of meiosis.
- In general, each replicated chromosome looks like the letter **X**, with two linear **sister chromatids** being attached in the middle at a structure called the **centromere**.
- **Meiosis I** involves the complex mechanism of crossing over in prophase I (described in the next section) and the separation of homologous chromosomes into two separate cells (each are now 1n).

- **Meiosis II** repeats the steps of meiosis I, the key difference being the physical separation of the sister chromatids (X → < >), and the migration of each into a gamete cell that will contain one copy of each chromosome (**1n** or **haploid** gametes).
- **Key Point**: Don't confuse the **homologous chromosomes**, which make the precursor cell 2n, with the copies of each that are produced before meiosis begins. As long as the sister chromatids are attached to each other at the centromere, they are considered a single replicated chromosome.

CHECK YOUR UNDERSTANDING

(a) On a sheet of scratch paper, make a diagram that compares the ploidy of diploid (2n) cells proceeding through mitosis and meiosis, and indicate where each change in ploidy takes place.

The Phases of Meiosis I

- Meiosis I contains two key events in the generation of genetic diversity from generation to generation.

- The first event is **crossing over** between homologous chromosomes in prophase I, and the second event is the **separation of homologous chromosomes** in anaphase I. These two steps generate incredible genetic diversity in the gametes, and recognizing how each contributes to this diversity should be a key focus in your study of Chapter 12. Both are discussed in detail here, but before we examine them in detail, let's review the steps of meiosis I.

- Review **Figure 12.4** for a visual summary of the details of meiosis.

- **Prophase I**—This step of meiosis takes the most time, primarily due to the complexity of crossing over. As it progresses, the chromosomes **condense** into tightly wound balls of DNA (X-shaped, visible with microscopy). Condensation is mediated by small **histone proteins** that bind to the chromosomal DNA and wrap the DNA around their surface. Crossing over occurs in prophase I and involves exchange of long stretches of DNA between homologous chromosomes. This is a key step in recombination of DNA for the production of genetically diverse gametes.

- **Metaphase I**—The condensed chromosomes migrate to the midline of the cell and form a line. Homologous chromosomes are adjacent to each other at the midline. Cytoskeletal proteins, known as microtubules, mediate this migration.

- **Anaphase I**—The second key step in generation of genetic diversity in gametes. The paired homologues are separated from each other, migrating to opposite ends of the cell. Following migration, each pole of the cell contains one complete set of chromosomes.

- **Telophase I**—The chromosomes complete their migration and are enclosed in two newly formed daughter cells (each is $1n$; confirm by counting centromeres in each cell).

The Phases of Meiosis II

- **Meiosis II** is really just a repeat of meiosis I, with a few important differences. At the start of meiosis II, each daughter cell contains haploid chromosome content ($1n$). Each replicated chromosome is composed of the two sister chromatids. These chromatids, attached to each other at the centromere, move into separate cells during anaphase II. Once the chromatids have segregated into gametes, they are called chromosomes. Because there is no crossing over in prophase II, this step is more rapid than prophase I.

CHECK YOUR UNDERSTANDING

(b) Describe the two steps in meiosis where most genetic recombination occurs.

_____

_____

_____

12.2 The Consequences of Meiosis

Chromosomes and Heredity

- A chromosome is a long, linear **macromolecule** composed of deoxyribonucleic acid (DNA). As discussed in **Chapters 13** and **14**, chromosomes hold information in the form of distinct regions of DNA that contain a code, found within the ordering of chemical subunits of the chromosome.

- Each information-carrying DNA region that carries the blueprint for a single **protein** is known as a **gene**.

- Two genes that encode the same protein but have small differences in coding information are called **alleles** of each other. There can be hundreds of alleles of a single gene. In a diploid ($2n$) organism, each gene is present two times, on each homologous chromosome.

- Proteins carry out functions within the cell; they form the cellular infrastructure (cytoskeleton), carry out steps in energy production, break down sugar into usable energy, and carry out cellular communication. _Proteins do things_, while genes are blueprints for the proteins.

How Does the Separation and Distribution of Homologous Chromosomes Produce Genetic Variation?

- The separation (or segregation) of homologous chromosomes in anaphase I generates a great deal of **genetic diversity** in the subsequent gametes.

- Diploid organisms, carrying two **homologs** of each chromosome, divide the diploid DNA content into two cells at the completion of meiosis I. This process is random, ensuring that each cell gets one replicated copy of each chromosome (a total of 23 in humans). In the production of gametes, the homologous chromosomes **segregate** into different cells at the completion of meiosis I. Notice that the chromosome number is now reduced to a haploid ($1n$) status.

- Because each gamete receives one of the two homologs of each chromosome, and there are 23 chromosomes in humans, we are capable of generating 8.4 million (223) genetically unique gametes. Add to that the additional diversity generated by crossing over in prophase I, and that value increases dramatically.

- **Sexual reproduction** offers incredible diversity to offspring. As we'll see next, organisms invest a great deal of energy into increasing the **genetic diversity** in their offspring. Why is such effort expended to ensure genetic diversity from generation to generation? Doesn't this increase the likelihood of mistakes occurring during gamete formation?

The Role of Crossing Over

- In prophase I, a complex process of **chromosomal recombination** is initiated. Let's focus on one pair of chromosomes, knowing that each pair of homologs will go through the same process.

- The homologous chromosomes each underwent DNA replication before meiosis began, and they are attached to the copy at a centromere. The homologous chromosomes come together (**synapsis**), recognizing each other by their matching DNA sequence and forming a tightly associated **tetrad** of chromosomes. *The tetrad contains both homologs and their copies, for a total of four strands of chromosomal DNA.*

- The four strands of DNA associate very closely with each other, and association is between corresponding regions of each chromosome. Crossing over is simply the exchange of sections of DNA from one strand of DNA to another (**Figure 12.7**). The site of crossing over, the **chiasma**, can sometimes be visualized microscopically as an X-shaped structure along the paired chromosomes.

- *Crossing over results in a shuffling of gene alleles between homologous chromosomes, which in turn results in unique chromosomes that were not previously present. These unique chromosomes contain a combination of paternal and maternal alleles of genes.* Your cells have two copies of each chromosome (one from mom and one from dad). In production of gametes, you will generate recombined chromosomes carrying alleles from your mother and from your father.

- *The regions exchanged are homologous;* that is, they carry the same genes, but subtle allelic differences in the genes contributed by each parent make these crossover products unique from the original chromosomes.

12.3 Why Does Meiosis Exist? Why Sex?

The Paradox of Sex

- Many types of organisms thrive through the process of **asexual reproduction**. A fascinating question in biology focuses on why sex is such a dominant reproductive strategy in multicellular organisms. Can you define the meaning of Maynard Smith's "twofold cost of males" in a population?

- Does the answer lie in the genes? Are organisms that reproduce sexually able to produce more diversity and adaptability in their offspring?

- Maynard Smith predicted that asexually reproducing organisms should reproduce faster and outcompete similar organisms that invest in sexual reproduction. What assumptions of this model might be problematic when you consider the influence of sexual/asexual reproduction on inheritance? What environmental factors might explain the success of sexual reproduction?

- Two models that address the paradox of sex are (1) the **purifying selection hypothesis** and (2) the **changing environment hypothesis**.

- Individuals reproducing asexually will transmit deleterious mutations to all of their offspring, while individuals reproducing sexually will transmit deleterious mutations to only half of their offspring (on average). The purifying selection hypothesis predicts that over time, natural selection will reduce the number of deleterious alleles in a population that is reproducing sexually. Individuals with deleterious mutations will display diminished fitness, will be less likely to reproduce, and will be less likely to pass their genes on to the next generation.

- The changing environment hypothesis predicts that populations reproducing sexually (with genetically diverse offspring) will have an improved ability to adapt and survive when environmental conditions change.

- Examine the examples offered in support of each of these hypotheses (**Figure 12.14**).

CHECK YOUR UNDERSTANDING

(c) How might asexual reproduction offer an advantage in some environments? Describe the two steps in meiosis where most genetic recombination occurs.

_____

_____

_____

_____

(d) Why is the freshwater snail (*P. antipodarium*) a good choice for experiments comparing sexual and asexual reproduction?

_____

_____

Experiment

Question: Why does sexual reproduction occur?

Hypothesis: In habitats where parasitism is common, sexually produced offspring have higher fitness than do asexually produced individuals.
Null hypothesis: There is no relationship between the presence of parasites and reproduction method.

Experimental setup:

1. Collect snails from a wide array of habitats.

2. Document percentage of males in each population, as an index of frequency of sexual reproduction. More males means that more sexual reproduction is occurring.

3. Note two types of populations: in one, males are common; in the other, males are almost nonexistent. Infer that sexual reproduction is either common or almost nonexistent.

4. Document percentage of individuals infected with parasites in sexually versus asexually reproducing populations.

Prediction: In populations where sexual reproduction is common, parasitism rates are high. In populations with only asexual reproduction, infection rates are low.
Prediction of null hypothesis: No difference in parasitism rate between populations that reproduce sexually versus asexually.

Results:

Conclusion: Sexual reproduction is common in habitats where parasitism is common. Asexual reproduction is common in habitats where parasitism is rare.

12.4 Mistakes in Meiosis

How Do Mistakes Occur?

- Meiotic mistakes are an inherent risk in the complex process of generating genetic diversity during gamete production.

- Some zygotes carry severe genetic abnormalities: abnormal numbers of chromosomes (**aneuploidy**) or fragmented chromosomes that were damaged at some point during meiosis. These zygotes typically do not survive to produce viable offspring.

- Down syndrome is a relatively common birth defect in humans (0.15 percent of live births), and is caused by an extra copy of chromosome 21, a trisomy-21 karyotype.

- Aneuploidy is caused by a mistake in meiosis, specifically a nondisjunction event (Figure 12.15). In nondisjunction, chromosomes that should segregate into separate cells are improperly segregated into the same cell. This generates one gamete that is lacking a chromosome and one gamete that has an extra chromosome. If either of these gamete types is involved in forming an embryo, the embryo will have an abnormal chromosome number (aneuploidy).

(e) Explain nondisjunction as it relates to meiosis; where does this problem predominantly arise?

_____

_____

_____

_____

B. INTEGRATING YOUR KOWLEDGE

Looking Back—Key Concepts from Earlier Chapters

The Cell Cycle—Chapter 11

While reading Chapter 12, you'll notice striking similarities between **mitosis** and **meiosis** and the basic cellular architecture enabling both processes. Examine **Chapter 11** and focus on cell components that are critical to **cellular division**. How might the **cytoskeleton** (filaments and microtubules), **cellular membranes**, and **chromosomes** be prepared for meiotic division? Would **DNA replication** precede this process?

Looking Forward—Key Concepts in Later Chapters

Mendel and the Gene—Chapter 13

Mendel worked out much of the meiotic process with no knowledge of what a chromosome really was. The **principles of segregation and independent assortment** both describe what happens to chromosomes in meiosis.

DNA and the Gene—Chapter 14

We'll learn about DNA replication in **Chapter 14**, identifying how signals in the cell activate the deliberate process of copying the **genome** before division can occur.

Principles of Development—Chapter 21

Meiosis is a key process in **reproduction** and **fertilization**. In **Chapter 21**, we examine **gametogenesis** in more detail and follow the gamete through fertilization.

C.   MASTERING CHALLENGING TOPICS

Mitosis versus Meiosis

Mitosis and meiosis share many characteristics—chromosome condensation, alignment of chromosomes at the cellular midline, migration of chromosomes to the poles, and formation of two new cells. What characteristics are observed only in mitosis or only in meiosis? How is meiosis I different from meiosis II?

Carefully examine these processes, focusing on the ploidy (or DNA content) at each stage and on how segregation at anaphase occurs. Some fundamental differences between these processes illustrate the purpose of each process. Mitosis occurs in all of our somatic cells, and it requires the careful transfer of all genetic information from mother cell to the daughter cells. Meiosis expends energy and engages in the dangerous process of rearranging our chromosomes to generate genetic diversity.

Genetic Recombination

Crossing over is a complex process that is often confusing at first. This is especially true when you consider the impact of crossing over on the independent assortment of genes. Diagram the formation of a tetrad and a crossover event between two homologs (use color to make the crossover more visible). Resolve the crossover by illustrating what would then happen to these homologs in anaphase I; notice how all four chromosomes are now unique, each carrying a slightly different DNA content.

Terminology

This chapter introduces many new terms and concepts that are crucial to your understanding of this, and future, chapters. Take time to learn these concepts, and you'll be prepared to move on to more complex concepts in genetics. Here are some terms to carefully examine and define as you study:

aneuploidy	chiasma	chromatids
diploid	gamete	gene
haploid	homologous	life cycle
chromosomes	nondisjunction	outcrossing
karyotype	trisomy	zygote
self-fertilization	sexual reproduction	
synapsed chromosomes		

D.   ASSESSING WHAT YOU'VE LEARNED

1. Which of the following occurs during meiosis II, but does not occur as part of meiosis I?
   a. crossing over between homologous chromosomes
   b. separation of homologous chromosomes
   c. separation of sister chromatids
   d. migration to middle of cell

2. Crossing over:
   a. occurs between sister chromatids on the same chromosome
   b. occurs between chromatids of homologous chromosomes
   c. occurs very rarely
   d. occurs during prophase II of the cell cycle

3. An individual plant that exhibits self-fertilization can produce offspring different from itself primarily as the result of:
   a. mutations that occur during mitosis
   b. independent assortment and crossing over during the formation of gametes
   c. crossing over (only) during the formation of chromosomes
   d. the fact that the number of chromosomes in pollen is always different from the number of chromosomes in egg cells

4. **Figure 12.13** in your textbook demonstrates which of the following?
   a. Organisms produced by asexual reproduction are more variable than those produced by sexual reproduction.
   b. Sexual reproduction is always more successful than asexual reproduction.
   c. Asexual reproduction is always more successful than sexual reproduction.
   d. Asexual organisms generally produce more offspring than sexually reproducing organisms.

5. Robert inherited alleles *A* and *B* from his father, and alleles *a* and *b* from his mother (both genes are on one chromosome). If Robert marries a woman who has only *a* and *b* alleles on the corresponding chromosomes, how is it possible that their child has *a* and *b* alleles on one chromosome and *A* and *b* alleles on the other?
   a. independent assortment
   b. crossing over of Robert's wife's chromosomes
   c. crossing over of Robert's chromosomes
   d. crossing over of Robert's father's chromsomes

6. A **proximate explanation** of a biological mechanism would describe:
   a. why something happens
   b. how something happens

7. Curtis Lively generated evidence in support of the "changing-environment hypothesis" when he found:
   a. sexually reproducing *Daphnia* populations with high numbers of deleterious mutations
   b. sexually reproducing *Daphnia* populations with high numbers of deleterious mutations
   c. asexually reproducing populations of snails to be less susceptible to parasitism
   d. sexually reproducing populations of snails to be less susceptible to parasitism

8. Bacterial conjugation mediates a form of genetic recombination in these prokaryotic cells. Conjugation offers recombination via the:
   a. bidirectional exchange of chromosomal DNA
   b. transfer of portions of chromosomal DNA from a donor bacterium to a recipient bacterium
   c. transfer of an intact chromosome from one bacterium to another
   d. a process almost identical to meiosis in sexually reproducing eukaryotic organisms

9. Diagram meiotic crossing over between two adjacent genes on the same chromosome (*A* and *B*, shown in Chart 1). The parental cell entering into meiosis is genotypically *aB/Ab*, and will generate four haploid gametes at the completion of meiosis. Diagram the genetic makeup of the gametes produced by this meiotic process, and compare these gametes to the original parental contributions.

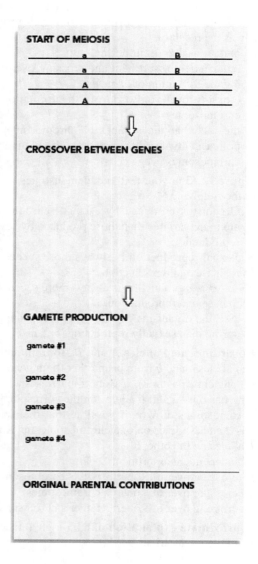

CHAPTER 12—ANSWER KEY

Check Your Understanding

(a) You can determine ploidy by counting the centromeres; a diploid cell with replicated chromosomes is still diploid, although it has doubled its DNA content.

   Mitosis—Before entering into mitosis, the $2n$ cell first replicates its genome. At anaphase, *sister chromatids* are separated and migrate to opposite poles of the cell. The cell then divides into two $2n$ daughter cells.

   Meiosis—Before entering into meiosis, the $2n$ cell first replicates its genome. At anaphase I, *homologous chromosomes* are separated and migrate to opposite poles of the cell. The cell then divides into two $1n$ cells (sister chromatids are still attached). At anaphase II, *sister chromatids* are separated and migrate to opposite poles of each cell. Then the cell divides into two $1n$ gametes.

(b) **Prophase I**—Homologous chromosomes (replicated) synapse into a tetrad and exchange homologous regions of DNA. This crossing over between chromosomes happens at least once for each tetrad.

   Anaphase I—The tetrad separates and homologous chromosomes migrate to opposite poles of the cell. Each centromere remains intact, with two sister chromatids attached, until anaphase II. The various combinations of homologs generated at the end of meiosis I offer dramatic diversity to the subsequent gamete.

(c) The most common site of nondisjunction is meiosis I, during which homologous chromosomes do not properly migrate to opposite poles of the cell. This can occur during meiosis II, but it is less common (see **Figure 12.15**).

(d) Reproduction in the absence of sex is faster, requires less energy, and does not require the fusion of germ cells from two sources. Asexually reproducing organisms can reproduce much more rapidly than sexually reproducing organisms. But, researchers have shown that asexual reproduction offers less adaptability (genetic variability) to the organism.

(e) This snail can reproduce both sexually and asexually. Populations of snails with high numbers of males tend to be reproducing sexually, while asexually reproducing populations have more female snails. The researcher (Curtis Lively) could thus analyze the ratio of males to females in populations of snails that are resistant or sensitive to disease (parasitism).

Assessing What You've Learned

1. c; 2. b; 3. b; 4. d; 5. c; 6. b; 7. d; 8. b

9. Once the tetrad is formed, strands are exchanged between the homologous chromosomes. If this occurs between the two genes as diagrammed below, then half of the resulting gametes will have a unique combination of alleles relative to each parent.

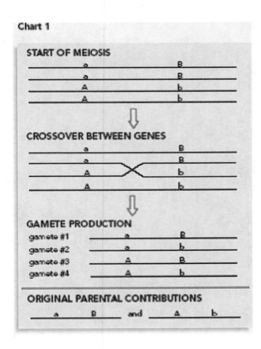

Chart 1

START OF MEIOSIS

CROSSOVER BETWEEN GENES

GAMETE PRODUCTION

gamete #1

gamete #2

gamete #3

gamete #4

ORIGINAL PARENTAL CONTRIBUTIONS

Mendel and the Gene

13

A. KEY CONCEPTS

13.1 Mendel's Experiments with a Single Trait

What Questions Was Mendel Trying to Answer?

- **Blending versus acquired characters**—These competing hypotheses of Mendel's time were both severely damaged by data produced from his simple experimentation with pea plants. Review these hypotheses, and the specific results that contradicted them.

- **Incomplete dominance** is a form of blending, with the offspring displaying the phenotype of both alleles. This occurs with some genes, but did not occur in Mendel's analyses with peas. We'll talk more about incomplete dominance later in the chapter.

Garden Peas Serve as the First Model Organism in Genetics

- Consider all of the characteristics you'd want in a **model organism**. If you were setting up the experiments, what category of organism would you select—a plant, insect, worm, fungus, mammal, or primate? How would you select from among all the different possible representatives of each group? What traits would be the most, or the least, valued?

- Mendel chose a common garden pea (*Pisum sativum*) and revolutionized our knowledge of genetics for all eukaryotic organisms. He used the garden pea because:

  1. It is easy to grow.
  2. It is inexpensive; little requirement for specialized equipment.
  3. Its matings can be controlled completely (**Figure 13.1**).
  4. It has easily observed traits (phenotypes).
  5. It has a rapid life cycle (seedling to reproductively mature).

(a) Mendel's choice of the pea plant (*Pisum sativum*) as his experimental organism was critical to his success. Describe how things might have turned out differently if he had chosen laboratory mice.

_____

_____

_____

_____

Here are some critical terms and concepts from Chapter 13:

- **Allele**—A single gene may have multiple alleles. Each allele is different from all other alleles, with some difference(s) in information content.

- **Dominant/recessive traits**—Most normal or wild-type traits are dominant; mutant traits are recessive. In other words, the mutation is hidden in organisms that carry both a wild-type and a mutant allele (different types of the same gene). Some mutations are dominant to the wild-type allele, but these are unusual.

- **F_1 generation**—The progeny of the parental organisms are called the "first filial" or F_1 generation. These progeny are genetically composed of one copy of each chromosome from the mother and one copy of each chromosome from the father.

- **Genotype**—The collection of alleles comprising the genome (total DNA content) of an organism.

- **Heterozygous**—In somatic cells, individuals carrying two different alleles of the same gene are heterozygous for that gene.

- **Homozygous**—In somatic cells, individuals carrying two copies of the same allele are said to be homozygous for that gene.

- **Parental generation**—In a standard mating, these would be the original parental organisms that donate gametes. In Mendel's experiments, parental plants were true-breeding, indicating several generations of self-fertilization.

- **Reciprocal crosses**—Two separate matings (crosses) performed with reversed male and female contributions. For example, pollen taken from a purple plant fertilizing an oocyte from a white plant (cross 1); and, in a second cross, pollen from a white plant fertilizing an oocyte from a purple plant. These crosses are valuable in identifying genes that are located on the sex-determining chromosomes (see the white-eyed fly example, **Figure 13.11**).

Inheritance of a Single Trait

- Gregor Mendel performed many simple crosses, focusing on one trait (such as flower color) in experiments known as **monohybrid crosses** (**Figure 13.4**). Consider for example, when a wild-type, purple-flowered plant (designated PP) is crossed by a mutant, white-flowered plant (designated pp). If the F_1 progeny are all purple, what does that indicate about the relationship between the two alleles?

- You will remember from **Chapter 12** that the product of gamete fusion is **diploid**. Thus the F_1 plant is heterozygous for the flower color gene (Pp), and since the F_1 progeny are all purple, the allele generating purple flowers (P) is **dominant** to the allele generating white flowers (p). The dominant allele is usually indicated using capital letters, while the **recessive** is indicated using lowercase letters. *Mendel did not observe blending!*

- Mendel crossed the purple-flowered F_1 progeny (or allowed **self-fertilization**), resulting in an F_2 generation. These plants displayed a **3:1 ratio** of purple- to white-flowered plants (¾ of all progeny were purple, and ¼ of all progeny were white). The recessive phenotype reappeared in the next generation, and thus must have been "hidden" in the F_1 plant.

- The segregation of alleles in gamete production is revealed in the ratios of F_2 progeny phenotypes. Examine the **Punnett square**; it is useful in the analysis of inheritance.

- Mendel described gene segregation in meiosis long before scientists even began to look at chromosomes in the production of sperm or oocytes. Mendel developed a model in which a plant carrying a genotype of Pp generates 50 percent P gametes and 50 percent p gametes. Thus, the parental organism carries two determinants for each trait, and these segregate separately in the generation of gametes. This is referred to as Mendel's **principle of segregation**.

- Mendel used probability calculations to predict the ratios produced by specific crosses.

- Using the preceding example, what is the probability of producing a white-flowered (pp) plant if a heterozygous (Pp) plant is allowed to self-fertilize?

- Mendel knew that in order to make this probability calculation, he needed to include the individual probabilities for the contribution of the recessive allele in each gamete.

- The probability for any individual gamete to carry the recessive p allele is .50 (50 percent). To generate the white-flowered plant, a p oocyte must be fertilized by a p pollen grain. These two independent events must occur simultaneously to generate a homozygous recessive plant; thus, their probabilities are multiplied together.

- Thus .25 or 25 percent of the progeny from a self-fertilized heterozygous plant will be white-flowered (pp).

- What is the probability (p) of the same heterozygous F_1 plant generating a genotypically homozygous F_2 plant with respect to the P gene (i.e., either PP or pp)?

- This type of calculation, called an **either-or** situation, requires one additional step. You first use the product rule to determine the individual probabilities for the generation of a PP plant (.25) or a pp plant (.25).

- These independent probabilities are then added together to calculate the probability of generating either one or the other in any given fertilization event. Thus, .25 + .25 = .50, or 50 percent. Mendel used simple calculations of this type to make testable predictions about inheritance, and to analyze his data with statistical tests.

13.2 Mendel's Experiments with Two Traits

The Principle of Independent Assortment

- As the inclusion of a second gene in a single cross, a **dihybrid cross** reveals the independent assortment of unlinked genes. One example given in the text (**Figure 13.5**) includes the cross of a plant that displays round (R), yellow (Y) peas by a plant that displays wrinkled (r), green peas (y). Thus the parental cross is $RRYY \times rryy$, and the F_1 progeny would be $RrYy$. The F_1 plants display the dominant phenotypes (round, yellow), but the F_2 plants display four combinations of phenotypes. Mendel found traits in these dihybrid crosses to segregate in a 9:3:3:1 pattern. Examine the Punnett square in the figure to confirm that this ratio makes sense.

- By careful analysis of his results, Mendel was able to establish the principle of **independent assortment**. *This principle states that genes segregate independently of each other in the production of gametes* (**Figure 13.5a**). This is true for most genes;

but some genes located on the same chromosome do not assort independently, and are described as linked (see **Section 13.4**).

- A simple way to "see" independent assortment is to diagram a simple meiotic division. Without knowing anything about DNA or chromosomes, Mendel was describing chromosomal assortment in meiosis.

- You can predict the genotypes and phenotypes of progeny from specific dihybrid crosses by using the product and sum rules. Compare the results of your calculations with the proportions predicted by a Punnett square.

Using a Testcross to Confirm Predictions

- The **testcross** is a simple analysis used in controlled mating experiments. It allows the investigator to focus on the meiotic events in only one of the two parents.

- A testcross is accomplished by crossing the organism of interest with a true-breeding organism that typically contains only recessive alleles for the genes being studied. Thus, for the genes being studied, the true-breeding organism will produce only one type of gamete.

CHECK YOUR UNDERSTANDING

(b) Using pea plants, you are analyzing two genes (A and B) using simple dihybrid crosses. You generate F_1 progeny that are heterozygous at both loci ($AaBb$), and interbreed these plants to generate F_2 progeny. What percentage of the F_2 progeny will carry the genotype $AABB$ or $aabb$? Try to answer this question using the Punnett square, then use probability calculations to determine the answer. Do they agree with each other?

_____

_____

_____

_____

_____

- Review the example given in text (an *RrYy* plant crossed by an *rryy* plant; **Figure 13.6**). This cross isolates the meiosis of the *RrYy* plant, because only *ry* gametes will be produced by the testcross plant. Thus, the progeny genotype/phenotype reveals what happened during meiosis of the *RrYy* plant. This powerful strategy has many applications in gene analysis.

13.3 The Chromosome Theory of Inheritance

- The **chromosome theory of inheritance** was born out of careful observations of meiosis and the movement of individual chromosomes during the stages of meiosis. Early researchers realized that the reduction in chromosome number during **gametogenesis** (gamete formation) was consistent with Mendel's principle of segregation. They also realized that the movement of homologous chromosomes was consistent with Mendel's principle of independent assortment (**Figure 13.8**), and that the chromosomes likely contained heritable units of information that were passed from generation to generation.

CHECK YOUR UNDERSTANDING

(c) Explain the value of a testcross. What advantage does it have for a geneticist? Compare a standard dihybrid cross, of *AaBb* by *AaBb*, to a testcross of *AaBb* by *aabb*.

_____

_____

_____

_____

_____

13.4 Testing and Extending the Chromosome Theory

The Discovery of Sex Chromosomes

- In the early years of observing chromosomes in meiotic cells, scientists struggled with observation of a **heteromorphic** pair of chromosomes that align with each other in anaphase I of meiosis (**Figure 13.10**). These chromosomes were called heteromorphic since they did not match (different sizes), and all other homologous chromosomes were precisely matched with a chromosome of the same size.

- These heteromorphic pairs represent the sex chromosomes—the X and Y chromosomes.

- It was observed that XY males generate X gametes and Y gametes, and that the composition of the sperm regarding X or Y content determined the sex of any fertilized oocyte.

- Thomas Hunt Morgan, a researcher at Columbia University, utilized the fruit fly (***Drosophila melanogaster***) as a model organism to examine mechanisms of inheritance. He established a line of flies carrying a recessive mutation that resulted in white eyes (rather than red). In monohybrid breeding experiments to examine this trait, Morgan found the expected 3:1 ratio of red-eyed to white-eyed flies in the F_2 generation. However, Morgan found that this trait did not segregate evenly among male and female progeny.

- Examine the reciprocal crosses shown in **Figure 13.11**. Notice that the F_1 and F_2 generations are different depending on which parental fly displays the white-eyed trait. *Note:* Whenever the results of reciprocal crosses generate different ratios of males/females that display the trait, the gene is located on a sex-determining chromosome (the X or Y).

- In pedigree analysis, whenever a phenotype within a pedigree is predominantly associated with males or females, one should suspect that the gene is located on a sex-determining chromosome (the X or Y).

- Review Morgan's data using a **Punnett square**; remember, if the gene is located on the X chromosome, the sex of parental cells used can dramatically influence the phenotypical ratios of the progeny.

- An example—If gene g^+ is located on the X chromosome, then the male genotype may be represented as g^+Y. In gamete production g^+ will segregate with the X chromosome and will be present in 50 percent of the gametes (carrying the X), while the other 50 percent of the gametes receive the Y chromosome. Use a Punnett square to analyze reciprocal monohybrid crosses for *g*. How are the F_1 and F_2 progeny different in these crosses?

CHECK YOUR UNDERSTANDING

(d) What is a reciprocal cross? In what type of analysis is it particularly valuable?

_____

_____

_____

_____

What Happens When Genes Are Located on the Same Chromosome?

- When examining dihybrid crosses, all examples of independent assortment we've discussed to this point depend on genes being located on separate chromosomes. What happens when two genes are located near each other, on the same chromosome?

- First, their assortments are not independent of each other; they are physically attached (or **linked**) to each other on that chromosome. Linkage can be identified by segregation patterns that do not show independent assortment of the two genes (something other than 3:1 for a monohybrid, or a 9:3:3:1 for a dihybrid cross).

- This is a difficult concept to visualize; review the meiotic process again, taking into account two genes located on the same chromosome. Those two linked alleles will segregate into the same gamete *every time*, unless a crossover event separates them onto homologous chromosomes. But any gametes that require a crossover event to be produced will be uncommon. Let's look at one example provided in the text (**Figure 13.12**).

- Two genes are utilized to illustrate gene linkage. Genes/alleles included in the example are w^+—red eyes, w—white eyes, y^+—gray body, and y—yellow body. The parental animals in this example donated a w^+y gamete and a wy^+ gamete to generate the F_1 displayed in part (a). Notice that the contributions of the parents are linked on the same chromosomes, and that in meiosis, the only way to generate a w^+y^+ gamete or a wy gamete is for crossing over to occur.

- Therefore, the parental genotype (for w and y) will be the most common genotype found in the gametes produced. Independent assortment will not occur between these two alleles; they are linked. The phenotypes of the F_2 progeny are not 9:3:3:1, but rather are dominated by the parental phenotype (wy^+ and w^+y).

- Linked genes have also been shown to be valuable in generating genetic maps of chromosomes. The key to this strategy was the recognition that the frequency of a crossover event between two linked genes is dependent upon the distance between those two genes. The closer the genes are, the less frequent crossover events will be. Thomas Morgan and his research students were the first to develop this strategy, developing genetic maps of *Drosophila* chromosomes.

13.5 Extending Mendel's Rules

Multiple Alleles and Polymorphic Traits

- There can be many **alleles** of a single gene, since minor differences in gene sequence are common. Some genes in the human genome have many alleles, an example being the (-globin gene. This gene encodes a component of hemoglobin (carries oxygen in red blood cells). Some of these allelic differences in gene sequence will affect the function of the encoded protein, generating a **trait or phenotype** specific for that allele. A polymorphic trait is any phenotype with more than two distinct forms.

Incomplete Dominance and Codominance

- **Mutant alleles** of a single gene are not always clearly dominant or recessive to the **wild-type alleles**. Sometimes the heterozygous organisms display an intermediate phenotype, between the wild-type and mutant phenotypes; this is described as **incomplete dominance**.

- A nice example of incomplete dominance is flower color (four-o'clock). When breeding purple by white, the F_1 plants will have lavender flowers, an intermediate trait. Why? Remember that genes encode proteins that perform a cellular function. In this case, the gene involved in flower color encodes an enzyme that synthesizes flower pigment. Plants that lack this enzyme, "pp" genotype, make no pigment. Plants that are heterozygous at this gene, "Pp" genotype, have only one functional copy of the gene and make half the normal amount of enzyme. Thus less pigment is made by the less available enzyme, and the flowers display an intermediate phenotype—lavender. In many allelic relationships, a functional allele can completely compensate for a nonfunctional allele, thus generating the familiar **dominant/recessive allele** terminology.

- A heterozygous organism, displaying the phenotype of both alleles of a single gene, is said to display **codominance**. Neither allele is dominant or recessive to the other. Examples of codominance in humans include ABO blood typing (A and B are alleles of a single gene) and the L gene that encodes a membrane glycoprotein.

Pleiotropy

- A pleiotropic gene infulences many traits, rather than just a single trait. An example is the gene responsible for Marfan syndrome.
  Genes Are Affected by the Physical Environment and Genetic Environment.

- Phenotypes produced by most genes and alleles are strongly affected by the individual's physical environment. Thus, an animal's phenotype is a combination of its genotype and its physical environement.
  Interactions with Other Genes Have a Profound Effect on Phenotpyes.
  Alleles of different genes affect each other in a way that influences the observed phenotype.

- Excellent examples include the genes for coat color in mammals. The *B* and *b* alleles control the production of pigment in hair cells, generating either black or brown hair. Thus *BB* and *Bb* animals have black pigments, and *bb* animals have brown pigments.
- However, the recessive allele of a second gene (*C*) can completely turn off pigment synthesis. So a *BBcc* animal is white (as are *bbcc* and *Bbcc*), illustrating that *cc* is **epistatic** to *BB* and eliminating its impact on phenotype.

Quantitative Traits

- Some traits are influenced by the contribution of multiple genes and display wide variability (e.g., skin color). These traits are referred to as complex due to the complicated pattern of gene and phenotype segregation observed in standard crosses.

CHECK YOUR UNDERSTANDING

(e) Allelic codominance is a genetic relationship that Mendel did not observe, but it was identified later as an exception and extension to common patterns of inheritance. Give an example of allelic codominance.

_____

_____

_____

13.6 Applying Mendel's Rules to Humans

In analyzing human inheritance, pedigrees are often a valuable tool in determining the mode of inheritance for disease-causing gene alleles. A dominant mutation will be passed from affected individuals to their offspring, which will also have the mutation.

Recessive mutations display a more complex pattern of inheritance: affected individuals can be born of phenotypically normal parents who each carry a copy of the disease allele. If two parents are heterozygous for alleles of a gene (*Aa*), what is the likelihood of their having an affected (*aa*) child?

If a recessive mutation is found on the X chromosome, a pedigree will show that most of the affected individuals are male (e.g., hemophilia). This is because males that inherit an X chromosome with that mutation will display the trait. They have only a single X, and thus the recessive gene will generate the phenotype.

B. INTEGRATING YOUR KNOWLEDGE

Looking Back—Key Concepts from Earlier Chapters

Inside the Cell—Chapter 7

As you explore the work of Mendel, review the cell biology behind **phenotype** ("show-type" or form). For example, what aspects of cell biology might result in the loss of specific coloration in a flower? The genes that encode pigment can be **mutated**, and their loss or defect is a very obvious change in flower coloration (abnormal phenotype).

Meiosis—Chapter 12

Without knowledge of the **meiotic process**, it is difficult to follow the key conclusions of Mendel's work. Review meiosis, examining how gene segregation, independent assortment, and gene linkage fit into the production of gametes. All of Mendel's work was really a characterization of chromosomal behavior in meiosis, though meiosis was not discovered until many years later.

Looking Forward—Key Concepts in Later Chapters

How Genes Work—Chapter 15

Mendel established the foundations of gene inheritance without knowing what a gene really was! **Chapter 15** examines the basics of **gene structure and function**, revealing details of these heritable units. Each chapter in Unit 3 builds upon Mendel's work, moving toward the utilization of genes in powerful engineering strategies that have revolutionized biology.

Plant Form and Function—Chapter 35

Mendel was able to control mating in plants by manually fertilizing each cross to ensure that he could identify both parents of each cross. **Chapter 35** offers additional information on **plant structure** and **plant reproduction**. Mendel also examined traits that include flower color, seed shape and color, and even plant height.

C. MASTERING CHALLENGING TOPICS

Linkage Analysis

A prominent experimental goal in genetic analysis is to determine the location of a specific gene—first identifying the chromosome and then the region of that chromosome where the gene is located. Disease-causing genes can sometimes be "mapped" to a specific place by using the medical records of families that carry the disease-causing allele. Finding the gene that causes a specific disease is a difficult task, but the rewards of finding the gene make the incredible effort worthwhile. There are many success stories of scientists identifying disease-causing genes (Huntington's disease, breast cancer). But how does this process work? Scientists use various strategies, one of which is fundamentally described in this chapter—linkage analysis.

Determining genetic linkage involves looking for two genes (or markers) in the genome that do not assort independently of each other. The alleles being studied are often found to segregate together from the parental genome. The easiest way to see linkage is to examine the testcross example. Examine the cross of an *AaBb* organism by an *aabb* organism. What predictions would you make if the genes are unlinked and assort independently of each other? How would linkage between the two genes change the results?

Linkage between *A* and *B* would be revealed by the ratios of progeny produced by this cross. Using these data, scientists can actually determine the distance between two linked genes. As the geneticists gradually refine their assessment of where the gene is located, they can then begin the process of actually finding the gene itself.

D. ASSESSING WHAT YOU'VE LEARNED

1. Draw a diagram of a diploid cell with two sets of chromosomes, placing alleles of two genes on the chromosomes as follows: gene *A* on chromosome 1, and gene *a* on chromosome 1′; gene *B* on chromosome 2, and gene *b* on chromosome 2′. Now diagram the 1*n* products of meiosis in these cells. In what four possible ways will these alleles segregate? *Hint:* The key to this segregation is anaphase 1.

2. If an organism is heterozygous for two traits that are linked, how many different genotypes are possible in the gametes? Assume that no crossing over occurs.
   a. 1
   b. 2
   c. 4
   d. 8

3. If an organism is heterozygous for two traits that are linked, how many genotypes are possible in the gametes produced from a single germ-line cell if even a single crossover event occurs during meiosis I?
   a. 1
   b. 2
   c. 4
   d. 8

4. Pepper color is a trait that is controlled by several genes. The *Y* allele generates red peppers while the recessive *y* allele generates yellow peppers (*yy*). If

a pepper plant that is heterozygous for these alleles (*Yy*) is mated by a genetically similar plant (*Yy*), what percentage of the progeny will be yellow?
   a. 50%
   b. 33%
   c. 25%
   d. 75%

5. A female housefly with red eyes mated with a male housefly with white eyes. Their offspring were:
   a. 90 females with white eyes
   b. 43 males with red eyes
   c. 47 males with white eyes

6. Based on this information, which of the following statements is most accurate?
   a. The gene for red eyes must be dominant.
   b. The gene for red eyes is inherited in a 3:1 ratio.
   c. The gene for red eyes must be on the X chromosome.
   d. The gene for white eyes must be on the Y chromosome.

7. As researchers study genes more closely, they are finding that many genes are pleotropic. This means that:
   a. The wild-type allele displays incomplete dominance over mutant alleles.
   b. The genes encode multiple unique proteins.
   c. The phenotype of mutant alleles depends upon environmental conditions.
   d. Mutations in these genes generate more than one distinct phenotype (trait).

8. When flipping three coins simultaneously, what is the probability of getting all heads?
   a. ½ or 50%
   b. ¼ or 25%
   c. ⅛ or 12.5%
   d. ¾ or 75%

9. A woman is about to deliver triplets. If each child is the product of an independent fertilization event, what is the probability that she will deliver all girls or all boys?
   a. ½ or 50%
   b. ¼ or 25%
   c. ⅛ or 12.5%
   d. ¾ or 75%

10. Using Punnett square analysis, examine reciprocal crosses of fruit flies carrying recessive copies of the *g⁺* gene (located on the X chromosome). First cross a mutant male fly (*gY*) with a wild-type female fly (*g⁺g⁺*). Next cross a wild-type male fly (*g⁺Y*) with a mutant female fly (*gg*). Do these crosses generate different ratios of F₁ progeny? Calculate the ratio/phenotypes.

Cross 1

Cross 2

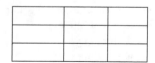

11. Walter Sutton and Theodore Boveri hypothesized that the chromosomes they visualized during meiosis were the hereditary determinants predicted by Gregor Mendel 30 years earlier. They utilized light microscopy to observe the behavior of chromosomes over time as meiosis proceeded. Define Mendel's principle of segregation and independent assortment, and the meiotic processes Sutton and Boveri would use to explain these principles (what did they see?).

12. Define a genetic map. How would a testcross be used to identify the genetic difference between two genes? What is the difference between physical distance and genetic distance?

CHAPTER 13—ANSWER KEY

Check Your Understanding

(a) Laboratory mice are much more difficult to study in large numbers; they reproduce in litters of 5–10. In studying pea plants, Mendel analyzed hundreds to thousands of progeny from each cross and then analyzed the data using statistical analysis. If he was examining only a few progeny, it would have been much harder to establish statistically relevant data.

Peas have many other advantages over mice for Mendel's analysis. Peas are easier to take care of, reproduce more rapidly, and display easily identified phenotypes; further, it is fairly simple to generate true-breeding lines, and they are edible. Mice are important experimental organisms in genetics; but for Mendel's experiments, peas were the better choice.

(b) Determine this answer by making a simple 4 (4 Punnett square like the one in **Figure 13.5**. The genotypes *AABB* and *aabb* each appear in 1/16 of the progeny. For either *AABB* or *aabb*, the frequency is 1/16 + 1/16 = 2/16, or .125 (12.5%).

Also make this determination by calculating individual probabilities of each genotype. Remember that for unlinked genes in a dihybrid cross, an *AaBb* organism generates an equal distribution of the four possible gametes (25% *AB*, 25% *ab*, 25% *Ab*, and 25% *aB*). Probability of generating an *AABB* plant will require the fertilization of an *AB* gamete by another *AB* gamete. These two independent events must occur simultaneously; thus, you will use the **product rule** to calculate probability of this fertilization event.

Probability for *AABB*: (.25)(.25) = .0625 (or 6.25%)

Probability for *aabb*: (.25)(.25) = .0625 (or 6.25%)

Now you can use the **sum rule** to calculate the probability of generating either

AABB or aabb: (.0625) + (.0625) = .125 (or 12.5%).

(c) The testcross allows for careful examination of meiosis in one parent and controls the gametes introduced by the other parent. For example, when crossing an *AaBb* plant by an *aabb* plant, all of the gametes from the second plant will be *ab*. Thus, meiosis in the *AaBb* plant dictates ratios of genotypes and phenotypes in the progeny. The dihybrid cross *AaBb* by *AaBb* generates a 9:3:3:1 ratio of phenotypes. The testcross *AaBb* by *aabb* will generate a 1:1:1:1 ratio of phenotypes.

(d) Reciprocal crosses are two crosses analyzing the same traits, but in each cross the genotype of the parents is reversed. For instance, a *bb* male fruit fly may be crossed by a *BB* female fruit fly. Then the reciprocal cross will be a *BB* male by a *bb* female. These crosses are particularly valuable in identifying sexlinked genes, or genes located on sex-determining chromosomes (X or Y). If reciprocal crosses generate different ratios of progeny genotypes, the gene involved is likely located on a sex chromosome. Review the story of the white-eyed male fruit fly (**Figures 13.11** and **13.13**).

(e) The ABO blood groups in humans illustrate allelic codominance. *A*, *B*, and *O* refer to phenotypes possible from the allelic composition of a single gene (*I*). Three alleles of this gene?I^A, I^B, and *I*—control the composition of cell-surface molecules on red blood cells. If you are genotypically *ii*, there are none of these cell surface molecules; if you are $I^A I^A$, you have the *A* cell-surface marker; but if you are $I^A I^B$, you have both *A* and *B* on the cell surface. These two alleles are codominant relative to each other; the heterozygotes display both phenotypes.

Assessing What You've Learned

1. Four haploid gametes are possible (*AB, Ab, aB,* and *ab*). If the chromosomes assort independently, each of these gametes will be produced in equal proportions.

2. b; 3. c; 4. a; 5. c; 6. d; 7. c; 8. b

Cross 1

	g	Y
g^+	g^+g	g^+Y
g^+	g^+g	g^+Y

Because g^+ is dominant to g, all of the flies will display a normal phenotype.

Cross 2

	g	Y
g	g^+g	gY
g	g^+g	gY

All male flies display the mutant phenotype. All female flies are wild type. The reciprocal crosses have different results! The mutant phenotype is associated with males only in the second cross. This gene must be located on the X chromosome.

10. The principle of segregation states that plants have two determinants (genes) that control each trait, and these determinants segregate into separate gametes (meiosis). The principle of independent assortment states that determinants controlling different traits (separate genes) are transmitted to gametes independently of one another, and they assort independently. Sutton and Boveri noticed darkly staining structures that appeared to be copied at the start of mitosis and meiosis. Then, during meiosis they would pair with a similarly sized structure at the early stages of meiosis and ultimately segregate into separate gametes. The structures were chromosomes, and this behavior was consistent with the principle of segregation. In addition, they noticed multiple pairs of chromosomes, each segregating independently as predicted by the principle of independent assortment.

11. A genetic map is a linear diagram of genes and their ordering on a chromosome. This can be determined using genetic crosses. A testcross is especially helpful in genetic mapping as meiotic events in *one* parent are clearly displayed in the progeny generated from that cross. Physical distances on chromosomes are the actual distances between genes, determined from DNA sequencing analyses.

14

DNA and the Gene: Synthesis and Repair

A. KEY CONCEPTS

14.1 DNA as the Hereditary Material

Transformation Experiments

- In 1928 Fredrick Griffith utilized the bacterium *Streptococcus pneumoniae* to evaluate its role in human infection and disease. Griffith knew there were **virulent** strains of *S. pneumoniae* that could cause death in the infected individual, and **nonvirulent** strains that were harmless as pathogens.

- What was different about these strains of the microbe? As Griffith examined these strains, he stumbled upon the capacity of one bacterium to alter the other, presumably through the transfer of an information-bearing molecule.

- **Figure 14.1**; the virulent *S. pneumoniae* can be distinguished from the nonvirulent strains simply by growing the cells in a petri dish (virulent = colonies that appear smooth; nonvirulent = rough or bumpy colonies).

- Griffith utilized mice to assay for the virulence of individual samples, injecting the samples into healthy animals and evaluating their tolerance of each sample.

- The discovery that samples containing heat-killed virulent bacteria could somehow transform nonvirulent cells to virulent was exciting (**Figure 14.2**). Somewhere within the heat-killed sample were molecules that carried the information for becoming virulent, and the nonvirulent cells were likely taking up those molecules.

- What experiments might you suggest to extend these results in identifying the information-carrying molecule that was transferred between strains? Be imaginative, but be sure to include controls as part of the analysis.

Is DNA the Genetic Material?

The Avery et al. Experiment

- In the 1940s, researchers followed up on Griffith's experiment by selectively removing macromolecules from the sample of heat-killed, smooth *S. pneumoniae* (**Figure 14.3**). The researchers used enzymes that selectively degrade DNA, RNA, and protein, removing other macromolecules (lipids, carbohydrates) with an extraction step.

- Notice that only the samples treated with a **DNase** (an enzyme that degrades DNA) lost their ability to transform the nonvirulent cells to virulent.

- These experiments strongly supported the concept of DNA being information carrying, altering the virulence of the exposed cells.

The Hershey-Chase Experiment

- A second key set of experiments used bacteriophage to determine if DNA was the central genetic material. Bacteriophages are made of a protein coat, with DNA inside the phage head structure. This virus attaches to the bacterium and *transfers material into the bacterial cell*, forcing the bacterium to synthesize hundreds of new viral particles (**Figure 14.4**). The material carries information, but is the transferred material made of protein or DNA?

- As illustrated in **Figure 14.5**, viral proteins were radioactively labeled (^{35}S) and viral DNA labeled (^{32}P). Separate bacterial cultures were infected with either type of radioactively labeled bacteriophage, and the bacteria were then separated from any bacteriophage present in the culture medium. The amount of radioactivity of the bacterial cells was then measured to determine if either the radioactive DNA or protein had been transferred into the bacterial cells. The scientists found that only radioactive bacteria were those that had been infected by a bacteriophage with radioactive DNA. Thus the transfer of DNA into the bacterium mediated the viral infection.

- This experiment strongly supported the hypothesis that DNA was the transforming factor and that DNA carried information on its structure. But questions remained. What type of code could be contained on such a simple molecular structure? How could this information-containing molecule be faithfully copied during cell division?

Experiment

Question: Do viral genes consist of DNA or protein?

DNA hypothesis: Viral genes consist of DNA.

Protein hypothesis: Viral genes consist of protein.

Experimental setup:

* Viral DNA is radioactive.

* Viral protein is radioactive.

E. coli

E. coli

Protein coats outside

Genes inside

Protein coats in solution

Viral genes within bacterial cells in pellet

1. Grow some viruses in presence of ^{32}P (P is in DNA but not in viral protein) and some viruses in presence of ^{35}S (S is in protein but not in DNA).

2. Allow viruses with labeled DNA to infect one culture of E. coli cells and viruses with labeled protein to infect another.

3. Agitate cultures in kitchen blender to separate empty viral protein coats from bacterial cells in each culture.

4. Centrifuge solutions of bacterial cells from each culture to force cells into a pellet. Record location of radioactive labels.

Prediction of DNA hypothesis: Radioactive DNA will be located within pellet.

Prediction of protein hypothesis: Radioactive protein will be located within pellet.

Results:

Radioactive DNA is in pellet

Radioactive protein is in solution

DNA

Protein

Conclusion: Viral genes consist of DNA. Viral coats consist of protein.

14.2 Testing Early Hypotheses about DNA Synthesis

- To better understand this chapter, you should review DNA structure, described in **Chapter 4**. The double helix and the nature of this macromolecular structure are vital review topics.

- Compare the three mechanisms of DNA replication that were proposed following Watson and Crick's discovery of the double-helix structure (**Figure 14.8**).

    1. **Semiconservative replication**—The double helix separates, and each old strand is copied to generate two new chromosomes. Thus *each new chromosome is composed of one strand of old DNA and one strand of newly synthesized DNA.*

    2. **Conservative replication**—During replication the original chromosome is copied but remains unchanged. Both of the original strands would remain

at the end of this replication, and the *new chromosome will be completely new.*

3. **Dispersive replication**—The replication process generates two new chromosomes, with *new and old sections of DNA mixed together randomly.* This model is complex and would require a great deal of cutting and splicing of DNA strands.

The Meselson–Stahl Experiment

- *Escherichia coli* is a prokaryotic bacterium that divides rapidly, is easy to culture, and is readily available. Many aspects of basic cellular function are shared among all living organisms; and though eukaryotic organisms are generally more complex, the underlying processes are similar. *E. coli* was an excellent **model organism** for early DNA replication studies.

- To allow for experimental separation of newly synthesized DNA from the original template DNA, Matthew Meselson and Frank Stahl grew *E. coli* bacteria in the presence of **"heavy" nitrogen** (^{15}N rather than the normal ^{14}N). Is nitrogen a part of the DNA structure? Where?

- After many generations of growing bacteria in the presence of ^{15}N-containing media, they changed the bacteria to normal ^{14}N media.

- How would this experimental design label DNA differently in each of the three models? Would each generate a unique result? How might you detect this difference?

- Meselson and Stahl separated the DNA by *density* using high-speed centrifugation and found that the semiconservative mode of DNA replication was indicated by their data. Review the data, remembering that the more dense the DNA, the further the band will travel downward in the centrifuge tube (**Figure 14.8**). The semiconservative replication model predicted that after one cell division in ^{14}N-containing media, all of the DNA would be of intermediate density—having one complete strand of DNA with ^{15}N and a new strand with ^{14}N. After a second cell division, half of the DNA would be of intermediate density and the other half would be lower density (all ^{14}N).

14.3 A Comprehensive Model for DNA Synthesis

- DNA is synthesized from nucleosides (dATP, dGTP, dCTP, and dTTP). The three components of each are a nitrogenous base, ribose, and a triphosphate.

- Once researchers were able to isolate **DNA polymerase I** from *E. coli*, enzymatic activity detected in an assay was directed by a **DNA template** strand, and the new DNA was an exact copy. Subsequent studies

(a) The Meselson and Stahl experiment focused on discerning between competing models of DNA replication. Describe the results the researchers would have generated if DNA replication proceeded by conservative replication.

_____

_____

_____

showed that another DNA polymerase, called **DNA polymerase III**, performed the bulk of DNA replication during bacterial cell division while DNA polymerase I performed a key role in the repair of mutation.

How Does Replication Get Started?

- DNA polymerase III catalyzes DNA synthesis in the 5′ to 3′ direction. This makes more sense when you examine the structure of a nucleotide. That is to say, new nucleotides are added to the 3-carbon hydroxyl (–OH) on ribose of an existing DNA strand.

- On a circular prokaryotic chromosome, DNA replication starts at specific sites and proceeds in both directions from the starting point. This **bidirectional DNA replication** proceeds in each direction with the advancement of a Y-shaped **replication fork** (**Figure 14.11**).

How Is the Helix Opened and Stabilized?

- The fork originates with the action of **DNA helicase**, an enzyme that opens the double helix to allow enzymes to attach to each strand. Small proteins known as single-strand DNA-binding proteins (SSBPs) bind the single-stranded DNA and keep it from reforming double-stranded DNA.

- DNA polymerization requires a primer (a 3′ –OH) on which to add nucleotides. The **primase** enzyme makes short RNA primers to start the DNA synthesis. This RNA is later replaced with DNA.

How Is the Lagging Strand Synthesized?

- How can this replication work if the DNA polymerase works only in the 5′ to 3′ direction? Examine **Figure 14.12**; it displays the continuous **leading strand** and the discontinuous **lagging strand** found at a replication fork. DNA polymerase III enzymes are associated with each replication fork, mediating the leading/lagging strand synthesis reactions.

(b) The synthesis of a DNA strand is an enzymatic process that proceeds quickly. Rapid enzymatic reactions are energetically favorable. Why is DNA synthesis an energetically favorable process?

_____

_____

_____

- Reiji Okazaki discovered the mechanism of lagging-strand synthesis. The short fragments generated at the site of lagging-strand synthesis are known as **Okazaki fragments**. These fragments are eventually connected to one another via the activity of DNA polymerase I and DNA ligase.

- Review the stepwise examination of DNA polymerization offered in **Figure 14.13**. Each of the key steps is presented in proper chronology and should make intuitive sense relative to the process we've introduced up to this point. Remember that this is a summary; DNA replication is a complex process requiring many proteins (some of which have yet to be identified). **Table 14.1** offers a valuable summary of the proteins involved in the process of DNA replication.

(c) Explain the need for RNA in the process of DNA replication.

_____

_____

(d) Define the differences in functional roles for DNA polymerase I and DNA polymerase III during DNA replication.

_____

_____

_____

14.4 Replicating the Ends of Linear Chromosomes

- The ends of *linear* chromosomes are called **telomeres**, and they contain many repeats of a six-base sequence TTAGGG (repeated ~1000 times).

- As the replication fork nears the end of a linear chromosome (like those found in human cells), the lagging strand makes an RNA primer for the end of the

REVIEW TABLE 14.1 **Proteins Involved in DNA Synthesis**

Name	Structure	Function
Opening the helix		
Helicase		Catalyzes the breaking of hydrogen bonds between base pairs and the opening of the double helix
Single-strand DNA-binding proteins		Stabilize single-stranded DNA
Topoisomerase		Breaks and rejoins the DNA double helix to relieve twisting forces caused by the opening of the helix
Leading strand		
Primase		Catalyzes the synthesis of the RNA primer
DNA polymerase III		Extends the leading strand
Sliding clamp		Holds DNA polymerase in place during strand extension
Lagging strand		
Primase		Catalyzes the synthesis of the RNA primer on an Okazaki fragment
DNA polymerase III		Extends an Okazaki fragment
Sliding clamp		Holds DNA polymerase in place during strand extension
DNA polymerase I		Removes the RNA primer and replaces it with DNA
DNA ligase		Catalyzes the joining of Okazaki fragments into a continuous strand

chromosome. But there is no way to replace this RNA with DNA, because there is no available primer for DNA synthesis (review **Figure 14.14**, step 4). The RNA is removed, leaving a section of single-stranded DNA (leading strand) at one end of each new chromosome.

- In most dividing cells, this residual single-stranded DNA is degraded and the telomere shortens by that length with each cell division. Telomere length then serves as a record of cell division and limits

the number of times a somatic cell can divide. But some cells, like germ cells, replicate the telomere completely.

- **Telomerase** is an enzyme that mediates the replication of telomeres, solving the problem of the single-stranded DNA described above. As is illustrated in **Figure 14.15**, telomerase adds more six-base repeats to the end of the leading strand. Then primase makes an RNA primer, and DNA polymerase uses the primer to synthesize the lagging-strand sequence.

14.5 Repairing Mistakes and Damage

How Does DNA Polymerase Proofread?

- During DNA synthesis, nucleotides are added to the new strand via their ability to form hydrogen bonds with the template nitrogenous base. Thus, using H-bonding interactions, an A in the template will direct the addition of a T to the new strand.

- To find the proofreading enzyme, researchers screened for mutant cells that made lots of mistakes during DNA replication.

- *E. coli* bacteria carrying a mutation in the epsilon (ε) subunit of *pol III* were found to make many errors during DNA replication. This subunit mediates **proofreading** during DNA replication by cutting out any nucleotides that were added incorrectly. The enzymatic removal of these nucleotides is termed **exonuclease** activity and acts in the 3' to 5' direction (**Figure 14.16**).

- DNA polymerase III makes one mistake every 100,000 nucleotides, and with proofreading activity, the error rate is reduces to 1 mistake every 10 million nucleotides (1×10^{-7}).

- If the *pol III* enzyme leaves a mistake behind, the cell has other DNA repair enzymes available to it. *MutS* is one example of a DNA repair enzyme; it identifies **mismatched nitrogenous bases** and replaces the incorrect nucleotide.

Nucleotide Excision Repair

- Mutagenic compounds generated by cellular metabolism (free radical compounds) can damage nitrogenous bases in DNA. Also, we are exposed to mutagenic compounds in our environment: ultraviolet light, poisons we inhale in smoke, and toxins found in mold-contaminated foods (aflatoxins). These compounds can change the chemical structure of nitrogenous bases, making it impossible to properly replicate this DNA.

- Ultraviolet light causes adjacent thymine bases to covalently bond with each other. These mutations are called **thymine dimers** or **cyclobutane pyrimidine dimers** (**Figure 14.17**).

Xeroderma Pigmentosum: A Case Study

- **Xeroderma pigmentosum (XP)** is an inherited disease in humans; it results in sensitivity to UV light, X-rays, and other DNA-damaging treatments. People affected by this disease rapidly develop skin lesions following exposure to sunlight.

- This disease is caused by mutation of one of several **excision repair** enzymes utilized by humans. Diminished excision repair enzymes result in a deficient cellular capacity to repair damaged DNA. As illustrated in **Figure 14.18**, excision repair involves the removal of a mismatched nucleotide and some adjacent nucleotides, followed by the addition of new, correctly matched nucleotides. This process requires three enzymatic activities; exonuclease, DNA polymerase, and DNA ligase.

- See **Figure 14.19** for an illustration of the structure of two key experiments characterizing DNA repair in cells isolated from XP patients. These studies simply examine the effect of UV light upon skin cells isolated from unaffected individuals and from XP sufferers. UV exposure was much more toxic to the XP cells.

B.   INTEGRATING YOUR KNOWLEDGE

Looking Back—Key Concepts from Earlier Chapters

Nucleic Acids and the RNA World—Chapter 4

The **double helix**, **antiparallel strands**, the 5′ and 3′ **ends** of each strand, and the role of **nitrogenous bases** in DNA structure—these are key concepts in Chapter 14, and their introduction in **Chapter 4** should be reviewed carefully as you begin to study DNA synthesis.

The Cell Cycle—Chapter 11

Finally, in Chapter 14 we begin to truly define the process of **DNA replication** that precedes cell division. This chapter fills in some of the holes from **Chapter 11**. Replication of a cell's genome is no small matter, and performing this task carefully is crucial to the successful transfer of information to the daughter cells.

Looking Forward—Key Concepts in Later Chapters

Bacteria and Archaea—Chapter 28

Much of the early work in DNA synthesis and isolation of enzymes in chromosome replication was first identified in **bacteria**. These cells are simple compared to eukaryotic cells, but the underlying mechanisms of cell function are shared. **Chapter 27** offers a critical introduction to bacteria and **prokaryotic cell biology**.

C.   MASTERING CHALLENGING CONCEPTS

One common theme in this chapter is the utilization of mutants to isolate and characterize the mutated gene. Scientists interested in a key cellular function can sometimes predict what a cell lacking that capacity will display as a mutant phenotype. One example from the chapter is the examination of DNA repair enzymes by isolating bacteria that made lots of mistakes during DNA replication. If the mutant phenotype is expected to be lethality, or a dramatic loss of cell division capacity, isolation of these mutants becomes difficult. Then temperature-sensitive mutations might be a better choice, because cells harboring this type of mutation may grow normally at the permissive temperature and display the mutant phenotype only at the restrictive temperature.

Humans afflicted with mutations that weaken the DNA repair mechanisms can display many phenotypes, including light sensitivity, brittle hair and nails, and skin that is easily sloughed off. Classic examples of this type of mutation are **Xeroderma pigmentosum**, **Cockayne syndrome**, and **trichothiodystrophy (TTD)**. Recently, Vermeulen and colleagues (2001) identified a group of people who suffer the symptoms of TTD whenever they run high fevers.* Upon carefully examining a specific DNA repair gene in these people, the researchers discovered a mutation that results in a **temperature-sensitive** phenotype. This means the mutant enzyme works at normal temperatures, but at slightly elevated temperatures it is unable to function because of its mutation. Wow! It is easy to think of bacteria or yeast as a cell type that might offer temperature-sensitive mutations to researchers, but this research actually identified them in humans!

D.   ASSESSING WHAT YOU'VE LEARNED

1. Which of the following best describes the structure of a DNA molecule?
   a. a deoxynucleoside triphosphate with an adenine, thymine, guanine, or cytosine base attached to it
   b. a sugar-phosphate backbone with adenine, thymine, guanine, and cytosine bases projecting toward the inside of the backbone

---

*W. Vermeulen, S. Rademakers, N. G. Jaspers, E. Appeldoorn, A. Raames, B. Klein, et al. (2001). A temperature-sensitive disorder in basal transcription and DNA repair in humans. *Nature Genetics* 27(3), 299–303.

c. a sugar-phosphate backbone with adenine, thymine, guanine, or cytosine bases projecting toward the outside of the backbone

d. a sugar-phosphate backbone made up of adenine, thymine, guanine, or cytosine bases hydrogen bonded to each other

2. The following DNA sequence occurs at the start of a DNA strand. Which of the following sequences would most likely bind to this sequence to initiate DNA replication?
   a. 3′–AATTGCAGATTCA–5′
   b. 5′–TTAACGTCTAAGT–3′
   c. 3′–TTAACGTCTAAGT–5′
   d. 3′–UUAACGUCUAAGU–5′
   e. 5′–UUAACGUCUAAGU–3′

3. Which of the following represents the correct order of events as they occur in the process of DNA replication?
   a. Helicase opens helix, *pol III* synthesizes RNA primer of leading strand, *pol I* excises primer and fills gap.
   b. Primase synthesizes primer, *pol I* excises primer, *pol III* elongates primer and synthesizes leading and lagging strands.
   c. Helicase opens helix, *pol III* elongates primer and synthesizes leading and lagging strands, DNA ligase links fragments of DNA.
   d. Primase synthesizes RNA primer, *pol III* synthesizes lagging and leading strands, helicase opens helix.

4. The classic experiment performed by Alfred Hershey and Martha Chase revealed:
   a. RNA was not the hereditary material.
   b. Microbes could exchange genetic information.
   c. In a viral infection, protein was transferred into the infected cell.
   d. In a viral infection, DNA was transferred into the infected cell.

5. Semiconservative DNA replication involves:
   a. the separation and cutting of template strands; these sections are then copied and recombined to make the new DNA strands
   b. the outward rotation of nitrogenous bases in a double helix, and the generation of completely new double-stranded DNA copy
   c. the separation of strands of DNA, and the synthesis of a new complementary strand on each old strand
   d. synthesis of an RNA copy of the DNA, and the generation of a new DNA strand using reverse transcriptase

6. Cancer can occur when DNA is damaged. Which of the following statements is true regarding the cause of damage to DNA molecules?
   a. The structure of DNA can be damaged only by environmental toxins such as cigarette smoke, aflatoxins, and other toxins in food.
   b. Excision repair enzymes are able to prevent all mutations in DNA.
   c. Small quantities of ultraviolet radiation have no effect on DNA structure.
   d. DNA molecules can break down in the absence of environmental toxins or radiation.

7. Imagine that you are a researcher working on developing a gene therapy for XP patients. Which of the following would be the best "gene" to incorporate into the skin cells of these patients, based on the data shown in Figure 14.19?
   a. a gene for the production of "normal" DNA strands
   b. a gene that would increase the production of DNA polymerase III
   c. a gene that would increase the production of DNA polymerase I
   d. a gene that would increase the production of thymidine in damaged cells

8. Once DNA strands are separated by DNA helicase, _____ block the reformation of a double helix by the template strands.
   a. RNA primer strands
   b. topoisomerase enzymes
   c. single-stranded DNA-binding proteins
   d. RNA polymerase complexes

9. James Cleaver showed diminished DNA repair activity in XP cells by:
   a. isolating DNA polymerase I from these cells
   b. culturing the cells in radiolabeled deoxyribonucleotide
   c. staining repaired DNA with a fluorescent dye
   d. examining damaged DNA with a microscope

10. Humans with a defect in the *mutS* gene display a defect in mismatch repair and develop:
    a. Alzheimer's disease
    b. hereditary colorectal cancer
    c. skin cancer
    d. limb deformities

CHAPTER 14—ANSWER KEY

Check Your Understanding

(a) Labeling over many generations with ^{15}N would generate heavy DNA throughout the bacterial culture.

Transfer of these cells to ^{14}N media would result in light DNA with all new DNA synthesis. If conservative replication occurred, then density gradient centrifugation would generate only heavy and light DNA bands; no intermediate bands would be observed. Review **Figure 14.8** for an illustration of this process.

(b) DNA strands are synthesized from nucleoside triphosphate molecules (dATP, dCTP, dGTP, and dTTP). The triphosphate group on these molecules contains covalent bonds that "release" energy when they are broken. Thus the breaking of the phosphoanhydride bond in a dNTP molecule during DNA synthesis makes this process energetically favorable.

(c) DNA polymerases are unable to synthesize DNA from a template strand unless there is an available primer on which to add new nucleotides. RNA polymerases are able to synthesize short RNA primers de novo, and they are utilized during DNA replication to prime new DNA synthesis. The RNA primers are later destroyed and replaced with the correct DNA sequence.

(d) DNA polymerase III performs the majority of template DNA copying during replication. It is a key enzymatic component of the replication fork. DNA polymerase I plays a role in filling in the gaps that are left between Okazaki fragments (and in DNA repair following mutation).

Assessing What You've Learned

1. b; 2. d; 3. c; 4. d; 5. c; 6. d; 7. c; 8. c: 9. b; 10. b

15

How Genes Work

A. KEY CONCEPTS

15.1 What Do Genes Do?

- Archibald Garrod was a physician who first began to think of inherited mutations and how they might influence cellular biochemistry (or **metabolic** processes). Garrod noticed a mutant phenotype that was prominent within a family, and he suspected that a mutation was passed from parent to child in the family.

- Garrod called the abnormality an **inborn error of metabolism**, and he suspected that the phenotype was the result of abnormal metabolic function. The disease, called **alkaptonuria**, caused affected individuals to display (among other things) black urine.

- How might one defective gene/protein generate abnormal metabolism? Think of metabolic processes as an assembly line, with stages along the line contributing consecutive pieces of the final product (**Figure 15.1**). Individual proteins (enzymes) perform a task at each stage, altering the compound to ultimately generate the final product. The loss of one enzyme results in the accumulation of unfinished product, all stuck at the same stage of processing. This buildup of an intermediate is exactly what happens in alkaptonuria; the accumulated intermediate (homogentisic acid) is disposed of in the urine, turning it black.

- Mutant alleles that generate a nonfunctional protein product are called **null mutations**, **knock-out mutations**, or **loss-of-function mutations**.

- In 1941, George Beadle and Edward Tatum announced discoveries from their study of mutant cells of the bread mold, *Neurospora crassa*. They utilized biochemistry and genetic analysis to characterize the steps involved in biosynthesis of various cellular compounds, including the vitamin pyridoxine.

- If you recall the assembly-line example of metabolic pathways described earlier, you'll see that each step in the pathway modifies the substrate molecule, eventually generating the product of the pathway (pyridoxine). These types of experiments led to the **one-gene, one-enzyme hypothesis**. A mutation in one gene generates one abnormal protein (or enzyme).

- Another team of researchers took these experiments further. Adrian Srb and Norman Horowitz used *N. crassa* to study the arginine biosynthetic pathway.

- Arginine is an amino acid that cells can make for themselves if it isn't in the growth medium. They isolated mutant cell lines that were unable to grow without arginine supplementation in the growth medium, indicating that the biosynthetic pathway was not functioning properly.

- *The arginine-dependent cell lines must carry mutations in genes that encode enzymes in arginine biosynthesis.*

- Srb and Horowitz knew some of the details about arginine synthesis, and they knew the identity of some of the intermediates (citrulline, ornithine). They added these intermediate forms to the growth medium.

- Why? If a single stage in the assembly line is blocked and you introduce an intermediate that occurs downstream from the blocked step, the intermediate will be successfully converted into product (arginine), allowing the cells to grow normally (review Figure 15.1).

- Using this strategy, Srb and Horowitz took individual mutant cell lines and identified where the mutation blocked the pathway. Did it work? Yes! Review **Figure 15.2** to examine their experiment more closely.

CHECK YOUR UNDERSTANDING

(a) How can a mutation in a gene destroy the activity of the encoded protein (or enzyme)?

_____

_____

_____

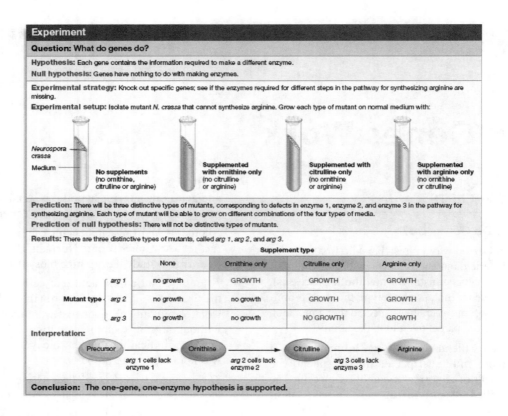

Experiment

Question: What do genes do?

Hypothesis: Each gene contains the information required to make a different enzyme.

Null hypothesis: Genes have nothing to do with making enzymes.

Experimental strategy: Knock out specific genes; see if the enzymes required for different steps in the pathway for synthesizing arginine are missing.

Experimental setup: Isolate mutant *N. crassa* that cannot synthesize arginine. Grow each type of mutant on normal medium with:

Neurospora crassa

Medium

No supplements (no ornithine, citrulline or arginine)

Supplemented with ornithine only (no citrulline or arginine)

Supplemented with citrulline only (no ornithine or arginine)

Supplemented with arginine only (no ornithine or citrulline)

Prediction: There will be three distinctive types of mutants, corresponding to defects in enzyme 1, enzyme 2, and enzyme 3 in the pathway for synthesizing arginine. Each type of mutant will be able to grow on different combinations of the four types of media.

Prediction of null hypothesis: There will not be distinctive types of mutants.

Results: There are three distinctive types of mutants, called *arg 1*, *arg 2*, and *arg 3*.

		Supplement type		
		None	Ornithine only	Citrulline only	Arginine only
Mutant type	*arg 1*	no growth	GROWTH	GROWTH	GROWTH
	arg 2	no growth	no growth	GROWTH	GROWTH
	arg 3	no growth	no growth	NO GROWTH	GROWTH

Interpretation:

Precursor → Ornithine → Citrulline → Arginine

arg 1 cells lack enzyme 1

arg 2 cells lack enzyme 2

arg 3 cells lack enzyme 3

Conclusion: The one-gene, one-enzyme hypothesis is supported.

CHECK YOUR UNDERSTANDING

(b) To initiate a "mutant hunt," the researcher must first isolate a large group of mutant populations of cells and then utilize them in further experimentation. Describe the process of isolating arg⁻ populations of *Neurospora*.

_____

_____

_____

_____

_____

15.2 The Central Dogma of Molecular Biology

- If one gene = one enzyme, then how is the protein encoded within the structure of DNA? DNA is double stranded, with antiparallel strands and the **nitrogenous bases** adenine (A), cytosine (C), thymine (T), and guanine (G). Francis Crick proposed that the sequence of nitrogenous bases in the DNA

strand was a kind of Morse code, defining the protein structure with a linear grouping of the nitrogenous bases (i.e., AGT = serine). His predictions were fundamentally correct.

RNA as the Intermediary between Genes and Proteins

- RNA is single stranded, and it is composed of A, U, G, and C (U stands for uracil; there is no thymine in RNA). In addition, the ribose found in RNA has a hydroxyl (–OH) on carbon 2 (DNA lacks this hydroxyl). Francois Jacob and Jacques Monod first proposed that RNA might serve as an intermediary (messenger) between DNA and protein synthesis machinery (**Figure 15.3**).

- **Ribosomes** are small structures found in the cytosol and associated with the **endoplasmic reticulum (ER)**. They synthesize proteins within the cell by reading the sequence of **messenger RNA (mRNA)**.

- Jerard Hurwitz and J. J. Furth isolated **RNA polymerase**, the enzyme that synthesizes mRNA by reading DNA sequence. The researchers examined cellular fractions for RNA synthesizing capacity. By adding the needed reagents for RNA polymerase activity (template DNA, ribonucleotides) they could measure RNA synthesis in each fraction. Newly synthesized RNA would be radioactive through incorporation of radioactive ribonucleotide triphosphates (not deoxyribonucleotides).

- How does RNA polymerase read the DNA to make the mRNA? In the coding sequence, the RNA polymerase will read a C and add the complementary G to the RNA strand (**Figure 15.5**). One unusual exception is that when the polymerase reads an A in the DNA, a U is added to the RNA (instead of thymine).

The Central Dogma

- **The central dogma of molecular biology defines information flow within the cell. This was first articulated by Francis Crick.**

$$DNA \rightarrow RNA \rightarrow protein$$

- DNA is a long-term form of information storage, allowing for the stable maintenance of the information and its passage from generation to generation.

- RNA carries the information from the DNA to the translation machinery (the ribosome). In eukaryotic cells, the DNA and the ribosomes are located in separate cellular compartments.

- RNA is unstable within the cell, lasting for only minutes to hours before it is destroyed.

- The proteins that are synthesized by the ribosomes carry out cellular functions.

- DNA is transcribed into RNA, and RNA is translated into protein.

- Not all RNA is translated; some RNA is functional within the cell. Ribosomes are made up of rRNA bound together with protein subunits, and transfer RNAs (tRNAs) facilitate translation. Functional RNAs are involved in (1) regulation of gene expression, (2) mRNA processing, (3) carrying amino acids to the ribosomes—tRNAs, and (4) ribosome function—rRNAs.

15.3 The Genetic Code

How Long Is a Word in the Genetic Code?

- George Gamow predicted that each code word in the genetic code would be three nucleotides long (**3 nucleotides = 1 codon**). How could he make such a prediction? *Key*: Twenty amino acids are used by ribosomes to synthesize proteins.

- Francis Crick and Sydney Brenner experimentally revealed that the code is indeed based on three nucleotides for each amino acid. Crick made this discovery by making **deletion/insertion mutations** in viral DNA. He found that the **reading frame** of a gene could be destroyed by mutation, but it could then be restored if the total number of deletions or additions were multiples of three. Review *the fat cat*

ate the rat as an illustration of reading frame (also, **Figure 15.7**).

How Did Researchers Crack the Code?

- Identifying the three-nucleotide sequence coding for each amino acid was a difficult proposition. Eventually the genetic code was deciphered, and it is shown in **Figure 15.8**. Each of 20 amino acids used by ribosomes to synthesize proteins have multiple codons that designate their addition to a protein.

- In the 1960s, Marshall Niernberg and Heinrich Matthaei were able to synthesize RNA and performed in vitro (in the test tube) translation experiments. Having control over the content of the RNA, they could then identify which amino acids were attached to each other during translation.

- They found that a polymer of U residues in an RNA strand (5′–UUUUUUUUUUUUUU–3′) resulted in the translation of that sequence into a polymer of phenylalanine. Polymers of A (5′–AAAAAAAAAA–3′) were translated to polymers of lysine, polymers of C were translated to polymers of proline, and polymers of G were translated to polymers of glycine.

- Thus, if RNA is synthesized as the complementing sequence of the gene encoding DNA, then the codon for phenylalanine must be **AAA**. A piece of the code is revealed! This type of strategy revealed a great deal of the genetic code.

- Nirenberg worked with Phil Leder to characterize more complex codons. They synthesized RNAs with the three-letter codon of interest and mixed ribosomes and tRNAs with amino acids attached (charged tRNAs). They then determined which amino acid was bound to the ribosome in response to a particular codon.

- There is only one **start codon** (AUG), which signifies the start of the protein-encoding sequence in an mRNA.

- There are three **stop codons** in the genetic code (UGA, UAA, and UAG).

Exceptions to the Central Dogma (Essay)

- The tobacco mosaic virus (TMV) contains no DNA; it is an RNA virus. Upon entering the cell, the viral RNA is directly translated into viral proteins.

- Some RNA viruses enter the cell and are reverse transcribed to DNA as the first step of viral infection. The process of converting RNA to DNA is called **reverse transcription**, and is mediated by an enzyme known as **reverse transcriptase**. This process is discussed in more detail in later chapters.

CHECK YOUR UNDERSTANDING

(c) George Gamow predicted that the genetic code would contain "words" that were three nucleotides in length. His predictions were based on mathematical considerations rather than experimental data. On what information were his predictions based? Was he correct?

_____

_____

_____

CHECK YOUR UNDERSTANDING

(d In cracking the genetic code, Marshall Nirenberg and Heinrich Matthaei performed an ingenious yet laborious series of experiments. Which two key experimental designs did they use to decipher individual codons?

_____

_____

_____

_____

B. INTEGRATING YOUR KNOWLEDGE

*Looking Back—Key Concepts
from Earlier Chapters*

Nucleic Acids and the RNA World—Chapter 4

RNA is more than just a messenger. The capacity of RNA to aid in **protein synthesis**, to have **catalytic activity** (it can cut RNA!), and to carry **complex messages** from the nucleus to the cytosol is truly remarkable. Review this introductory material on RNA to enhance your study of Chapter 15.

Review the structure of **double-stranded DNA**; it will be important to your study of gene structure in this chapter.

Mendel and the Gene—Chapter 13

Phenotype and the influence of mutation on physical characteristics of organisms were crucial to Mendel's work. With an understanding of genes and the central

dogma, we can now better understand what phenotype really entails. A **mutant gene** produces a **mutant protein** (or no protein at all), and the loss of this protein's function generates an abnormal phenotype. The mutant gene really is just an incorrect blueprint for a protein.

*Looking Forward—Key Concepts
in Later Chapters*

Transcription and Translation—Chapter 16

In **Chapter 16**, we will examine the process of **protein synthesis** and learn how the information carried on a messenger RNA (mRNA) is **translated** into a functional protein.

**Control of Gene Expression and Cell
Differentiation—Chapters 17, 18, and 21**

If the **human genome** is composed of 30,000 genes, and every cell in your body contains each of these genes, how are these utilized differently by different cells? Each cell type in the human body acts differently due to **differential gene expression**. The genes expressed in a cell dictate the shape, growth, and function of that cell. Regulating gene expression is central to how we utilize the information in the genome.

C. MASTERING CHALLENGING TOPICS

This chapter contains some complex experimental data, generated by laboratory strategies that are new to the reader. Anticipate some confusion when first reading through these experiments; but realize that you can make sense of these experiments with a little extra review of the material. The characterization of the arginine biosynthetic pathway offers a good example. This experiment utilizes *Neurospora crassa* (a bread mold), a mutational screen, and finally a carefully designed series of phenotypic analyses. What is the relevance of each step in the experiment? Should you ignore these "minor details" and move on to memorizing the conclusions? No! The experimental strategy is critical to this discussion; work through the details one at a time to ensure that you understand why this experiment worked and what it was designed to accomplish. Read the text, the figure legends, and the study guide closely to ensure that you've thoroughly examined the experiment. Also review background material referenced in previous chapters. The relevance of each of the listed terms is described. *Scientists explore biology through careful experimentation. Recognizing the "how" of important discoveries is critical to your development as a scientist.* Don't give up on the experiments included in the text. They are included expressly to teach each student more than just a set of conclusions about important discoveries, and they offer details of the experiments themselves.

D.   ASSESSING WHAT YOU'VE LEARNED

1. Archibald Garrod hypothesized that alkaptonuria was:
   a. caused by a DNA mutation
   b. caused by an inherited defect in an enzyme
   c. an example of recessive mutation in a gene
   d. caused by abnormal chromosomal sequence

2. Which of the following serves as a template for making proteins from DNA?
   a. tRNA
   b. mRNA
   c. rRNA
   d. DNA polymerase

3. Which of these describes the function of RNA polymerase?
   a. amplifies the "message" by making multiple copies of an mRNA molecule after it has been transcribed from DNA
   b. converts a protein sequence to mRNA
   c. transcribes DNA to mRNA
   d. translates DNA into protein

4. Which of the following is an *exception* to the central dogma of molecular biology?
   a. production of RNA in the nucleus of a eukaryotic cell
   b. double-stranded DNA from a virus inserts into the host genome
   c. single-stranded DNA from a virus inserts into the host genome
   d. single-stranded RNA from a virus is used as a template for DNA that inserts into the host genome

5. In a skin cell, which of the following changes may lead to skin cancer?
   a. a change in a single nucleotide in an mRNA molecule
   b. a change in a single nucleotide in a DNA molecule
   c. a change in multiple amino acids of a single protein molecule
   d. a change in the conformation of an enzyme molecule

6. Consider the following two mRNA molecules. Original RNA sequence:
   a. ACGUGUACCG
   b. ACGUUGUACC
   c. The difference between the two molecules would most likely result in which type of change in the resulting protein molecule?
   d. frameshift mutation, resulting in a change in the amino acid sequence
   e. deletion mutation, resulting in a change in the amino acid sequence

   f. no change in the protein
   g. intron

7. George Beadle and Edward Tatum revealed that:
   a. One genetic mutation corresponded to one defective protein (enzyme).
   b. DNA is the heritable material.
   c. RNA is the heritable material.
   d. The arginine biosynthetic pathway is defective in bread molds.

8. Which of the following was considered important evidence that RNA was an intermediary between DNA and protein biosynthetic processes?
   a. RNA is located in the nucleus (with DNA) and the cytosol (with ribosomes).
   b. RNA is synthesized, but it is then fairly rapidly degraded.
   c. DNA does not colocalize with ribosomes in the cytosol.
   d. All of the above are correct.

9. If the genetic code was based on a 2-digit system, how many different amino acids could be coded for by this system?
   a. 8
   b. 16
   c. 32
   d. 64

10. Which of the following statements distinguishes DNA and RNA?
    a. T is found only in DNA.
    b. DNA lacks a 2-hydroxyl on its ribose.
    c. DNA does not leave the nucleus.
    d. All of the above apply.

CHAPTER 15—ANSWER KEY

Check Your Understanding

(a) Enzymes catalyze each step in the biosynthetic pathway, modifying the intermediate forms to eventually generate the desired product. Each enzyme is a protein composed of a chain of amino acids. Changing the amino acid content of an enzyme, by altering the gene on which it is encoded, can profoundly influence the function of that protein.

(b) A culture of wild-type *N. crassa* would first be mutagenized using chemical treatment or exposure to radiation. The researchers would identify clonal populations of cells that grow in rich medium but cannot grow in medium that lacks the amino acid arginine. Cells unable to grow in the selective medium likely contain a mutation in one of the arginine biosynthesis genes. Review **Figure 15.2** to see how these isolated mutant cell lines can then be utilized to identify the genes involved in arginine biosynthesis.

(c) Gamow was correct. He knew that 20 amino acids were used by ribosomes to build cellular proteins. If the genetic code were only two nucleotides in length, then there would be only 4^2, or 16 codons available. But if they were three nucleotides in length, then the available number of distinct codons would be 4^3, or 64.

(d) The first strategy was to synthesize polymers of RNA; poly(U), poly(A), poly(G), and poly(C). They added these RNA polymers to a mixture of reagents capable of translating RNA into protein. Uracil polymers (UUUUUUUU) resulted in the synthesis of chains of phenylalanine, evidence that the codon for phenylalanine is UUU. The second strategy was to make specific codons and then measure which charged tRNA bound to the ribosome as it read that specific RNA sequence.

Assessing What You've Learned

1. b; 2. b; 3. c; 4. d; 5. b; 6. a; 7. a; 8. d, 9. b; 10. d

16

Transcription and Translation

A. KEY CONCEPTS

16.1 Transcription in Bacteria

RNA Polymerase Structure and Function

- **RNA polymerase** synthesizes mRNA transcript (and other types of RNA transcripts) by reading the DNA template of an activated gene.

- RNA polymerase reads the **template strand** of the DNA to synthesize RNA. This terminology is important to keep straight. The **non-template strand** of DNA matches the sequence of the synthesized RNA generated by transcription (except U is substituted for T). Because of this, the non-template strand is also called the **coding strand**.

- An RNA polymerase enzyme attaches **nucleoside triphosphates** together, selecting the correct nucleoside by its hydrogen-bonding capacities and synthesizing the RNA strand in the 5′ to 3′ direction (**Figure 16.1**).

- How is a deoxynucleoside different from a nucleoside? Review the structural differences defined in previous chapters.

- The presence or absence of an –OH (hydroxyl) functional group generates major differences in the characteristics of the polymers (DNA vs. RNA).

- The 3-dimensional structure of RNA polymerase revealed grooves or channels through which the template strand of DNA passes during catalytic activity (**Figure 16.2**).

Initiation: How Does Transcription Begin?

- The prokaryotic RNA polymerase is composed of a core complex (holoenzyme or "whole" enzyme) that has the ability to synthesize RNA. The holoenzyme is made up of four globular proteins named α, α', β, and β'.

- Initiation of transcription requires an additional protein, called **sigma (σ)**, that binds to a holoenzyme not yet associated with DNA. Sigma can "read" DNA and identify the start site of a gene for transcription. Thus sigma is critical to the proper initiation of transcription at the correct place on the chromosome, recognizing the DNA sequence that precedes the coding region of a gene (the **promoter**).

- What sequences does the RNA polymerase/sigma complex commonly bind? **David Pribnow** examined promoter sequences in bacterial and viral genomes. He allowed the bacterial RNA polymerase/sigma complex to bind to DNA and then characterized the bound DNA. **Figure 16.3**; Pribnow identified two elements found within most bacterial/viral promoters: the **–10 box** (**Pribnow box**) and the **–35 box**.

- The –10 box is located 10 bases upstream (in the 5′ direction) from the transcription start site. It contains a 5′–TATAAT–3′ consensus element.

- The –35 box is located 35 bases upstream from the transcription start site and contains a 5′–TTGACA–3′ consensus element.

- Sigma identifies the –10 and –35 promoter sites, properly orienting the RNA polymerase core complex for transcription at the gene start site. The polymerase then opens the DNA double helix and begins to catalyze the polymerization of nucleotides, moving down the DNA template as RNA is synthesized.

- Sigma dissociates from the holoenzyme once transcription is successfully initiated.

- Because sigma has so much influence over which genes are expressed, it was anticipated that there would be multiple forms of sigma in microbial cells. This has been shown to be the case. There are sigma factors whose activity is induced by different environmental conditions (heat, starvation). Thus, cells adapt to these environmental changes by switching sigma factors, altering patterns of gene expression.

HOW TRANSCRIPTION BEGINS

Promoter (on non-template strand)

−35 box   −10 box
CTG**TTGACA**ATTAATCATCGAACTAG**TATAATA**GTACGC
+1 site

Upstream DNA

Sigma

Downstream DNA

Active site

RNA polymerase

1. Initiation begins
Sigma binds to promoter region of DNA.

Template strand

Non-template strand

+1 site

RNA

2. Initiation continues
Sigma opens the DNA helix; transcription begins.

NTPs

Upstream DNA

3. Initiation is complete
Sigma releases; mRNA synthesis continues.

RNA

RNA exit site

Zipper
Rudder

Downstream DNA

Elongation and Termination

• Once RNA polymerization is initiated, the enzyme **elongates** the RNA transcript by reading the template strand and synthesizing RNA in the 5′ to 3′ direction (moving down the template DNA strand in the 3′ to 5′ direction). This enzyme proceeds at a rate of ~50 nucleotides added/second.

• Notice there are grooves in the holoenzyme through which pass the template and non-template DNA strands, nucleotides, and the newly synthesized RNA strand (**Figure 16.3**, step 3).

• Transcription is terminated when specific sequences, called **transcription termination signals**, are encountered in the DNA template. The process of RNA polymerase dissociating from the DNA can occur by several different mechanisms. One type of termination signal results in the RNA forming a hairpin structure (the RNA folds back on itself to make a short stretch of double-stranded RNA). Formation of the hairpin will "pull" the newly synthesized RNA out of the holoenzyme, terminating additional transcription. Another termination mechanism requires the action of a separate protein (rho factor) that binds the RNA at a termination sequence and initiates ribosomal disassembly.

16.2 Transcription and RNA Processing in Eukaryotes

• Eukaryotic transcription shares many fundamental characteristics with prokaryotic transcription, but it is more complex and includes some additional processing steps.

• Eukaryotic cells contain three RNA polymerases, though mRNA synthesis is mediated by only one of these, RNA pol II. Thus all protein synthesis in a eukaryotic cell depends on **RNA pol II**. RNA polymerase I makes ribosomal RNA (rRNA); RNA polymerase III is responsible for some rRNA and for transfer RNA (tRNA—see Section 16.4).

• The promoters of eukaryotic cells are highly variable, though many do contain a **TATA box** approximately 30 base pairs from the transcription start site.

• Many small proteins assemble with the RNA pol II to locate promoter elements and to regulate transcriptional activity; these are called **basal transcription factors**. These factors act in a manner that is very similar to the sigma factor of prokaryotic cells.

- Basal transcription factors were discovered by in vitro (in the tube) transcriptional studies. Purified RNA polymerase was combined with template DNA and nucleotides, resulting in RNA synthesis. However, researchers discovered that by adding specific proteins to the test tube, the RNA polymerase activity became more specific for the promoters of genes. Eventually several proteins that regulate RNA polymerase activity were identified and classified as basal transcription factors.

The Startling Discovery of Eukaryotic Genes in Pieces

- Prokaryotic genomes contain genes in one **continuous** DNA sequence.
- Characterization of eukaryotic genomes has revealed that genes are often **discontinuous**, having been broken into smaller fragments (**exons**) with noncoding sequences (**introns**) between each piece.
- This trait was first discovered by Philip Sharp, who is credited with developing and testing the **genes-in-pieces hypothesis.** His research team generated DNA:RNA hybrid molecules to reveal the discontinuous nature of genes on the chromosomal DNA. They first heated double-stranded DNA molecules to denature them into single strands, and then they mixed the denatured DNA with mRNA strands. The mRNA strands should anneal to the complementary strand of DNA sequence (i.e., the original site of that gene). Sharp discovered that the hybrid molecules frequently contained loops of DNA rather than a continuous annealed structure (**Figure 16.6**). Why? The mRNA must be missing some of the sequence! They proposed that some sequence was removed as the primary RNA was processed to a mRNA.
- As more scientists began to examine this phenomenon, it became clear that eukaryotic genes are discontinuous in the genome, and that following transcription the gene-encoding fragments in the **primary RNA transcript** were spliced together to make the mRNA product.

Exons, Introns, and RNA Splicing

- The coding fragments found in the chromosomal location of a gene are called **exons** (expressed), and the noncoding fragments interspersed between exons are called **introns** (intervening sequences).
- How are the exons spliced together? This complex process is mediated by *s*mall *n*uclear *r*ibo*nucleo*-*p*roteins (**snRNPs**, pronounced "snurps"). As their name suggests, these complex enzymes are composed of both protein and RNA. **Figure 16.7** illustrates how snRNPs identify introns by their sequence at each end

(the splice junction site), catalyze the removal of the intron, and splice surrounding introns together.
- An interesting structure is formed during the splicing process. An adenosine (A) base inside the intron offers a site to which the 5′ end of the intron is attached (**Figure 16.7b**, step 2). This forms a looped structure, called the lariat, and it is quite unusual.
- The complex of snRNPs that mediate this process is called a **spliceosome**.

Adding Caps and Tails to Transcripts

- After splicing of the primary transcript, two additional steps of RNA processing are performed in eukaryotic cells—the addition of a **5′ cap** and the addition of a **poly(A) tail** (**Figure 16.8**).
- The cap is a modified nucleoside (7-methyguanylate) triphosphate that is added to the front or **5′** end of the transcript. This cap aids mRNA translation by enabling binding of some translation regulatory proteins. It also increases the stability of the mRNA by decreasing the rate at which the mRNA is degraded by cellular RNA degrading enzymes (ribnucleases or RNases).
- The 3′ poly(A) tail is another modification of the RNA transcript that aids mRNA stability in the cell. Approximately 100–250 consecutive adenosine nucleosides are added to the 3′ end of the RNA.

CHECK YOUR UNDERSTANDING

(d) Robert Roeder's research team utilized adenovirus DNA to characterize transcription in eukaryotic cells. They used an in vitro assay, adding purified RNA polymerase to adenovirus DNA, but their early experiments resulted in transcription at random places on the DNA. They suspected that something was missing from the RNA polymerase. How did they alter their experimental strategy to address this problem, and were their changes helpful?

_____

_____

_____

_____

16.3 An Introduction to Translation

- Once the mRNA moves to the **cytoplasm**, the information contained within its sequence can be translated by the **ribosome** to generate the functional protein.

- Proteins are synthesized from individual **amino acids**, and each amino acid in the protein is coded for in the mRNA. Each codon (three nucleotides in mRNA) directs the ribosome to add one amino acid to the protein.

Ribosomes Are the Site of Protein Synthesis

- Ribosomes are composed of both protein and RNA, called ribosomal RNA (rRNA).

- Roy Britton used **pulse-chase experiments** to examine whether ribosomes were indeed the site of protein synthesis. *Britton wanted to determine if protein synthesis was occurring at the site of ribosome localization.* He labeled proteins being synthesized for a short period of time by adding a 15-second pulse of radioactive sulfate to the culturing medium of *E. coli* (immediately chased with non-radioactive sulphur). This would result in the incorporation of the radioactive tag with amino acids containing sulphur (methionine and cysteine), and a short window of time in which all proteins being synthesized would be radioactive. But how did Britton detect these proteins? How did he determine that they were associated with ribosomes?

- Ribosomes can be isolated via **density gradient centrifugation**, in which molecules are separated from each other according to their density. Britton found the pulse of radioactivity to localize to fractions containing ribosomes! As the radioactive pulse was chased with nonradioactive sulphur in the *E. coli* cultures, radioactivity disappeared from ribosomal fractions. This shows that the radioactive sulphur is associated with ribosome activity (the proteins being made), not with the ribosome structure itself.

- Because prokaryotic cells have no nucleus, translation occurs immediately following, or even during transcription (**Figure 16.9a**).

- Eukaryotic cells transcribe RNA in the nucleus, process the RNA to mRNA in the nucleus, and then export the mature mRNA to the cytosol for translation (**Figure 16.10**).

16.4 The Role of Transfer RNA

- **Adapter-molecule hypothesis**: In imagining how a codon of nucleotides in an RNA molecule could somehow direct the recruitment of specific amino acids, **Francis Crick** (of Watson and Crick) developed one hypothesis. He hypothesized there was an as yet undiscovered adapter molecule that "read" the mRNA codon directly and carried with it an amino acid for protein synthesis. The predicted adapter molecule was discovered (see next item) and named **transfer RNA (tRNA)**.

- Researchers studied protein synthesis by identifying the essential ingredients of active protein synthesis. These ingredients are mRNA, ribosomes, amino acids, guanosine triphosphate, and adenosine triphosphate. One additional and indispensible ingredient came from cellular fractions and was eventually shown to be a small RNA. The molecules discovered by this research team were the transfer RNAs.

- A tRNA bound to its corresponding amino acid is called an **aminoacyl tRNA**; the addition of amino acids to tRNAs is mediated by a specific enzyme (**aminoacyl tRNA synthetase, Figure 16.12**). There are many types of tRNAs, each carrying a specific amino acid and recognizing a codon on the mRNA. They are the adapters between the amino acids and the mRNA that Crick had predicted.

- A key experiment is highlighted in **Figure 16.13**, revealing the transfer of an amino acid from a tRNA to a growing peptide chain. The ribosome mediates this transfer. How was the location of the amino acid determined in this experiment?

What Do tRNAs Look Like?

- It is hard to understand what a tRNA really does during protein synthesis without examining tRNA structure and putting the tRNA, ribosome, and mRNA together in a diagram. Carefully examine **Figure 16.14**. Notice the 3′ end to which amino acids are added; the CCA sequence is found at 3′ end of all tRNAs. Throughout the structure are a series of **stem-loop** structures that give shape to the tRNA. These structures are formed by complementary base pairing among adjacent, antiparallel strands. A third important characteristic is the **anticodon** region.

- The anticodon actually forms an **antiparallel association** (by hydrogen bonding) with the codon on the mRNA strand. If the aminoacyl tRNA matches the codon being presented, the amino acid on the 3′ end is presented to the ribosome for addition to the protein being synthesized.

How Many tRNAs Are There?

- Since the genetic code includes 61 amino acid codons, scientists originally thought there should be 61 distinct tRNAs that match each codon. However, most cell types contain only ~40 different tRNAs. How can this be?

- Francis Crick proposed the **wobble hypothesis**, predicting that the first two nucleotides in a codon are more important than the third (wobble) base for some aminio acids. Recall that some amino acids are coded for by several different codons, but that all start with the same two bases (CAA and CAG both code for glutamine).

- It was experimentally shown that *some* tRNAs are capable of binding to codons that match at only two residues, while the third residue mismatch is tolerated. Thus 40 tRNAs are capable of mediating the translation of all 61 codons.

16.5 Ribosomes and the Mechanism of Translation

- Protein is synthesized by the ribosome as it reads the mRNA. This process occurs both in the **cytoplasm** and on the surface of the **endoplasmic reticulum**.

- The ribosome was first isolated using **density gradient centrifugation**. A crude mixture of proteins can be divided into components, separating them by their relative density. Ribosomes are composed of protein and rRNA.

- Density gradient centrifugation was used to isolate intact ribosomes and showed that the *E. coli* ribosome was composed of two major parts, now known as the **50S** and the **30S** subunits (S is a unit of sedimentation density). Each subunit is composed of proteins and rRNA strands. The rRNA found in ribosomes is important to the function and structure of the ribosome; it is never translated.

- Ribosomes **self-assemble**, meaning that if the individual components are placed together, they immediately assemble into a ribosome.

- As shown in **Figure 16.15**, there are several key structural sites in the ribosome for tRNA localization. These are A, P, and E sites; their role in protein synthesis is summarized next.

Initiation

- Remember that mRNA contains sequences in addition to an open reading frame (coding for protein); there are noncoding 5′ and 3′ RNA sequences. Thus the ribosome must locate the translation start site correctly to ensure that the correct information is translated into protein. How would you predict the process might occur? Think back to transcription; how was the correct starting site identified?

- Translation always begins with a **methionine** codon (**AUG** in the RNA sequence). The 30S subunit is recruited to a consensus sequence just 6 nucleotides upstream of the translation start site (**Figure 16.16**). Scientists have named it the **Shine-Dalgarno sequence** or the **ribosome binding site**. The consensus sequence at this site is 5′–AGGAGGU and is recognized by a small rRNA in the 30S subunit via complementary strand hydrogen bonding.

- Once the 30S subunit is bound to the mRNA, the anticodon of an **initiator tRNA** (methionine) binds to the AUG. The initiator methionine is unique from other methionine tRNAs because the methionine carried on this tRNA is modified with a formyl group. The resulting methionine is called *N*-formylmethionine. Eukaryotes use a normal methionine to initiate translation. Initiator tRNA association is a complex process, mediated by a series of small proteins known as **initiation factors** (**IF**).

- The **50S subunit** binds next, associating with the 30S subunit that is already properly oriented to the mRNA translation start site.

INITIATING TRANSLATION IN BACTERIA

1. Ribosome binding site sequence binds to a complementary sequence in the small subunit of the ribosome, with the help of protein initiation factors.

2. Initiator aminoacyl tRNA binds to start codon.

3. Large subunit of ribosome binds. Translation begins.

Elongation

- The ribosome-mediated process of protein synthesis can be summarized as regulated formation of peptide bonds between amino acids. *What is a peptide bond*? Do you recall peptide bonds as discussed in **Chapter 3**? If not, review that material to be sure you recognize the structure of amino acid and the reactive groups interacting to form a peptide bond (amine and carboxyl groups). *What characteristics of an amino acid make it unique?* Review the R-groups or side chains on amino acids; it is their makeup that generates the unique characteristics of each protein.

- Review **Figure 16.17** and the three steps of elongation in protein synthesis:

  1. *Filling the A site*—An aminoacyl tRNA moves into the A site of the assembled ribosome/mRNA complex. The anticodon of the rRNA forms hydrogen bonds with the codon in the A site. The adjacent P site contains the previously added tRNA with its amino acid bound by a peptide bond to the growing polypeptide chain.
  2. *Peptide bond formation*—The amino acid in the A site forms a peptide bond with the polypeptide chain, its amine group reacting with the carboxyl group of the P-site amino acid. As this bond is formed, the P-site amino acid breaks its covalent attachment to the P-site tRNA.
  3. **Translocation**—The P-site tRNA diffuses out of the complex; the ribosome advances forward one codon ($5' \to 3'$), transforming the A site to the P site and making room for the next aminoacyl tRNA.

- Energy for elongation is derived from high-energy phosphate bonds in GTP molecules. Small proteins known as elongation factors also aid in the elongation process.

Termination

- When the A site encounters a stop codon, instead of a new aminoacyl tRNA, small proteins (**release factors**) enter the site and initiate release of the synthesized protein. The bond between the last amino acid in the protein and the tRNA in the P site is hydrolyzed. The ribosome then releases the protein, separates from the mRNA, and the 30S and 50S subunits dissociate from each other.

- Universal stop codons: **UAA**, **UAG**, **UGA**.

Post-Translational Modifications

- The completed protein product is often further modified to mediate proper folding and to effect protein function. These modifications can be complex, and they vary depending on the specific protein involved. As these modifications are completed after translation, they are called **post-translational modifications**.

- A class of proteins that mediates folding by interacting with newly synthesized proteins is called **molecular chaperones**. They are cellular proteins that function to bind to other proteins and enable their proper folding. Many of these proteins are stress activated, expressed in high levels following some type of cellular stress that might threaten proper protein folding. An excellent example is the **heat-shock-activated chaperone proteins**. High temperatures can dramatically influence protein folding; heat-shock proteins protect against this threat.

- Proteins can also be covalently modified as a part of their proper synthesis (see Chapter 7 for details). These modifications occur in the rough endoplasmic reticulum and the Golgi apparatus. Modifications can include the addition of sugar molecules (**glycosylation**) or phosphate molecules (**phosphorylation**). Some proteins are processed by enzymatic **cleavage** of the final protein. Proteinases mediate this process and may be at the site of action for that protein, or they may be regulated by other signals.

CHECK YOUR UNDERSTANDING

(e) Define the process of translation initiation in bacteria.

_____

_____

_____

_____

_____

16.6 What Is the Molecular Basis of Mutation?

- A **mutation** is any permanent change in an organism's DNA sequence. DNA mutations affect phenotype only when the mutation is expressed (DNA → RNA → protein), and the resulting protein functions abnormally. Not all mutations will affect the protein's ability to function; thus they will not generate a phenotype.

Point Mutations

- There are many types of DNA mutations. One of the most common is a **point mutation**, which changes a single nucleotide. These can result from errors in DNA replication (**Figure 16.20**, **Table 16.3**), or exposure to mutagenic toxins.

- **Sickle-cell anemia** is a human disease resulting from a point mutation in the gene for hemoglobin. A simple T → A mutation changes a codon for the amino acid glutamic acid into a codon for valine (**Figure 16.21**). This type of point mutation is called a **missense or replacement mutation** since the codon has been changed from one amino acid to another. The result is an abnormal hemoglobin protein that crystallizes at low oxygen concentrations, causing abnormally shaped red blood cells and circulation problems.

- Some mutations do not change the amino acid added to the protein (notice that CTT and CTC codons both specify glutamic acid). These are known as **silent mutations**.

Chromosome-Level Mutations

- Larger **chromosome-level mutations** can affect regions of chromosomes, or even entire chromosomes.

- **Aneuploidy** is a situation where a cell has an abnormal number of chromosomes, either through chromosome loss or gain.

- **Polyploidy** is where there has been a duplication of the entire chromosome complement (a cell might be $4n$ rather than $2n$).

- **Chromosome inversions** are mutations where a region of DNA becomes flipped backward within its normal chromosomal location.

- **Chromosome translocations** are mutations where a region of one chromosome is moved to another chromosome (**Table 16.4**).

B. INTEGRATING YOUR KNOWLEDGE

Looking Back—Key Concepts from Earlier Chapters

Nucleic Acids and the RNA World—Chapter 4

Scientists are only beginning to understand the various roles for **RNA macromolecules** in cellular function. The discovery that RNA can be catalytic, functioning in the splicing of RNA strands, was amazing. Recent findings reveal that rRNA in the ribosome catalyzes the formation of **peptide bonds**. In this chapter we introduced several different examples of RNA function in the cell. RNA is an interesting molecule displaying an interesting diversity of activities.

Protein Folding and Processing—Chapter 3 and Chapter 7

Proteins are complex structurally; they must fold properly in order to function. The process of synthesizing new protein is much more complex than simply generating a linear array of amino acids attached via a peptide bond. The role of chaperones and post-translational modification should be reviewed. The endoplasmic reticulum and Golgi apparatus play a key role in cellular protein processing.

Looking Forward—Key Concepts in Later Chapters

Control of Gene Expression—Chapters 17 and 18

The next two chapters focus on the mechanisms controlling gene expression. Regulating gene expression at the chromosome is much more efficient than regulating RNA processing or ribosome activity. Much of the biology defining a cell type (e.g., a liver cell or a brain cell) is the direct result of **differential gene expression**.

Bacteria and Archaea—Chapter 28

Overuse of **antibiotics** in the United States, and the resulting rise in **drug-resistant bacteria**, are in the news today. Humans utilize molecules that attack critical cellular functions in bacteria (like protein synthesis). Why don't these drugs harm our cells? Study the differences between bacteria and human cells; can you hypothesize how these drugs could be harmless to us?

Viruses—Chapter 35

Many early experiments in RNA polymerase activity were performed using viruses. **Bacteriophage viruses** infect prokaryotic cells, and **adenoviruses** infect eukaryotic cells. Experiments using each of these viruses to characterize RNA polymerase are shown in **Chapter 35**. Why was viral infection of a cell selected for these classical experiments? Upon entry into the cell, the virus takes over the native RNA polymerase for expression of the viral genes. Thus scientists can examine the relatively few viral genes in their experiments, when compared to the large number of cellular genes. Examine this chapter for a detailed discussion of viral pathogenesis.

C. MASTERING CHALLENGING TOPICS

X-Ray Crystallography (Figure 16.15)

As we examine ribosomes and protein synthesis, it is important to realize how researchers determine the three-dimensional shape of specific macromolecules.

Chapter 4 introduced the process of solving the structure of a protein using X-ray diffraction analysis. Scientists have used this strategy to characterize DNA and protein interactions, protein and protein interactions, and even structural changes occurring with the activation of a specific protein (allosteric regulation).

You can find many crystallography images and resources on the Internet. Explore this experimental strategy with some simple searches. Here are two sites we recommend:

http://ncbi.nlm.nih.gov/Structure/MMDB/
mmdb.shtml

http://boatman.med.wayne.edu/~xray/
education.html

D. ASSESSING WHAT YOU'VE LEARNED

1. What is the maximum number of amino acids that could result from the following mRNA sequence?
   a. 5′–AAUCCGUAAAUGAGACCGUCGAUG-
      AAUUAGCG–3′
   a. 0
   b. 6
   c. 7
   d. 10

2. Which of the following best describes the effect of a mutation in the tryptophan aminoacyl tRNA synthetase that changed the conformation of the tRNA binding site?
   a. The amino acid tryptophan would not bind to the aminoacyl tRNA synthetase.
   b. The aminoacyl tRNA synthetase would not be produced.
   c. The tryptophan tRNA would not be produced.
   d. The enzyme would not be activated.

3. Which of the following are required as an energy source for translation?
   a. mitochondria
   b. aminoacyl tRNA synthetase
   c. GTP and ATP
   d. GTP only

4. What is the best description of the "loops" in the tRNA molecule?
   a. unpaired DNA bases
   b. regions created as the result of hydrogen bonding between base pairs
   c. regions of the molecules that remain as single-stranded RNA due to the absence of complementary base pairs
   d. regions provided for binding to the mRNA codon

5. The tRNA shown in **Figure 16.14** in your textbook depicts the typical "L" shape of this molecule. Which of the following will most likely result if some hydrogen bonds in the region labeled "hydrogen bonds" in the figure are broken using some type of chemical treatment?
   a. The tRNA will not be able to bind the amino acid properly.
   b. The tRNA will not have the correct anticodon.
   c. The anticodon will remain the same, but the amino acid bound to the tRNA will be different.
   d. The tRNA will not fit into the ribosome properly.

6. Refer to **Figure 16.16** in your textbook. Which of the following statements is *not* true regarding translation in bacteria?
   a. Translation requires the presence of initiation factors and both the small and large subunits of the ribosome.
   b. The peptide bonds between adjacent amino acids are formed within the large ribosomal subunit.
   c. The first amino acid in every protein is methionine.
   d. Translation proceeds from the 5′ end toward the 3′ end of the mRNA.

7. If there were a mutation in a ribosome that prevented tRNA molecules from moving into the E site, how many amino acids would be found in the resulting peptide chain formed on this ribosome?
   a. a maximum of 2
   b. a minimum of 2
   c. a maximum of 1
   d. none

8. Which of the following statements is correct?
   a. Protein enzymes catalyze the formation of peptide bonds between adjacent amino acids.
   b. Peptide bond formation occurs within the small subunit of the ribosome.
   c. Peptide bond formation occurs in the absence of any catalyst.
   d. A compound other than a protein enzyme is able to catalyze the formation of peptide bonds.

9. Some proteins can act as infectious proteins, or prions. These proteins have an abnormal conformation and cause other "normal" proteins to change to the mutant form, resulting in a number of disorders, including mad cow disease. This process is similar to which of the following normal processes in the cell?
   a. transcription
   b. translation
   c. mRNA processing
   d. post-translational modification

10. Aneuploidy is best defined as:
    a. a single nucleotide change
    b. an abnormal chromosome number
    c. a doubling of chromosome number
    d. a mutation

CHAPTER 16—ANSWER KEY

Check Your Understanding

(a) Pribnow examined the DNA sequence of a number of promoters for prokaryotic genes. He identified sequences that were always present in a promoter sequence, anticipating that this would reveal the RNA polymerase binding region. These sequence motifs are called consensus or conserved sequences.

(b) The promoters of prokaryotic genes have conserved sequences at the −10 and the −35 sites (distance to transcription start site). The −10 box contains a TATAAT sequence. The −35 box contains a TTGACA sequence. Both of these sites are critical for proper trascriptional activation.

(c) Bacteria have sigma factors present in the cell that activate the expression stress response genes. Upon heat shock, for example, a sigma factor that activates heat-shock protein expression is activated. Thus genes that are important for stress response have unique promoters that are recognized only by the stress-activated sigma factors.

(d) Roeder and co-workers hypothesized that their RNA polymerase preparation was "too pure" and lacked a protein that normally helped to identify the proper promoter elements (a eukaryotic sigma factor?). When they repeated the experiments with crude cell extracts, the researchers found that transcription activity was promoter specific. Eventually they purified the components of the crude cell extracts that delivered this specificity—basal transcription factors.

(e) The ribosome must assemble around the mRNA to begin the process of translation. Upstream from the AUG is the Shine-Dalgarno sequence. Initiation factors bind to the Shine-Dalgarno sequence and enable the binding of a ribosomal small subunit (30S) to the region. Next a charged tRNA (aminoacyl tRNA with *N*-formylmethionine) binds to the AUG with its anticodon sequence. Finally, the large subunit binds and completes the formation of the initiation complex.

Assessing What Youv'e Learned

1. b; 2. c; 3. c; 4. c; 5. d; 6. c; 7. a; 8. d; 9. d; 10. b

17

Control of Gene Expression in Bacteria

A. KEY CONCEPTS

- *Escherichia coli* serves as an excellent model organism for the study of prokaryotic gene regulation. These bacterial cells adapt rapidly to **environmental changes** and can often continue to grow. The adaptation is usually based on altered gene expression, activating genes needed to respond to a specific change in conditions.

- Why not express all the genes all the time? Sets of genes are expressed for a specific cellular function, and most genes are expressed sporadically as they are needed. Continual expression of all genes would be devastating to cellular biochemistry, and would certainly cause the rapid destruction of the cell.

- Regulation of gene expression is fundamental to the identity of a cell, controlling cellular characteristics and physiology. When we discuss eukaryotic gene regulation, you will see how gene regulation controls cell identity in a multicellular organism.

17.1 Gene Regulation and Information Flow

- The central dogma of molecular biology involves several steps that can be regulated before an active protein is produced within the cell (**Figure 17.1**).

  1. **Transcriptional regulation**—Control of gene expression at the chromosome, allowing RNA polymerase binding and activation only under cellular conditions favoring expression of that gene.
  2. **Translational (post-transcriptional) regulation**—Regulatory mechanisms present in cells serve to regulate the stability of mRNA messages, to process a primary RNA transcript to an mRNA, and even to increase the efficiency of translation of that transcript.
  3. **Post-translational regulation**—The completed protein product sometimes undergoes additional processing steps to generate activity. These processing steps include phosphorylation,

glycosylation, and cleavage of the protein. See **Chapter 16** for more discussion of these mechanisms.

- These types of regulation occur in both prokaryotic and eukaryotic cells. Prominent within these mechanisms is the regulation of transcription.

Metabolizing Lactose—A Model System

- Sugars are metabolized in cells to generate energy in the form of ATP. The pathway that initiates the breakdown of glucose is called **glycolysis** (breaking down glucose and gathering energetically valuable electrons in the process).

- The fastest and easiest way to feed glycolysis is to import glucose molecules from outside the cell and utilize them in glycolysis. If glucose molecules are not available, the bacterial cell must convert available molecules to glucose or some other energetically useful molecule. To perform this conversion, enzymes are required. These enzymes are typically expressed only when glucose is unavailable but another convertible molecule is available.

- Lactose is a **disaccharide** (glucose and galactose molecules covalently attached to each other) that can be cleaved to the two constituent **monosaccharides** by an enzyme called **β-galactosidase**.

- In the absence of lactose, no β-galactosidase enzyme is expressed in *E. coli* cells. Lactose is an **inducer** of the β-galactosidase gene (and other genes in the lactose operon).

- What about cells having both lactose and glucose available at the same time? Is one sugar preferred over the other? **Jacques Monod** and colleagues tested the hypothesis that *E. coli* would prefer glucose over lactose (**Figure 17.2**). They measured the expression of β-galactosidase under three conditions: media with only glucose, media with only lactose, and media with both glucose and lactose. They found that β-galactosidase is expressed only when glucose is absent and lactose is present, consistent with Monod's hypothesis.

CHECK YOUR UNDERSTANDING

(a) List all aspects of the central dogma of molecular biology that involve the regulation of gene expression.

_____

_____

_____

_____

17.2 Identifying the Genes Involved in Lactose Metabolism

How Were the Genes Found?

- Monod, working with **Francois Jacob**, initiated a hunt for mutant *E. coli* that are unable to metabolize lactose. This strategy allowed the scientists to screen for *E. coli* that are unable to utilize lactose as a source of carbon.

- Bacteria can be easily grown in **petri dishes**. Thousands of cells are spread onto each plate and allowed to grow into clonal colonies. Each cell in a **colony** (a pile of cells) shares the same genetic makeup, because all have grown from a single cell. If the original cell carried a mutation in β-galactosidase, then all cells in the colony will contain this mutation.

- Examine replica plating, displayed in **Figure 17.3**. This technique is used to make "copies" of populations of cells. These copies can then be tested in lactose medium, and researchers will still have copies of each clonal population for analysis in later experiments.

- **Indicator plates** were used in this mutant hunt. Monod and Jacob generated petri dishes containing an o-nitrophenyl-β-D-galactoside (**ONPG**) **indicator molecule** with a structure similar to that of lactose (two subunits bound by a β-1,4 glycosidic bond). Thus any enzyme that breaks down lactose would be expected to break down ONPG. Once broken down, the ONPG subunits generate a yellow coloration in the media, indicating enzyme activity. The researchers could thus examine thousands of *E. coli* colonies, identifying those that lacked the ability to break down ONPG by the absence of yellow coloration near the colony.

- Monod and Jacob were interested in these mutant cells; they hypothesized that the cells carried mutations in the genes that regulate or mediate lactose utilization.

Different Classes of Lactose Metabolism Mutants

- Monod and Jacob recognized that several enzymes were likely important in the utilization of lactose. They isolated several classes of *E. coli* mutants, each containing a different defect in this process (**Table 17.1**).

  1. **β-glycosidic** mutants—Loss of this enzyme activity results in an inability to cleave lactose into glucose and galactose. The gene encoding β-galactosidase was named *lacZ*.

  2. **Galactoside permease mutants**—A protein located in the cell membrane mediates the transport of lactose into the *E. coli* cell. In the absence of this enzyme, cells cannot accumulate lactose for cellular metabolism. The gene encoding galactoside permease was named *lacY*.

  3. **Constitutive mutants**—This class of mutants has lost the ability to regulate expression of the genes needed for lactose utilization. Unlike wild-type cells, they display ONPG cleavage in the *absence* of lactose, indicating that β-galactosidase is being expressed continuously (constitutive expression). One mutation that resulted in the constitutive expression of β-galactosidase was found in a gene named *lacI*. The "I" is for the phenotype where an inducer (lactose) is unnecessary for the expression of β-galactosidase. This makes perfect sense because the wild-type lacI⁺ protein **represses** gene expression, and the loss of this repression results in abnormal activation of gene expression. Many genes have been named after the phenotype observed when that protein is nonfunctional.

- Further studies revealed that these three genes are located near each other on the bacterial chromosome (**Figure 17.4**). In prokaryotic organisms, genes with a shared purpose (like lactose metabolism) are sometimes controlled by a single promoter; these groups of genes are called **operons**.

- A single promoter regulates the expression of all three *lac* operon genes, and that promoter is activated only under specific conditions (presence of lactose, absence of glucose).

- Once scientists were able to generate a genetic map of the chromosomal region near the *lacZ* gene, they found that *lacY* and a third gene—*lacA* (an acetyltransferase)—were all located immediately adjacent to each other and under the control of a single promoter. The *lacI* gene was found nearby but under control of its own promoter. Ultimately it was shown that the lacI encoded protein (the *lac* repressor) regulated gene expression from the lactose operon (*lacZ, lacY, lacA*).

CHECK YOUR UNDERSTANDING

(b) In the presence of lactose and glucose, why do *E. coli* cells prefer glucose as a source of energy?

_____

_____

CHECK YOUR UNDERSTANDING

(c) Isolation of cells having mutations in specific types of genes is a common research strategy. Researchers studying utilization of lactose performed a screen for cells with a mutation in the β-galactosidase gene. How was this screen designed?

_____

_____

_____

_____

17.3 Mechanisms of Negative Control: Discovery of the Repressor

- The lactose operon is regulated by **positive** and **negative control** mechanisms. Positive control is used when something must bind the promoter to activate expression; negative control is used when something bound to the promoter blocks expression of the operon.

- Leo Szilard (1950) suggested that Monod's data indicated the lacI⁺ protein was a negative regulator of *lacZ* and *lacY* expression. Thus lacI⁺ was a **repressor** of the lactose operon.

- Monod had shown that *lacI* expression was **constitutive**; in other words, it was always "on" regardless of the presence or absence of lactose in the medium. How then could lacI⁺ release its repression of the operon in the presence of lactose? Szilard proposed that the lacI⁺ protein interacted directly with

lactose, and that when lacI⁺ was bound to lactose, it released the promoter element (**Figure 17.7**).

- To test this hypothesis, the researchers developed an experimental strategy for transferring DNA from one microbe to another (using the bacterial process of **conjugation**). A recipient group of cells carried the *lacI⁻* mutation; they expressed *lacY* and *lacZ* continually, regardless of whether lactose was present or absent. When DNA from a *lacI⁺ lacY⁻ lacZ⁻* strain was introduced, the lacI⁺ protein from the introduced DNA restored negative control of the recipient strains' lactose operon. This experiment and others showed Szilard's hypothesis was correct. The introduced *lacI⁺ was restoring negative regulation at the lactose operon.*

The lac *Operon*

- There are three central hypotheses to the Jacob–Monod model of lac operon regulation:
  1. The lacZ, lacY, and lacA genes are adjacent to one another, and they are transcribed together as one polycistronic mRNA. Thus their expression is coordinately regulated—the lac operon.
  2. The lacI protein represses expression of the lac operon genes by binding to the operator motif within the promoter of the operon and repressing transcription.
  3. Lactose is an inducer in this system, binding to the repressor protein and blocking its ability to bind the operator. The inducer activates transcription of the operon.

A New Twist on Negative Control: Comparing the trp *and* lac *Operons*

- The **tryptophan (trp) operon** contains a group of five genes encoding enzymes that synthesize this amino acid. The operon has a single promoter that is regulated by the cellular concentration of tryptophan. Low cellular concentrations of tryptophan induce expression of the operon.

- Regulation of the *trp* operon has been studied carefully, and when compared to the lactose operon, displays important similarities and some striking differences (**Figure 17.9**).

- The *trp* operon displays negative regulation like the lactose operon, but the mechanism is quite different. The *trp* repressor protein binds the operator DNA only when it is bound to tryptophan (the corepressor). This fits with the role of the *trp* operon; when there is plenty of tryptophan around, the promoter should be blocked from expressing these genes.

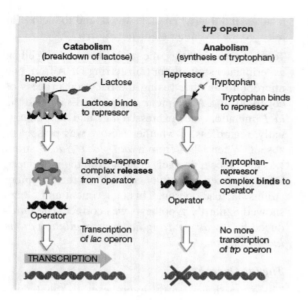

17.4 Mechanisms of Positive Control: Catabolite Repression

- **Catabolism** is the metabolic breakdown of large molecules to smaller, energetically useful molecules or catabolites. If the cell has plenty of these breakdown products, the catabolic pathway is typically stopped. This inhibition of a metabolic pathway is called **catabolite repression** (e.g., glucose inhibiting the lactose operon diagrammed in **Figure 17.12**).

- The absence of glucose actually helps to activate expression of the lactose operon via the action of another DNA-binding protein called the **catabolite activator protein (CAP)**.

- Inactivation of the *lacI*+ repressor allows for some expression of the *lac* operon; but binding by the positive regulator protein, CAP, generates a massive induction of expression.

- CAP is **allosterically** regulated by binding to a molecule called cyclic adenosine monophosphate (cAMP), a modified nucleotide present in high concentrations only when glucose levels are depleted. CAP bound to cAMP induces expression of the lactose operon by binding the **CAP site** in the promoter region to recruit RNA polymerase to that site for transcription.

- The concentration of cAMP is directly controlled by an enzyme—**adenylyl cyclase**. This enzyme synthesizes cAMP from ATP and is inhibited by high concentrations of glucose.

- CAP protein is an example of positive control of an operon; when bound to the promoter region, CAP *increases* operon expression.

- **Figure 17.13** is a good summary of positive and negative regulation of the *lac* operon.

CHECK YOUR UNDERSTANDING

(d) Define allosteric regulation and give an example.

_____

_____

_____

_____

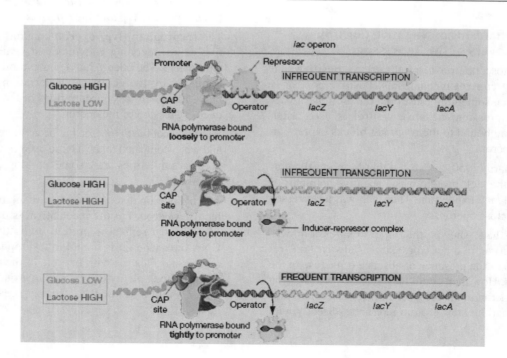

(e) Catabolite repression is commonly seen in *E. coli*. It involves inhibition of some operons by the presence of a product of that operon's metabolic action. How does catabolite repression work in the lactose operon? Predict other cellular metabolic pathways that might experience catabolite repression.

_____

_____

_____

17.5 The Operator and the Repressor—an Introduction to DNA-Binding Proteins

How Was the Operator Found?

- The **operator sequence** within the *lac* operon promoter is recognized by the repressor protein (lacI⁺). But how was this region of the promoter identified as being important?

- Initial experiments used mutant bacterial cell lines that displayed a lack of operator activity (no repression). The site was mapped to just downstream of the promoter region (between the promoter and the transcription start site).

- Scientists used an assay called **DNA footprinting** to precisely identify the site of repressor binding. This assay is summarized in **Box 17.2** and **Figure 17.15**. The researchers first radioactively labeled DNA fragments containing the *lac* promoter and mixed these fragments with repressor proteins to allow for repressor binding. They then added an enzyme that degrades DNA, knowing that any DNA that is bound up by the repressor will be protected from digestion. Upon analyzing the products, they identified one DNA sequence that was consistently bound tightly by the repressor protein—the operator sequence.

- The lactose operon promoter actually contains several consecutive operator sequences, each able to bind a repressor protein. Their sequence and organization are summarized in **Figure 17.16**. Notice that the DNA sequence of the operator is symmetrical (contains **dyad symmetry**); this is common in DNA sequences that bind regulatory proteins.

How Does the Repressor Bind to the Operator?

- **Domain**—Within a protein, domains are regions that display unique structural characteristics (α-helix, β-sheet).

- **Motif**—A motif is a conserved domain that is shared by a related group of proteins.

- Many structural motifs are observed in DNA-binding proteins. A common motif observed in the *lac* repressor is called the **helix-turn-helix motif**. This structure fits easily into the major groove of the double helix, and it actually "reads" the DNA sequence via hydrogen-bond interactions between amino acids in the protein and nitrogenous bases in the DNA (**Figure 17.17**). The region of the protein that interacts with the nitrogenous bases is known as the **recognition sequence**.

- The active *lac* repressor is a tetramer of four lacI⁺proteins, each of which can bind to an operator sequence to block the opening of the double helix for transcription. One tetramer is capable of binding two operator sequences and will cause the DNA between the operators to "kink" or "loop" (**Figure 17.18**).

- When a repressor interacts with an inducer (lactose or IPTG, an analog of lactose), the inducer binds to central region of the repressor, inducing a change in shape of the repressor tetramer. This **allosteric** influence of lactose on the repressor makes the repressor release the operator sequence, relieving negative regulation of the promoter.

B. INTEGRATING YOUR KNOWLEDGE

Looking Back—Key Concepts from Earlier Chapters

How Genes Work—Chapter 15

With this chapter, you are adding to your knowledge of how genes work at the level of molecular biology. Regulating the expression of genes is a crucial aspect of their relevance to cell growth and adaptability. Regulation can be simplified to the interaction of proteins with specific sequences in the **promote**r region that precedes the **open reading frame (ORF)**. This interaction can increase or decrease the ability of **RNA polymerase** to transcribe that gene.

Looking Forward—Key Concepts in Later Chapters

Principles of Development—Chapter 21

Cellular differentiation in multicellular organisms is fundamentally due to the careful control of gene expression. Your introduction into regulated gene expression in this chapter builds a foundation for your future studies of **tissue differentiation, developmental biology**, and even the cause of diseases like **cancer**. Research into how genes are regulated has broad relevance to many areas of biology.

Animal Form and Function—Chapter 41

It is incredible to imagine a simple microbe "making careful decisions" about which genes to express at any given time. How did this come about? Varied environmental pressures *exert a powerful* selection on microbes, and those that survive from generation to generation must display the ability to adapt to change.

C. MASTERING CHALLENGING TOPICS

Scientists have been examining regulation of the lactose operon for decades. They have identified a complex but fascinating combination of regulatory proteins and DNA regulatory sequences. It is easy to follow how this regulation works when examining one component at a time; but when viewing both positive and negative regulation, things get complicated.

Examine **Figure 17.13** in your textbook, walking through the process that occurs for each of the conditions listed. Which conditions might result in weak expression of the operon? What interactions must occur to have full activation of the operon? This exercise is vital to study of this operon; ideally, it will resolve any confusion you may have, helping you see how positive and negative regulation of gene expression work together.

D. ASSESSING WHAT YOU'VE LEARNED

1. A hypothetical bacterium isolated from a Martian sea uses silicose as its main energy source. This silica-based sugar is one of many sugars present on Mars, and tests have shown that silicose is present everywhere on the planet. Which of the following would be the most efficient type of control for the production of silicase, the enzyme this bacterium uses to metabolize silicose?
   a. constitutive transcription of silicase gene
   b. negative control of transcription of silicase gene
   c. catabolite repression transcription of silicase gene
   d. inducible operon in control of transcription of silicase gene

2. The role of the operator sequence in an operon is to:
   a. orient RNA polymerase at the transcriptional start site
   b. bind a transcriptional repressor protein
   c. bind an transcriptional inducer molecule
   d. encode a transcriptional regulator (positive or negative)

3. The allosteric regulation of protein activity could be described as:
   a. regulation via allelic duplication
   b. an increase in the expression of a protein

c. inactivation by proteolytic cleavage
d. a structural change in a protein, resulting in increased or decreased activity

4. The pattern of β-galactosidase expression illustrated in **Figure 17.2** indicates that:
   a. Lactose is the preferred source of energy in a bacterial cell.
   b. Glucose is the preferred source of energy in a bacterial cell.
   c. Glucose and lactose are utilized equally in bacteria.
   d. β-galactosidase is critical to glucose utilization.

5. The *trp* operon encodes genes for tryptophan biosynthesis and is regulated by:
   a. tryptophan, the *trp* repressor, transcriptional attenuation
   b. the concentration of glucose, transcriptional attenuation
   c. the cell cycle (growth), the *trp* repressor
   d. tryptophan, the *trp* repressor

6. Which of the following statements is true regarding the reaction shown in **Figure 17.12b**?
   a. β-galactosidase tightly regulates cellular glucose concentration.
   b. Increasing the concentration of galactose will block β-galactosidase activity.
   c. Increasing the concentration of glucose will decrease the production of β-galactosidase.
   d. Increasing the concentration of glucose will induce production of β-galactosidase.

7. Which of these conditions would activate DNA binding by the catabolite activated protein (CAP)?
   a. low levels of glucose in the cell
   b. high levels of glucose in the cell
   c. low levels of lactose in the cell
   d. high levels of lactose in the cell

8. DNA footprinting is a valuable technique in the identification of:
   a. open reading frames within genes
   b. the site of specific DNA–protein interactions
   c. genomic loci containing operons
   d. positive regulators of gene expression

9. When comparing the *trp* and *lac* repressor molecules, which of the following is true?
   a. *trp* repressor DNA binding is induced by tryptophan, while *lac* repressor DNA binding is repressed by lactose.
   b. *trp* repressor DNA binding is repressed by tryptophan, while *lac* repressor DNA binding is induced by lactose.

c. *trp* repressor DNA binding is induced by tryptophan, and *lac* repressor DNA binding is induced by lactose.

d. *trp* repressor DNA binding in repressed by tryptophan, and *lac* repressor DNA binding is repressed by lactose.

10. The diarrhea induced by *Vibrio cholera* in the human gut is the result of a bacterial toxin that induces:

a. destruction of blood vessels

b. activation of adenylyl cyclase activity in the human intestinal cells

c. destruction of cell membranes

d. infiltration of bacteria into human intestinal cells

CHAPTER 17—ANSWER KEY

Check Your Understanding

(a) Gene expression is regulated at several places:
- transcription (regulation of promoter activity)
- post-transcriptional processing (RNA splicing, mRNA stability)
- translation initiation (availability of initiation factors)
- post-translational processing (modifications of proteins enhancing activity)

(b) Glucose can be acted on directly by glycolysis and other metabolic processes to produce energy. Lactose must be converted to glucose. Thus the production of energy from glucose is more efficient and is favored by bacteria.

(c) Cultures of cells were exposed to a mutation-inducing compound and then screened for the ability to cleave a lactose analog (ONPG). Digestion of ONPG by β-galactosidase results in the release of a bright yellow molecule. Thus cells that lack β-galactosidase activity are observed as white cells in the presence of ONPG, even under conditions that activate β-galactosidase expression.

(d) The allosteric regulation of a molecule involves alteration of its structure to regulate its activity. The binding of lactose to the lactose repressor, and cAMP to the CAP protein, are examples of structural changes that regulate the capacity of these proteins to bind DNA.

(e) Catabolite repression of the lactose operon involves binding of CAP to the promoter region. The presence of this protein dramatically enhances gene expression by helping to recruit the RNA polymerase to that site. CAP can bind to the promoter only when it is associated with cAMP, an allosteric regulator of its binding activity. The presence of high concentrations of glucose in *E. coli* cells generates a very low level of cAMP in the cytosol; the operon is thus repressed until the glucose is depleted.

There are many examples of catabolite repression in bacterial cells. One prominent example is a group of operons that mediate amino acid biosynthesis. In the absence of a specific amino acid, the operons are activated; but when the amino acid is readily available, the operon is repressed.

Assessing What You've Learned

1. a; 2. b; 3. d; 4. b; 5. a; 6. b; 7. a; 8. b; 9. a;10. b

Control of Gene Expression in Eukaryotes

18

A. KEY CONCEPTS

18.1 Mechanisms of Gene Regulation— an Overview

- Regulation of gene expression in eukaryotic cells is more complex than that observed in prokaryotic cells. As summarized in **Figure 18.1**, eukaryotic cells display regulation at six places within the gene expression pathway: chromatin remodeling, transcription initiation, RNA processing, mRNA stability, translation, and post-translational modification of proteins.

- Eukaryotic DNA is tightly associated with **histone** proteins, generating a DNA–protein complex known as **chromatin**. To express a specific gene, the histones

at the site of that gene's promoter must be dislodged through a process known as **chromatin remodeling**.

- Eukaryotic gene expression involves **RNA processing**; introns are spliced out of the primary transcript, a 5′ cap is added, and a 3′ tail is added to the mRNA. The exons of some genes can be spliced together in different ways to generate unique protein products through a process known as **alternative splicing**. In addition, the stability of a mRNA in the cytosol can be regulated by cellular signals, sustaining or abbreviating the time during which mRNA is available for translation.

18.2 Eukaryotic DNA and the Regulation of Gene Expression

- **Chromatin** is the molecular complex formed between DNA and proteins in the cell nucleus. Proteins in this complex are called histones, and they are positively charged due the high density of **lysine** and **arginine** amino acids in chromatin's structure. DNA is negatively charged due to the phosphate residues in the phosphodiester backbone; therefore, the DNA–histone complex forms spontaneously.

- **Histones** look like a round disk around which the DNA double helix can wrap, protecting DNA and compressing the long molecule. One chromatin complex with DNA wrapped around its structure is called a **nucleosome**. Nucleosomes are visible in preparations of chromatin and can be visualized using an electron microscope (**Figure 18.2**).

- **H1 histone** is important to chromatin struture; it is not part of the disk-shaped complex, but associates with the "linker" DNA between each nucleosome complex.

- The **30-nm fiber** forms as the nucleosomes form a helical structure, wrapping around an axis and further condensing the chromosome. H1 histones interact with each other to mediate the formation of this structure.

Evidence that Chromatin Structure Is Altered in Active Genes

- Harold Weintraub and Mark Groudine tested the hypothesis that transcription of a gene would not occur if it were tightly associated with histones in nucleosome complexes. They tested this idea by examining genes that are expressed differentially in varied tissues.
  1. **β-globin** is expressed at high levels in **reticulocyte** cells (a type of blood cell).
  2. **Ovalbumin** is expressed at high levels in the **female reproductive tract**, but it is not expressed in reticulocyte cells.

- To evaluate any histone association with these two genes in reticulocytes, the researchers employed an assay to degrade DNA that is *not* associated with protein (**Figure 18.3**). The DNase enzyme cannot attach and digest DNA if it is bound tightly to histone proteins.

- Chromatin (purified from reticulocyte cells) is incubated with DNase I and breaks down the DNA not protected by protein association.

- But how can you identify the β-globin and ovalbumin chromosome fragments specifically? You must **probe** the DNA fragments produced for the intact genes (a Southern blot). The DNase fragments generated from reticulocyte cells contained intact ovalbumin genes and degraded β-globin genes. The β-globin gene was not associated with histones, consistent with their hypothesis.

- A second set of evidence was generated in studies with budding yeast (*Saccharomyces cerevisiae*). Mutations in histone genes resulted in abnormal patterns of gene expression, which is consistent with the hypothesis that chromatin has a key role in regulating gene expression.

How Is Chromatin Altered?

- If histone association with a chromosome occurs only at the site of inactive genes, activation of a gene must somehow be preceded by histone removal. How does this work?

- In the budding yeast (*Saccharomyces cerevisiae*), a group of proteins called the **chromatin-remodeling complex** is associated with the loss of histones from the chromosome. These complexes add acetyl ($-COCH_3$) or methyl ($-CH_3$) groups to the histones. These additions disrupt histone association with the DNA and diminish the positive charge on the histone (**Figure 18.4**).

- **Histone acetyl transferases (HATs)** add acetyl groups to histones to allow gene expression; **histone deacetylases (HDACs)** remove acetyl groups to allow chromatin reformation.

- The regulation of histone acetylation and deacetylation is a key area of research in gene regulation. There is much left to be discovered on this topic.

Chromatin Modifications Can Be Inherited

- In mitosis, the pattern of chromatin assembly on chromosomes is passed from mother cell to daughter cell. This serves an important role in maintaining cell identity and functionality by regulating which genes can be transcribed in that cell type. This is a form of epigenetic inheritance—where traits are influenced by factors independent of changes in DNA sequence.

CHECK YOUR UNDERSTANDING

(a) What were the two key intial experiments that revealed chromatin to play a role in regulating eukaryotic gene expression?

_____

_____

_____

_____

18.3 Regulatory Sequences and Regulatory Proteins

Some Regulatory Sequences Are Near the Promoter

- Most **regulatory sequences** are found 5′ or upstream of the open reading frame. The **promoter** is the site where RNA polymerase binds to initiate transcription.

- In eukaryotic cells, the **TATA sequence** is found just upstream of the transcription start site. It is bound by the **TATA-binding protein** (or **TBP**). TBP acts in a similar manner to the sigma factor found in prokaryotic cells, locating the promoter for the RNA polymerase enzyme and aiding in the initiation of transcription. Once transcription is initiated, the TBP dissociates (like sigma factor).

- In the 1970s, Yasuji Oshima and co-workers used the budding yeast, *S. cerevisiae*, to first identify an additional **positive regulator** that aids in regulating gene expression. They focused on genes required to utilize galactose; the genes are transcribed only when galactose is present in the growth medium.

- Yeast cell lines with a mutation in *GAL4* displayed inability to properly activate expression of galactose utilization genes. The protein encoded by this gene was shown to be a positive regulator of galactose utilization genes (similar to the CAP protein in *E. coli*). The GAL4 protein contains a helix-turn-helix motif (discussed in **Chapter 17**), and a **zinc-finger domain** common in DNA binding proteins. But what does the GAL4 protein bind?

- The GAL4 binding site was identified as a short DNA sequence 20 base pairs long and located 5′ of galactose utilization genes (**Figure 18.5**). The binding site is outside of the RNA polymerase binding site (basal promoter) and is called a **promoter-proximal element**. The GAL4 protein binds this site and increases transcriptional activity at the adjacent gene.

Some Regulatory Sequences Are Far from the Promoter

- **Enhancer sequences** are another type of regulatory sequence, but they are unique for their ability to influence gene expression from a long distance (up to 100,000 base pairs from the gene). They can be located upstream or downstream of the gene they regulate, or in intron sequences within the gene.

- Susumu Tonegawa and co-workers selected an **immunoglobulin**-heavy chain gene for study. They recognized that regulation of this gene would be dynamic, and they focused attention on the regulatory elements they hypothesized to be present in the gene's introns.

- Review **Figure 18.6**, which summarizes the work of Tonegawa and coworkers as they characterized regulatory elements associated with this immunoglobulin gene. They suspected that a region within an intron was influencing the expression of this gene. They selectively removed regions of this intron using restriction enzymes (see **Chapter 19**), and then assayed the remaining gene locus for expression. This selective cutting and reconnecting (ligating) is a fundamental process in the manipulation and cloning of gene sequences.

- Gene expression was measured in this assay using a **Northern blot**. This type of analysis allows the researcher to measure the expression of specific genes by quantifying the amount of mRNA present in a cellular extract. RNA samples are separated on an agarose gel (similar to the Southern blot), transferred to a nylon membrane, and then hybridized with a radioactive probe for a specific gene.

- The native and recombinant forms of the immunoglobulin gene were assayed for gene expression. They identified a region in the intron that was required for expression of the immunoglobulin gene; *when it was spliced out, no expression was observed.*

- This was the first demonstration of a regulatory element that regulated transcriptional activation from a considerable distance away. In addition, this element was *downstream* of the promoter.

- This element and others like it are called enhancers. They are grouped together based on the following characteristics:

  1. They can regulate a promoter from long distances upstream or downstream (up to 100,000 bases).

  2. Enhancers can regulate the promoter with similar efficiency when flipped backward at their normal location.

  3. Enhancers can regulate the promoter with similar efficiency when moved to a new location near the gene.

  4. Enhancers increase transcriptional activity at the regulated promoter.

  5. Enhancers are specific to eukaryotic gene expression.

- **Silencers** are similar to enhancers, with varied location and distance to the regulated gene; but they repress rather than activate gene expression.

What Role Do Regulatory Proteins Play?

- Julian Banerji further characterized enhancer sequences, identifying their role in generating a **tissue-specific activation** of gene expression (**Figure 18.8**). The researchers utilized the immunoglobulin gene discussed earlier, revealing that enhancer activity found in the intron was activated only in B cells (where immunoglobulins are made).

- The researchers then spliced the immunoglobulin enhancer sequence into the regulatory regions of the β-globin gene (normally not expressed in B cells). The β-globin transgene was now expressed in B cells, revealing the capacity for the enhancer element to activate a tissue-specific gene expression.

- Transcriptional **regulatory proteins** that are found in these differentiated tissues must bind enhancer, silencer, and regulatory elements found on the chromosome. Indeed, researchers have identified several classes of regulatory proteins that are specific for different types of tissue. Muscle cells express muscle-specific genes because they express muscle-specific regulatory proteins that recognize the regulatory elements of muscle genes.

(b) The GAL4 protein offers important clues into the mechanisms of gene regulation in eukaryotic cells (*S. cerevisiae*). What role does this protein play in activating gene expression under conditions where galactose is present? Is there an analogous protein in bacteria?

_____

_____

_____

_____

_____

(c) Summarize what is known about enhancer sequences.

_____

_____

_____

_____

18.4 Transcription Initiation

- Many DNA-binding proteins that regulate the initiation of eukaryotic transcription have been identified.

- The two classes of regulatory proteins are (1) **basal transcription factors** that associate with the core transcription complex and the promoter sequence, and (2) **regulatory transcription factors** that bind to enhancers and silencers (promoter-proximal elements).

- **Figure 18.10** summarizes the current model of transcription initiation in eukaryotic cells. This model summarizes the work of many researchers. Key to this model is the physical contact between regulatory proteins and the core transcription complex. For this to occur, the DNA must actually fold back on itself, allowing enhancer-binding proteins to come into contact with the core complex.

- Review **Figure 18.10**; initiation of transcription is summarized in the following steps:

  Step 1. Regulatory transcription factors recruit the chromatin-remodeling complex and histone acetyl transferases (HATs).

  Step 2. Acetylation of histones results in a weakening of histone-DNA interactions and generates

"naked DNA" for TFIID binding and transcription initiation.

  Step 3. Additional regulatory transcription factors bind enhancers and promoter-proximal elements. These substances aid in recruiting basal transcription factors to assemble at the promoter, forming the **basal transcription complex** (~10 proteins). The TBP in the complex identifies the TATA box and orients the complex with respect to the transcription start site. Its binding is the first event in assembly at the promoter sequence (**Figure 18.11**).

  Step 4. The RNA polymerase enzyme and associated proteins join the basal transcription complex at the transcription start site to form the **core transcription complex**. Enhancer-binding proteins continue to interact with the large complex through the looping of DNA between their respective locations.

- A key question in the regulation of eukaryotic gene expression focuses on the regulatory mechanisms of these enhancer/silencer-binding proteins. How do they communicate with the core transcription complex?

18.5 Post-Transcriptional Control

Alternative Splicing of mRNAs

- Eukaryotic genes are often broken into exons that are separated by noncoding introns. The primary transcript contains both introns and exons; in the process of post-transcriptional processing, the introns are spliced out to yield the intact open reading frame for translation.

- Some genes exhibit **alternative splicing** patterns, depending on the tissue type or environmental conditions. But what does alternative splicing involve, and how are the protein products different?

- Alternative splicing simply means that the cell splices different exons together, depending on the cell type or condition, generating a unique protein with each splicing pattern. The process of alternative splicing is regulated by cell-type-specific proteins that regulate the splicesome itself.

- One example from text is the mammalian muscle-specific protein **tropomyosin**, encoded in a large gene with 14 exons. Exons that are spliced together depend on the identity of the cell in which the primary transcript is generated (**Figure 18.12**). Smooth muscle cells and striated muscle cells generate alternatively spliced forms of the tropomyosin protein, conferring one important difference in these two cell types. There are many examples of alternative splicing; each generates two or more unique proteins from a single gene.

- At least 35 percent of human genes undergo alternative splicing; though we have only around ~25,000 genes, it is anticipated that we express over ~100,000 different protein products.

mRNA Stability and RNA Interference

- All mRNAs in the cytoplasm have a relatively short life span, though some mRNAs have a much shorter life span than others. Also, the life span of specific mRNAs is regulated by cellular signals. An example is the mRNA for casein—a prominent protein in milk generated during lactation. In non-lactating females the casein mRNA has an average life span of 1 hour, while in lactating females the life span is ~30 hours.

- One mechanism for the regulation of mRNA life span is known as RNA interference (**Figure 18.13**).

- Key steps to RNA interference:

    1. A small RNA with sequence complementarity to the mRNA of interest is transcribed.

    2. The sequences spontaneously forms a hairpin with itself and is trimmed in nucleus.

    3. In cytosol, small RNA is cut into smaller fragments (typically 22 bases in length). Now it is called a mature microRNA.

    4. One strand of microRNA associates with enzyme complex called RISC (RNA-induced silencing complex).

    5. The RISC/microRNA complex now targets mRNAs that have sequence complemetary to microRNA in the enzyme complex. The matching mRNAs are destroyed by RNA degrading activity of the RISC enzyme.

- RNA interference is a new discovery, but it is estimated that 20–30 percent of animal and plant genes are regulated by this process.

How Is Translation Controlled?

- In the cytosol, additional regulatory mechanisms control gene expression.

- Cytosolic factors can bind mRNAs to block their translation until needed. Oocytes (eggs) contain high concentrations of maternal mRNAs whose translation is induced following fertilization. Translation of the mRNAs is blocked by regulatory proteins that bind to them until the release of the mRNA is induced by cellular signals.

- The initiation of translation is mediated by translation initiation factors (TIFs). These can be inactivated by phosphorylation during times of cellular stress, blocking the translation of all but a few transcripts. Viruses can induce this type of response as they try to force the cell to synthesize huge amounts of viral proteins. If the endoplasmic reticulum becomes overwhelmed by proteins in need of processing, the unfolded protein response signal can be activated. This signal induces the phosphorylaton of key initiation factors, blocking the initiation of most cellular translation.

Post-Translational Control

- The activity and stability of proteins is tightly regulated by post-translational events that can include protein glycosylation, cleavage, and phosphorylation (introduced in Chapter 16). Glycosylation (attachment of sugar) of proteins is critical to protein folding and structure. Phosphorylation or cleavage of a protein offers a rapid method for activating or inactivating protein activity.

- The signal transducers and activators of transcription (STATs) are an excellent example of post-translational regulation via protein phosphorylation

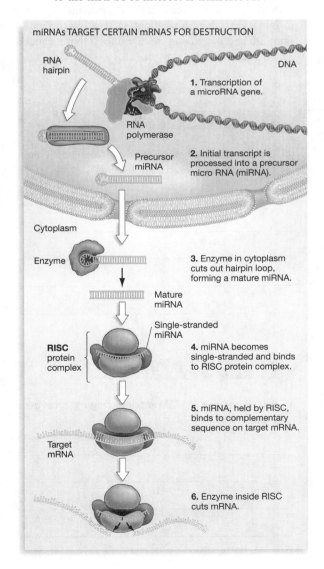

miRNAs TARGET CERTAIN mRNAS FOR DESTRUCTION

RNA hairpin

DNA

RNA polymerase

1. Transcription of a microRNA gene.

Precursor miRNA

2. Initial transcript is processed into a precursor micro RNA (miRNA).

Cytoplasm

Enzyme

3. Enzyme in cytoplasm cuts out hairpin loop, forming a mature miRNA.

Mature miRNA

Single-stranded miRNA

RISC protein complex

4. miRNA becomes single-stranded and binds to RISC protein complex.

5. miRNA, held by RISC, binds to complementary sequence on target mRNA.

Target mRNA

6. Enzyme inside RISC cuts mRNA.

(Figure 18.14). In this case the target of the induced signaling cascade is the activation of expression of select genes.

- **STAT proteins** are **signal transducers** and **activators of transcription**. These proteins are present in cytosol in a dormant state, and they are activated rapidly following receipt of an extracellular signal (e.g., phosphorylation of STAT). Following activation. these proteins can migrate to the nucleus and activate the transcription of specific genes. Genes activated by STAT proteins are often involved in regulation of cell division.

- Mutant forms of some STAT proteins have been shown to bind DNA in the absence of any external signal, and to activate unregulated cell growth (cancer).

- The active form of STAT3 protein is a **dimer** of two activated STAT3 proteins, and formation of the dimer is induced by phosphorylation of a specific amino acid in the protein.

A 50-Year Retrospective: How Does Gene Expression in Bacteria Compare with That in Eukaryotes?

- The four primary differences between prokaryotic and eukaryotic gene expression are (1) packaging of DNA in eukaryotes, (2) splicing of mRNA in eukaryotes, (3) complexity of transcriptional control in eukaryotes, and (4) operons in prokaryotes.

18.6 Linking Cancer with Defects in Gene Regulation

- Humans are composed of trillions of cells, working together to produce a healthy organism. Each day many old, damaged, or infected cells are selectively removed from your body and replaced by the mitotic division of healthy neighboring cells. The maintenance of your size and shape requires a dynamic balance of cell death and cell division. **Cancer** is one disease that results when this balance is lost, and a group of cells grows uncontrollably, ignoring the signals sent to control its division.

- Cancer is frequently the result of mutation. Mutation-causing toxins (**mutagens**) often are shown to be **carcinogenic** (cancer causing) and are labeled as **carcinogens**.

- But how is a normal cell transformed into a cancer cell? As you may suspect, cancer often results from mutations in genes that (1) stop or slow the cell cycle or (2) trigger cell growth/division. Genes that stop or slow the cell cycle are known as **tumor suppressor genes**. When cell activity is lost due to mutation (loss-of-function mutations), cells are released from this negative control of the cell cycle. Genes that trigger cell division are called **proto-oncogenes**. Mutation can modify a proto-concogene into an oncogene (a gain-of-function mutation), where the encoded mutant protein now induces uncontrolled cell division.

p53—Guardian of the Genome

- **p53** is a critical regulatory protein that has been shown to be abnormal in many types of cancer. Its normal cellular function is to regulate the progression of the cell cycle. Loss-of-function mutations in the *p53* gene are present in half of all human cancers.

- Warren Maltzman and Linda Czyzyk revealed that *p53* expression is dramatically activated following exposure to a chemical mutagen.

- *p53* was later shown to act as a **tumor suppressor gene**, arresting cell division following DNA damage (**Figure 18.15**) and allowing the cell to properly repair any damaged genes. The loss of *p53* function allows the cell to replicate DNA despite the presence of mutations, thus dramatically elevating the mutation rate and increasing the likelihood that a cancer-causing mutation may result.

CHECK YOUR UNDERSTANDING

(d) Explain how alternative splicing can generate unique protein products from a single gene.

_____

_____

_____

_____

_____

CHECK YOUR UNDERSTANDING

(e) Explain the role of STAT proteins in regulation of gene expression.

_____

_____

_____

_____

B.   INTEGRATING YOUR KNOWLEDGE

Looking Back—Key Concepts from Earlier Chapters

How Genes Work—Chapter 15

Structures of genes within chromosomes, and the nature of the **central dogma**, are complicated by the revelation of **exons** and **introns** in eukaryotic DNA. Each gene may produce a distinct protein, but the information may be broken up into several distinct units.

Control of Gene Expression in Bacteria— Chapter 17

Now that you've examined regulation of gene expression in eukaryotic cells, review **homologous processes of gene regulation** in prokaryotes. In what ways do regulatory mechanisms from these organisms parallel each other? What are key differences between these organisms?

Looking Forward—Key Concepts in Later Chapters

Principles of Development—Chapter 21

Cellular differentiation in multicellular organisms is fundamentally due to the careful control of gene expression. Your introduction into regulated gene expression in this chapter builds a foundation for your future studies of **tissue differentiation, developmental biology**, and even the cause of diseases like **cancer**. Research into how genes are regulated has broad relevance to many areas of biology.

The Immune System in Animals—Chapter 49

In this chapter, we've examined the regulated expression of immunoglobulins **in B cells. The immune system is one of the most fascinating examples of cellular differentiation to generate cells with a specific function. Your body is constantly replenishing the immune system from undifferentiated** stem cells **in bone marrow, stimulating their maturation to an array of cells that fight infection.**

C.   MASTERING CHALLENGING TOPICS

Processing of mRNA as a form of gene regulation is a recent addition to our understanding of eukaryotic biology. The genes and enzymatic mechanisms that control alternative splicing are not yet well understood. The concept of differentially splicing a primary transcript depending upon cell type and environment is interesting, yet confusing at the same time. How did genes gain this adaptability, enabling a kind of "mix-and-match" strategy of exon combinations? This has become one of the most important areas of study in eukaryotic gene regulation. Scientists had anticipated that the genome project would reveal humans to have ~100,000 genes, basing their predictions on cellular complexity, mRNA diversity, and genome size. Completion of the genome in 2001 (see **Chapter 20**) revealed that humans have only ~30,000 genes. At first this did not seem to make sense, but subsequent research has shown that alternative splicing is very common in human cells. So, though we have only 30,000 genes, we may still be able to generate 100,000 different mRNAs by differentially splicing exons together. Alternative splicing is far from an interesting side note in eukaryotic gene expression; it is just as important as promoter-proximal elements. Certainly there will be many more surprises as scientists characterize the regulation of gene expression.

D.   ASSESSING WHAT YOU'VE LEARNED

1. Which of the following DNA sequences is most likely to serve as a restriction enzyme recognition site? (*Hint:* Find the palindrome!)
   a. GATTAG
   b. CAATTG
   c. AAATTTCCC
   d. CTAGGG

2. Which of the following cells were included as *positive* controls in the experiment described in the previous question?
   a. normal human B cell, control mouse cell, and intact gene in mouse cell
   b. normal human B cell and intact gene in mouse cell
   c. control mouse cell only
   d. normal human B cell only

3. Which of the following is *not* true about eukaryotic regulatory sequences?
   a. Some increase transcription, and others suppress it.
   b. They can function if their 5′ to 3′ orientation is reversed.
   c. They always have the same function in different cell types.
   d. They can function if they are moved to a new location.

4. Which of the following regulatory elements is matched correctly to its chemical composition?
   a. promoter-proximal element; DNA
   b. enhancer; protein
   c. RNA polymerase; DNA
   d. silencers; protein

5. Which of the following is involved in transcription of prokaryotes, but *not* in eukaryotic transcription?
   a. enhancers
   b. RNA polymerase

   c. promoters

   d. operators

6. In smooth and striated muscle cells, DNA for tropomyosin is the same, yet the mRNA produced is different. Which of the following is the cause of this phenomenon?

   a. The presence of cell-type-specific compounds influences the splicing patterns of the mRNA.

   b. The tropomyosin gene is very large and contains a many introns and exons.

   c. Distinct tropomyosin proteins are produced in the two cell types.

   d. A shorter region of the tropomyosin DNA in striated muscles cells is transcribed into mRNA, resulting in a smaller mRNA transcript.

7. If a hormone bound to a cell in order to induce transcription of a given gene by using the STAT pathway, which of the following would most likely occur?

   a. An inactive STAT protein would bind to the enhancer.b.

   b. An activated STAT protein would bind to the gene, inhibiting transcription.

   c. The STAT proteins would become nonphosphorylated and remain in the cytoplasm.

   d. The STAT proteins would become phosphorylated and move into the nucleus.

8. Which of the following could be a mechanism to help prevent cancer?

   a. increased expression of a protein similar in structure and function to p53

   b. STAT proteins that are easily dimerized

   c. STAT proteins that are constitutively active

   d. moving genes involved in cell growth close to regions of a chromosome that are actively transcribed

9. Which of the following is an application of recombinant DNA technology?

   a. animal cloning

   b. in vitro fertilization

   c. rice plants that produce rice with β-carotene

   d. using bacteria to control the insect infestation of food crops

10. Which of these changes to the histone structure generates dissociation from DNA?

   a. proteolytic cleavage

   b. acetylation (negative charge)

   c. glycosylation (sugar addition)

   d. all of the above

CHAPTER 18—ANSWER KEY

Check Your Understanding

(a) Weintraub and Groudine used a DNase enzyme to analyze whether specific genes were associated with histones. If a gene was bound tightly to histones, the DNase is inhibited from cutting the gene into as many pieces as would be generated by a gene that is not bound to histones. They showed active genes to display diminished histone association. A second experiment was performed in the yeast *Saccharomyces cerevisiae*. Yeast cells carry mutations in histone genes that display abnormal gene regulation. Thus the histones must play an important role in regulating gene expression.

(b) GAL4 is a zinc-finger, DNA-binding protein that is analogous to the CAP-binding protein utilized in the lactose operon. In the presence of galactose, this protein binds upstream from the transcription start site of select genes and positively regulates activation of expression.

(c) Enhancers elevate the level of transcription for a specific gene. Enhancers can generate tissue- or cell-type-specific regulation of gene expression.

(d) Enhancers can work even if their normal $5' \rightarrow 3'$ orientation is flipped.

(e) Higher eukaryotic organisms divide their genes (open reading frames) into small fragments called exons, which are dispersed along the chromosome. The primary RNA transcript for the gene must then be spliced together to restore the continuous reading frame for the ribosome. There are now many examples of genes that splice the exons together differently, depending on cellular conditions. Consequently, exons are added or deleted from the final product depending on the splicing decisions. The result is a dramatically altered protein product.

(f) STAT proteins are signaling proteins normally located in the cytosol. These STAT proteins signal the receipt of external growth signals that bind to membrane-associated receptors. The activated receptors mediate phosphorylation of STAT proteins, and the phosphorylated form migrates to the nucleus to activate gene expression. This signaling mechanism is rapid, because the STAT proteins are readily available to carry this signal to the nucleus. The STAT proteins display an altered structure when phosphorylated, and this structural change is thought to activate the nuclear migration and activation of gene expression. Mutations in STAT proteins that generate similar changes in structure have been shown to generate unregulated cell division (cancer).

Assessing What You've Learned

1. b; 2. b; 3. c; 4. a; 5. d; 6. a; 7. d; 8. a; 9. c; 10. b

19

Analyzing and Engineering Genes

A. KEY CONCEPTS

This chapter introduces the strategies scientists use to study individual genes and to determine the function of their encoded proteins. The ability to manipulate genes has revolutionized the study of biology. **Genetic engineering** involves the manipulation of DNA sequences (cutting and recombining gene fragments) to enable experimental study of those sequences.

19.1 Using Recombinant DNA Techniques to Manufacture Proteins: The Effort to Cure Pituitary Dwarfism

- **Pituitary dwarfism** results from the abnormal production of **human growth hormone (hGH)**, encoded by the gene *GH1*. This disease can be inherited from generation to generation, or it can occur spontaneously.

- The mutant form of the *GH1* gene is **recessive** to the dominant allele; thus humans can carry the mutant allele without displaying any outward abnormal phenotype.

- Humans affected by pituitary dwarfism grow slowly, reaching a maximum adult height of about 4 feet.

- hGH can be purified from pituitary glands dissected out of **human cadavers**. Isolating the hormone in this way is expensive, but it is effective in treating pituitary dwarfism.

- In the 1980s researchers discovered contamination of this hormone supply with the infectious agent that causes **neurodegenerative** disorders (**Kuru and Creutzfeld-Jacob disease**). Kuru was first observed in cannibalistic tribes in **New Guinea**, passing from person to person via consumption of human tissue. The discovery that some patients receiving hGH therapy had acquired the infectious agent caused this source of hGH to be banned as a supply source for human treatment.

Using Recombinant DNA Technology to Produce a Safe Supply of Growth Hormone

- Recombinant DNA strategies for producing human growth hormone focused only on *cloning the human gene*, introducing the gene into the *Escherichia coli* bacterium, and having these *microbes synthesize the hormone* (a protein). This supply of hormone would be comparatively inexpensive and disease free.

- **Overview of genetic cloning (Figures 19.2 through 19.5)**—To isolate the *hGH* gene from human cells, researchers first isolated GH1 mRNAs from pituitary gland tissue, copied all of these mRNAs into **cDNA** using **reverse transcriptase**, and inserted these cDNA fragments into circular plasmid DNA.

- The process of copying a mRNA into a cDNA is summarized in **Figure 19.1**. The reverse transcriptase enzyme is unique to a type of RNA virus (retrovirus) that is able to convert its RNA genome into DNA inside an infected cell. This enzyme is now used to make single-stranded DNA copies of mRNA, called cDNA (complementary DNA). To convert the single-stranded cDNA into double-stranded cDNA, a DNA polymerase is typically used.

- DNA fragments can be attached to each other using an enzyme called **DNA ligase**. The double-stranded DNA can be ligated into a circular plasmid, as shown in **Figure 19.2**, to generate a recombinant molecule. In our example, cloning the GH1 gene, the double-stranded cDNA would be ligated (attached) into the plasmid.

- The plasmids are then used to carry the cDNA into living *E. coli* cells. Researchers can identify bacterial cells carrying the plasmid by their ability to grow in the presence of antibiotics (antibiotic resistance genes are engineered onto the plasmid).

- The researchers then screen all of the bacteria in search of those carrying the *hGH* cDNA (GH1) within a plasmid. The scientists must use a DNA probe to examine each colony of cells. Key steps of this process:

1. *Using plasmids*—**Plasmids** are small, circular segments of DNA carried in prokaryotic cells (separate from the chromosome; see **Chapter 7**). Scientists use these plasmids as **vectors** for DNA cloning and manipulation; they have engineered many new plasmids just for use in research, typically adding an **antibiotic resistance** gene and other components to the plasmid sequence. Thus simply adding antibiotics to the media can determine the presence of plasmid within a cell; cells without the plasmid will die, while those with plasmid grow normally. Thus scientists can amplify millions of copies (clones) of a piece of DNA simply by growing bacterial cells that carry that plasmid.

2. *Using restriction endonucleases* (**Figure 19.2**)— Scientists can choose from a large group of enzymes for use in cutting DNA at specific places. These restriction enzymes are made by bacteria to defend against viral infection. The enzymes cut the viral DNA but ignore similar sequences in the bacterial DNA, and the bacteria block those sites by methylating nitrogenous bases at that site. Restriction enzymes are a key tool in molecular biology; they give researchers a method of cutting DNA at specific places. Thus scientists can selectively isolate a specific piece of DNA from a plasmid or other DNA segment. Restriction endonucleases cut at specific **palindrome** sequences in the DNA (they read the same backward and forward). **Figure 19.2** (step 4) requires the utilization of a restriction enzyme to ensure that the cDNA is inserted into the correct site. The pieces of DNA being attached to each other must have complementary "sticky ends," which are the short, single-stranded sequences generated by a restriction enzyme digestion. After ligating the pituitary cDNAs into plasmids (now called a **library** of cDNAs), the scientists needed to amplify these plasmids for further screening. Amplification is done easily in bacterial cells.

3. *Transformation of Bacteria*—Transferring a plasmid into a bacterial cell requires the DNA to cross the cell wall. To do this, scientists use **chemical treatment** or even a **pulse of electricity** to open holes in the cell wall. Once the DNA is inside, the cells are grown in the presence of antibiotics to select for cells carrying a plasmid and to kill cells that are not carrying plasmid. Typically, a tube of cells is transformed with the collection of plasmids and then placed on a petri dish containing media plus antibiotic. Individual cells carrying the plasmid will grow into a pile of cells, known as a clonal colony of cells.

4. *Screening the cDNA library for GH1*—To probe for the bacterial cells carrying the GH1 gene, scientists had to generate a **probe** for the gene. Scientists use many types of probes to search DNA or RNA for specific sequences; but regardless of the type, each probe searches for its target by looking for nucleic acids to which it can **hybridize**. Double-stranded DNA is bound together through hydrogen bonding of its **nitrogenous bases**. DNA probes the hydrogen bond with **complementary sequence** in exactly the same way, and it will search through long DNA and RNA sequences to hybridize to its exact match. **Figure 19.5** displays the screening of thousands of individual bacterial colonies with a *GH1* DNA probe. An imprint of each colony is generated on a positively charged filter paper (DNA is negatively charged). The DNA from each imprint is treated to make it single stranded, and it is then permanently attached to the filter. The filter is then probed with a radioactively labeled *GH1* probe to identify which colony on the original plate contains the *GH1* gene.

5. Once the correct plasmid-bearing bacteria are identified, the researchers can use the cloned gene to synthesize large quantities of human growth hormone (*hGH*). This is accomplished by placing the cloned gene under control of a strong bacterial promoter and putting a plasmid carrying this construct into bacteria. Bacterial cells then make a large quantity of the hGH protein that can be purified and used to treat patients.

Southern Blotting (Box 19.1)

- The screening of colonies (step 4 above) is very similar to Southern blotting (described in **Box 19.1**). The key difference is that Southern blotting involves no bacteria, only the analysis of pure chromosomal DNA isolated from some source. The DNA is typically digested with a restriction enzyme, run on an agarose gel, denatured to single-stranded DNA, transferred to positively charged filter paper, and then probed with a gene-specific radioactive probe. This technique is a powerful strategy for analyzing chromosomal DNA (**Figure 19.8**).

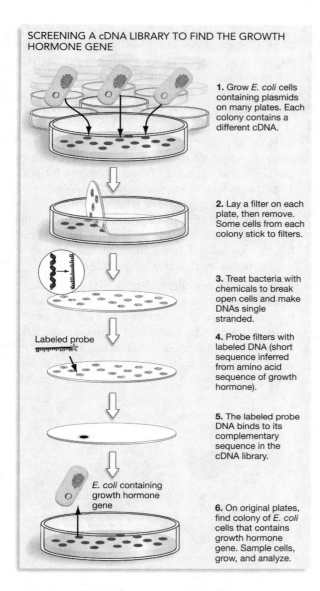

SCREENING A cDNA LIBRARY TO FIND THE GROWTH HORMONE GENE

1. Grow *E. coli* cells containing plasmids on many plates. Each colony contains a different cDNA.

2. Lay a filter on each plate, then remove. Some cells from each colony stick to filters.

3. Treat bacteria with chemicals to break open cells and make DNAs single stranded.

Labeled probe

4. Probe filters with labeled DNA (short sequence inferred from amino acid sequence of growth hormone).

5. The labeled probe DNA binds to its complementary sequence in the cDNA library.

E. coli containing growth hormone gene

6. On original plates, find colony of *E. coli* cells that contains growth hormone gene. Sample cells, grow, and analyze.

19.2 Another Approach to Cloning: The Polymerase Chain Reaction

- The **polymerase chain reaction (PCR)** is simply a methodology for amplifying short fragments of DNA from a few copies to billions of copies. Kary Mullis developed PCR in the early 1980s. The strategy exploits a heat-resistant DNA polymerase cloned from a bacterium that grows readily in hot springs (*Thermus aquaticus*).

- The technique is extraordinary due to its simplicity and its value to several areas of genetic analysis: gene cloning, forensic analysis, and rapid analysis of relatedness or ancestry, to name a few.

- The technique is limited only by the need for a target DNA sequence, which is used to design primers for that specific target (**Figure 19.6**).

(a) Why is recombinant hGH better as a therapeutic agent than hGH isolated from human cadavers?

_____

_____

(b) In the process fo using a DNA probe to detect single-stranded DNA sequences on a filter paper, how does the probe "search" for the target sequences?

_____

_____

_____

_____

- A PCR reaction contains just a few key ingredients:
  1. DNA template
  2. Two DNA primers matching the template target site
  3. dNTPs (nucleoside triphosphates)
  4. Buffer
  5. Heat-resistant DNA polymerase (known as *Taq* DNA polymerase)

- A PCR reaction requires about 30 cycles, with each cycle containing time points at three different temperatures:
  1. A denaturing step to separate the strands of the target sequence (95°C)
  2. An annealing (primer binding) step to allow the primers to attach to the target site via hydrogen bonding of complementary sequence (45–70°C)
  3. An extension step to allow the thermostable DNA polymerase to extend the primer and synthesize new DNA in the 5' to 3' direction (72°C)

- How do these changes in temperature amplify DNA? Examine **Figure 19.7**, which illustrates several keys to understanding PCR.
  1. Denaturation of the template means to separate the double-stranded DNA to single strands (using temperature).
  2. PCR primers are designed for a DNA target sequence, annealing on each side of the target and offering a 3'-OH group for DNA polymerization.
  3. The primers anneal or bind to opposite strands of the DNA template. Remember, if DNA synthesis

is always 5' to 3', this process works only if the primers activate DNA synthesis toward each other (steps 3 and 4).

4. The extension step allows the thermostable DNA polymerase (*Taq*) to synthesize a new strand of DNA, starting at the 3' end of each primer.

5. In one cycle, we've doubled the number of copies of the target sequence. Imagine repeating this process for 30 cycles!

PCR in Action: Studying Fossil DNA

• Svante Pääbo and colleagues used PCR to compare DNA from *Homo neanderthalensis* and *Homo sapiens*, testing a hypothesis that these species may have interbred in the past. They first amplified small pieces of *H. neanderthalensis* mitochondrial DNA, isolated from ancient bones (estimated to be 30,000 years old). With the same DNA primers, they used PCR to amplify the same regions of *H. sapiens* mitochondrial DNA. They then used dideoxy sequencing on the amplified DNA fragments to compare them (see next section). They found that *H. sapiens* and *H. neanderthalensis* mitochondrial DNA sequences were very different; thus, it is unlikely these two species ever interbred.

19.3 Dideoxy DNA Sequencing

• Fredrick Sanger developed dideoxy sequencing, a key strategy for DNA analysis (summarized in **Figure 19.9**).

• Gene sequence information is of tremendous value to a scientist studying that gene. The DNA sequence can be examined to determine the amino acids in the encoded protein. In addition, the DNA sequence can be used to identify other genes with related DNA sequence and presumably related protein function.

• The DNA sequencing procedure is based on dideoxynucleosides. Review the structure of a normal **deoxynucleoside** and a **dideoxynucleoside**. Why is the missing 3' –OH group so important in DNA polymerization? How might the addition of a dideoxynucleoside—to the 3' end of the DNA strand being synthesized—affect the polymerization process?

• Once a dideoxynucleotide is added during DNA polymerization, the extension of that strand is stopped. Though the DNA polymerase is still present at the site, there is no available 3' –OH group to bond with an incoming nucleotide. Thus the reaction is stopped precisely at that site.

• All polymerization reactions are initiated with a primer that anneals to the region of DNA to be sequenced. Each extension from this primer proceeds for some distance, unless a dideoxynucleoside is

added to the chain. Thus, you can "see" where A, T, G, or C nucleotides are added by examining the fragments generated during a series of polymerization reactions containing the dideoxynucleotides.

• Four separate reactions are performed. Why? The dideoxynucleosides are added differentially to each of the four reactions (one has ddTTP, one has ddCTP, etc.). Following the reaction, the mixture of components is run into a gel matrix that denatures the DNA and separates DNA fragments by size. These fragments are detectable on X-ray film, as the polymerization reaction is performed with radioactively labeled dNTPs (**Figure 19.9**). Try reading the DNA sequence on the auroradiogram shown in this figure. Can you determine the sequence, and the 5' to 3' direction, of the template strand?

• Automated sequencing has made Sanger DNA sequencing easier and faster to perform. Instead of four reactions to sequence from one primer, only a single reaction tube is required. The key modification is the use of fluorescently labeled ddNTPs, with each type of dideoxy nucleotide being labeled with a differently colored fluorescent dye (termination

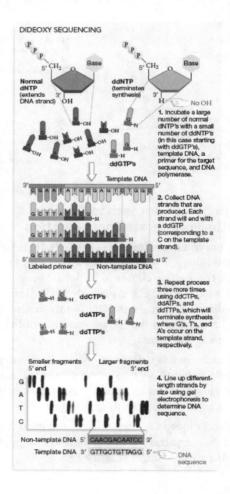

with a "ddATP" generates a red fragment, etc.). The mixture of fragments is analyzed in a capillary tube with a detector at the bottom of the tube. As each colored fragment emerges from the column, the detector identifies the fluorescent color and sends the data to a computer (**Figure 19.10**).

19.4 Finding Genes by Mapping: The Huntington's Disease Story

- Nancy Wexler and colleagues' search for the mutation behind **Huntington's disease** is an excellent example of how molecular biology has dramatically enhanced research strategies. This disease is an inherited **neurodegenerative disorder** resulting from a dominant mutation. It includes symptoms such as loss of motor function, changes in personality traits, and decreased intelligence. Onset of symptoms is usually at about 40 years of age.

- Once scientists were certain that this disease passed from generation to generation, they set out to use **pedigree analysis** to find the gene that is mutated in these families. But how does this work? How does a pedigree give information about the gene's identity?

- Researchers must first determine the mode of inheritance (dominant or recessive, X-linked or autosomal). By examining the pedigree data, scientists can generate reliable predictions of the mode of inheritance, though human pedigrees often offer a limited amount of information. Examine **Figure 19.12**; what mode of inheritance is consistent with this pattern of inheritance?

  1. In this example, *all affected individuals were born to affected parents* (no "skipping of generations"). There is no example of unaffected parents generating an affected child (which would be consistent with a recessive mutation).
  2. Also, consistent with autosomal mutations, there seems to be fairly equal distribution of affected males and females.
  3. The mutation displays an **autosomal dominant** mode of inheritance.

How Was the Huntington's Disease Gene Found?

- The DNA samples available from affected families can be analyzed with genetic markers to follow the inheritance of specific DNA regions from generation to generation.

- Researchers look for cosegregation (or linkage) of chromosomal markers with the disease phenotype, hoping to find which chromosome, and where on that chromosome, the disease-causing mutation is located (**Figure 19.11**). But to establish cosegregation of the markers and the mutation, researchers must analyze DNA from many members of affected families.

Nancy Wexler and colleagues examined the DNA of people living near Lake Maracaibo, Venezuela. People in that region have a high incidence of early onset Huntington's disease (symptoms appearing in early adulthood). By comparing patterns of inheritance in these families, and searching for regions of chromosomes that are passed from one affected individual to his or her children, the location of the mutation was eventually discovered.

- Genetic markers are simple differences in chromosomes, generally within noncoding regions. Genetic markers used include restriction enzyme cut sites (RFLP analysis), hypervariable repetitive sequences, and more recently the detection of single nucleotide polymorphisms (SNPs). Each individual inherits these polymorphic sites from his or her parents, and thus a specific DNA region inherited from a parent can be identified by the genetic markers associated with it (described as a haplotype). Through this type of genotyping, we can "see" a chromosomal region as it is passed on through several generations. We see that region by identifying its associated marker in individuals within the pedigree.

- **Restriction fragment length polymorphisms (RFLPs)** are differences between chromosomes that are measured as the loss or gain of a restriction enzyme cut site. Digestion of the chromosome at this site generates a specific group of DNA fragments that can be detected by a probe for that DNA region. This procedure works somewhat like the Southern blots described in **Box 19.1**. The presence or absence of restriction enzyme cut sites generates an altered pattern of bands on the resulting filter. Thus individuals with slightly different DNA sequences are revealed through this RFLP analysis. Wexler and colleagues used RFLP analysis to examine the patterns of RFLP marker segregation through families carrying the Huntington's disease mutation. Eventually they identified a region on chromosome 4 that was always passed from one affected individual to his or her affected offspring.

- Imagine performing this analysis on DNA samples from a family affected by a disease-causing mutation. These analyses might find an RFLP that is *always* present in affected individuals and is *never* present in unaffected individuals! This result may suggest that the disease-causing gene is closely linked to the chromosomal location of that RFLP. Researchers use probability-based statistical analyses to establish linkage between a marker and the specific disease.

- Once the general location on chromosome 4 was established, the researchers had to examine all the genes at that site to determine which gene is mutated in Huntington's sufferers.

- Pinpointing the defect: Once the location of the Huntington's disease gene was narrowed to ~500,000 base pairs on chromosome 4, the next step was to isolate all the genes in this region and examine them for mutations.

- Each mRNA generated from this site was isolated, and the gene sequence was determined from its cDNA.

- Scientists found one gene within the isolated chromosomal region that was abnormal in people with Huntington's disease. It had an expanded **trinucleotide repeat** (CAG) at the 5′ (front) end of the coding region. Normal humans have about 11–25 of these repeats, and affected indi-viduals have 40 or more. The codon codes for **glutamine**, and affected individuals have a protein that is inactivated due to the effect that all the extra glutamines have on its configuration.

- The protein encoded by this gene was called **huntingtin**, and the gene was named *IT15*.

What Are the Benefits of Finding a Disease Gene?

1. Improved understanding of phenotype: Identifying the specific protein whose defecient function generates a disease phenotype gives researchers enhanced insight into that disease. For example, Huntington's disease is induced by the formation of aggregates of the huntingtin protein (IT15). These aggregates induce apoptosis in neurons. Thus, successful treatment of this disease will likely require a method for inhibiting the formation of these aggregates.

2. Therapy: A key route to developing effective therapies for a disease includes the development of an animal model. If the gene defect is known, animals can sometimes be engineered to display the disease phenotype. These model animals are an excellent starting point for the development of therapies.

3. Genetic testing: Once the disease-causing gene is known, reliable tests for the defective gene can be developed and offered to the public. These tests can help people to determine if they are carriers for a disease before they choose to have children. Prenatal testing also can be made available.

19.5 Can Gene Therapy Cure Inherited Diseases in Humans? Research on Severe Immune Disorders

- Gene therapy involves the introduction of a gene that will replace or augment a mutant gene that is causing an abnormal phenotype. Three criteria must be met for this process to be successful.

CHECK YOUR UNDERSTANDING

(c) What characteristics in a pedigree would suggest to you that a specific mutation was autosomal and recessive?

_____

_____

_____

1. The wild-type allele for the gene in question must be sequenced and its regulatory sequences understood.
2. There must be a method for introducing a copy of the normal allele into affected individuals.
3. The normal allele must dominate a recessive mutant allele, or replace a dominant mutation completely.

How Can Novel Alleles Be Introduced into Human Cells?

- The greatest difficulty in gene therapy strategies is stable introduction of the therapeutic gene into many cells in a human patient.

- The two prominent vectors for introducing these genes into human cells are both **viral vectors**. Because they have been stripped of genes that might cause damage to the recipient cells, these viruses serve only as a vehicle for delivery of the transgene.

- The gene is transferred into the viral sequence, and the virus is used to infect the patient's cells.

- The two most commonly used viruses are
  1. **Retroviruses**—These contain an RNA genome and convert themselves to DNA upon infection of the host cell. This type of virus **integrates** into the chromosomal DNA and remains there indefinitely while expressing the new gene (**Figure 19.13**).
  2. **Adenoviruses**—These viruses contain a DNA genome and are a common type of virus that can infect cells in the respiratory tract through inhalation. These viruses do not integrate into the chromosome, and thus are carried in infected cells for only a short period.

Using Gene Therapy to Treat X-Linked Immune Deficiency

- The disease known as **severe combined immunodeficiency (SCID)** is devastating; it profoundly weakens

the immune system. In this disease, white blood cells are unable to properly mature into **T cells**.

- A research team led by Marina Cavazzana-Calvo developed a strategy to cure one category of SCID, in which a gene called *SCID-X1* (a gene on the X chromosome) is mutated. The gene encodes a **growth factor receptor** called γc that receives a growth-activating signal from outside the cell.

- The team developed a retrovirus carrying the wild-type copy of this gene and tested the retrovirus on cells from dogs and mice.

- The first attempt at using this strategy on humans included two male infants, each affected with the mutant *SCID-X1* gene. No bone-marrow donors were available for either child.

- The research team removed **bone-marrow cells** from each of the boys and then infected the cells with the retrovirus they had constructed (**Figure 19.15**). The cells were then reintroduced into each patient, and gradually each child began to produce his own T cells. Researchers are now carefully watching T-cell production in these young boys to ensure that the transgene continues to provide a normally functioning γc receptor.

19.6 Biotechnology in Agriculture: The Development of Golden Rice

- Most strategies for genetic engineering in agriculture focus on reducing herbivore damage, reducing competition between the crop plant and weeds (making the crop more herbicide resistant), and improving the quality of the food product.

Synthesizing β-Carotene in Rice

- β-carotene is a dietary component that can be rapidly converted to vitamin A. In countries with diets dominated by rice, people commonly suffer vitamin A deficiency. There is a small amount of β-carotene in the rice husk, but it is removed in the polishing step of rice processing.

- Ingo Potrykus and colleagues set out to make a rice crop that was enriched for β-carotene to relieve this dietary problem.

- As shown in **Figure 19.16**, the synthesis of β-carotene can be accomplished through the conversion of a molecule called geranyl geranyl diphosphate (GGPP) in an enzyme-mediated process. Potrykus recognized that GGPP was present in the endosperm of rice seeds. His team set out to introduce into the rice endosperm the enzymes needed to convert GGPP to β-carotene. The result would be a rice seed with an accumulation of this dietary precursor to vitamin A;

but to accomplish this, he would need to make a transgenic plant.

The Agrobacterium *Transformation System*

- Researchers who do genetic engineering in plants often use a bacterium (***Agrobacterium tumefaciens***) to mediate the process of transferring genetic information into plant cells. This bacterium often infects plants, and as part of the infection process, the bacterium induces growth of a gall (a tumor-like mass) in the plant (**Figure 19.17**).

- The gall-inducing genes are found on a plasmid in the bacterium (Ti plasmid). The plasmid has several components, but the key part is a DNA sequence (T-DNA) that is transferred into the host plant's cells to integrate into the chromosome.

- Researchers recognized the value of DNA being transferred into the plant cell this way, and they have exploited the process to carry other gene sequences along with the T-DNA sequence.

- Potrykus used *Agrobacterium* to carry the β-carotene genes into cells from a rice plant, eventually generating a transgenic plant—now called golden rice (it is yellow from the high concentration of β-carotene).

CHECK YOUR UNDERSTANDING

(d) When compared to retroviral gene therapy strategies, what disadvantages are there for using adenoviruses instead? Advantages?

_____

_____

_____

(e) Loss of the gene *GH1* results in the type I form of pituitary dwarfism, in which affected individuals are benefited simply by receiving an external source of human growth hormone (hGH). In SCID, another example from this chapter, T cells are unable to properly mature, even in the presence of the growth-stimulating signal. Why is the introduction of the external signal sufficient in one case but not the other? Explain.

_____

_____

_____

_____

B. INTEGRATING YOUR KNOWELDGE

*Looking Back—Key Concepts
from Earlier Chapters*

Mendel and the Gene—Chapter 13

Sex-linked traits were introduced in **Chapter 13** with a discussion of inheritance patterns displayed by a **white-eye trait** in *Drosophila melanogaster*. Thomas Morgan was able to make sense of the unusual segregation pattern of X-linked traits in reciprocal crosses. In humans, examining patterns of inheritance over several generations can reveal X-linked traits.

Control of Gene Expression in Eukaryotes—Chapter 18

Restriction enzymes are a key tool in the characterization and recombination of DNA sequences. Microbes make these enzymes as a defense against viral infection (they degrade the viral DNA during infection). The enzymes are key in analyses such as **Southern blotting**, **RFLP analysis**, and **gene cloning** into vectors.

Looking Forward—Key Concepts in Later Chapters

Viruses—Chapter 35

The life cycle of a virus can be quite complicated, sometimes including a period of integration into the host genome (**lysogeny**) before replicating itself and emerging from the host cell. Researchers have harnessed the lysogenic capacity of some viruses to carry genes into the human genome in **gene therapy** strategies.

The enzyme **reverse transcriptase** is produced by **retroviruses** to copy their RNA genome into DNA as part of the infectious process. Reverse transcriptase is an unusual enzyme that has become tremendously valuable in the production of **cDNA** for research.

C. MASTERING CHALLENGING TOPICS

Imagine how tedious the early research into Huntington's disease really was. Researchers worked together to find regions in the human genome that displayed variability regarding the presence or absence of a specific restriction enzyme cut site. The researchers then would develop a probe for this region that could be used to screen blots of digested DNA. They prepared DNA samples and screened them for each of these polymorphic regions to determine the individual's RFLP makeup. The data for this type of analysis would be complicated to look at on a gel, but eventually each individual would be typed for each locus. However, then what would the researchers do?

Somehow within the data there is linkage between a marker and the disease-causing allele. With Huntington's disease the mutation was dominant, and thus each

person with the mutant allele would display the disease eventually (after age 35).

Examine the pedigree displayed in **Figure 19.12**, which displays only the RFLP makeup of each individual at the chromosome 4 marker. Do you see the 100 percent correlation (cosegregation) between Huntington's disease and the C haplotype? There is no correlation with the A, B, or D haplotypes. It is this type of correlation that scientists hope for as they examine *thousands* of individual genetic loci. If a disease were caused by a recessive mutation, would the process of mapping the gene be significantly influenced? Explain.

Today the methodology is faster, using PCR to amplify polymorphisms and enabling the simultaneous screening of many loci. But even with these improvements, the mapping of a disease gene is truly a daunting task requiring a great amount of patience and endurance.

D. ASSESSING WHAT YOU'VE LEARNED

1. Genetic engineering became possible with the discovery of which of the following compounds?

   a. restriction enzymes and DNA ligase
   b. ddNTP molecules
   c. DNA polymerase
   d. reverse transcriptase

2. Which of the following is the main reason it was necessary to develop a method of using recombinant DNA technology to produce human growth hormone for treating individuals with pituitary dwarfism?

   a. There was no available source of functional growth hormone.
   b. The drug was extremely expensive.
   c. The growth hormone supply led to hereditary diseases.
   d. The growth hormone supply was contaminated with infectious agents.

3. Try reading the gel presented in **Figure 19.9**. Determine the first 10 nucleotides on the 3′ end of the *template* sequence by reading the X-ray film. Which of the following sequences are correct?

   a. 3′–GGA GAT CAA T–5′
   b. 3′–CCT CTA GTT A–5′
   c. 3′–GTA TCC AGA G–5′
   d. 3′–CAT AGG TCT C–5′

4. The plasmid used in the experiment illustrated in **Figure 19.2** of your text must include which of the following, before being inserted into *E. coli* cells?

   a. a marker gene
   b. restriction recognition site
   c. cDNA synthesized from mRNA of gene being examined
   d. all of these

5. Refer to **Figure 19.5** to answer this question. Which of the following would be the most likely result if step 3 of the procedure were omitted?

   a. The X-ray film would not show any black spots.
   b. The X-ray film would show multiple black spots (one for each colony on the plate).
   c. Half of the colonies would demonstrate radioactivity on the X-ray film.
   d. The results would be identical to step 5 in the figure.

6. Restriction enzymes are formed in bacteria to break down viral DNA. When restriction sites are placed on a human gene, and human DNA and viral DNA with the same restriction site are cut with the restriction enzyme, the "sticky ends" of the two can recombine, and the virus may be used to incorporate the gene into a human chromosome.

   Pick the statement that best describes the paragraph above.

   a. This is not true; bacteria are not able to cut viral DNA in the way described.
   b. The DNA from a human and a virus would not exhibit the same sequence at the restriction site.
   c. A virus would not be able to insert genetic material into a human chromosome.
   d. This is an example of the utilization of recombinant DNA technology in gene therapy.

7. How is the pedigree of an individual generally constructed?

   a. by sequencing the genome of the individual
   b. by interviewing the individual regarding the sex and physical traits of his or her family members
   c. by sequencing the genes of the individual's parents
   d. by analyzing the genetic fingerprint of the individual and of all of his or her family members

8. One example of a disease caused by the expansion of a repetitive DNA sequence (codon repeats) is:

   a. leukemia
   b. pituitary dwarfism
   c. fragile-X syndrome
   d. Down syndrome

9. Which of the following is a disadvantage of using biotechnological techniques in agriculture?

   a. Genetically modified corn grows poorly and is less nutritious.
   b. Transgenic herbicide-resistant plants may transfer their resistance transgene to closely related weeds (through pollination), making them resistant as well.
   c. Genetically modified crops are often slightly poisonous.

   d. It is expensive to produce seed from transgenic crop plants because the T-DNA impedes seed production.

10. Golden rice offer a potential dietary benefit because the rice grains are engineered to synthesize β-carotene from:

   a. vitamin A
   b. vitamin C
   c. geranyl geranyl diphosphate
   d. glucose

CHAPTER 19—ANSWER KEY

Check Your Understanding

(a) Isolation of hGH from cadavers is very expensive and carries with it the risk of transferring disease (prion disease).

(b) The probes used in Southern blotting, Northern blotting, and the blots used in RFLP mapping are composed of single-stranded DNA that has been labeled in some way (typically with radioactivity). The single-stranded DNA will anneal tightly with any DNA sequences that are composed of the reverse, complementary sequence of the probe sequence. The probe and the target will anneal to one another in a manner similar to double-stranded DNA annealing to itself. Researchers can then localize the probe attached to the blot by exposing an X-ray film to the filter paper.

(c) Recessive mutations often appear in pedigrees, with affected children being born to unaffected parents. Also, recessive mutations typically skip generations within the pedigree, and sometimes they may not reappear unless two related individuals have children together. If the sexes of affected individuals in the pedigree are equally distributed, the mutation is likely in an autosomal gene.

(d) Adenoviruses can be inhaled by the patient, and this offers an excellent delivery mechanism for gene therapy strategies involving the lungs and respiratory system. But these viruses don't integrate into the host-cell genome and don't offer long-term transgene expression.

(e) Type I pituitary dwarfism is caused by mutation in a gene responsible for the synthesis of hGH. SCID, in the gene therapy example, is caused by a deficient growth factor receptor. Thus, in SCID-X1, the disease results from cells being unable to *receive* the signal, while the pituitary dwarfism is caused by an inability to *send* the signal.

Assessing What You've Learned

1. a; 2. d; 3. b; 4. d; 5. a; 6. d; 7. b; 8. c; 9. b; 10. c

20

Genomics

A. KEY CONCEPTS

20.1 Whole-Genome Sequencing

- The first complete sequencing of the genome for a prokaryotic cell was completed in 1995 (*Haemophilus influenzae*), and the first eukaryotic genome followed soon after in 1996 (*Saccharomyces cerevisiae*).

- The human genome sequencing project was essentially completed in 2001 (3 billion nucleotides). Genome sequencing data are publicly available through the GenBank website, which now contains more than 100 billion nucleotides of DNA sequence.

- The capacity for sequencing facilities to generate sequence data has advanced rapidly in recent years. Review **dideoxy sequencing**, a key research tool introduced in **Chapter 19**. Scientists still use this strategy but have changed some aspects of detection to speed the process by automation.

- Fluorescent tags are now used in DNA sequencing. These are attached to the dideoxynucleoside triphosphate molecules (**ddNTPs**), and they label each DNA fragment generated in the sequencing reaction *at the termination event*. Thus if all the labeled ddNTPs (ddATP, ddGTP, ddCTP, ddTTP) are added to a single sequencing reaction that also contains dNTPS, they will occasionally be incorporated as the polymerase synthesizes the complementary strand. With the incorporation of a ddNTP, the polymerization is arrested, generating a DNA fragment that contains the fluorescent tag on one end.

- The ddNTPs are each labeled with a different-colored **fluorescent tag**, allowing researchers to determine which ddNTP addition terminated the reaction (e.g., ddATP = red, ddGTP = blue, etc.).

- Instead of running four separate reactions labeled with radioactivity, the researchers use one reaction and analyze the fragments by their fluorescence. A computer reads the color of each fragment generated in the sequencing reaction (analyzed by size) and automatically displays the DNA sequence found in the template strand.

How Are Complete Genomes Sequenced?

- The availability of automated sequencing facilities dramatically speeds up the process of genome sequencing, but researchers still require DNA template and primers to analyze the sequence of specific regions of a chromosome. How does this process work, and what methods improve the speed of this process?

- Chromosomes are very large strands of DNA, and they must be cut into smaller fragments (using **restriction enzymes**) for sequencing. One **vector** for carrying fragments of a chromosome is called a **bacterial artificial chromosome (BAC)**. Vectors are DNA strands used in cloning to carry DNA fragments inside their circular structure. A BAC is unusual in that it can carry very large fragments of DNA (up to 200,000 base pairs).

- What does it mean for a vector to "carry" a piece of DNA? In cloning, scientists use *Escherichia coli* to make many copies of the DNA being studied. They simply place the microbe in a flask of media, and one cell carrying the vector can rapidly grow to billions (each with the vector). Thus we can now isolate the vector from the bacteria and have many copies to work with.

- In genome analysis, the entire genome of an organism is cut into fragments and transferred into BACs for amplification in bacteria. This collection of BACs is called a **BAC library**.

- The BAC inserts are then analyzed individually. Each is purified, chopped into smaller 1-kb fragments using sonication, and inserted into another circular vector, called a plasmid. The resulting collection of plasmids is called a **shotgun library** because it contains small fragments of the original large piece of DNA. Sequencing reactions are performed on each plasmid, and the continuous sequence of the entire chromosomal fragment carried on the BAC is determined by computer analysis.

- How does the computer *align* these sequences? The computer exploits one detail in this analysis, and it is well illustrated in steps 5 and 6 of **Figure 20.2**.

SHOTGUN SEQUENCING A GENOME

Genomic DNA — ~160 kb fragments

BAC library — BAC — Main bacterial chromosome

1-kb fragments

"Shotgun clones"

Shotgun sequence
TAGCGATCGATTTAGACTCGATAA
TAGACTCGATAAGGATGCGATACTACG

TAGCGATCGATTTAGACTCGATAAGGATGCGATACTACG

Draft sequence

1. Cut DNA into fragments of ~160 kb, using sonication.

2. Insert fragments into bacterial artificial chromosomes; grow in *E. coli* cells to obtain large numbers of each fragment.

3. Purify each 160-kb fragment, then cut each into a set of 1-kb fragments, using sonication, so that 1-kb fragments overlap.

4. Insert 1-kb fragments into plasmids; grow in *E. coli* cells. Obtain many copies of each fragment.

5. Sequence each fragment. Find regions where different fragments overlap.

6. Assemble 1-kb fragments from each 160-kb fragment by matching overlapping ends.

7. Assemble sequences from different BACs (160-kb fragments) by matching overlapping ends.

The fragments that are generated **overlap** with each other; upon sequencing, these overlapping sequences are valuable in identifying the correct order of the sequenced fragments.

Which Genomes Are Being Sequenced, and Why?

- The first genome to be completely sequenced was that of a bacterium, *Haemophilus influenzae* (1995). Since that time, many microbial genomes have been sequenced since they are relatively small genomes (*H. influenzae* genome is 1.8 million base pairs).

- In 1996, the first eukaryotic genome sequencing project was completed. The *Saccharomyces cerevisiae* genome project sequenced all *16 chromosomes of this organism* (12 million base pairs).

- At the writing of this text, complete genome sequences are available for 510 bacterial species, 52 archaeobacterial species, and 47 eukaryotic species.

- The selection of organisms for sequencing projects is a challenging process. Organisms are selected based on the anticipated value of their genome sequence to the public and to the scientific community.

Which Sequences Are Genes?

- Imagine the task of finding the genes in a genome sequence. What sequences identify the presence of a gene? Computers can search a sequence for reading frames with large regions that lack a stop codon.

These regions are called **open reading frames (ORFs)** and are usually a gene (**Figure 20.3**).

- In bacteria and archaea there are no introns, and genes can be identified by highly conserved promoter sequences associated with a distinct open reading frame.

- In eukaryotic organisms, the process of finding genes is complicated by introns and diverse regulatory sequences. One strategy is first to identify mRNA sequences using a **cDNA library** and then to hunt for the genes within the genomic sequences. In the process of generating an mRNA, the introns have been spliced out; but the sequence of exons should match exactly with short sections of the genomic sequence.

- Also, researchers use known genes from other organisms to search for homologous (similar) genes in the organism being studied.

CHECK YOUR UNDERSTANDING

(a) In your own words, describe the process of aligning DNA sequences from a "shotgun cloning" strategy. Can you think of a good analogy?

_____

_____

_____

(b) In higher eukaryotic cells, genes are often broken up into exons, with introns separating these coding fragments. How do researchers find these genes in the chromosomal sequence when performing genomic analyses?

_____

_____

_____

_____

20.2 Bacterial and Archaeal Genomes

The Natural History of Prokaryotic Genomes

- The size of the genome correlates with the complexity of microbial metabolic processes.

  1. *Mycoplasma* bacteria have a very small genome and are unable to grow in the absence of a **host cell**. They parasitize the host cells, growing in their cytosol and absorbing metabolites for their own use. With no host cell, *Mycoplasma* bacteria cannot survive (**obligate parasites**).
  2. Cells with large genomes are much more complex metabolically; they can survive in varying environments by converting available nutrients to the required compounds.

- There is a great deal of genetic diversity among prokaryotes.

- Some genes are found multiple times within the prokaryotic genome. The value of this **redundancy** is not known, but it may be due to slight differences in function or a requirement for a rapid induction of gene expression following activation.

- Many genes or chromosome regions appear to have originated with other microbes. This transfer of large fragments of DNA from one microbe to another is called **lateral gene transfer**.

Evidence for Lateral Gene Transfer

- Two key pieces of evidence for lateral gene transfer are (1) when a section of the chromosome in a species is more similar to genes in distantly related microbes than to those in closely related microbes, and (2) when sections of the chromosome are rich in either GC or AT nucleotides relative to the rest of the chromosome.

- ***Thermotoga maritima*** is a unique bacterial species thriving deep under the sea near hot-spring vents, where the temperature is very hot. In the same environment, some archaea species also grow well. Genome analysis revealed sections of the *T. maritima* genome that closely match the archaea genome. It appears that DNA was transferred from an archaeobacterial microbe to this eubacterial microbe, possibly improving this microbe's unusual ability to grow in a harsh environment.

- **Pathogenicity islands** have been discovered in some disease-causing strains of *E. coli*. These are chromosome regions that contain genes for aiding in infection; they also allow for localization of the bacterium to unique areas in the gut, and they generate devastating side effects such as severe diarrhea. Origins of these loci have been traced to other bacteria, such as *Shigella* spp., and even to the viruses that infect bacteria.

(c) Using what has been learned from completed prokaryotic genome projects, what key points can now be made about genomes and prokaryotic natural histories?

_____

_____

_____

_____

20.3 Eukaryotic Genomes

Natural History: Types of Sequences

- **Exons** comprise less than 5 percent of the human genome, and repeated sequences (noncoding) comprise more than 50 percent of the genome. In prokaryotic cells, usually 90 percent of the genome is comprised of genes. Why is there such a dramatic difference?

- **Repetitive sequences** in the human genome are due to simple sequence repeats (noncoding), transposable elements, and families of genes.

- Sources of repetitive sequences include the **transposons**. These DNA sequences, found in our chromosomes, occasionally "jump" from one place to another (they are transposable). Our chromosomes have many transposon sequences.

- **Long interspersed nuclear elements (LINES)**, comprise one category of transposons. These elements have a promoter followed by two genes—**reverse** transcriptase and **integrase** (remember reverse transcriptase from our discussion of cDNA libraries).

- Expression of the LINE locus results in synthesis of both proteins; the reverse transcriptase acts on the mRNA to generate a cDNA of the LINE sequence. Integrase then nicks a new site in a chromosome and mediates integration of the cDNA fragment into that site. So the LINE transposon has moved, leaving a copy behind; a new copy is now present in the genome. See **Figure 20.5** in your text to review the mechanism of transposition.

Natural History

Repeated Sequences and DNA Fingerprinting

- **Microsatellites** are repeating sequences that are found in the genome with repeats of 1 to 5 bases.
- Minisatellites are repeating sequences found in the genome with repeats of 6 to 500 bases.
- Where do these repeats originate? One prominent hypothesis is that these repeats are produced by an abnormal process called **slipped mispairing** that can occur during DNA replication. Details of that model are beyond the scope of this text, but it can be thought of as a stuttering of the DNA polymerase, accidentally repeating small sequences.
- Repeating sequences are hypervariable, meaning that the number of repeats is highly variable from one person to the next and even between homologous chromosomes. Thus each person has a unique set of repeat sequences—a genetic "fingerprint" that is unique to that one person.
- The highly polymorphic nature of these repetitive regions can be traced to crossing over during meiosis (**Figure 20.6**). Repetitive sequences would have the capacity to align improperly during prophase I, and if crossing over were to occur in this misaligned area, the chromosomes generated would have an expansion or a contraction of the size of the repetitive region. Examine the figure closely to understand this point.
- DNA fingerprinting is commonly used today in forensic science as authorities investigate and prosecute criminals. It is simply the determination of a person's haplotype at a number (~12) of highly polymorphic sites in the genome. Except for identical twins, no two people will have the same haplotypes for all of these sites.
- DNA fingerprinting has many other applications in the analysis of human ancestry, determining relatedness between people.

Gene Duplication and the Origin of Gene Families

- Data from analysis of genomes has shown that duplicated genes are a key source of new genes. Evidence

for duplication are the groups of similar genes in one region of chromosome. Similarities include sequence, exon/intron organization, and regulatory elements. Over time, duplicated genes may change via mutation with no deleterious effect on the cell, because the original copy of the gene is still producing a functional protein. Sometimes the mutations in the duplicate gene may generate a protein with a new function.

- Genetic loci that are closely related to each other, and physically close to each other on the chromosome, are called **gene families**.
- One example is the **globin gene family**, which is defined as a tandem array of genes, each member highly similar to other members of the family (**Figure 20.9**).
- Some members of the gene family are nonfunctional **pseudogenes**. They are remnants of a functional copy of the gene; but due to mutation, they have lost the ability to be transcribed.
- Examples of organisms with genomic modifications to benefit their survival:
  1. *Anopheles gambiae* (a mosquito species) has 58 genes for fibrinogen-like proteins, used to prevent blood clotting at the site of feeding. Fruit flies eat rotting fruit (not blood) and have only 13 genes for fibrinogen-like proteins.
  2. *Plasmodium* spp. are single-celled, parasitic eukaryotes that are responsible for malaria in humans. In the blood, foreign cells are detected by our immune system cells through examination of cell-surface proteins and sugars. Plasmodium cells have between 600 and 800 copies of genes that code for cell-surface proteins. This is hypothesized to be a strategy for evading the immune system.
  3. *Prochlorococcus* spp. are prokaryotic bacteria commonly found in the ocean. These microbes are photosynthetic, gathering light to produce cellular energy. Researchers have found a correlation between the number of chlorophyll genes and the quantity of available light for each species' distinct habitat (depth in the ocean). Species living deeper in the ocean have more chlorophyll genes and a more sensitive light-catching antenna complex.

Insights from the Human Genome Project

- The analysis of genome data sets has revealed a great deal about the proportions of different categories of genes in each organism (**Figure 20.10**). Notice that most genes identified in each genome have no known function. There is a great deal more to learn.
- **Table 20.1**—A major surprise from genome sequencing is that there is not a clear increase in gene number

as cellular complexity increases. Humans have about the same number of genes as fruit flies and round-worms do (approximately 25,000).

- One explanation for this confusing result involves alternative splicing of RNA transcripts (discussed in Chapter 18). This mechanism allows for different proteins to be made from a single gene, depending upon how the RNA transcript is spliced together. Early attempts to quantify the number of alternative products for each human gene have generated a value of ~3 alternative products/gene.

- The human and chimp genomes are very similar (98.8 percent identical) with most proteins showning only one or two different aminio acids. Yet undoubtedly, humans and chimps display major differences in their biology. How can this be? One prominent hypothesis holds that the key differences between chimps and humans are in the regulation of gene expression. Small changes in regulatory elements can profoundly affect the timing and duration of gene expression.

CHECK YOUR UNDERSTANDING

(d) What sequence-based evidence is there that a long interspersed element (LINE) originated as a transposon?

_____

_____

_____

20.4 Functional Genomics and Proteomics

- With all the information contained in the genome, scientists must now begin to methodically analyze how these genes and their encoded proteins work together to generate a living cell—an area of study called functional genomics.

- Activation of gene expression is an important function of cell differentiation, stress response, and cell division. Scientists have used Northern blotting as a way to detect changes in gene expression, analyzing one gene at a time. **DNA microarrays** allow researchers to simultaneously measure the expression of every gene in the genome.

- Imagine comparing all of the genes expressed in normal liver cells, compared to cancerous liver cells. The differences in gene expression may offer important insight into the nature of mutations that generate a cancer cell.

- Researchers create microarrays by putting thousands of tiny spots onto a glass slide (or chip), with each

spot containing short strands of single-stranded DNA (**Figure 20.11**). Then fluorescently labeled cDNA strands (copied from mRNA, introduced earlier in the chapter) can be used to probe the glass slide. The spots that bind to the cDNA probe reveal the expressed genes within that cell type. In a typical experiment, cDNAs from normal cells (control) are labeled one color, and cDNAs from abnormal cells are labeled another color. If the two cell types display the same level of expression for a gene, equal amounts of each color of cDNA will bind to that gene's spot on the slide (green = red). However, if the abnormal cells display an increase in that gene's expression, one color will increase relative to the other. Notice **Figure 20.12**, displaying spots of various colors and revealing changes in gene expression.

- The microarray is an incredibly valuable tool for the simultaneous, quantitative analysis of all gene expression in a group of cells.

What Is Proteomics?

- Simultaneous analysis of thousands of proteins, in a manner similar to the DNA microarrays, would greatly enhance scientists' ability to characterize how proteins work.

- One type of proteomic assay involves spotting proteins onto a solid matrix (like glass) and then exposing these proteins to a pure preparation of a specific protein or molecule. Detection of protein-to-protein binding and even catalytic activity may then be measured, thus identifying which proteins in the array interact with the introduced protein/molecule.

20.5 Can Genomics Help Improve Human Health and Welfare?

Identifying Potential Drug Targets

- Genome projects offer huge databases of information for studying living organisms. One application of this knowledge is the pursuit of disease prevention and treatment strategies. For instance, one focus has been identifying the genes that make some microbes deadly to humans.

- **Virulence** is the tendency of a parasite to harm its host. *Escherichia coli* is a species containing both benign (harmless) and virulent strains. Researchers have used genomics to study the genetic basis of virulence in *E. coli* strain 0157:H7, a common cause of severe food poisoning from undercooked meat. Researchers found 0157:H7 to contain the normal *E. coli* genome with an extra region of chromosome derived from the *Shigella* bacterial genome, which enables the *E. coli* to cause a severe food poisoning. *Shigella* spp. are a common cause of dysentery.

- Researchers hypothesize that a bacterial virus carried this portion of the *Shigella* genome into *E. coli*; this is a valuable example of lateral gene transfer.

- Identifying the key proteins that make such a microbe virulent aids in the development of drugs to treat those afflicted with this disease.

- Researchers are using the genome sequence of *Plasmodium* to develop drug targets for proteins found only in the parasitic microbe, not in the human host.

Designing Vaccines

- Vaccines against disease-causing microbes are being developed using genome databases. One example is *Neisseria meningitidis*, which causes meningitis (see "Mastering Challenging Topics").

- This microbe grows rapidly in the infected human patient (often children) and can cause permanent tissue damage or death before antibiotics can effectively treat the infection. Thus preventative vaccination is a better route to protecting children from this disease.

- Many cell-surface proteins on this microbe are coated with polysaccharides (sugars) and thus do not make good antigens for a vaccine. The *N. meningitidis*

genome data have offered information on all the proteins in that infectious bacterium, thus enabling a more thorough testing of those proteins as antigens.

- Of 350 open reading frames tested, researchers identified seven encoded proteins that generate a robust immune response in mice (**Figure 20.13**).

Finding Genes Associated with Inherited Disease: The HapMap Project

- The genome project has enabled the development of a large catalog of sites in the human genome that often contain single nucleotide polymorphisms (SNPs). This database is incredibly valuable to the process of mapping genes that are responsible for specific traits (including diseases). This initial catalog was enhanced by the HapMap Project, designed to seek out additional SNP sites. There are now ~10 million SNP sites in the human genome, enabling high-resolution mapping of chromosomes. This will be a major asset to scientists trying to identify the genes responsible for diseases like diabetes and cancer.

CHECK YOUR UNDERSTANDING

(e) Explain the relevance of *E. coli* O157:H7 to evidence for lateral transfer of DNA between prokaryotic organisms.

_____

_____

_____

B. INTEGRATING YOUR KNOWLEDGE

Looking Back—Key Concepts from Earlier Chapters

How Genes Work—Chapter 15

With information available from genome analysis, we can look back at our introductory examination of genes in a new light. Key findings include the discovery of **families of genes** clustered together in the genome, **pseudogenes** scattered on our chromosomes, and the amazing complexity of **gene structure**.

Analyzing and Engineering Genes—Chapter 19

Dideoxy sequencing is a critical concept to understand, because we build on that technology with an improved fluorescence labeling strategy. The chemistry of Sanger dideoxy DNA sequencing is still the basis of all sequencing used today, though some details have changed.

Looking Forward—Key Concepts in Later Chapters

Bacteria and Archaea—Chapter 28

Bacteria display an incredible diversity of genetic sequence, and the evidence that genetic information is exchanged between even distantly related microbes is truly fascinating. Microbes appear to readily gather available genetic information via **lateral transfer of genes**, sometimes benefiting their ability to adapt and survive.

C.   MASTERING CHALLENGING TOPICS

Rational Drug/Vaccine Design

Once scientists know all the genes of a particular organism, how can they use this information to develop better antibiotics, vaccines, or even medicines that prevent the progression of disease? Remember that the proteins within cells perform tasks and generate structure in a cell. Knowing a gene's DNA sequence allows researchers to predict the amino acid sequence in the protein that would be synthesized by that gene and offers clues to the gene's structure and function. Examining the genome offers vital insight into the "complete picture" of cellular function. Thus strategies for treating a specific disease will now utilize genomics data sets to choose better targets for drug action.

One example included in the text discusses the production of a vaccine against a dangerous microbe (*Neisseria meningitidis*) that is one cause of infections in spinal fluid. A vaccine includes inactivated components of an infectious organism, and it is injected into a patient to reveal identifiable characteristics of the pathogen to the immune system. Thus, as soon as that organism enters the body, the immune system recognizes the danger and rapidly reacts. Genomic analysis of *N. meningitidis* has revealed all of the proteins encoded with that genome, so that researchers can now test all of these proteins for their ability to stimulate the immune system effectively. By understanding this genome sequence, scientists can make a better vaccine, one that will save many lives. The value of sequencing the human genome is undeniable; but certainly we are only beginning to grasp the best strategies for using this information. What applications do you see?

D.   ASSESSING WHAT YOU'VE LEARNED

1. Using fluorescent tags in dideoxy sequencing is an improvement on using radioactive tags for the following reasons (choose all that apply):
   a. Automated laser scanners can read results and compile data on a computer.
   b. Radioactive sequencing is inaccurate.
   c. Each ddNTP nucleotide is a different color, and thus the sequence can be determined from electrophoresis of a single sample (four are needed in radioactive labeling).
   d. Fluorescent nucleotides are less expensive than radioactive nucleotides are.

2. Scientists studying the "normal" benign *E. coli* and the pathogenic *E. coli* strain O157:H7 have made which of the following determinations?
   a. The O157:H7 strain has fewer genes than the benign strain.
   b. The O157:H7 strain has more introns than the benign strain in key virulence-determining genes.
   c. The O157:H7 strain has the same genetic sequence as the benign strain, so differences in virulence must be the result of post-transcriptional modification of the mRNA of virulence determinants.
   d. The O157:H7 strain has gene sequences that are similar to *Shigella* bacteria that are not found in the benign strain.

3. Lateral gene transfer is indicated when (choose all that apply):
   a. A microbe has the genetic components for anaerobic growth.
   b. A microbe has an isolated region with a distinct guanine and cytosine content.
   c. A microbe has a genomic region containing genes that are similar to a distantly related microbe.
   d. A microbe has distinct lateral gene transfer motifs in its DNA sequence.

4. *Prochlorotrichus* spp. offer an excellent example of:
   a. genomic adaptation
   b. parasitism
   c. microbial conjugation
   d. microbial virulence

5. Transposable elements:
   a. have the ability to replicate themselves and move from one location to another in a genome
   b. occur very rarely in humans
   c. are always active, if present
   d. make up as much as one-half of a bacterium's genetic material

6. Please refer to **Figure 20.5** in your textbook. If there were a mutation in the reverse transcriptase formed in this transposon-containing cell that prevented it from binding to the LINE mRNA, how would the resulting situation differ from that shown in the figure?
   a. The LINE mRNA would not be translated.
   b. A DNA copy of the LINE transposon would not be made.
   c. The reverse transcriptase would not be produced.
   d. The DNA would not be nicked.

7. Overall gene number and cellular/organismal complexity do not correlate as researchers had predicted. One explanation for this phenomenon is:
   a. our misunderstanding of biological complexity
   b. our inability to identify genes in genome sequence
   c. the importance of alternative splicing in multicellular organisms
   d. the poor quality of genome sequence data

8. The globin gene family shown in **Figure 20.9** in your textbook is thought to have arisen from which of the following processes?
   a. polyploidy
   b. transposable elements
   c. unequal crossover during meiosis
   d. alternative splicing

9. Shotgun sequencing involves:
   a. cutting genomic DNA into small overlapping fragments, sequencing each individually, and using computer software to align the sequences
   b. isolating bacterial artificial chromosomes (BACs) containing huge genomic fragments, and sequencing each BAC using nonspecific "shotgun" primers
   c. generating genomic DNA fragments using a high-pressure gas "shotgun" syringe, and sequencing each fragment generated
   d. radioactive sequencing rather than fluorescent sequencing

10. A "gene family" is:
    a. a group of functionally related genes that are located together and function together (i.e., lactose utilization)
    b. genes transferred together from one microbe to another, through conjugation
    c. most often observed as a pathogenicity island
    d. a group of genes that are similar in sequence, and are thought to be duplicated from a common ancestral gene

CHAPTER 20—ANSWER KEY

Check Your Understanding

(a) Of course, this will be the student's answer and own analogy. Imagine having 100 freight trains, each 500 cars long, with exactly the same ordering of freight cars. If you were to break each train into fragments of 20–30 freight cars, each separated at different sites, could you examine the order of each of the sections and regenerate the original sequence of cars? Yes! You could read the sequence of each section and then put the sections together into the correct order by aligning sequences with information from overlapping sections. DNA sequencing with "shotgun cloning" uses exactly this strategy; the large BAC is broken up, sequenced in small fragments, and the original BAC sequence is determined by aligning the smaller sequences.

(b) This task is difficult, and several strategies are employed. The most productive is to search the DNA sequence for matches with known cDNA sequences isolated from cDNA libraries. The small matching regions will reveal the exons within the larger sequence.

(c) There is a general correlation between the size of a genome and the metabolic capacities of the organism.
   • Prokaryotic organisms display an incredible array of genetic diversity.
   • Prokaryotic organisms commonly display genetic redundancy (multiple copies of genes).
   • Prokaryotes display a high degree of lateral transfer of genetic information.

(d) LINES contain a transcription start site, a reverse transcriptase gene, and an integrase gene. These two genes are known to mediate the transposition of a DNA sequence from one genomic site to another.

(e) Most *E. coli* are harmless to humans—they are even a common inhabitant of the human gut. The *E. coli* O157:H7 strain can be lethal to humans (especially children), causing a dehydrating "bloody diarrhea." Comparative analysis of the genomes of this pathogenic strain of *E. coli* and other nonpathogenic *E. coli* revealed a distinct region in the O157:H7 strain. This region included genes that closely resemble those of a known pathogenic bacterium (*Shigella*). The *Shigella* genes appear to have been transferred into the *E. coli*, giving this strain an enhanced pathogenic potential.

Assessing What You've Learned

1. a, c; 2. d; 3. b, c; 4. a; 5. a; 6. b; 7. c; 8. c; 9. a; 10. d

21

Principles of Development

A. KEY CONCEPTS

21.1 Four Essential Developmental Processes

- The development of a complex multicellular organism from a single cell is a fascinating process. Four categories of cellular processes drive development: cell division, cell migration, alterations in gene expression and cell morphology, and cell-to-cell communication.

Cell Proliferation and Programmed Cell Death

- Cell division (mitosis) is regulated at several different checkpoints, affecting the activity of the central kinase in this process, called **mitosis promoting factor**. Review Chapter 8 for details on the cell cycle.

- Cells can stimulate or inhibit the growth of adjacent cells. Receptor proteins in the cell membrane typically mediate these **social controls** over the cell cycle.

- The preprogrammed loss of specific cells during development is known as **programmed cell death (PCD**; see also **Box 21.1)**. A classic example is the development of the hand. In early development a pancake-like structure forms at the end of each limb. The death of specific cells in this structure, and the retention of others, forms the digits of the hand. PCD is an essential process in the development of a complex three-dimensional organ or tissue.

- A more general term that defines the regulated process of cell destruction is **apoptosis** ("falling away"). Apoptosis is essentially PCD, but occurring in different circumstances. Apoptosis is induced in cells that are functioning abnormally, typically due to mutation, viral infection, or some type of environmental insult. Apoptosis involves a protease and DNase-mediated process of destroying cellular protein and DNA, and breaking the cell into small vesicles that can be consumed by neighboring cells. The elimination of abnormal cells protects the multicellular system.

Cell Movement and Expansion

- In animal development, the process of gastrulation involves the organization of a mass of cells into three types of embryonic tissues. These three types of tissue are called the ectoderm (forms skin), mesoderm (forms muscle), and endoderm (forms gut). A key part of this organization of tissue is cell migration.

Cell Differentiation

- The process of cellular differentiation is really a transformative process. The cell becomes something different structurally and functionally, though its DNA remains unchanged. How does this transformation occur? The fundamental mechanism of cell differentiation is altered gene expression. Inactivation of some genes and activation of others tranforms the cell and mediates the process of differentiation.

- Cell differentiation is a progressive, sequential process as cells move from one cell type through a series of changes that culminate with fully differentiated sets of cells. Some cells have the capacity to de-differentiate (e.g., differentiated plant cells that have the potential to generate a new plant).

- **Stem cells** remain in an undifferentiated or partially differentiated state, and they are a resevoir for the production of differentiated cells in an adult organism.

Cell–Cell Interactions

- Cell-to-cell communication is critical to the process of organizing billions of cells into an organ or an organism. Review **Chapter 8** and the topic of signaling that is mediated by the extracellular matrix.

CHECK YOUR UNDERSTANDING

(a) What three tissues, from which are derived all adult tissues, are present in animal embryos?

_____

_____

Define the tissues that are generated by these three tissues in animals.

_____

_____

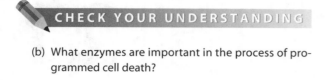

(b) What enzymes are important in the process of programmed cell death?

_____

_____

_____

21.2 The Role of Differential Gene Expression in Development

- Patterns of gene expression define one differentiated cell type in comparison to another. That makes sense because ultimately proteins do things in cells, and gene expression controls which proteins are present in any given cell.

Are Differentiated Animal Cells Genetically Equivalent?

- A key question in the study of development and cellular differentiation focuses on the effect of differentiation on the genome—the DNA content of a cell. Is there an irreversible change in the DNA of a cell as it differentiates? A key technology used to test this question has been nuclear transfer experiments, in which the nucleus of a differentiated cell is transplanted into an unfertilized egg. Researchers hypothesized that if there is no change to the DNA content of a differentiated cell, then the transferred nucleus should enable the normal development of a complete individual.

- In the 1950s, scientists were first able to isolate the nucleus of one frog cell and transfer it to an unfertilized frog egg by a process called **nuclear transfer**. Researchers found that the transfer of a nucleus from a blastula- or neurula-stage embryo into an egg was capable of generating a normal frog tadpole.

- However, isolation of nuclei from tadpoles for use in similar experiments was unsuccessful. Something was different about the DNA in these differentiated tissues.

- Ian Wilmut (1997) performed nuclear transfer experiments with sheep mammary gland cells and enucleated sheep eggs. Though the success rate was very low, this process was successful in generating cloned sheep (**Figure 21.3**).

- Tissue differentiation does not involve any irreversible change in DNA content, and it lies primarily within the differential regulation of gene expression in different cell types.

What Is the Most Important Level of Control over Gene Expression?

- In **Chapter 18**, we learned that there are several places within the central dogma of gene expression where regulation occurs. In studies of development, it has become clear that transcriptional regulation is the key site of regulation used to drive cell differentiation.

- In eukaryotic cells, transcriptional control is accomplished through transcription factors that affect chromatin remodeling and bind to promoter-proximal elements.

21.3 What Triggers Differential Gene Expression?

Master Regulators Set Up the Major Body Axes

- **Pattern formation** in a developing embryo is the establishment of the three-dimensional, spatial organization that initiates many developmental processes. Cells receive molecular signals that indicate where they are within the embryo, and that information directs gene expression and cell-fate determination.

- The three body axes in an organsim are
    1. the anterior (head) to posterior (tail) axis,
    2. the ventral (belly) to dorsal (back) axis, and
    3. the left to right axis (**Figure 21.5**).

- The fruit fly, *Drosophila melanogaster*, has been a key model organism for the study of pattern formation.

The Discovery of bicoid

- Christiane Nüsslein-Volhard and Eric Wieschaus performed a large mutant hunt for genes that influence pattern formation in *Drosophila melanogaster*. They **mutagenized** flies and then examined the fly progeny for defects in development. They looked specifically for mutant flies that generated progeny with abnormal pattern formation.

- The *bicoid* mutation generated a dramatic phenotype, generating larvae that lacked structures on the anterior end (**Figure 21.6**). The normal head/thorax region develops into posterior structures, making the larvae *bicoid*, which means "two-tailed."

- The *bicoid* gene was shown to display **maternal effect inheritance**. This term means that the phenotype in the larvae is due only to the mother's genotype. If the mother lacks a functional copy of the *bicoid* gene, then the embryo will be two-tailed. The mother is responsible for providing functional *bicoid* mRNAs to the egg, and these cytoplasmic determinants aid in larval development.

- The *bicoid* gene was isolated via **gene mapping** techniques. Review the examination of **polymorphisms** and **genetic linkage** discussed in **Chapter 13** to illustrate how this process works.

Where Is the bicoid *Product Found?*

- Once the gene was isolated, researchers could generate probes for localizing *bicoid* mRNA (in situ hybridization) and protein (immunolocalization) within the embryo.

- A technique called in situ hybridization is used by researchers to localize specific mRNAs (**Figure 21.7**). If specific mRNAs are critical for the development of select cells during develoment, then that mRNA should be concentrated in those cells.

- In situ hybridization is a technique similar to Northern blotting, in which a labeled probe is used to screen unlabeled mRNAs for its complementary sequence. In this case the target mRNAs are located within existing tissue, thus they are said to be in situ (Latin for "in the normal or natural position").

- Using in situ hybridization analysis, *bicoid* mRNA was found in the anterior of the egg cytoplasm (**Figure 21.7**). This mRNA is from the mother and is present at fertilization as a crucial cytoplasmic determinant.

- Antibodies specific for the *bicoid* protein were used to determine when and where the protein was synthesized (**Figure 21.8**). Immediately following fertilization, *bicoid* protein is synthesized, and its concentration within the zygote is organized as a gradient.

- There are very high concentrations of the bicoid protein in the zygote's anterior and low concentrations at its posterior.

- The **gradient hypothesis** proposed by Nüsslein-Volhard stated that this gradient of *bicoid* protein was key in establishing anterior and posterior axes within the embryo. Thus, high concentrations of *bicoid* protein were hypothesized to signal anterior development, and low concentrations of *bicoid* protein were hypothesized to signal posterior development.

- To test the gradient hypothesis, Nüsslein-Volhard purified *bicoid* mRNA and injected it into different places in developing embryos. To determine if purified bicoid mRNA was functional when injected into an embryo, the researchers injected the mRNA into the anterior of embryos derived from mutant *bicoid* females. The injected embryos developed normally and did not display the two-tailed phenotype.

- When the researchers injected *bicoid* mRNA into wild-type (nonmutant) embryos, anterior structures always developed at the site of injection. These results strongly supported the gradient hypothesis.

- The *bicoid* protein was shown to act as a **transcription factor** within *Drosophila* embryos, binding DNA and regulating the expression of genes vital to development at the anterior of the embryo (**Figure 21.9**). Identifying what genes are regulated by *bicoid* was vital to unraveling some of the mysteries of early development.

Regulatory Genes Provide Increasingly Specific Positional Information

- Many other genes were identified via the mutant hunt conducted by Nüsslein-Volhard and Wieschaus, and each of these genes is vital to early developmental processes. Some of them can be grouped into one of three general categories of genes involved in embryo **segmentation** during *Drosophila* development (**Figure 21.10**).

- **Gap genes**—Loss-of-function mutation in gap genes results in a loss of several consecutive segments in the fly embryo. One key example of a gap gene is named *hunchback*. Mutation of *hunchback* results in embryonic phenotypes similar to those described for *bicoid* (lacking segments in the anterior of the embryo). However, *hunchback* functions via embryonic gene expression and is not a maternally derived cytoplasmic determinant like *bicoid*.

- **Pair-rule genes**—Loss of function in pair-rule genes results in the loss of **alternating** segments in the embryo, indicating that these genes are critical to the formation of every other segment formed during embryogenesis.

- **Segment polarity genes**—These genes are critical to formation of components of each segment; the loss of these genes generated embryos lacking portions of each segment.

- Segmentation genes are expressed in a precise order during embryogenesis:

  gap genes → pair-rule genes → segment polarity genes

- How does this pattern of consecutively activated genes get started? Do the genes activate each other? The maternal cytoplasmic determinant *bicoid* is known to activate expression of *hunchback* (a gap gene whose expression is located in the embryo anterior; similar to the *bicoid* expression pattern), so this is an example of a cytoplasmic determinant activating a member of the gap genes.

- But do the gap genes activate the pair-rule genes? Yes, many of the gap genes code for transcription factors that bind to regulatory sequences in pair-rule gene promoters, activating their expression at a precise time during development. As you might now suspect, members of the pair-rule gene family activate the segment polarity genes.

- This complex array of stepwise gene activation events is called a **regulatory cascade**, and it can be quite complex to consider in its entirety (**Figure 21.11**).

These waves of gene expression are critical to establishing proper **pattern formation**. They continue through development, eventually being restricted to specific groups of cells as specific tissues begin to form. But how are distinct groups of cells given unique signals during development? In other words, what signals establish the individual segments?

Homeotic Genes: What Are They, and Where Do They Fit In?

- Edward Lewis (1940s) performed a mutant hunt with *Drosophila melanogaster*, isolating flies that displayed abnormal body patterns in the *adult fly*. Lewis isolated many mutant flies and recognized that some of the mutant phenotypes were due to abnormal development of individual segments. Some mutants had segments that developed into a phenotype that matched an adjoining segment, generating legs where antennae should be, an extra set of legs, or even an extra set of wings.

- This phenomenon of one segment developing incorrectly into the phenotype of another segment is called **homeosis** (like condition).

- The genes responsible for proper identity in each segment were named **homeotic complex**, or *Hox*

genes. Eight were identified by gene cloning and sequencing, each gene being responsible for the proper development of a specific segment or group of segments in the embryo (**Figure 21.12**).

- All eight of these genes are located on one chromosome. The five genes expressed in the anterior of the embryo are called the *Antennapedia* complex; three that are expressed in the posterior of the embryo are called the *Bithorax* complex.

- The expression of the *Hox* genes is sequential, moving in order along the chromosome (*lab* → *AbdB*) and from the anterior to the posterior of the embryo (*lab* is expressed first and in the far anterior of the embryo; *AbdB* is expressed last and at the far posterior of the embryo). Examine **Figure 21.13a** closely to follow the parallels between chromosome location and the timing/distribution of gene expression.

- Each of the proteins encoded by the *Hox* genes contains a **DNA-binding region**; these regions are thought to act as transcription factors, activating genes critical for proper development of the relevant segments. The DNA sequence encoding these regions in each protein is called the **homeobox** (a 180-bp segment within each of the *Hox* genes).

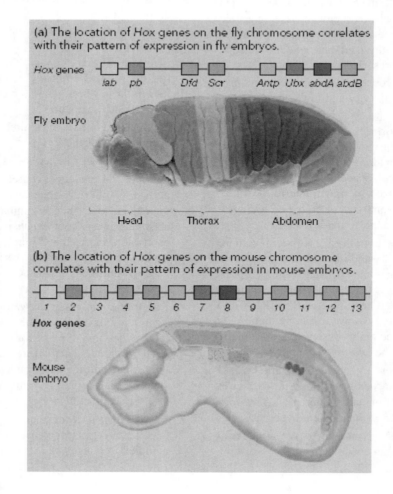

(a) The location of Hox genes on the fly chromosome correlates with their pattern of expression in fly embryos.

(b) The location of Hox genes on the mouse chromosome correlates with their pattern of expression in mouse embryos.

- How are the *Hox* genes activated during development? There is clear evidence that the segmentation genes play a critical role in activating the proper set of *Hox* genes in each segment. Flies carrying mutant segmentation genes display abnormal activation of the *Hox* genes.

Cell–Cell Signals and Regulatory Genes Are Evolutionarily Conserved

- Homeobox genes, clustered into groups as in fruit flies, are found in the genomes of all animals examined to date. In these other organisms, they are expressed sequentially, and they appear to be transcriptional regulators with a function similar to that observed in fruit flies. Examine **Figure 21.13b**, revealing the *Hox* genes and their head-to-tail pattern of expression in a developing mouse embryo (same sequence as *Hox* genes in a fruit fly).

- The *Hox* genes play a key role in determining the head-to-tail axis of a developing embryo.

- The fruit fly *Antp* (*Antennaepedia*) gene and the mouse *Hox6b* are homologous to one another and affect the development of similar structures.

21.4 Changes in Developmental Pathways Underlie Evolutionary Change

- Minor changes in the activity of key regulatory proteins (like the homeobox proteins) can have profoundly affect the development of an embryo. Mutations in these genes are thought to play a key role in change of form over time—the evolution of new body plans.

- **Tetrapods** (Latin, meaning "four-legged") include lizards (with four legs) as well as snakes. The genetic evidence is clear that these are closely related animals. Yet why do snakes lack functional legs? Some snakes have tiny pelvic bones, consistent with the belief that they are descendents of organisms that had functional legs.

- In tetrapods, *Hoxc6* and *Hoxc8* are expressed together in vertebrae that eventually develop ribs. In areas where *Hoxc6* is expressed without *Hoxc8*, forelimbs develop. In snakes, *Hoxc6* expression and *Hoxc8* expression co-occur, and functional forelimbs do not form. This indicates that the pattern of expression of these genes in snakes may be the cause of the loss of functional forelimbs.

- Sonic hedgehog is a transcription factor that controls hindlimb development in tetrapods. In snakes, this gene shows changes that have altered the functionality of the gene product.

- It is clear that changes in these transcription factors have played a primary role in the generation of snakes that lack functional limbs.

B.   INTEGRATING YOUR KNOWLEDGE

Looking Back—Key Concepts from Earlier Chapters

Control of Gene Expression in Eukaryotes—Chapter 18

In Chapter 21 we have examined how vital the regulation of gene expression is in development. Many of the key regulatory molecules in early development are **transcription factors**, binding promoter elements to turn on genes at the proper place in the embryo and at the proper time (review the segmentation genes in *Drosophila*). Your understanding of **Chapter 18** content, transcription factors, and the regulation of gene expression is critical to your understanding of the homeobox and segmentation genes.

Looking Forward—Key Concepts in Later Chapters

An Introduction to Animal Development—Chapter 22

Now that we have carefully developed a framework in which we can examine development, we will study animal development in more detail. **Chapter 22** introduces more detail into our understanding of early animal development, defining specific events that mold the embryo (**gastrulation**) and the genes that regulate this process.

CHECK YOUR UNDERSTANDING

(c) Within the *Drosophila* model of development, what early events establish the signals that initiate pattern formation?

_____

_____

_____

_____

CHECK YOUR UNDERSTANDING

(d) In *Drosophila*, how does the observed pattern of consecutively activated segmentation genes, as illustrated in **Figure 21.13a**, get started?

_____

_____

_____

C. MASTERING CHALLENGING TOPICS

In Situ Hybridization Techniques (Figure 21.7)

A critical research tool introduced in this chapter is in situ hybridization, which enables scientists to detect the location of gene expression in an organism. The technique is similar to that used for a Northern blot, whereby mRNA transcripts are detected with a radioactive probe (DNA or RNA) that is complementary to a specific mRNA species (one gene). In Northern blotting, the pool of mRNAs are run on an agarose gel and transferred onto a membrane before the probe is used to screen through the transcripts. In situ hybridizations involve the addition of probe directly to the tissue.

Tissues to be examined using in situ hybridization are "fixed," freezing cellular macromolecules in place with a chemical treatment (formaldehyde). The probe enters into the cells of the tissue being tested and hybridizes to the mRNA present in the fixed cells. The site of probe hybridization is detected by placing an emulsion (a kind of liquid X-ray film) over the tissue, which is typically attached to a slide. The dots are generated as the emulsion is exposed to radioactivity and then developed in a manner similar to that used for photographic film. Researchers can then determine (with proper controls) the localization of gene expression within a tissue sample, or even within an entire organism, depending on the size of the organism. In **Figure 21.9**, the site of *Vg1* expression within a fly egg is revealed using in situ hybridization. How was this information valuable to scientists studying the process of embryo maturation in flies? Could this information have been gathered in another way?

D. ASSESSING WHAT YOU'VE LEARNED

1. List the advantages of using *Drosophila melanogaster* as a model organism for identifying the genes and processes responsible for cell-fate decisions in early development.
   a. _____
   b. _____
   c. _____
   d. _____

2. If the mother of larvae displaying the mutant *bicoid* phenotype carries only the mutant gene, how was the mother able to develop normally in the first place?
   a. The bicoid phenotype can be suppressed by culturing the embryos at cool temperatures.
   b. The *bicoid* mRNA is a maternal effect gene, donated to an egg by the mother regardless of the egg genotype.
   c. The mother likely also carried a *bicoid* suppressor mutation.
   d. All of the above apply.

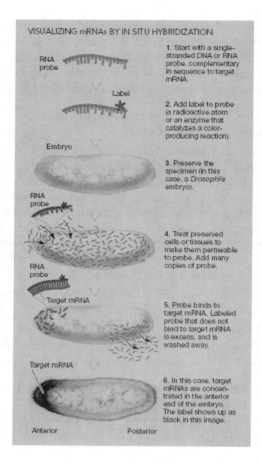

3. What was the experimental focus of the screen performed by Nüsslein-Volhard and Wieschaus?
   a. to evaluate the mutagenic properties of varied treatments
   b. to discover the *bicoid* gene
   c. to generate two-tailed fruit-fly embryos
   d. to generate fruit flies with mutations in genes that control fruit-fly development

4. Which of the following is *not* an accurate description of the *bicoid* gene?
   a. Mutation generates larvae without anterior development.
   b. Encoded protein is a DNA-binding protein.
   c. The gene displays a maternal effect mode of inheritance.
   d. A male that is homozygous for this mutation cannot generate viable offspring.

5. Which of the following is true of maternal effect inheritance?
   a. A gene is transcribed in the offspring and affects the phenotype observed in the mother.
   b. A gene is transcribed in the mother and affects the phenotype observed in the offspring.
   c. The genes of the mother negate the genes of the father.
   d. The genes of the father negate the genes of the mother.

6. What do you call the 180-base-pair DNA sequence that was revealed in each gene in the *Antennapedia complex* and the *Bithorax complex*?
   a. homeobox
   b. homeotic complex
   c. homeosis
   d. homeotic genes

7. Defects in pair-rule genes cause:
   a. the loss of paired sets of structures, like eyes
   b. in a segmented developmental region, effects on alternerating sections (every other section)
   c. defects in the anterior of the embryo
   d. defects in the posterior of the embryo

8. The *hunchback* gene is a type of:
   a. pair-rule gene
   b. gap gene
   c. segment polarity gene
   d. *bicoid* family member

9. When researchers injected *bicoid* mRNA at different sites of the early fruit-fly embryo, what happened?
   a. development of legs at the site of injection
   b. early lethality of embryo
   c. developemnt of anterior structures at site of injection
   d. development of dorsal structures at site of injection

10. Stem cells are:
    a. cells in a mature organism that have the ability to develop into fully differentiated cells
    b. found exclusively in the bone marrow
    c. cells that occur only in early embryogenesis of an animal
    d. a hypothetical cell type that scientists have not yet been able to isolate

CHAPTER 21—ANSWER KEY

Check Your Understanding

(a) **ectoderm** (generates skin and nerve cells); **vegetal pole—endoderm** (generates gut and organs); **mesoderm** (generates muscle and organs).

(b) The process of programmed cell death involves the removal of unwanted cells as the developing embryo is sculpted into a functional organism. To remove a cell without damaging neighboring cells, the contents of the cell must be digested. Enzymes that destroy cellular macromolecules (DNA, RNA, protein, and lipid) mediate this process: DNases, RNases, proteases, lipases.

(c) The maternally derived cytoplasmic determinant, *bicoid*, is expressed in the anterior of the fertilized embryo and generates a gradient of *bicoid* protein (high concentration in anterior to low concentration in the posterior). This gradient of concentration establishes a polarized pattern of gene expression.

(d) *Bicoid* is a transcription factor, activating the expression of some segmentation genes along its concentration gradient in the early embryo. The key example is the gap gene *hunchback*, whose expression is directly activated by *bicoid*. Then some of the gap genes activate the pair-rule genes, and some of the pair-rule genes activate the segment polarity genes, driving embryonic development forward.

Assessing What You've Learned

(a) *Drosophila* generate large numbers of progeny very rapidly (the time from mating to reproductive maturity is about 3 weeks).

(b) *Drosophila* exhibit complex developmental processes (multicellular organism), serving as a better model than simpler multicellular organisms.

(c) *Drosophila* has well-characterized genetics (**Chapters 12** and **13**).

(d) Fruit flies are easily maintained and bred in a laboratory setting.

(e) Fruit flies can be mutagenized fairly easily (X-rays or chemical mutagens), and flies with mutant phenotype can be detected by visual inspection of progeny.

(f) Fruit-fly embryos are deposited into a rich nutrient source (in the laboratory, culturing medium), and are therefore easy to isolate for study.

Assessing What You've Learned
2. b; 3. d; 4. d; 5. b; 6. a; 7. b; 8. b; 9. c; 10. a

22

An Introduction to Animal Development

A. KEY CONCEPTS

22.1 Gametogenesis

Sperm Structure and Function

- **Gametogenesis** is the formation of gametes (sperm and egg) in the reproductive organs.
- The mammalian sperm cell, which has haploid DNA, contains four major compartments (head, tail, neck, and midpiece) and several critical structural components (**Figure 22.2**).
    1. **Acrosome**—A lysosome-derived structure at the head of the sperm and containing a mixture of hydrolytic enzymes involved in degrading specific components of the egg cell wall. The acrosome is located between the nucleus and the anterior cell wall of the sperm head.
    2. **Centriole**—In the sperm's neck region, the centriole is a small, hollow, cylindrical structure comprised of microtubule fibrils.
    3. **Mitochondria**—Each sperm structure contains many tightly packed mitochondrial organelles within the midpiece region. The capacity for sperm motility depends on ATP production from these mitochondria.
    4. **Flagellum**—A proteinaceous "propeller" structure composed of microtubules, the flagellum generates the thrust necessary for sperm motility to the egg.
- **Pollen grains** serve as plant sperm and are small enough to be carried by the wind or to powder the extremities of insects. Once these grains come into contact with an ovule, a pollen tube grows from cells in the pollen grain, a single pollen cell then divides mitotically, and the resulting haploid sperm nuclei migrate down the pollen tube to the ovule (see **Chapter 23**).

Egg Structure and Function

- The mammalian egg cell is much larger than the sperm cell (**Figure 22.3**) and contains a cytosol enriched with nutrients as well as several structural components regulating the processes of fertilization and early development.
    1. **Yolk granules**—Small, dense structures in the egg cytosol, packed with protein and/or lipid for energy and macromolecular biosynthesis during early development.
    2. **Vitelline envelope**—A fibrous structure surrounding the cell membrane and lying beneath any associated gelatinous layer (egg jelly, zona pellucida, etc.).
    3. **Cytoplasmic determinants**—These protein and mRNA molecules are key regulators of early development in many animals (excluding mammals). Segregation of these molecules during cleavage controls key processes leading to cell differentiation.
    4. **Cortical granules**—These small vesicles are localized to the periphery of the cytosol. They contain enzymes critical to fertilization (see "blocking polyspermy" in **Section 22.2**).

22.2 Fertilization

Fertilization in Sea Urchins

- The sea urchin has proven to be an incredible research system for the characterization of fertilization. Researchers can isolate large quantities of sperm and eggs quite easily. Large numbers of sperm and egg are released from adult sea urchins, and fertilization occurs externally in the surrounding seawater.
- As summarized in **Figure 22.5**, the fertilization of a sea urchin egg involves a complex chain of events.
    1. The sperm is recruited to the egg by an attractant that is present in the **egg coat** and slowly diffuses out into the surrounding seawater. The **attractant** is a small protein (peptide) that is recognized by receptor proteins present in the sperm.

2. Upon direct contact between the sperm head and the egg cell, the acrosome releases its contents from the sperm head.
3. This mixture of hydrolytic enzymes degrades the gelatinous egg coat.
4. At the same time the hydrolytic enzymes are released, a structure is built on the sperm head by the **polymerization of actin** proteins together. This structure is called the **acrosomal process**, and it mediates the fusion of sperm-cell membranes with the egg-cell membranes.
5. Once the membranes have fused, the sperm nucleus can enter the egg cytosol, eventually fusing with it to generate the zygote with a *2n* chromosomal content.

How Do Gametes from the Same Species Recognize Each Other?

- When exposed to a mixture of sperm from different sea urchin species, an egg can be successfully fertilized only by sperm of its own species. If the mechanisms of fertilization are similar between these species, how is it possible for an egg to discriminate between different sperm so effectively?

- In 1910, Frank Lille identified a compound, present on the surface of sea urchin eggs, that appeared to bind sperm in a species-specific manner. The molecule was called **fertilizin**. His simple assay for fertilizin–sperm interaction was to mix the two together and measure any clumping of the sperm cells. Lille predicted there was a corresponding protein on the surface of the sperm cells and that it bound to fertilizin in a "lock-and-key" manner.

- In the 1970s researchers made another breakthrough in the study of species recognition. They isolated a protein on the surface of the *sperm* and named the protein **bindin**. Each species had a slightly unique form of this protein, which bound only to eggs from the same species.

- Kathleen Foltz and William Lennarz tested the hypothesis that fertilizin protein on the egg was the binding partner of bindin on the surface of the sperm. As summarized in **Figure 22.6**, they used proteases to release cell-surface proteins from sea urchin eggs and then isolated any of these proteins that could bind sperm or bindin proteins of its own species. Key to this strategy was the prediction that in the collection of proteins released from the egg would be the fertilizin receptor proteins (attached to bindin proteins in a species-specific manner).

- The egg-sperm interaction was species specific due to the presence of the bindin proteins in sperm and the fertilizin receptors in eggs. These proteins bind tightly to enable fertilization but are very selective, binding to each other only if each protein originated from the same species (species lock and key).

Why Does Only One Sperm Enter the Egg?

- One key question of fertilization is how only one sperm is successful in fertilizing the egg, despite the presence of hundreds of sperm nearby, all of them intent upon fertilizing the egg.

- A Ca^{2+}-based signal is induced rapidly after fertilization and inhibits fertilization by any additional sperm by generating a **fertilization envelope**. As summarized in **Figure 22.7a**, fertilization results in the release of Ca^{2+} from intracellular stores. The increase in cytosolic Ca^{2+} results in the fusion of cortical granules with the egg-cell membrane. These granules are small, membrane-bound vesicles located near the cell membrane; upon fusion with the membrane, they release their contents into the extracellular matrix.

- What do these cortical granules contain that could inhibit additional fertilization events?

  1. **Proteases** that break down proteins associated with the egg-cell wall. These include the receptor protein (fertilizin), and the loss of these structures inhibits receptor binding to additional sperm.
  2. A high concentration of **solutes**; these alter the concentration of solutes at the cell exterior and induce a sudden influx of water from outside the cell (**osmosis**).

- The influx of water generates a water-filled space between the vitelline envelope and the cell membrane, generating a barrier to further fertilization events. The thick matrix of cell-wall materials that have been pushed away from the cell membrane by water influx is called the **fertilization envelope** (**Figure 22.7b**).

Fertilization in Mammals

- Mammalian fertilization is unique from that of invertebrates like the sea urchin in several ways, including the absence of the species-specific binding between sperm and egg. Also, the sperm and egg meet within the female body, at the **oviduct (fallopian tube)**.

- Paul Wassarman examined the proteins that mediate binding between sperm and eggs during mammalian fertilization. He isolated **glycoproteins** (proteins with sugar additions to the structure) from the zona pellucida of mouse eggs, and identified one (**ZP3**) that binds to sperm heads. ZP3 did not display species-specific binding, as was observed with sea urchin proteins that shared a similar function.

- The interaction between the sperm and the glycoprotein in the egg's zona pellucida generates an acrosomal reaction. This reaction blocks polyspermy, but via a mechanism that is distinct from that observed in sea urchins.

- At fertilization, mammalian eggs do not produce a fertilization envelope, but they do release enzymes that modify the receptor glycoproteins. This causes the release of any additional sperm that are bound to the zona pellucida.

CHECK YOUR UNDERSTANDING

(a) What mechanisms are used by the sea urchin eggs to avoid polyspermy?

_____

_____

(b) Do mammals block polyspermy via a mechanism that is identical to the one used by sea urchins? Explain your answer.

_____

_____

_____

22.3 Cleavage

- **Cleavage** refers to the rapid cell divisions that follow fertilization (without growth) to generate a blastula. The cells generated by cleavage are called **blastomeres.** Cleavage is similar to normal mitotic cell division except it is more rapid, and it involves no increase in size. The two cells generated by cleavage will each be half the size of the original parental cell.

Variation in Cleavage Patterns

- Cleavage occurs in different patterns depending upon the species of the embryo (**Figure 22.8**).
- During **radial cleavage**, the cells divide at right angles (90°) to form columns or tiers.
- During **spiral cleavage**, the cells divide at non-right or oblique angles, and these cells pile up as cleavage progresses.
- During **discoidal cleavage**, which occurs in birds and fish, cleavage produces a cluster of cells on top of the yolk.
- **Superficial cleavage** is specific to fruit flies and other insects. In this process, many nuclei are generated in the nucleus by multiple mitotic events that are not followed by cytokinsesis. The result is a large embryo with hundreds of nuclei in the cyotoplasm. These nuclei migrate to the edge of the cell and are subsequently surrounded with cell membrane (**Figure 12.8d**).

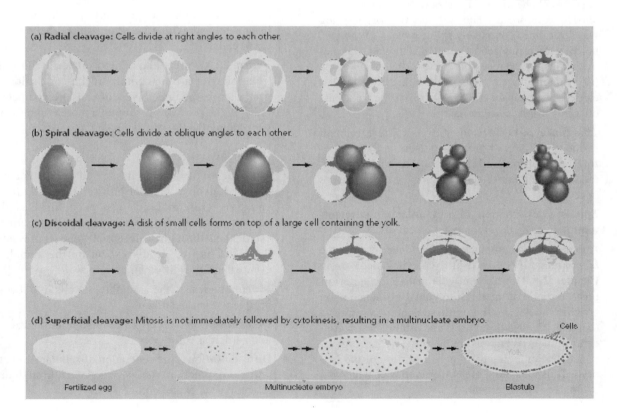

(a) **Radial cleavage:** Cells divide at right angles to each other.

(b) **Spiral cleavage:** Cells divide at oblique angles to each other.

(c) **Discoidal cleavage:** A disk of small cells forms on top of a large cell containing the yolk.

(d) **Superficial cleavage:** Mitosis is not immediately followed by cytokinesis, resulting in a multinucleate embryo.

Cells

Fertilized egg   Multinucleate embryo   Blastula

- Rotational cleavage occurs in human embryonic development. The oviduct or fallopian tube serves as a site for fertilization of the human oocyte and is the site of cleavage events. The oviduct is a small structure that connects the ovary to the uterus. As described in **Box 22.1** and diagrammed in **Figure 22.9**, cleavage generates a blastocyst that is composed of a thin layer of **trophoblast** cells on the exterior surrounding a ball of cells known as the **inner cell mass (ICM)**. A water-filled cavity in the center of this structure is known as the **blastocoel**. The embryo hatches from the zona pellucida 3.5 days after fertilization and attaches (implants) to the wall of the uterus. The trophoblast cells combine with maternal cells to form the **placenta**, and the ICM continues to develop as the embryo.

What Role Do Cytoplasmic Determinants Play?

- In 1894, H. E. Crampton established a link between the pattern of cleavage in early development and the establishment of shell-coiling direction later in development. Working with a species that produced a left-handed coil, Crampton was occasionally able to find mutant snails that display a right-handed coil. Surprisingly, when examining early development in these snails, Crampton discovered that the mutant cells underwent cleavage events with a switched orientation.

- An unusual phenomenon was observed with more extensive experiments. The coil direction of offspring was determined completely by the *mother's genotype*, not the offspring genotype. If the mother was homozygous for the recessive allele (*dd*), then all of the offspring displayed left-handed shells. If the mother carried the dominant allele (*DD* or *Dd*), then all of the offspring displayed right-handed shells.

- This **maternal effect** on embryonic development is due to mRNAs or proteins present in the egg before fertilization, and therefore it affects development independent of sperm genotype or the zygote genotype.

- This is a key example of a **cytoplasmic determinant** regulating developmental processes.

- Another example of maternal effect inheritance is the *bicoid* mutation in the fruit fly. Review bicoid from Chapter 21. This protein establishes the anterior-posterior axis of the embryo via its influence upon specific nuclei in the multinucleate fly embryo (**Figure 22.8d**). Its expression establishes a concentration gradient of bicoid protein, the highest concentration being the anterior portion of the embryo. Thus the bicoid mRNA/protein is considered a cytoplasmic determinant.

22.4 Gastrulation

- During **gastrulation** a complex migration of cells occurs, organizing the blastula and further establishing the key embryonic cell types (endoderm, ectoderm, and mesoderm). Gastrulation begins as cleavage processes are completed.

- Within the blastula are the following (**Figure 22.10**):
    1. **Animal pole—ectoderm** (generating skin and nerve cells)
    2. **Vegetal pole—endoderm** (generating gut and organs)
    3. **Central region—mesoderm** (muscle and organs)

- Review **Figure 22.11** to examine the formation of distinct layers of cells during gastrulation in frog embryos. Early experiments characterizing these processes utilized simple staining of subsets of cells during cleavage and then followed the cells through gastrulation. This mapping of cell fate has proven incredibly useful in the study of early developmental processes.

- **Step 1:** The **blastocoel** is the water-filled hole in the center of the blastula.

- **Step 2:** Gastrulation begins with the formation of a **blastopore** (hole) on the surface of the blastomere. Cells migrate away from a specific site on the embryo surface and form an **invagination**. This structure eventually develops into the anus.

- **Step 3 and 4:** Cells migrate inward, shrinking the blastocoel and gradually forming three distinct layers of cells.

- **Figure 22.12** offers photographs of sea urchin gastrulation over a time course. At the conclusion of gastrulation, the three layers of cells are competent to begin the process of forming a more complex organization of cells—the organs.

22.5 Organogenesis

- The process of forming tissue and organs is complex and involves cellular proliferation and differentiation. Differentiation can be defined as the generation of cellular diversity (structure, function), established through differential gene expression. Review **Chapter 21** discussion on the signaling events that drive differentiation forward.

- **Morphogenesis**—the assembly of cells into recognizable tissues and organs.

- **Determination**—a cell's irreversible commitment to a particular cell type.

- **Differentiation**—the generation of cellular diversity (structure, function) that is established through differential gene expression.

- Chronology within development:

  pattern formation → morphogenesis
  → determination → differentiation

Organizing Mesoderm into Somites: Precursors of Muscle, Skeleton, and Skin

- At the completion of vertebrate gastrulation, the mesoderm is a thin layer of cells located between the endoderm and ectoderm. **Figure 22.13** offers an excellent summary of early organogenesis in chordates.

- As development proceeds, the endoderm forms the beginnings of the **digestive tract**.

- At the same time, the mesoderm begins to form the **notochord** on the dorsal side of the embryo. The notochord is maintained in some adult chordates, but in most it is a transient structure that is important in organizing the body plan during organogenesis and is destroyed by programmed cell death in later stages.

- Signals from the notochord stimulate the adjacent ectoderm along the dorsal side of the embryo to fold and form the **neural tube**. This process involves dramatic changes in the morphology of specific ectoderm cells. The cytoskeletal proteins in these cells stretch the cells vertically, widening them at their ventral end and narrowing them at their dorsal end. This causes the cells to fold upward in the dorsal direction, eventually forming at tube. The neural tube is a precursor to the brain and spinal chord.

- The fully formed neural tube emits a molecular signal that organizes the neighboring mesodermal cells into blocks of cells known as **somites** (**Figure 22.14**). The neural tube is soon surrounded by somites, mediated by selective adhesion events (mediated by cadherins).

- The somites develop into a variety of adult tissues inclluding vertebrae, ribs, body wall, some muscle, and even skin (**Figure 22.15**). The fate of each group of somites is activated by cell–cell signals initiated in adjacent tissues (ectoderm, mesoderm, notochord, neural tube).

- Transplantation studies have shown that initially, all somites have the potential to become any of the somite-derived tissues. Transplanting somites from one site to another alters the differentiation fate of those cells. However, later in development, the somite's fate becomes irreversibly **determined**, and somatic cells that are transplanted to other sites in the embryo retain their initial differentiation fate.

Differentiation of Muscle Cells

- Cells in the somites receive molecular signals from several surrounding tissues (neural tube, ectoderm, and mesoderm). Cells located on the periphery of each somite are destined to become muscle cells.

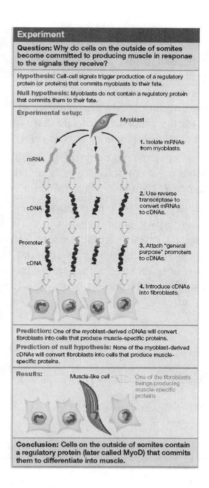

CHECK YOUR UNDERSTANDING

(c) Compare and contrast the terms *morphogenesis, determination,* and *differentiation* as they relate to early development.

- Researchers examined the molecular characteristics of myoblasts, somite-derived cells whose cell fate is determined to be muscle (**Figure 22.18**). The researchers isolated mRNA from myoblasts, converted the mRNA into cDNA through the process of **reverse transcription**, and reintroduced the **cDNAs** into fibroblast cells. They sought to identify genes whose expression in the fibroblasts results in the development of muscle-like characteristics. They were hunting for the genes that activate cell-fate determination in the development of muscle cells.

- Researchers discovered one gene that could convert the fibroblasts into muscle-like cells, and they named the gene *MyoD* ("myoblast determination").

- *MyoD* codes for a transcription factor that binds to the regulatory sequences of muscle-specific genes and activates their expression. Activation of *MyoD* expression in somites is stimulated by signals from neighboring embryonic tissues, and it results in the development of muscle from some of the somites.

CHECK YOUR UNDERSTANDING

(d) How can researchers establish when a somite cell fate is determined?

_____

_____

_____

_____

CHECK YOUR UNDERSTANDING

(e) Describe the methodology used for the cloning of genes involved in muscle-cell differentiation. How were researchers able to assay each isolated gene for the capacity to activate muscle-specific differentiation?

_____

_____

_____

_____

_____

B. INTEGRATING YOUR KNOWLEDGE

Looking Back—Key Concepts from Earlier Chapters

Control of Gene Expression in Eukaryotes—Chapter 18

The regulation of **cadherin expression** is a critical component of development, enabling **selective adhesion** between specific groups of cells and disabling the adhesion between other cells. Beneath these complex developmental processes is the critical regulation of gene expression.

Analyzing and Engineering Genes—Chapter 19

Isolating the genes that regulate and mediate the developmental process enables scientists to perturb these processes in order to better characterize the molecular mechanisms behind development. Using genetic engineering, scientists can produce an abnormal pattern of gene expression, delete the expression of specific genes, and/or observe the influence of mutant proteins on the system. Examine the methodology used to clone the *GH1* gene, and then review the experiments used to clone and characterize *MyoD*.

Looking Forward—Key Concepts in Later Chapters

Animal Reproduction—Chapter 48

Key to this chapter are the processes of egg fertilization and embryo development. Examine **Chapter 48**, recognizing the complexity of these processes and the molecular mechanisms that drive development and differentiation forward.

C. MASTERING CHALLENGING CONCEPTS

Reverse Transcription and cDNA Libraries

Our discussion of the work of Harold Weintraub and colleagues (*MyoD* cloning) introduces an important research tool, the utilization of a **cDNA library** to hunt for genes. The researchers set out to find genes involved in muscle-cell determination, and they prepared mRNAs from a collection of myoblasts. Certainly, within the pool of mRNAs are some transcripts that signal and maintain the muscle-specific determination of these cells. But finding the relevant gene is not easy. It requires a functional screening of all the mRNAs to find those associated with cell determination.

To screen mRNAs, the researchers copied each single-stranded RNA molecule into a copied DNA or cDNA. An enzyme called **reverse transcriptase** is utilized for this purpose. Retroviruses (a category of RNA viruses) use reverse transcriptase to convert their RNA genome into DNA during the process of viral infection. The enzyme can be purified for research purposes, converting mRNAs into cDNAs.

Weintraub and colleagues generated cDNA for each of the mRNAs isolated from myoblast cells and then set out to screen each of these cDNAs for an ability to alter the determination status of fibroblast cells. Their goal was to find, amidst that collection of thousands of different cDNAs, those capable of activating "muscle-like" determination in the fibroblasts. To do this, each of the cDNAs would have to be expressed within a group of fibroblasts to identify any influence of that gene on fibroblast biology. The researchers used DNA plasmids, each carrying a strong promoter, and transferred the cDNAs into their structure. The result is a collection of plasmids with cDNA inserts in their structures, each of which expresses the gene encoded on the cDNA. The plasmids were then transferred into fibroblasts, and the plasmid-carrying fibroblasts were analyzed for changes that might indicate muscle determination.

In this instance, the research team was successful in identifying one plasmid carrying the *MyoD* cDNA, which dramatically altered the fibroblasts. The cDNA library had been successfully screened, and a critical insight in cell determination was achieved.

D.   ASSESSING WHAT YOU'VE LEARNED

1. This category of cleavage is associated with a polynucleate embryo, whose nuclei migrate to the margins of the embryo and are subsequently surrounded by cell membrane (**Figure 22.8**).
   a. radial
   b. spiral
   c. discoidal
   d. superficial

2. When is a cell considered differentiated?
   a. when a cell first becomes irreversibly committed to a particular fate
   b. when a cell begins its pattern formation
   c. when a cell manufactures proteins that are specific to a particular cell type
   d. when a cell is part of recognizable tissues or organs

3. Which of the following is *not* necessary for fertilization to take place?
   a. The sperm and egg must be in the same place at the same time.
   b. The sperm and egg must fuse.
   c. The fusion of the two nuclei must trigger the onset of development.
   d. Intercourse must occur between members of the opposite sex of the same species.

4. In sea urchins, the gelatinous coat surrounding the egg contains a molecule that attracts sperm. Which of the following is true regarding this phenomenon?
   a. The attractant is a small peptide that diffuses out of the jelly layer into the surrounding seawater.
   b. The attractant is not species specific, and this can result in sperm from one species of sea urchin trying to fertilize an egg from a different species.
   c. Sperm respond to the attractant by swimming toward any area that contains it, regardless of the concentration.
   d. The release of the attractant also triggers the polymerization of actin into microfilaments that form a protrusion, which extends until it makes contact with the egg-cell membrane.

5. Research over the past 70 years has revealed a wide array of mechanisms that block polyspermy in various animal species. In sea urchins, fertilization results in the erection of a physical barrier to sperm entry via the formation of a fertilization envelope. Which of the following are involved in the mechanism that creates the fertilization envelope?
   a. The entry of the sperm causes calcium ions to be released from stored areas inside the egg.
   b. Cortical granules located just inside the membrane fuse with the egg-cell membrane and release their contents to the exterior.
   c. An influx of water separates the envelope matrix from the cell.
   d. The contents of the cortical granules include proteases that digest the exterior-facing fragment of the egg-cell receptor for sperm.
   e. All of the above apply.

6. Which of the following is true of cleavage?
   a. Cleavage divisions partition the egg cytoplasm during additional cell growth.
   b. These divisions proceed slowly in the initial creation of a multicellular embryo.
   c. When cleavage is complete, the embryo consists of a sphere of cells that is ready to undergo gastrulation.
   d. Cleavage occurs exclusively in plant cells.

7. What were researchers able to determine by transferring cells from one somite to another?
   a. A cell in a somite can become any of the somite-derived elements in the body.
   b. Somites are responsible for controlling apoptosis.
   c. Somites are transient structures that appear relatively briefly during development.
   d. Mesodermal cells in a somite are destined for a variety of structures.

8. Weintraub and co-workers hypothesized that myoblasts contain at least one regulatory protein that commits them to their fate. Their research found the gene that encodes this protein, which they named the protein MyoD. All of the following are true of MyoD *except*:
   a. MyoD encodes a transcription factor.
   b. *MyoD* is the gene that regulates activities of *ced-3* and *ced-4* genes.
   c. MyoD protein binds to enhancer elements located upstream of muscle-specific genes.
   d. MyoD protein activates expression of the *MyoD* gene.

9. Cells located at the periphery of a somite are destined to become:
   a. bone
   b. muscle
   c. neurons
   d. skin

10. A classic example of maternal genotye playing a key role in embryonic phenotype is the _____ gene.
    a. bicoid
    b. MyoD
    c. hunchback
    d. frizzled

CHAPTER 22—ANSWER KEY

Check Your Understanding

(a) A Ca^{2+}-based signal is induced rapidly after fertilization and inhibits fertilization by any additional sperm by generating a **fertilization envelope**. As is summarized in **Figure 22.7a**, fertilization results in the release of Ca^{2+} from intracellular stores. The increase in cytosolic Ca^{2+} results in the fusion of cortical granules with the egg-cell membrane. These granules release proteases that cleave egg-receptor proteins, and solutes that induce water flow into the space between the cell membrane and the cell envelope.

(b) Mammalian embryos do not form a fertilization envelope like sea urchins. However, at fertilization, they do release proteases that degrade receptor proteins in the zona pellucida.

(c) *Morphogenesis* is the migration and assembly of cells into distinct structures/ tissues. *Determination* refers to the molecular "commitment" to a cell type, activating cellular changes that result in cellular differentiation. Activation of *MyoD* within myoblasts is an excellent example of cellular determination. Differentiation refers to the generation of diversity in cellular structure and function, and it is established through differential gene expression.

(d) As illustrated in **Figure 22.16**, somites may be removed from one site and transplanted to another to determine if the transplanted somite is capable of differentiating into other cell types. Early in somite maturation, each transplanted somite cell is capable of differentiating into any of the somite-derived cell fates. Later during somite development, this is no longer possible, indicating the somite fate is now irreversibly determined.

(e) Harold Weintraub and colleagues isolated mRNAs from determined muscle precursor cells called myoblasts. The mRNAs were reverse transcribed into cDNAs and cloned into expression plasmids. This procedure enables the expression of these cloned fragments in any cell type. The researchers screened through their cDNA clones, transferring each into fibroblast cells and examining the cells for muscle-like differentiation following expression of the cDNA clone. The muscle-like changes within these cells would be identified as changes in cellular morphology and the expression of genes that are active only in muscle.

Assessing What You've Learned

1. d; 2. c; 3. d; 4. a; 5. e; 6. c; 7. a; 8. b; 9. b; 10. c

23

An Introduction to Plant Development

A. KEY BIOLOGICAL CONCEPTS

- This chapter extends our exploration of development to that of plants. A focus of this chapter is upon *Arabidopsis thaliana*, a flowering mustard plant that has been studied extensively (**Figure 23.1**). This plant grows to reproductive maturity rapidly, and it can be readily cultured in a greenhouse. The genome of this plant has been sequenced. Researchers can readily make transgenic *A. thaliana* plants using the techniques discussed in **Chapter 19** (*Agrobacterium*-based transfer). Information gained from the study of *A. thaliana* has been applied to many other plants. This organism thus serves as an excellent model organism for plants, in the same way that fruit flies, worms, yeast, and mice have served as model organisms for animals.

23.1 Pollination and Fertilization

- Gamete formation and fertilizationare more structurally complex in plantsthan they are in animals. In flowering plants, the egg is fertilized inside an **ovule**—a structure at the base of the flower (**Figure 23.2**). For fertilization to take place, the pollen must extend a long pollen tube that reaches inside the ovule. Once the egg is fertilized, the ovule and embryo together develop into a **seed** through a process known as **embryogenesis**. Notice the haploid (1*n*) and diploid (2*n*) phases of this life cycle. In Figure 23.2, these are highlighted in red and blue.

- Haploid **pollen grains** (male gamete) are produced on the anther of a mature flower via meiosis followed by mitosis. The two cells generated by this mitosis have different roles in fertilization. One cell mediates formation of the pollen tube, while the other cell forms sperm and fertilizes the egg (**Figure 23.3a**). These cells are surrounded by a tough coat that protects the pollen grain from dehydration. Pollen is distributed by the wind or by birds and insects that come into contact with the flower.

- To fertilize an egg, the pollen must be deposited on the stigma of a mature flower. Plants can determine if a particular pollen grain is the appropriate species via specific protein–protein interactions between the pollen and the stigma (**Figure 23.3b**). This is very similar to bindin and fertilizing proteins in animal fertilization (Chapter 22).

- If the protein–protein interactions are specific, the pollen grain **germinates**. This involves the elongation of a pollen tube that reaches down the stigma and toward the egg; it also involves the generation of sperm that will be carried by the pollen tube to the egg (**Figure 23.3c**).

- The plant ovule is fertilized by two sperm nuclei carried by the pollen tube. This **double fertilization** is key to the development of both the embryo and the **endosperm** tissue that is important to providing nutrition for the young plant (**Figure 23.4**). The embryo has a 2*n* or *diploid* chromosome number (sperm plus egg) and the endosperm has a 3*n* or *triploid* genotype (sperm plus two maternal cells).

- The endosperm surrounds the embryo and provides nutrition that includes proteins, carbohydrates, and fats for embryonic developmental growth. Endosperm is analogous to the yolk found in animal eggs, providing nutrition to embryo cells that are rapidly growing and developing.

CHECK YOUR UNDERSTANDING

(a) Describe the composition of a pollen grain.

_____

_____

_____

23.2 Embryogenesis

- Embryogenesis occurs inside the ovule following fertilization. Compare with cleavage, gastrulation, and organogenesis (introduced in **Chapter 22**).

> ### CHECK YOUR UNDERSTANDING
>
> (b) Define the cellular products of double fertilization, and the ploidy of each.
>
> _____
>
> _____
>
> _____
>
> _____

What Happens during Embryogenesis?

- Following fertilization, an *Arabidopsis* embryo undergoes asymmetric mitotic cell division, generating a **basal** cell (large) and an **apical** cell (small). The basal cell replicates several times in an axis that is perpendicular to the axis of the embryo, generating a column of cells called the **suspensor** at the base of the embryo. The apical cell ultimately develops into the mature embryo (**Figure 23.5**).

- This early step in development establishes the **apical-basal axis** (tip to base of the embryo).

- The apical cell divides to generate a **globule** (ball) of cells at the tip of the suspensor structure. The exterior layer of cells on this globule are morphologically distinct from those in the center.

- **Cotyledons** (precursors to leaves) then begin to develop as two extensions at the apical tip of the embryo. The **hypocotyl** (precursor to stem) is attached to the base of the cotyledons. The cotyledon and hypocotyl together are called the **shoot**. Basal to the hypocotyl is the **root** structure, which will become the belowground portion of the seedling.

- The **meristem** cells of the embryo are like stem cells in an animal embryo. These are groups of cells that are undifferentiated, and they divide to generate a source of cells for building specific differentiated structures.

- The early embryo contains a **shoot apical meristem (SAM)** at the apical side of the embryo, and a **root apical meristem (RAM)** at the basal side of the embryo.

- A cross section of the embryonic stem reveals three tissues critical to the development of the adult plant (**Figure 23.6**). These tissues are the exterior layer of protective cells, or **epidermis**, the cells below the epidermis, or **ground tissue**, and the **vascular** tissue in the center of the tissue. These tissues could be compared to ectoderm, mesoderm, and endoderm in animals.

- The ground tissue can develop into several different cell types that will be involved in photosynthesis or other functions.

- The vascular tissue develops into a transport system for the movement of water and nutrients throught the entire adult plant.

Which Genes and Proteins Set Up Body Axes?

- Gerd Jürgens and colleagues used a mutant hunt to isolate genes responsible for establishing the apical-basal axis (**Figure 23.7**). They predicted that plants carrying mutations in these genes would display abnormal development along this axis. They predicted that the gene products of these genes would be involved in regulating and signaling each cell's site along the apical-basal axis. The mutants they isolated were placed into three groups: **apical mutants** lacking cotyledons, **central mutants** lacking hypocotyl structures, and **basal mutants** lacking hypocotyl *and* root structures.

- The mutants that lacked hypocotyls and roots contained a mutation in a gene they named *MONOPTEROUS*. The encoded protein appears to be a transcription factor, and its expression is regulated by **auxin**. Auxins are a class of plant hormones (phytohormones) that regulate plant cell growth and development.

- Researchers draw parallels between auxin and *bicoid* as molecules that establish polarity within the embryo through their differential concentration along the axis of the embryo.

23.3 Vegetative Development

- The adult plant contiues to adapt and grow in response to environmental conditions. Trees extend branches in the direction of available sunlight. *Arabidopsis* plants that grow against a type of physical barrier will initiate growth away from that barrier. How does this work? There must be complex molecular signaling processes that regulate these processes.

Meristems Are Responsible for Continuous Growth and Development

- In the adult plant, meristems (SAM, RAM) play a central role in mediating new growth and development. The adult meristems are found at the tip of each shoot and at the tip of each root (**Figure 23.8a**).

- Examine the labels in **Figure 23.8b**, which shows the groups of cells present in a SAM. Notice the cells emerging from the meristem and beginning to differentiate into epidermal, ground, and vascular tissue.

- Auxin has the ability to stimulate leaf growth from a SAM. This appears to be a key cell–cell signal regulating vegetative growth from an adult plant.

- The new leaf, originating from the SAM, must develop all the tissues of a leaf in proper orientation. Therefore, the leaf must establish axes in all three dimensions (**Figure 23.9**).

- The *PHANTASTICA* (*PHAN*) gene has recently been shown to regulate establishment of the upper-lower axis in leaf development. Not surprisingly, the protein encoded by this gene encodes a transcription factor. This protein is expressed in the upper layer of the developing leaf. It induces the expression of genes that cause cells to form the upper layers of the leaf, and it suppresses genes that induce differentiation as lower leaf cells.

- To better characterize the funtion of *PHAN* in plants, scientists engineered tomato plants that carried a defect in this gene and then examined their leaf morphology. Their results are consistent with *PHAN* playing an important role in tomato leaf growth, similar to what had been observed in *Arabidopsis* plants. Review the methodology of generating transgenic plants discussed in **Chapter 19**.

23.4 Reproductive Development

- In animals, **germ cells** are set aside early in development, and somatic cells are involved in building most

(c) Define some of the effects of auxin on plant development.

_____

_____

_____

_____

of the animal body plan. The gonads (testes, ovary) are sites of germ cell origination only and are established early in development. In plants, meristem cells generate flowers and their pollen and egg cells. Meristem cells can go through many mitotic divisions during the normal life of the plant, with each division offering the risk of incorrect DNA replication (mutation). Thus, plants can pass on mutations as meristem cells with mistakes in their DNA are used to make new germ cells. Animal cells are thought to pass on fewer new mutations because they isolate and protect their germ cells.

The Floral Meristem and the Flower

- SAMs can convert from vegetative development to reproductive development, from making new leaves to making flowers.

- A SAM that has begun reproductive development is known as a **floral meristem**.

- The flower is composed of four layers of different cell types, and the pattern of layering is known as **whorls within whorls**. A whorl is a shape that coils or curls, like the swirls in icing on a cake. As shown in **Figure 21.11**, the four layers of whorls generate the flower structure. The layers are (1) a protective layer of **sepals** on the outside, (2) colorful **petals** inside of the sepals, (3) pollen-producing **stamen**, and (4) egg-producing structures, or **carpels**, in the center of the flower.

- In the 1800s, researchers identified mutant *Arabidopsis* plants that resulted in one category of reproductive organ replacing another (e.g., a flower that is missing the stamen layer and instead has two layers of petals). These mutants are called **homeotic** mutants. They can be compared to segmentation mutants in *Drosophila*, where a segment of cells develops abnormally due to defective signaling (**Chapter 21**).

- The powerful experimental strategies that are possible in *Arabidopsis* have enabled the discovery of mutations in specific genes as the source of the homeotic phenotype.

- Elliot Meyerowitz isolated a group of mutant *Arabidopsis* plants displaying abnormalities in flowering (homeotic mutants). He identified the genes responsible for each of the observed phenotypes. These genes can be divided into three categories, with each category of mutant phenotype displaying the absence of two adjacent whorls. The three phenotypes are carpel/stamen only, sepal/carpel only, and petals/sepals only (**Figure 22.12**).

- The research team predicted that each category of mutant phenotype was the result of a specific gene defect, and therefore three genes are the key regulators of flower development.

- Meyerowitz explained these three categories of genes/mutant phenotypes using the **ABC model** of flowering (**Figure 23.13**). This model includes three central components:

    1. Each of the three genes regulating flowering (*A*, *B*, and *C*) is expressed only in two adjacent whorls.
    2. If each gene is expressed in only two adjacent whorls, then four different combinations of gene products are possible (*A*, *AB*, *BC*, *C*). Each combination activates a unique molecular signal.
    3. These unique signals are capable of activating different whorls of differentiated meristem tissue: sepal (*A*), petal (*AB*), stamen (*BC*), and carpel (*C*).

- An additional prediction regarding this model is that the presence of protein A inhibits the expression of protein C. Also, the expression of protein C inhibits the production of protein A. They are mutually inhibitory gene products—if one is expressed first, the other is suppressed. This relationship would explain the mutant phenotypes observed. Look at **Figure 23.13**. If a plant lacked a functional copy of gene *C*, what phenotype would be generated?

- To test their model, Meyerowitz and colleagues next had to clone the genes that encode each of these proteins. They mapped the chromosomal locus of each protein and cloned the associated gene using techniques described in **Chapter 19**. They then examined the pattern of expression of each of these genes using in situ hybridization, described in **Chapter 21**. Recall that this type of analysis uses a probe made of single-stranded DNA to infiltrate slices of tissue and show where the mRNA that complements that probe is located. The mRNA and the probe stick to each other because they have complementary sequences. This analysis supported Meyerowitz's model. Gene *A* was expressed in the outer two whorls, gene *B* in the middle two whorls, and gene *C* in the inner two whorls.

- Compare these results and the ABC model with the *Hox* genes and their pattern of expression during animal development (**Chapter 22**). The parallels between these developmental processes are striking.

- Each of the three genes was also shown to encode DNA-binding regions in its protein product, sharing functional homology with the homeobox sequences found in the *HOM-C* genes. The plant DNA-binding motifs are known as MADS-boxes.

CHECK YOUR UNDERSTANDING

(d) Using the ABC model, explain the mutant phenotype of a flower from an *Arabidopsis* plant that lacks the *C* gene.

_____

_____

_____

_____

B. INTEGRATING YOUR KNOWLEDGE

*Looking Back—Key Concepts
from Earlier Chapters*

Analyzing and Engineering Genes—Chapter 19

In this chapter you learned about the use of gene mapping, cloning, sequencing, and in situ hybridization to characterize key genes that regulate the development of flowers. These techniques allowed researchers to test very specific hypotheses that had been developed from their studies of mutant plants. Notice the application of molecular biology to a fundamental question in developmental biology. Another example is the production of transgenic plants, which carry engineered mutations in the *PHAN* gene. This work enabled an improved understanding of leaf development.

*Looking Forward—Key Concepts
in Later Chapters*

**Plant Sensory Systems, Signals
and Responses—Chapter 39**

In this chapter auxin was introduced as avital, developmentally important hormone in plants. Auxin was the first hormone to be discovered. It is actually not one molecule, but a class of structurally similar hormones. It regulates root growth, shoot growth, and various other key developmental processes.

(a) The ABC model

Idea 1: Three genes pattern the flower. Each gene is expressed in two adjacent whorls.

Idea 2: A total of four different combinations of gene products occur.

Idea 3: Each of the four combinations of gene products triggers the development of a different floral organ.

Assumption: Presence of gene A product inhibits expression of gene C. Presence of gene C product inhibits expression of gene A.

(b) Testing the ABC model

Interestingly, most of these influences can be traced to the impact of auxin in regulating the expression of specific genes.

C.   MASTERING CHALLENGING CONCEPTS

Plant development presents a dizzying array of terminology used to describe the complex architecture of plants. It is challenging to a student learning about plants for the first time. As you work through this chapter, a helpful methodology for keeping track of terms is to diagram a flowchart of terms and structures that are relevant at each stage of development. Then move forward in developmental time and write out the structures that are critical at that stage. You can connect embryonic structures and terms with their corresponding adult structures and terms by color-coding or simply by interconnecting the terms with a drawn line. The result is similar to a concept map, but you are making a map of plant development. You can branch off in another direction to add another group of terms, for instance, reproductive organs. You will find that as new terms are introduced, you can add them to your map and they are much easier to remember. They are no longer individual points of information, but part of a web of information.

D.   ASSESSING WHAT YOU'VE LEARNED

1. Compare and contrast an animal sperm cell to a plant pollen grain.

   _____

   _____

   _____

   _____

   _____

2. In flowering plants, an event known as double fertilization occurs when:
   a. Two sperm enter the ovule; one fertilizes the egg, and the other fertilizes a $2n$ cell to generate endosperm.
   b. Two sperm cells fertilize one normal egg cell to make a $3n$ embryo; the embryo later becomes $2n$ during a unique mitotic division.
   c. Two eggs are simultaneously fertilized by two sperm nuclei.
   d. None of the above apply.

3. Species-specific selection of pollen occurs via:
   a. allowing only species-specific pollen to come into contact with the stigma
   b. toxin-based destruction of foreign pollen
   c. specific protein–protein binding between pollen and stigma
   d. Plants do not block fertilization of their egg by foreign pollen.

4. Germination of a pollen seed refers to:
   a. rapid mitotic divisions and growth of the embryo
   b. establishment of meristem cells
   c. growth of roots and cotyledons
   d. growth of a pollen tube and production of sperm

5. In early embryonic development, the _____ cell divides into a support structure called the suspensor.
   a. basal
   b. ground
   c. apical
   d. vascular

6. The hypocotyl is an embryonic precursor to the:
   a. flower
   b. root
   c. stem
   d. leaf

7. Plant embryogenesis is comparable to which of the following developmental processes in animals?
   a. segmentation and determination
   b. fertilization
   c. cleavage, gastrulation, organogenesis
   d. meiotic cell division

8. Using in situ hybridization experiments, the Meyerowitz research group examined the expression of flower development in gene *B* and found its expression to be limited to which of the following whorls in a normal developing flower?
   a. sepal
   b. petal
   c. carpel
   d. stamen

9. The *PHAN* gene is a transcription factor that regulates:
   a. axis development in leaf
   b. apical-basal axis development in embryo
   c. radial axis development in flower
   d. development of the suspensor structure

10. Following fertilization of the egg, the plant ovule develops into:
    a. exterior of the seed
    b. the flower
    c. a meristem
    d. the shoot apical meristem (SAM)

CHAPTER 23—ANSWER KEY

Check Your Understanding

(a) Pollen grains (male gamete) are composed of two cells, essentially a cell within another cell. The outer cell mediates the formation of the pollen tube. The inner cell forms the sperm that fertilize the plant egg after moving down the extended pollen tube (**Figure 23.3a**). These cells are surrounded by a tough coat that protects the pollen grain from dehydration.

(b) Double fertilization is a unique fertilization process in plants. One sperm fertilizes the egg to generate the plant embryo (diploid or $2n$). A second sperm fertilizes a diploid ovule cell to generate endosperm (triploid or $3n$). The endosperm feeds the developing embryo by producing large amounts of carbohydrate, fat, and protein for the embryo. Seeds are primarily composed of endosperm and a small plant embryo.

(c) Auxin is a complex plant hormone that has wide-ranging effects on plant development. In this chapter, we have defined auxin as an inducer of MONOPTEROUS expression, a key protein in the establishment of the embryonic body axis. In addition, auxin stimulates the shoot apical meristem (SAM) to generate new leaves.

(d) In the ABC model, protein products from each gene work together to specify the type of tissue that develops: sepal (*A*), petal (*AB*), stamen (*BC*), or carpel (*C*). The *C* gene is normally expressed in the inner two whorls of the developing flower, and it normally suppresses expression of the *A* gene in those whorls. Thus the whorls in a *C*-deficient plant would have the following protein products (from outer layer to inner layers): A, AB, AB, A. The flower of this plant would develop with sepals on the outer layer and the innermost layer, with petals in the middle layers. See **Figure 12.13**.

Assessing What You've Learned

1. The plant sperm cell is a pollen grain. This structure contains very little of the structural specialization displayed by the animal sperm cell. The pollen grain contains two haploid nuclei, whereas the animal sperm contains only one. The plant pollen grain relies on attachment to the fertile plant and growth of the pollen tube to the egg cell. Two nuclei are delivered. The relevance of these two nuclei involves the formation of endosperm ($3n$ in DNA content) along with the zygote ($2n$).

2. a; 3. c; 4. d; 5. a; 6. c; 7. c; 8. b and c; 9. a; 10. a

24

Evolution by Natural Selection

A. KEY CONCEPTS

- The theory of evolution by natural selection is one of the key theories in biology. This theory was controversial because it conflicts with the theory of special creation.

- The pattern recognized by the theory of evolution is that species are related and change through time. Natural selection is the mechanism that explains this pattern. Modern biologists continue to study evolution, amassing yet more evidence in support of this theory.

24.1 The Evolution of Evolutionary Thought

- The theory of evolution by natural selection was revolutionary because it overturned one idea about how nature works, replacing it with a new and completely different idea.

- The theory that species are unchanging types created by God had dominated thinking about organisms since Plato articulated it more than 2000 years before.

- Another Greek philosopher, Aristotle, proposed the **great chain of being** that built on Plato's ideas by organizing species into a sequence from smallest and/or least complex to largest and/or most complex (**Figure 24.1**).

- Plato's and Aristotle's concept of species as unchanging types with insignificant or misleading variation typify **typological thinking**. Typological thinking is also found in the Bible's book of Genesis and led to Linnaeus's hierarchical classification system in the 1700s.

- Jean-Baptiste de Lamarck was the first to propose a theory of **evolution** when he argued that species had changed through time (1809). However, Lamarck thought organisms originated via spontaneous generation at the base of the great chain of being and then moved up the chain as they evolved over time.

- Lamarck proposed that organisms progressed from lower to higher forms by inheriting **acquired characters**—characters developed by individuals in response to environmental challenges.

- Darwin and Wallace (**Box 24.1**) did not believe that evolution followed any linear progressive pattern. They also argued that change through time relies on variation among individuals in a population.

- Darwin and Wallace's theory represents a switch to **population thinking**. Natural selection occurs when individuals with certain traits leave more offspring than other individuals do. Thus, instead of seeing variation as unimportant, this theory critically relies on variation in a **population** (individuals of the same species living in the same area at the same time).

CHECK YOUR UNDERSTANDING

(a) Why do traits have to be heritable in order for evolution by natural selection to occur?

_____

_____

_____

_____

24.2 The Pattern of Evolution: Have Species Changed through Time?

- Darwin described evolution as **descent with modification**—meaning that change over time produces modern species from ancestral species.

- The suggestion that species change through time and are related by common ancestry was a radical departure from the independently created and immutable species concept described by the theory of special creation.

Evidence for Change through Time

- Traces of organisms that lived in the past are called **fossils** (**Figure 24.3**). Those that have been found and described in the scientific literature make up the **fossil record**.

- Most fossils are found in **sedimentary rocks** that form in layers from deposits of sand, mud, or other materials at beaches or river mouths. Fossils from lower sedimentary layers are older than those from upper layers . These layers are associated with different periods in the **geologic time scale**.

- Researchers can now use radioactive isotopes to date rocks, and they have assigned absolute ages to the various geologic time periods. Geologic data show that Earth is about 4.6 billion years old, and fossil data show the earliest signs of life in rocks about 3.4 billion years old.

- Observations from the fossil record and observations of living species convinced biologists that species do indeed change through time.

Extinction

- Researchers have discovered many fossils unlike any known living organisms. These species are **extinct**; they no longer exist. More species have gone extinct than exist today, and species have gone extinct throughout Earth's history.

- Darwin interpreted extinction as evidence that species are dynamic and can change. He reasoned that if species have gone extinct, then the variety of species on Earth has changed through time.

Transitional Forms

- Fossils often resemble living species found in the same region. This "law of succession" means that extinct fossil species are typically succeeded, in the same region, by similar species (**Figure 24.4a**). Darwin interpreted this as evidence that extinct forms are the ancestors of modern forms, and species change over time.

- Improvements in the fossil record led to the discovery of many **transitional forms** with traits that are intermediate between older and younger species.

- The whale lineage includes fossil transitional forms that show a gradual change over a 12 million-year period from land-dwelling hippopotamus-like animals to modern whales (**Figure 24.4b**). Many other transitions are similarly well documented and provide strong evidence for species change through time.

Vestigial Traits

- **Vestigial traits** are functionless structures similar to functioning structures in related species. Biologists interpret these vestigial traits as evidence that trait structure and function change over time (e.g., whale hip bones, human appendix, pseudogenes; **Figure 24.5**).

- Vestigial traits are strong evidence that species change over time. The existence of these traits is not adequately explained by the theory that a perfect supernatural being created unchanging species.

Current Examples

- Changes in modern species in response to environmental change are also well documented (e.g., evolution of antibiotic resistance in bacteria, herbicide resistance in weeds, etc.). Thus change in species over time continues today and can be measured directly.

- Evidence from both the fossil record and from living species supports the idea that life is ancient and that species have changed through time and continue to change today.

CHECK YOUR UNDERSTANDING

(b) Transitional forms and vestigial traits have both been used as evidence that species change over time. Comment on the strengths and weaknesses of these two types of evidence.

_____

_____

_____

_____

_____

Evidence That Species Are Related

Geographic Relationships

- Darwin, as ship naturalist, collected plants and animals on his *Beagle* voyage. The Galápagos mockingbirds were superficially similar, but different islands had distinct species. Darwin proposed that the species were similar because they had descended with modification from a common ancestor (**Figure 24.6**).

- This pattern of distinct yet similar species is commonly found on island groups worldwide. Related species share a **phylogeny** or family tree, and species' ancestor–descendant relationships can be graphically depicted in a **phylogenetic tree**.

- The presence of similar species in the same geographic area is strong evidence that these species are descended from a common ancestor that lived in the same region.

Homology

- **Homology**, in biology, refers to resemblance due to common descent (similar traits derived from a common ancestor).

- The most fundamental homology shared by organisms is **genetic homology**—similarity in the DNA sequences of different species.

- For example, fruit fly and human eyes are very different, but the genes that determine where eyes develop are similar in these two groups (**Figure 24.7**). Biologists conclude that a common ancestor of fruit flies and humans had a similar gene that was involved in forming a primitive light-gathering organ.

- Evidence that all Earth's organisms descend from a single common ancestor include the most fundamental of all genetic homologies—the universal genetic code. Almost all organisms use the same 64-mRNA codons to specify the same amino acids (the few exceptions differ from this universal code for only one or two codons).

- Other universal homologies are the presence of a plasma membrane made of a phospholipid bilayer with interspersed proteins, use of RNA polymerase to transcribe the information coded in DNA into RNA, use of ribosomes to synthesize proteins, use of ATP as an energy currency, and use of DNA polymerase to copy DNA.

- Biologists also study **developmental homology**—a similarity in embryo **morphology** (form) and/or pattern of tissue differentiation.

- Early in embryonic development, all **vertebrates** (animals with backbones) have gill pouches and tails (**Figure 24.8**). Biologists hypothesize this pattern exists because the common ancestor of all vertebrates was a fishlike animal with gill pouches and a tail.

- Gill pouches are lost later on in the embryonic development of most terrestrial vertebrates (e.g., chicks, humans, cats) and can be considered a vestigial trait. In humans, the embryonic tail is another vestigial trait.

- An example of developmental homology in tissue differentiation is that the same group of embryonic cells develops into the jaw of both fish and mammals. Once again, the similarity in trait is explained by the trait's presence in a common ancestor.

- Genetic homologies give rise to developmental homologies that in turn cause **structural homology**, meaning similarity in morphology.

- For example, all vertebrates share the same general limb structure (**Figure 24.9**). It is not logical to use the same underlying pattern for limbs of such differing function, but the similarity is easily explained if all mammals descended from a common ancestor.

- The many examples of similarity in traits among different organisms support the idea that species are related to each other by descent from a common ancestor. If species were created independently of one another, there would be no necessity for such ubiquitous similarity.

CHECK YOUR UNDERSTANDING

(c) Why is genetic homology "the most fundamental" level of homology?

_____

_____

_____

Evolution Is Change through Time

- Evidence that species change through time and are related is summarized in **Table 24.1**. By the late 1800s, most biologists were convinced that evolution was (and remains) the best explanation for the patterns observed in the natural world.

24.3 The Process of Evolution: How Does Natural Selection Work?

- The idea of evolution predates Darwin, but Darwin and Wallace were the first to propose a process—natural selection—that satisfactorily explains the observed pattern of descent with modification.

- A population experiences evolution by natural selection when (1) individuals vary in their traits; (2) some of these variations are heritable; (3) some individuals survive and reproduce better than other individuals; and (4) differential survival and reproduction are influenced by the heritable traits of individuals.

- In other words, some heritable traits help an individual to survive or reproduce better than other individuals. Such traits will, over time, increase in frequency in the population, causing **evolution**—the genetically based change in a population's traits over time.

- Population thinking and an appreciation for the importance of variation among individuals are the keys to understanding evolution. Evolution by natural selection occurs when heritable variation within a population leads to differential success in survival and reproduction.

- For example, in some moth populations the wing color depends, in part, on individual genotypes since different alleles cause more or less melanin pigment production. In environments with light-colored tree bark, moths that produce less pigment survive better than darker moths because bird predators have a harder time finding light-colored moths (**Figure 24.10**).

- In the above example, differential bird predation causes natural selection on wing color. This natural selection causes evolution—a change in the population's allele frequencies over time.

(a) If heritable variation...

Color varies among individuals primarily because of differences in their genotype

(b) ... leads to differential success...

Birds find and eat many more dark-winged moths than light-winged moths

(c) ... then evolution results.

	Frequency of A_1 allele	Frequency of A_2 allele	
Start	0.5 (10 of 20 alleles present)	0.5 (10 of 20 alleles present)	Allele frequencies have changed in the surviving moths
End	0.625 (5 of 8 alleles present)	0.375 (3 of 8 alleles present)	

- **Darwinian fitness** is the ability to produce more offspring, as compared to other individuals in the population. Researchers can measure the relative fitness of an individual by counting how many offspring it produces relative to other individuals in a population.

- Fitness has meaning only in reference to a particular environment. For example, while light-wing moths have higher fitness in environments with light tree bark, dark-wing moths would have higher fitness in an environment with dark bark.

- **Adaptations** are heritable traits that increase an individual's fitness relative to individuals lacking that

trait. Adaptations are also meaningful only in reference to an organism's environment, because traits that increase fitness in one environment may not be advantageous in another environment.

(d) List the traits that might help some individuals survive or reproduce better than other individuals for each of these organisms: rabbit, turtle, oak tree, bacteria.

_____

_____

_____

_____

24.4 Evolution in Action: Recent Research on Natural Selection

- The theory of evolution by natural selection is testable. Here are some examples of how biologists test for evolution by natural selection in actual populations. Hundreds of other examples are available.

How Did Mycobacterium tuberculosis Become Resistant to Antibiotics?

- *Mycobacterium tuberculosis* causes **tuberculosis (TB)**, which in the nineteenth century killed up to a third of all adults in big cities. TB still kills more adult humans than does any other virus or bacteria.

- Sanitation, nutrition, and antibiotics succeeded in greatly reducing deaths due to TB in industrialized countries between 1950 and 1990. Unfortunately (but predictably), TB is on the rise again as antibiotic-resistant strains evolve and spread.

- Consider the evolution of antibiotic resistance in a young man diagnosed with TB. After more than 33 weeks of treatment with rifampin and other antibiotics, his lung fluid and chest X-rays were normal and he was considered cured.

- Two months later, the man was readmitted to the hospital with TB and treated with more antibiotics, including rifampin (a common antibiotic for TB). But new lung-fluid cultures grew rifampin-resistant bacteria. Antibiotics did not kill the bacteria; the man died.

- DNA from the rifampin-resistant bacteria was compared to DNA from the original antibiotic-sensitive bacteria cultured when the man was first admitted to the hospital with TB. The only difference between the two DNA sequences was a single point mutation (C → T).

- This mutation caused a single amino acid change in the *rpoB* gene that codes for a component of RNA polymerase, the protein that transcribes DNA to mRNA.

- Rifampin acts by binding to RNA polymerase in *M. tuberculosis* and preventing mRNA transcription (eventually killing the cell). But because rifampin does not bind effectively to the new mutant RNA polymerase, it cannot kill cells with this mutation.

- Researchers deduced that one or a few cells with this mutation were present in the original population. Cells with this mutation don't usually survive and reproduce as well as cells with the normal *rpoB* gene (the mutant RNA polymerase does not work as well).

- When antibiotics were added to the bacteria's environment, bacteria with the normal *rpoB* gene died; the bacterial population declined so much that the man appeared cured. However, cells with the mutant *rpoB* gene now had a selective advantage—they survived, reproduced, and eventually became so numerous that they killed their host (**Figure 24.11**).

- Natural selection had caused evolution—a change in a population's allele frequencies over time.
  1. Variation existed in the initial bacterial population.
  2. The variation was heritable.
  3. Bacteria differed in their reproductive success.
  4. Bacteria with certain traits produced more offspring than other bacteria; when rifampin was present, bacteria with the mutant *rpoB* gene were more fit than other bacteria—so the mutant *rpoB* gene is an adaptation to antibiotic-containing environments.

- Individual bacterial cells did not evolve; they merely survived or died and reproduced more or less. It was the population that evolved over time due to the effects of natural selection on individuals. Natural selection acts on the individual, but it is the population that evolves.

- Evolution of antibiotic resistance in bacteria is increasingly common, as is the evolution of resistance to insecticides, fungicides, antiviral drugs, and herbicides in insects, fungi, viruses, and plants.

Why Are Beak Size, Beak Shape, and Body Size Changing in Galápagos Finches?

- Biologists can study natural selection even when the alleles and molecular mechanisms responsible for variation in the population are not known.

(e) How did the M. tuberculosis researchers establish heritability of rifampin resistance?

_____

_____

_____

_____

- Peter and Rosemary Grant led long-term research on changes in body size and beak size and shape in medium ground finches (**Figure 24.12**) on Isle Daphne Major of the Galápagos Islands.

- Because these finches live on a small island, researchers could catch, weigh, measure, and mark all individuals. They found that body size and beak size and shape varied among individuals and that variation in these traits was heritable (e.g., parents with deep beaks tend to have offspring with deep beaks).

- In 1977, a severe drought eliminated 84 percent of the island's finch population. Most of the finches died of starvation since few plants were able to produce seeds.

- This sudden environmental change created a **natural experiment** that allowed researchers to compare groups created by a natural change in conditions (as opposed to a manipulation under controlled conditions). This particular natural experiment allowed comparison of the finch population before and after the drought and mass starvation.

- Survivors had deeper beaks on average than did nonsurvivors. Grant's group hypothesized that individuals with large, deep beaks were more likely to survive because they could eat the tough fruits of *Tribulus cistoides*.

- Natural selection caused an increase in the finch population's average beak depth in just one generation (**Figure 24.13**). Alleles responsible for development of deep beaks must have increased in frequency in the population as an adaptation for cracking large, tough fruits and seeds.

- More recent work (**Box 24.2**) has identified one of the genes for which allele frequencies may have changed during this period of natural selection.

- In 1983 there was much more rain than average, causing lots of plant growth, and finches fed primarily on abundant small, soft seeds. Small finches with small, pointed beaks had particularly high reproductive success during this period, causing yet another change in the population's characteristics.

- The Grants continue documenting evolution in response to continuing environmental changes on Daphne Major (**Figure 24.14**). Long-term studies like these document how populations experience natural selection in response to environmental change.

CHECK YOUR UNDERSTANDING

(f) A pattern of repeated speciation events is often found in island chains. What about islands promotes speciation?

_____

_____

_____

24.5 The Nature of Natural Selection and Adaptation

Selection Acts on Individuals, but Evolutionary Change Occurs in Populations

- Individuals do not change during natural selection. Those that are selected simply produce more surviving offspring than other individuals do, leading to a change in the genetic makeup of the population.

- Darwinian evolution is different in this respect from Lamarck's proposed evolution by inheritance of acquired characters. Individuals may change over their lifetime in response to environmental change, but these changes are not passed on to offspring.

- **Acclimation** occurs when an individual changes in response to changes in its environment (e.g., making more red blood cells at high altitudes). These phenotypic changes are not heritable and, consequently, cannot cause evolution.

- Adaptation can occur only when a population changes in response to natural selection.

Evolution Is Not Goal-Directed or Progressive

- Mutations, as in the tuberculosis example, occur randomly. Likewise, adaptations do not occur because organisms need them. They just happen, and when they do, natural selection may lead to their becoming more common.

- Adaptations are not goal directed, and neither is evolution. Although groups that appear later in the fossil record are often more morphologically complex than related groups that appeared earlier, this tendency is not universal. Traits can be lost or gained due to evolution by natural selection.

- For example, populations that become parasitic tend to evolve simpler morphologies (e.g., tapeworms lost their gut). These and other data show that evolution is not progressive. This evidence contrasts with Lamarck's idea that evolution caused progress toward higher levels on a "ladder of life" (**Figure 24.15**).

- According to modern evolutionary theory, there is no such thing as "higher" or "lower" organisms. There are only more ancient and less ancient groups with different adaptations that allow the groups to thrive in different environments.

Animals Do Not Do Things for the Good of the Species

- Purely self-sacrificing behavior is not observed in nature, because it is selected against. If a self-sacrificing behavior were to arise due to a mutation, that allele would swiftly be selected against as individuals with more selfish behaviors reproduced and left more offspring (**Figure 24.16**).

Not All Traits Are Adaptive

- Adaptation is not a perfect process. For example, vestigial traits do not increase fitness; they exist simply because they were present in the ancestral population.

- Other nonadaptive traits include holdovers from structures present early in development and neutral mutations where a change in the DNA sequence does not change the amino acid sequence of the protein encoded by that gene (due to the redundancy of the genetic code).

Genetic Constraints

- Many alleles that affect body size also affect other aspects of size. In the finches studied by the Grants, for example, beak depth was correlated with beak width. Thus deep, narrow beaks were not able to evolve even though birds with narrow beaks were better able to twist open the tough *Tribulus* fruits.

- When selection on alleles for one trait causes a correlated but suboptimal change in another trait, the number of possible evolutionary outcomes is limited. This type of constraint is called **genetic correlation**.

- Lack of genetic variation can also constrain evolution because natural selection can work only on extant variation in a population.

Fitness Trade-offs

- A **fitness trade-off** is a compromise between traits. For example, there may be a trade-off between size of egg or seed and number of offspring, where larger eggs or seeds are correlated with fewer offspring. Fitness trade-offs occur because selection acts on individuals with many traits.

Historical Constraints

- Because all traits evolve from previously existing traits, adaptations are constrained by history.

- For example, the bones in our middle ear derive from bones that used to be part of the jaw and braincase of mammal ancestors. Natural selection had to work on the traits available—unlike an engineer, who is free to imagine the best possible way to solve each structural challenge.

CHECK YOUR UNDERSTANDING

(g) Because most genes are inherited and used by both sexes, selection for traits that would be beneficial for males but harmful for females (or vice versa) is constrained. Is this a genetic constraint? a fitness trade-off? both? neither?

_____

_____

_____

_____

_____

B. INTEGRATING YOUR KNOWLEDGE

*Looking Back—Key Concepts
from Earlier Chapters*

Evolution and Natural Selection— Chapters 1 and 4

The process of evolution by natural selection was introduced in **Chapter 1** along with some examples of how researchers study evolution. **Chapter 4** discussed the related process of chemical evolution and gave an example of experimental selection. Here in Chapter 24, natural selection and research on its action in modern organisms are described in greater detail.

Inheritance of Traits—Chapters 13 through 16

Chapter 24 emphasizes that only heritable traits can be acted on by natural selection to cause evolution in a population. The genetic basis of inheritance was explained in earlier chapters.

Mutation—Chapters 14 and 16

Chapter 24 describes how a single-point mutation led to antibiotic resistance in *Mycobacterium tuberculosis*.

For additional information on the molecular basis of mutation, look back to **Chapters 14** and **16**.

Transcription—Chapter 16

Chapter 24 introduces you to an antibiotic, rifampin, that can disrupt RNA polymerase's ability to transcribe DNA. See **Chapter 16** for an explanation of DNA transcription in bacteria.

*Looking Forward—Key Concepts
in Later Chapters*

Natural Selection and Mutation—Chapter 25

More information on natural selection and mutation can be found in **Chapter 25**, which explores all the possible causes of evolution.

Shared Ancestry—Chapters 26 and 27

Chapter 26 describes how two species can evolve from one ancestral population, and **Chapter 27** explains how relationships between species are studied and summarizes the history of the diversification of life on Earth.

Bacterial Diseases—Chapter 28

Mycobacterium tuberculosis is only one of many disease-causing bacteria. **Chapter 28** presents more information on bacterial diseases.

C. MASTERING CHALLENGING TOPICS

Natural selection sounds relatively simple, yet it occurs in many different settings and with drastically different results. Don't lose sight of the basics when learning the details of individual instances. Instead, try to relate the details to the four steps leading to natural selection: variation, heritability, differential reproductive success, and reproductive success based on those variable heritable traits. Test your understanding by imagining how natural selection might have acted to produce one of your favorite species. What variable traits does that species have? Are those traits heritable? Would those traits affect survival and reproduction? How might that species have changed through time as individuals with the successful heritable traits had more offspring than those with less successful traits?

Remember that natural selection acts on the individual: an individual is either selected for or selected against. *On average*, some traits lead to greater success than do other traits (random exceptions will occur; they just won't matter in the long run if the population is large enough—more on this in the next chapter). This process makes the *population* evolve. Individuals do not change their heritable traits, but the frequency of those traits changes in a population over time in response to natural selection on individuals.

D. ASSESSING WHAT YOU'VE LEARNED

1. Aristotle's great chain of being is an example of:
   a. the first theory of evolution
   b. the law of succession
   c. population thinking
   d. typological thinking

2. Which of the following is a good restatement of Darwin's phrase "descent with modification"?
   a. Species change over time, and those that exist today descended from other, preexisting species.
   b. Humans are different from a preexisting ancestor species that lived in trees.
   c. Ancestral species are more primitive and lower down on the phylogenetic tree than descendant species alive today.
   d. Natural selection changes species over time.

3. Which of the following are supported by evidence from the fossil record? (Choose all that apply.)
   a. the concept that species can go extinct
   b. the hypothesis that individuals evolve by acquiring new traits
   c. the idea that the Earth is 4.6 billion years old
   d. the existence of transitional forms
   e. the existence of developmental homology

4. Which of the following is a statement of the law of succession?
   a. Species in the fossil record were succeeded, in the same region, by similar species.
   b. Species in the fossil record were always succeeded, in the same region, by species more closely adapted to the current environment.
   c. Species in the fossil record were succeeded, in the same region, by larger and more complex species.
   d. Species in the fossil record were succeeded, in the same region, by species that were more successful.

5. An example of a vestigial trait in humans is:
   a. kidneys
   b. cataracts
   c. wisdom teeth
   d. There are no vestigial traits in humans.

6. The use of RNA polymerase by all organisms is an example of:
   a. a vestigial trait
   b. genetic homology
   c. developmental homology
   d. structural homology

7. Which of the following is considered Darwin's greatest contribution to science?
   a. Darwin recognized the pattern of evolution (descent with modification).
   b. Darwin recognized the process of evolution (natural selection).
   c. Darwin recognized "The original species."
   d. Darwin recognized the heritable basis of traits (genetics).

8. Which of the following is the best definition of Darwinian fitness?
   a. Darwinian fitness is the ability of an individual to survive.
   b. Darwinian fitness is the ability of a population of organisms to persist.
   c. Darwinian fitness is the ability of an individual to survive and reproduce.
   d. Darwinian fitness is the ability of an individual to stay healthy by eating well and exercising.

9. Which of the following is a modern definition of evolution?
   a. Evolution is a change in an individual's ability to survive and reproduce as it grows older.
   b. Evolution is a change in allele frequencies in a population over time.
   c. Evolution is an increase in the fitness of a population.
   d. Evolution is a change in an individual's fitness over time.

10. What are the four requirements for evolution by natural selection in a given population?
    a. (i) Population varies; (ii) all variation is heritable; (iii) some individuals produce more offspring than others do; (iv) populations differ in production of offspring.
    b. (i) Individuals in a population vary; (ii) the variation is neutral; (iii) offspring are produced that can survive; (iv) individuals produce offspring.
    c. (i) Individuals in a population vary; (ii) variation is heritable; (iii) more offspring are produced than can survive; (iv) all individuals have different alleles.
    d. (i) Individuals in a population vary; (ii) variation is heritable; (iii) some individuals produce more offspring than others do; (iv) certain traits lead to greater reproductive success than other traits do.

11. Which statement restates the fact that rifampin resistance in *Mycobacterium tuberculosis* is an adaptation?
    a. Rifampin resistance is a heritable trait that increases fitness relative to susceptible individuals in certain environments.
    b. Rifampin resistance is caused by a point mutation in the *rpoB* gene.

c. A mutation occurred in the *rpoB* gene of one cell, and this mutant cell continued to grow and divide.

d. Drug-resistant strains now account for about 10% of the *Mycobacterium tuberculosis*–caused infections throughout the world.

12. Resistance to a wide variety of insecticides, fungicides, antibiotics, antiviral drugs, and herbicides has recently evolved in hundreds of insects, fungi, bacteria, viruses, and plants. Why?
    a. Mutations are on the rise.
    b. Humans are altering the environments of these organisms, and the organisms are evolving by natural selection.
    c. No new species are evolving, just resistant strains or varieties. This is not evolution by natural selection.
    d. Humans have better health practices, so these organisms are trying to keep up.

13. Why did the Grants observe larger beak depths (in the medium ground finch), on average, following a drought?
    a. Deeper beaks help finches find water.
    b. Larger birds always survive better than smaller birds.
    c. During the drought, natural selection led to increased beak depth in the finch population.
    d. The Grants did not observe deeper beaks after the 1977 drought.

14. Which of the following statements is true according to evolutionary theory?
    a. Acclimation can lead to evolution in a population.
    b. Flowering plants are more advanced or "higher" organisms than algae are.
    c. Natural selection always creates more complex species from simpler ancestor species.
    d. Individuals do not change during evolution by natural selection.

15. Which of the following are examples of evolutionary constraints?
    a. genetic correlation between beak depth and beak width in medium ground finches
    b. lack of genetic variation in some populations
    c. the bones of the mammalian inner ear
    d. all of the above

CHAPTER 24—ANSWER KEY

Check Your Understanding

(a) Traits that are not heritable may increase an individual's ability to survive or reproduce. But because the trait is not passed on to the offspring, it does not change the population frequency for that trait; thus evolution does not occur.

(b) Transitional form evidence typically comes in the form of intermediate fossils. Fossils are critical because they provide our most direct evidence of extinct life-forms. However, fossils are rare and the fossil record is biased toward organisms that are common, large, have hard parts, and so on. Vestigial traits however, especially ones like pseudogenes, can be found in a wide array of living organisms with less bias toward certain taxonomic groups.

(c) Although the underlying genetics is not always known or understood, developmental and structural homologies are based on genetic similarities—similarity in the genes controlling the structure or development. In Chapter 27 you learn that similarities in distantly related organisms with similar lifestyles are considered to be analogous rather than homologous if the evolutionary and genetic bases of the traits differ.

(d) *Rabbit*—ability to run fast; *turtle*—shell thick enough to protect against predators, but not too heavy to carry around; *oak tree*—leaves that taste bad to herbivores; *bacteria*—ability to grow and divide quickly.

(e) Researchers identified the DNA mutation that caused the antibiotic-resistant trait. Any trait in a single-celled organism that is coded for by a specific DNA sequence is by definition heritable because DNA carries the genetic information. In multicellular organisms, mutations that occur only in nonreproductive cells (e.g., skin cancer) are not heritable.

(f) Islands promote speciation because the organisms on each island experience different environments and therefore different types of natural selection. In the next chapter you will learn some other reasons why islands can be sites of rapid speciation (genetic drift).

(g) The goal of this question is to provoke thought about genetic constraints versus fitness trade-offs. This situation is a type of genetic correlation, but the correlation is between traits in different sexes rather than in the same individual. Thus, it can be evaluated only across a population or over many generations of male and female offspring. The fitness trade-off is also across generations rather than in an individual. For example, a trait that increases the success of male offspring but decreases the success of female offspring is selectively neutral for an individual that has equal numbers of male and female offspring.

Assessing What You've Learned

1. d; 2. a; 3. a, d; 4. a; 5. c; 6. b; 7. b; 8. c; 9. b; 10. d; 11. a; 12. b; 13. c; 14. d; 15. d

25

Evolutionary Processes

A. KEY CONCEPTS

- Natural selection is the only process that consistently leads to adaptation. However, there are three other processes that can change allele frequencies and cause evolution in a **population**—a group of individuals from the same species that live and breed together.

- The four mechanisms that can cause evolution are natural selection (increases frequency of alleles that improve reproductive success), genetic drift (randomly changes allele frequencies), gene flow (due to migration in and out of a population), and mutation (introduction of new alleles).

- Effects of these evolutionary processes can be compared to a null model, which shows what happens to allele frequencies when none of the evolutionary mechanisms are affecting a population.

25.1 Analyzing Change in Allele Frequencies: The Hardy–Weinberg Principle

- Evolutionary processes are often first analyzed using mathematical models. Model predictions are then tested with experimental or observational data. The model may then be further modified if necessary, and the results used to solve problems in conservation biology or human genetics.

- G. H. Hardy and Wilhelm Weinberg independently developed a mathematical model to calculate what happens to allele frequencies when no evolution is taking place. Biologists compare data to this null model in order to determine whether evolution is taking place in a population.

- Hardy and Weinberg started by considering all the gametes produced in one generation by individuals in a population and placing them all in a **gene pool**. They then calculated what would be observed if the next generation were made by randomly picking gametes from this gene pool.

- The researchers first examined what happens for a particular gene found in a population. Let A_1 and A_2 represent the only two alleles for this gene. The frequency of A_1 is designated p, and the frequency of A_2 is symbolized by q. There are only two alleles, so the two frequencies must add up to 1: $p + q = 1$.

- Three genotypes are possible: A_1A_1, A_1A_2, and A_2A_2. When gametes are randomly picked from the gene pool, the resulting frequency of the A_1A_1 genotype in the new generation is p^2 (the probability of drawing two A_1 alleles $= p \times p$), and the frequency of A_2A_2 will be q^2. The frequency of A_1A_2 will be $2pq$ (see **Figures 25.1** and **25.2** for two ways of depicting this).

- Because all individuals in the new generation must have one of the three genotypes, the sum of the three genotype frequencies must equal 1 (100% of the population): $p^2 + 2pq + q^2 = 1$. When allele frequencies are calculated for this new generation, the frequency of A_1 is still p and the frequency of A_2 is still q.

- No evolution occurred; this result is called the **Hardy–Weinberg principle**: (1) If allele frequencies are p and q, then genotype frequencies will be p^2, $2pq$, and q^2 for generation after generation; (2) allele distribution under Mendelian inheritance does not change allele frequency.

✏️ CHECK YOUR UNDERSTANDING

(a) It is easy to determine the frequency of homozygous recessive genes (q) because individuals with the homozygous recessive genotype show the recessive phenotype (whereas heterozygotes and homozygous dominants may have the same phenotype). There is some evidence that straight hairlines are recessive and widow's peaks are dominant in humans. If Hardy–Weinberg assumptions are met for this trait in a human population, and 81% of the population have straight hairlines (A_2A_2), then:

q^2 must equal _____, q must equal _____, and p must equal _____. Thus, in this human population, the frequency of homozygous dominants (A_1A_1) must be , and the frequency of heterozygotes (A_1A_2) must be _____.

The Hardy–Weinberg Model Makes Important Assumptions

- The Hardy–Weinberg model assumes that mating is random with respect to the gene being examined and that none of the four evolutionary mechanisms are acting.

- The Hardy–Weinberg principle holds when there is (1) no natural selection—all parents contribute equally to the next generation; (2) no genetic drift—alleles are drawn from the gene pool in exactly the frequency that they are found in the population; (3) no gene flow; (4) no mutation; and (5) random mating.

CHECK YOUR UNDERSTANDING

(b) Can you think of a situation in which the Hardy–Weinberg expected genotypes could be found even though the Hardy–Weinberg assumptions were not strictly met?

_____

_____

_____

_____

How Does the Hardy–Weinberg Principle Serve as a Null Hypothesis?

- A null hypothesis specifies what we should observe when there are no differences among treatment groups. When biologists want to test whether natural selection, nonrandom mating, or some other evolutionary mechanism is acting on a particular gene, the Hardy–Weinberg principle serves as a null hypothesis.

- Whenever biologists observe genotype frequencies that do not conform to Hardy–Weinberg proportions, it indicates that evolution or nonrandom mating is occurring in that population.

Are MN Blood Types in Humans in Hardy–Weinberg Equilibrium?

- Most human populations have two alleles for the MN blood group. This gene codes for a protein on the surface of red blood cells, and human genotype (*MM*, *MN*, or *NN*) can be determined from blood samples.

- Using a large sample, geneticists can estimate genotype frequency for a population by dividing the number of individuals observed for each genotype by the total number in the sample. Allele frequencies can then be calculated from the genotype frequencies (**Table 25.1**).

- Allele frequencies can then be used to calculate the expected genotype distribution according to the Hardy–Weinberg principle; if mating is random and no evolution is occurring, genotype frequencies (*MM* : *MN* : *NN*) should equal $p^2 : 2pq : q^2$.

- Statistical tests can determine whether any differences between observed and expected values are so small as to be due to chance (for example, which individuals happened to be included in the blood sample group) or large enough to reject the Hardy–Weinberg null hypothesis for a particular allele in a given population.

- In all cases, observed and expected genotype frequencies are almost identical for MN blood groups. Because genotypes occur in Hardy–Weinberg proportions, evolutionary processes do not currently affect MN blood group, and mating must be random with respect to this trait.

Are HLA Genes in Humans in Hardy–Weinberg Equilibrium?

- The *HLA-A* and *HLA-B* genes code for proteins that help immune system cells recognize and destroy invading bacteria and viruses. Alleles for these genes are codominant, and they code for proteins that recognize slightly different disease-causing organisms.

- Researchers hypothesized that heterozygotes might be more fit than homozygotes because their immune systems would recognize more disease-causing organisms. To test this hypothesis, they determined the genotypes of Havasupai tribe members and used these data to estimate population allele frequencies.

- Researchers calculated the expected genotype frequencies using the Hardy–Weinberg principle and compared the observed and expected frequencies. The Havasupai population has more heterozygotes and fewer homozygotes than expected (**Table 25.2**), so at least one of the Hardy–Weinberg assumptions must be violated for these alleles in this population.

- Mutation, migration, and genetic drift are negligible in this case, so either mating is not random with respect to the *HLA* genotype and/or heterozygous individuals have higher fitness.

- College students can distinguish genotypes at *HLA*-like loci by body odor and are more attracted to the smell of genotypes different from their own. In other research, Hutterite women with the same *HLA*-related alleles as those of their husbands were found to have difficulty getting pregnant and had higher rates of spontaneous abortion.

- Research continues, but it appears that nonrandom mating, as well as low fitness of homozygous fetuses, may cause the higher than expected frequency of heterozygotes observed in the Havasupai tribe.

25.2 Types of Natural Selection

- Natural selection occurs when individuals with certain heritable phenotypes survive and reproduce better than others do. The alleles responsible for this increased reproduction then increase in frequency in succeeding generations, causing evolution.

- The pattern of natural selection documented in the Havasupai tribe is known as **heterozygote advantage**. When heterozygous individuals have higher fitness than homozygous individuals have, natural selection maintains **genetic variation**—the number and relative frequency of alleles in a population.

- Other patterns of natural selection decrease genetic variation. Low genetic variation can be dangerous for a population because if the environment changes, alternative alleles that confer high fitness under the new conditions may not be present. If severe environmental change occurs, the decrease in average population fitness can even cause extinction.

Directional Selection

- Directional selection occurs when natural selection increases the frequency of one allele, as in the drug-resistance example (**Chapter 24**). This type of selection reduces population genetic diversity over time.

- If directional selection continues for long enough, the favored alleles eventually reach a frequency of 1.0 (100%) and are said to be fixed, whereas those alleles that are no longer found in the population are lost.

- Often several genes influence a trait, causing a bell-shaped distribution of population phenotypes. Directional selection for these traits acts on many genes at once and causes the bell curve to shift in one direction (**Figure 25.3**).

- This type of selection was observed in cliff swallows during a cold snap when small birds starved to death.

- Directional selection is not necessarily constant throughout a species' range or over time. Sometimes opposing patterns of directional selection on related traits maintain genetic variation in a population (e.g., in cliff swallows, selection for large body size versus selection for maneuverability provides a fitness trade-off).

Stabilizing Selection

- **Stabilizing selection** occurs when individuals with intermediate traits reproduce more than others do; this type of selection maintains intermediate phenotypes in a population (**Figure 25.4**). Stabilizing selection decreases a population's genetic variation over time but does not change its average trait value.

- Human birth weight is under stabilizing selection because babies of average size survive better than do babies with low or high weights.

Disruptive Selection

- **Disruptive selection** is the opposite of stabilizing selection; it occurs when intermediate phenotypes are selected against and extreme phenotypes are favored. Disruptive selection maintains genetic variation but does not change the mean value of a trait (**Figure 25.5**).

- For example, due to the types of food available, African seedcrackers with very short or very long beaks survive better than do those with intermediate phenotypes.

- Disruptive selection can cause speciation if individuals with one extreme of a trait start mating preferentially with individuals that have the same trait.

CHECK YOUR UNDERSTANDING

(c) You observe three lab populations of euglena (a photosynthetic protist that can phagocytose and digest bacteria) that differ in their environments. You determine the mean chloroplast number per euglena over several generations. Population A has a habitat that varies from dark with lots of bacteria to light with few bacteria. This population shows an increasingly bimodal distribution of chloroplast number per organism over time, but the mean number of chloroplasts per euglena does not change. Population B has a more constant habitat with both bacteria and light present throughout. The mean chloroplast number does not change, but you observe decreased variation in chloroplast number. Population C has a habitat with lots of light but no bacteria. The mean chloroplast number increases in this population. Which population(s) is/are experiencing selection, and what type of selection appears to be occurring?

_____

_____

_____

_____

25.3 Genetic Drift

- **Genetic drift** is change in allele frequencies due to chance (**sampling error**). It causes allele frequencies to drift up and down randomly over time.

- For example, 8 men and 6 women colonized Pitcairn Island in 1789. Possible offspring from six couples can be simulated by flipping coins to determine which parent alleles happens to be in the gametes that contribute genetic material to the next generation (**Table 25.3**).

- Just by chance, in the coin flipping you could get more heads than tails, causing one allele to increase in the frequency and another allele to decrease in frequency—evolution by random chance.

- *Genetic drift is random with respect to fitness.* Allele frequency changes are due to chance and are not adaptive.

- *Genetic drift is most pronounced in a small population.* It is easier to deviate significantly from a 50 percent heads ratio when you flip a coin 10 times than when you flip a coin 1000 times. In the first case, a single coin-flip has a large impact on the heads "allele" frequency. Chance events have a larger impact on allele frequencies in small populations than in large populations.

- *Over time, genetic drift can lead to the random loss or fixation of alleles.* "Lucky" alleles may become fixed, whereas "unlucky" alleles may be lost, causing a decline in the population's genetic variation.

- Computer simulations have been used to model patterns of population allele change resulting from random allele combination (**Figure 25.6**). The simulations demonstrate that genetic drift is much more pronounced in small than in large populations.

- Over long periods of time, genetic drift can also affect large populations, especially for alleles that do not affect fitness and thus are not subject to natural selection (e.g., pseudogenes and silent mutations that do not affect the gene product).

Experimental Studies of Genetic Drift

- Warwick Kerr and Sewall Wright studied genetic drift in fruit flies using leg-bristle morphology (normal straight bristles versus "forked," bent-looking bristles) as a **genetic marker**—an easily identifiable trait with known alleles.

- Four females and four male flies were placed in each of 96 cages such that the initial frequencies of normal and forked-bristle alleles were 0.5 in each cage. In each succeeding generation, four male and four female offspring were randomly chosen to continue each population.

- After 16 generations, 29 of the 96 populations had only forked leg bristles, and 41 had only normal bristles. In these populations one allele had become fixed (**Figure 25.7**). Only 27 percent of the populations still had both alleles.

- Leg bristles do not affect fitness in the lab, migration was not allowed, and mutation rate for the bristle gene is very low; therefore, the observed changes in population allele frequencies must have been due to genetic drift.

CHECK YOUR UNDERSTANDING

(d) Why did Kerr and Wright choose only 4 female and 4 male fruit flies from each population to breed in each generation?

_____

_____

_____

Genetic Drift in Natural Populations

- A sampling process occurs during fertilization in every population in each generation, but sampling error primarily affects population allele frequency in small populations. Other sampling events can also affect populations.

- Genetic drift is currently a topic of great concern to conservation biologists because, in many species, population sizes are decreasing due to habitat destruction and other human activities. Small populations found on nature reserves or in zoos are especially susceptible to genetic drift.

How Do Founder Effects Cause Drift?

- A **founder event** occurs when a group leaves a population, emigrates to a new area, and starts to establish a new population. If the founding group is small, its allele frequencies probably differ from those of the source population. This sampling effect on the new population's allele frequencies is called a **founder effect**.

- Founder events and founder effects are especially common in the colonization of isolated habitats like islands, mountains, caves, and ponds. For example, a raft of at least 15 green iguanas colonized the island of Anguilla.

How Do Population Bottlenecks Cause Drift?

- A sudden decrease in population size, called a **population bottleneck**, can lead to a **genetic bottleneck**—that is, a sudden reduction in the number of alleles in a population.

- Genetic bottlenecks are commonly caused by disease outbreaks and natural catastrophes.

- Genetic drift often occurs in the resulting small population (e.g., the serious vision deficit, *achromatopsia*, that increased in frequency following the typhoon and famine-induced population bottleneck on Pingelap Atoll).

25.4 Gene Flow

- **Gene flow** is the movement of alleles from one population to another. It occurs any time individuals leave one population, join with another, and reproduce.
- Gene flow reduces genetic differences between the source and recipient populations.
- Following the volcanic explosion of Mount St. Helens in the state of Washington, prairie lupine seeds recolonized the ash plain surrounding the mountain. Researchers determined allele frequencies in several new prairie lupine populations and observed changes in allele frequencies over time.
- Initially, new populations showed strong founder effects, but population allele frequency differences decreased over time. Researchers proposed that populations became more similar as bees caused gene flow by carrying pollen between populations of lupines (**Figure 25.8**).
- Gene flow can increase the average fitness of individuals in a population where genetic diversity was lost due to a nonadaptive process like genetic drift; but gene flow can decrease the average fitness of individuals in a population where natural selection had produced adaptation to a specific habitat.
- Gene flow is random with respect to fitness; depending on the situation, it could increase or decrease average fitness. Either way, gene flow always decreases genetic differences between populations.

25.5 Mutation

- Whereas most evolutionary mechanisms reduce genetic diversity, **mutation** restores genetic diversity and creates new alleles. A mutation occurs when copy errors by DNA polymerase cause random changes in DNA sequence that may lead to an altered amino acid sequence.
- Because errors are inevitable, mutation is always adding new alleles into populations at all gene loci.
- Mutation is a random process; because most organisms are well adapted to their current habitats, random changes usually result in **deleterious alleles**—that is, alleles that lower fitness because their gene products do not work as well as do the other naturally selected alleles.

- However, mutation does occasionally produce an advantageous allele. Beneficial alleles created by mutation can then increase in frequency in a population due to the effects of natural selection.

Mutation as an Evolutionary Mechanism

- For a given allele (in humans), there is only about one mutation per 10,000 gametes. Thus, mutation rates are too low to significantly affect allele frequencies by themselves. Another evolutionary mechanism must act on alleles created by mutation in order for significant evolutionary change to occur.
- Although mutation is extremely slow, organisms contain many genes (over 20,000 in humans), so almost every individual in every population has a new allele due to mutation. These mutations are the only way genetic diversity can be increased.

What Role Does Mutation Play in Evolutionary Change?

- Richard Lenski and colleagues tested the role mutation plays in evolutionary change by using single *Escherichia coli* cells to found many populations of *Ara+* and *Ara−* bacteria (The *Ara* gene allowed researchers to easily determine relative growth in competition experiments). *E. coli* is asexual, so mutation was the only source of genetic variation.
- Cells were grown under identical conditions. Each day some cells from each population were transferred to new growth media. At regular intervals, researchers froze sample cells from each population.
- Natural selection acted on these populations, but no gene flow was allowed, and populations were large enough that genetic drift was unlikely to be important.
- At the end of the experiment, frozen "fossil" cells from previous generations were thawed and grown with cells from more recent generations. Those newer generations produced more cells during these competition experiments, demonstrating their adaptation over time to the experimental environment (**Figure 25.9**).
- Relative fitness increased by irregular intervals over the experiment, for an almost 30 percent total increase in fitness. Each fitness increase must have been due to a new beneficial mutation that then rapidly increased in frequency through succeeding generations due to natural selection.
- Mutation provides the genetic variation upon which natural selection can act to produce evolution. **Table 25.4** summarizes evolutionary mechanisms and describes their effects on fitness and genetic variation.

(e) Could you do an experiment like Lenski's (**Figure 25.9**) on mice? In your answer, explain why or why not.

_____

_____

_____

_____

25.6 Nonrandom Mating

- The Hardy–Weinberg model assumes mating occurs at random, but in nature mating is rarely random. Inbreeding and sexual selection are two processes that violate the random-mating assumption.

Inbreeding

- In small, isolated populations, mating between relatives—known as **inbreeding**—may be common. Because relatives are more likely to share alleles, inbreeding violates the Hardy–Weinberg assumptions.

- Inbreeding decreases the frequency of heterozygotes and increases the frequency of homozygotes in each generation (**Figure 25.10**). Although inbreeding affects genotype frequency, it does not change allele frequencies; thus it does not cause evolution.

- **Inbreeding depression** is the decreased fitness that occurs as more deleterious recessive traits are expressed in the phenotypes of homozygous recessive individuals (see **Figure 25.11**). Homozygotes also have lower fitness for traits that show a heterozygote advantage.

- Inbreeding depression can increase the rate at which natural selection removes disadvantageous recessive alleles by exposing these alleles in the homozygous phenotype.

- Inbreeding and sexual selection both violate the assumption of random mating. Inbreeding affects only genotype frequencies (not allele frequencies); in contrast, sexual selection (a type of natural selection) affects both allele and genotype frequencies, but only for genes involved in mate competition.

- As **Table 25.5** shows, inbreeding depression is pronounced in humans. Many species, including humans, have mechanisms to avoid inbreeding.

Sexual Selection

- Mate choice often plays an important role in speciation. **Sexual selection** is selection for enhanced ability to attract mates. It is a type of natural selection.

(f) Sometimes mate choice is nonrandom, such that genetically different individuals are more likely to mate. This is called outbreeding. What would happen to allele frequencies and genotype frequencies in a population experiencing an outbreeding mating pattern?

_____

_____

_____

- The Bateman-Trivers theory states that sexual selection typically acts more strongly on males than on females, so that the mate-attracting traits tend to be more elaborate in males. This pattern is explained by the **fundamental asymmetry of sex**—females usually invest more in their offspring than do males.

- Eggs are large and energetically expensive. Females typically produce relatively few offspring during their lifetime, and their fitness is primarily limited by the ability to gain resources necessary to produce and rear young (not by the ability to find a mate). Females should therefore be choosy about the males they mate with.

- Sperm, on the other hand, are cheap to produce, so males can father many offspring. Male fitness is primarily limited by ability to acquire mates (not by the ability to acquire resources for offspring production). Therefore, males should compete with each other for mates, and sexually selected traits (used in courtship or competition for mates) should be found primarily in males.

Sexual Selection via Female Choice

- Female birds invest a great deal of time and energy in offspring, and they should be choosy about the males they mate with. In several bird species, females prefer healthy, well-fed males.

- Red and yellow pigments called carotenoids cause bright feather and beak coloration in many bird species. Carotenoids also stimulate the immune system and prevent tissue damage from free radicals. However, animals cannot make carotenoids and must obtain them in their diet.

- Birds with the brightest beaks and feathers must be well fed and healthy, because carotenoids are used in beaks and feathers only when they are not needed by the immune system.

- Researchers fed brother zebra finches diets that differed in carotenoid content. The males eating the carotenoid-rich diet developed more colorful beaks. Females preferred these bright-beaked males to their duller-beaked brothers, supporting the hypothesis that females choose to mate with healthy, well-fed males. (**Figure 25.12**).

- Other experiments on other bird species also support the conclusion that bright male coloration and other courtship displays indicate male healthiness. Females that choose to mate with such a male presumably have offspring with alleles that help them fight disease and feed efficiently.

- Males that provide resources for egg production or care of young are also preferred mates in many animal species. For example, brown kiwi females produce enormous eggs (**Figure 25.13**), and they choose to mate with males that incubate and care for the offspring.

- In conclusion, females often choose mates based on (1) characteristics that signal good genes, (2) male resources or parental care, or (3) both. However, in some species, females have little choice in mating, and competition among males causes sexual selection.

CHECK YOUR UNDERSTANDING

(g) Male European blackbirds have a bright orange bill due to the presence of carotenoids. Researchers injected some males with sheep red blood cells. These males then displayed a strong immune response. Based on your reading above, predict what happened to the beak color of these males.

_____

_____

If females were to choose between these blackbirds and a control group of blackbirds (injected with just a salt solution), which males do you think they would choose and why?

_____

_____

_____

_____

Sexual Selection via Male–Male Competition

- Male elephant seals fight over beach patches where females come to bear their pups. Males that win

battles obtain larger territories. A **territory** is an actively defended area where the owner has exclusive use—in this case, exclusive access to females inhabiting that part of the beach.

- Male elephant seals are more than four times larger than females. Large male size has evolved because fights for beach territories are almost always won by the larger males. The alleles of large territory-owning males have a fitness advantage due to sexual selection.

- A few male elephant seals father the vast majority of offspring produced each year, so sexual selection is intense in this species and is driven by male–male competition rather than by female choice. There is less variation in female reproductive success (**Figure 25.13**).

What Are the Consequences of Sexual Selection?

- Most females that survive to adulthood are able to mate and produce offspring, but many males do not.

- Because reproductive success usually varies more in males than in females, sexual selection is typically more intense in males. Thus, males tend to have more traits that function primarily in courtship or male–male competition. Evolution of these traits is driven by sexual selection.

- **Sexual dimorphism**, or male–female differences, can be observed in many species (**Figure 25.15**). These sexually selected traits include the weapons used in male–male competition as well as ornamentation and behavior used in courtship displays.

- Nonrandom mating due to sexual selection violates Hardy–Weinberg assumptions and can cause certain alleles to increase in frequency. Like other forms of selection (and unlike inbreeding), nonrandom mating can cause evolution.

B. INTEGRATING YOUR KNOWLEDGE

*Looking Back—Key Concepts
from Earlier Chapters*

Evolution—Chapters 1, 6, 24

This chapter builds on discussions of evolution in earlier chapters by presenting you with processes that produce evolution in a population.

Mutation—Chapters 4, 14, 15, 18

In **Chapter 4** you first learned how some researchers use mutation in selection experiments. In **Chapters 14, 15**, and **18** you learned more about how mutations occur. In this chapter, the role of mutation as the ultimate source of genetic variation is highlighted. Mutation provides the variation that natural selection acts upon.

Genetics—Chapters 13, 15, 19

Chapters 13 and **15** provide background information critical to your understanding of how allele frequencies change in populations and how genes work to determine organism phenotype. The discussion in Chapter 25 of the Hardy–Weinberg principle shows how researchers study genetics at the population level. **Chapter 19** describes how researchers can artificially alter individual (and population) allele frequencies.

Looking Forward—Key Concepts in Later Chapters

Evolution—Chapters 26, 27, 51, and others

Because evolution is one of the unifying themes of biology, it reappears throughout your textbook. Larger-scale evolutionary patterns are examined in **Chapters 26** and **27**. **Chapters 28** through **35** discuss the evolution of specific groups of organisms, and **Chapter 51** examines the evolution of behavior.

Adaptation—Chapters 27–35, 36, 41, and others

As with evolution, you can find mention of adaptations (traits that increase fitness) throughout this textbook. **Chapters 28** through **35** describe the major organismal groups—something that could hardly be accomplished without mentioning the salient characteristics, or adaptations, of each group. **Chapters 36** and **41** have more examples of plant and animal adaptations.

Populations—Chapter 52

Evolution occurs in populations. Although much of this textbook examines other biological levels (species, organism, cell), **Chapter 52** focuses on populations again, in a thorough discussion of their ecology.

C. MASTERING CHALLENGING TOPICS

Understanding the Hardy–Weinberg equations can be challenging for those lacking an intuitive understanding of the math involved. It may help you to work through concrete examples like that given in the "Check Your Understanding" question (a). You might try mapping this example in a Punnett square figure like that shown in **Figure 25.2** in addition to doing the calculations. You could also use two candy colors or two different coins to go through the process of picking gamete alleles and calculating genotype frequencies, as shown in **Figure 25.1**.

Keep in mind that in a large population, if 70 percent of the alleles in the gene pool are A_1, then 70 percent of the gametes should be A_1 (under Hardy–Weinberg conditions). But if you try to replicate this process on your own, picking blindly, you may get a different percentage in the gametes due to random chance—in which case you have just shown the effects of genetic drift!

Now think about this process of picking alleles to go into the next generation when the other evolutionary processes are acting. How do the processes affect the allele frequencies? For example, if you are doing this exercise with orange and green candies, and you prefer the flavor of the green ones, you may find that you eat the green ones. The orange alleles survive better and are thus more likely to make it into the next generation. Presto—natural selection!

D. ASSESSING WHAT YOU'VE LEARNED

1. What is a population?
   a. A population is any group of many individuals of the same species.
   b. A population is a group of individuals that live in the same area and regularly interbreed.
   c. A population is a group of interacting species that live in the same area.
   d. A population is two or more groups that regularly interbreed.

2. The Hardy–Weinberg principle acts as a null model because it describes the relationship between allele and genotypic frequencies under conditions where:
   a. None of the four evolutionary forces are acting, and mating is random.
   b. New species are differentiating.
   c. Evolution by natural selection increases the fitness of individuals.
   d. Individuals in a population are not mating randomly.

3. Which of the following is *not* one of the Hardy–Weinberg assumptions?
   a. no genetic drift
   b. no mutation
   c. no random mating
   d. no natural selection
   e. no gene flow

4. Which of the following scenarios best illustrates heterozygote advantage?
   a. A population has more heterozygotes than homozygotes.
   b. For a given locus, heterozygous individuals are more fit than are homozygous individuals.
   c. Individuals in a population are nonrandomly mating without inbreeding depression.
   d. Parents with similar alleles produce more offspring than do parents with dissimilar alleles.

5. Which of the following examples of natural selection in action would tend to increase the genetic variation in the population?

   a. In a pond-dwelling fish species, higher average summer temperatures selects for fish with greater heat tolerance over time.

   b. A small population of butterflies is blown by a windstorm and colonizes a new mountaintop.

   c. Natural selection simultaneously selects against heavy field mice (they starve during winter) and light field mice (they freeze during winter).

   d. Males with bright red feet produce more offspring than do those with duller feet in one bird species.

   e. In alpine skypilots, large flowers set more seed above the timberline and small flowers set more seed below the timberline.

6. In mammals, young born from pregnancies that are much longer or shorter than the species average tend to have higher rates of infant mortality. This is an example of _____ selection.

   a. disruptive

   b. directional

   c. sexual

   d. stabilizing

7. Why aren't there any female elephant seals with over 90 offspring?

   a. because males are larger than females and can therefore take care of more offspring

   b. because females must compete for male copulations

   c. because female elephant seals have one pup per year, and they do not live 90 years

   d. because the beaches would get too crowded

   e. all of the above

8. Evolution by natural selection is a powerful force of change. What keeps organisms from becoming perfectly adapted to their environment?

   a. Environments change, so different alleles become more fit.

   b. Genetic drift causes random changes in allele frequencies.

   c. Mutations introduce new variation into a population.

   d. All of the above processes impede, constrain, or prohibit evolution by natural selection.

9. Which of these statements most fully characterizes the fundamental asymmetry of sex?

   a. Female fitness is limited most by the ability to get resources for producing eggs and rearing young, whereas male fitness is limited by the ability to attract females.

   b. Female fitness is limited most by the ability to attract males, whereas male fitness is limited by the ability to get resources for provisioning the female.

   c. Female fitness is limited most by the ability to get resources for producing eggs and rearing young, whereas male fitness is limited by the ability to get resources for provisioning the female.

   d. Female fitness is limited most by the ability to attract males, whereas male fitness is limited by the ability to attract females.

10. Genetic diversity is required for natural selection to act, but natural selection can reduce or eliminate diversity. What process can restore genetic diversity to a population?

    a. genetic drift

    b. mutation

    c. sexual selection

    d. stabilizing selection

11. Which of the following biological processes regularly causes adaptation?

    a. mutation

    b. migration

    c. natural selection

    d. genetic drift

    e. inbreeding

12. What is an important consequence of gene flow in natural populations?

    a. Gene flow increases the mutation rate among sedentary organisms.

    b. Gene flow moves individuals from one habitat to another on a seasonal basis.

    c. Gene flow tends to separate allele frequencies among populations.

    d. Gene flow tends to reduce genetic differences among populations.

    e. All of the above are true.

13. In what kind of population is random genetic drift most pronounced?

    a. Drift is greatest in large populations.

    b. Drift is greatest in small populations.

    c. Drift is greatest in migrating populations.

    d. Drift is greatest in fixed populations.

14. Which of the following is *not* associated with inbreeding?
    a. Allele frequencies change in a population.
    b. The frequency of homozygote genotypes increases in a population.
    c. Individuals in a population may experience reduced fitness.
    d. In plants, individuals may be self-fertilizing.
    e. The rate of evolutionary change due to natural selection may increase.

15. Under extreme inbreeding like that shown in **Figure 25.10**, if genotype frequencies in one generation were 0.20 A_1A_1, 0.60 A_1A_2, and 0.20 A_2A_2, what would be the genotype frequencies in the next generation?
    a. 0.5 A_1A_1, 0.0 A_1A_2, and 0.5 A_2A_2
    b. 0.20 A_1A_1, 0.60 A_1A_2, and 0.20 A_2A_2
    c. 0.45 A_1A_1, 0.10 A_1A_2, and 0.45 A_2A_2
    d. 0.35 A_1A_1, 0.30 A_1A_2, and 0.35 A_2A_2
    e. none of the above

CHAPTER 25—ANSWER KEY

Check Your Understanding

(a) q^2 equals 0.81, q equals 0.9, and p equals 0.1. The frequency of homozygous dominants (A_1A_1) must be 0.01, and the frequency of heterozygotes (A_1A_2) must be 0.18.

(b) Theoretically this situation could occur if exactly the same allele proportions were migrating into a population as were migrating out. It could also occur if natural selection against some allele were exactly balanced by mutations and/or gene flow bringing that allele back into the population.

(c) All three populations appear to be experiencing selection. Populations A, B, and C appear to be experiencing disruptive, stabilizing, and directional selection respectively.

(d) They chose only 4 males and 4 females because they wanted to observe the effects of genetic drift, and genetic drift has a greater impact on allele frequencies in small populations.

(e) Lenski's experiment could not be performed on mice, because you cannot freeze mice and then use them to compete against later generations. Also, observing 2000 mice generations would take way too much time.

(f) Outbreeding would not change allele frequencies either, but it would increase the frequency of heterozygotes relative to homozygotes.

(g) Because carotenoids are used to stimulate the immune system, you should predict that the beak color would dim. This is, in fact, what occurred. Females probably would choose the brighter-beaked males (the control group) since, in nature, a dim beak might indicate either sickness or a low quality diet. This prediction follows from the zebra finch results, but would have to be tested experimentally.

Assessing What You've Learned

1. b; 2. a; 3. c; 4. b; 5. e; 6. d; 7. c; 8. d; 9. a; 10. b; 11. c; 12. d; 13. b; 14. a; 15. d

26

Speciation

A. KEY CONCEPTS

- Populations that are no longer connected by gene flow may diverge genetically due to mutation, genetic drift, and/or natural selection. This genetic divergence may eventually lead to **speciation**, the creation of new species. When speciation occurs, two or more distinct species arise from one ancestral group (**Figure 26.1**).

26.1 How Are Species Defined and Identified?

- A **species** is a distinct type of organism—an evolutionarily independent population or group of populations.

- Gene flow reduces genetic differences among populations, so evolutionary independence begins with lack of gene flow. Once gene flow stops, mutation, selection, and drift can act on populations independently, and genetic divergence can occur. Genetic divergence, in turn, may lead to speciation.

- The biological species, morphospecies, and phylogenetic species concepts are three different methods used by biologists to evaluate whether populations show enough genetic divergence (are evolutionarily independent enough) to be considered separate species.

The Biological Species Concept

- The **biological species concept** considers populations to be evolutionarily independent if they are reproductively isolated from each other (no gene flow).

- **Prezygotic isolation** occurs when individuals from different populations are unable to mate (e.g., breed at different times or have incompatible genitalia). **Postzygotic isolation** occurs when individuals from different populations can interbreed, but the hybrid offspring have low fitness (**Table 26.1**).

- Applying the biological species concept of reproductive isolation is difficult for populations that are geographically separated (difficult to know if the geographic separation will continue to prevent interbreeding in the future). This concept cannot be applied to asexual or fossil species.

The Morphospecies Concept

- Evolutionary independence can also be estimated by looking at population morphology, because different appearances often evolve when populations experience different natural selection, mutations, and genetic drift. These different forms will persist only if gene flow is restricted.

- The morphospecies concept can easily be applied to most populations and species, but it is rather subjective (**Figure 26.2**). How much do populations need to differ to be considered different species? Which traits should be evaluated for difference?

The Phylogenetic Species Concept

- The **phylogenetic species concept** defines species as evolutionary distinct populations and is based on a reconstruction of evolutionary history.

- Data on genes, shape, physiology, behavior, and other traits of organisms can be used separately or together to estimate evolutionary relationships among organisms and to create phylogenetic trees depicting the relationships.

- On phylogenetic trees, an ancestral population plus all its descendants (but no other groups) is called a **monophyletic group**.

- A phylogenetic species is the smallest monophyletic group on a tree created from data on related populations (**Figure 26.3**). On a tree of populations, each tip is a phylogenetic species.

- The phylogenetic species concept has the advantages of (1) being applicable to all populations; and (2) being logical—because monophyletic groups, by definition, differ in some traits and are isolated from gene flow with other groups (they have their own independent evolutionary history).

- Unfortunately, good phylogenetic trees have been created for only a few related populations. Use of the phylogenetic species concept may lead to recognition of more species.

- In practice, biologists use all three species concepts (**Table 26.2**). Sometimes the species they recognize differ, depending on the species concept used.

(a) Which species definition (i.e., biological, morphospecies, or phylogenetic) would you use for (1) a new fossil species, (2) a well-studied bacterial species, and (3) a well-studied mammalian species? Why?

_____

_____

_____

_____

_____

Species Definitions in Action: The Case of the Dusky Seaside Sparrow

- Several seaside sparrow (*Ammodramus maritimus*) subspecies with slightly different appearances have been named using the morphospecies concept. Populations with their own identifying traits, but not distinct enough to be a separate species, are called **subspecies**.

- The seaside sparrow subspecies were believed to be genetically isolated because the populations are geographically isolated and young birds breed near their hatching ground (**Figure 26.4**).

- Development of salt marshes where these birds live eliminates their natural habitat, causing population decreases. In 1980 the dusky seaside sparrow subspecies (*A. m. nigrescens*) had only six males left. In an effort to preserve this population, biologists bred these males with females from a nearby population (*A. m. peninsulae*) in hopes of reintroducing the hybrid offspring.

- Other biologists compared gene sequences from different seaside sparrow populations and determined the phylogeny shown in **Figure 26.4b**. They found that seaside sparrows belong to two monophyletic groups: one on the Atlantic Coast and one on the Gulf Coast.

- Thus, *A. m. nigrescens* is not distinct from other Atlantic seaside sparrows and did not need to be

individually preserved. However, the Atlantic Coast and Gulf Coast groups should be separately preserved. Because *A. m. peninsulae* belongs to the Gulf group, the hybrid offspring should not be reintroduced to the wild.

(b) Why does the fact that the seaside sparrow phylogeny shows only two distinct groups suggest there may be more gene flow within those two groups than previously thought?

_____

_____

_____

26.2 Isolation and Divergence in Allopatry

- Physical separation of populations occurs either when a group colonizes a new habitat (dispersal) or when a new physical barrier divides a population (**vicariance**). Reduction or elimination of gene flow due to physical isolation can lead to **allopatric speciation** (**Figure 26.5**).

- Populations living in different areas are said to be in **allopatry**. The study of the geographic distribution of species and populations is called **biogeography**.

Dispersal and Colonization Isolate Populations

- Peter and Rosemary Grant observed as five large ground finches from a nearby island colonized Daphne Major (the island where the Grants were studying medium ground finches). Because finches normally stay on the same island, the colonists were allopatric to their original or source population.

- Colonization events often cause speciation because the physical separation reduces gene flow, and colonists are likely to experience genetic drift (both during the founder event and afterward, if the colonizing group is small). If the new habitat differs from the original habitat, natural selection may also cause divergence of the new population.

- The Grants caught, weighed, and measured large ground finches on Daphne Major over the next 12 years and found the average bill size of the new Daphne Major population was larger than that of the original population. The colonists were a nonrandom sample of the original population.

- Colonizing populations often differ from source populations, causing a founder effect. Genetic drift and natural selection following colonization appear to have caused repeated speciation in many groups of island organisms.

(c) Colonization followed by genetic drift and natural selection has also led to speciation in aquatic organisms and organisms that live at the top of some mountains. Explain how this could happen.

_____

_____

_____

_____

_____

Vicariance Isolates Populations

- Sometimes an existing population is separated by a new physical barrier. For example, uplift of a new mountain range or a change in a river course may separate populations. This separation is known as a vicariance event.

- For example, vicariance events occurred during the last ice age as glaciers fragmented forest and grassland habitats. Physical isolation of populations in these fragmented habitats is thought to have led to the evolution of many modern species.

- Vicariance also occurred when the supercontinent called Gondwana broke up due to **continental drift**—the movement of continental plates (**theory of plate tectonics**; see **Figure 26.6**).

- As Gondwana broke up into the present-day continents, large flightless ratite birds and other species were split into isolated populations. Smaller populations likely experienced genetic drift, and natural selection must have affected populations as the continents moved and environments changed. Today, five different ratite species exist, each on a different continent or island.

- Genetic isolation followed by genetic divergence due to mutation, selection, and genetic drift often causes speciation. Thus, a series of vicariance events may cause a series of speciation events.

(d) Why is genetic drift more likely to be an important evolutionary factor in colonization events than in vicariance events?

_____

_____

26.3 Isolation and Divergence in Sympatry

- Populations or species that live in the same geographic region (close enough to mate) live in **sympatry**. Researchers used to think that speciation could not occur among sympatric populations, because gene flow would tend to eliminate any evolutionary differences.

Can Natural Selection Cause Speciation Even When Gene Flow Is Possible?

- In some situations, natural selection *can* overcome limited gene flow to cause **sympatric speciation**. Often this happens when populations become isolated by habitat preference.

- For example, the soapberry bug (found in the southern United States) feeds on plants in the Sapindaceae family by piercing fruits with their beaks and eating the seeds.

- In the mid-twentieth century, three Asian sapindaceous species were brought to North America. These introduced species spread, and soapberry bugs began using them as hosts even though their fruits are smaller than those of the American species.

- Soapberry bug beak length is correlated with host fruit size in native plants; bugs on species with big fruit have longer beaks than do those on species with small fruit. Presumably this beak/fruit size correlation allows bugs to feed more efficiently.

- Researchers measured beak length of bugs found on native and nonnative hosts to determine whether beak lengths had evolved shorter lengths to match the smaller introduced fruit.

- Bugs on large-fruited native plants have longer beaks than those found on small-fruited nonnative plants (**Figure 26.7**). It seems that disruptive selection has occurred on beak length since the appearance of the new Asian host plants.

- Because soapbugs mate near their host plants, populations on native plants may continue to diverge from the populations on nonnative plants to eventually form a new species. Gene flow between the two populations is decreased; natural selection causes divergence.

- Many insects and parasites are associated with specific hosts. When some individuals in a population switch hosts, they often experience different selection pressures and reduced gene flow with the original population. Thus, host switching is an important cause of speciation in sympatric populations.

CHECK YOUR UNDERSTANDING

(e) Has natural selection caused speciation yet in the case of the two soapbug populations? Why or why not?

_____

_____

_____

_____

_____

How Can Polyploidy Lead to Speciation?

- Mutation *alone* does not usually cause evolutionary change, because it is an inefficient evolutionary mechanism. But one type of mutation does cause speciation in sympatric plant populations. This type of mutation causes **polyploidy**—the presence of more than two sets of chromosomes.

- Polyploid mutations can cause sympatric speciation by reducing or eliminating gene flow between mutant and normal (wild-type) plants.

- Gene flow is reduced because if a normal diploid ($2n$) parent with haploid (n) gametes mates with a mutant **tetraploid** ($4n$) parent with diploid gametes, the offspring are **triploid** ($3n$). Even if this triploid survives, it cannot reproduce, because the three chromosome copies will be unable to undergo meiosis correctly (**Figure 26.8**).

- Mutations that result in a doubling of chromosome number produce **autopolyploid** individuals, whereas **allopolyploid** individuals are produced by a mating between parents of two different species.

Autopolyploidy

- Biologists found several polyploid maidenhair ferns. In the normal fern life cycle, individuals alternate between haploid (n) and diploid ($2n$) life stages (**Figure 26.9**). But the polyploid fern instead alternated between diploid and tetraploid ($4n$) life stages.

- The parent of the polyploid plants had a defect in meiosis, causing production of diploid instead of haploid gametes. Maidenhair ferns can self-fertilize, so these gametes combined to form tetraploid offspring that could in turn self-fertilize or mate with the parent plant.

- Because the polyploids are genetically isolated, they may experience different evolutionary forces (e.g., there are fewer polyploids than wild types, so the polyploids are more likely to experience genetic drift). If so, the two populations may diverge, creating two species where once there was just one.

Allopolyploidy

- Hybridization between two diploid species occasionally creates polyploidy.

- Many hybrid offspring are sterile because their chromosomes do not pair normally during meiosis. However, if a mutation doubles the chromosome number, then homologs can synapse normally and gametes can be produced. These gametes may be able to self-fertilize, producing a tetraploid offspring (**Figure 26.10**).

- The sequence of events just described seems to have caused speciation from three parent species of *Trogopogon*, a weedy European plant introduced to North America in the early 1900s. Two new tetraploid species appear to have multiple independent originations following allopolyploid events.

- Many diploid plant species have closely related polyploid species, indicating that this type of instantaneous sympatric speciation by autopolyploid and allopolyploid mutation has been relatively common in plants.

- Plant reproductive cells don't separate from somatic cells until late in development. If sister chromatids don't separate during an early mitotic division, a tetraploid cell is produced that may later undergo meiosis to form diploid gametes.

- Many plants self-fertilize, so diploid gametes can fuse, creating a new, genetically isolated tetraploid population. Cross-species hybridization is also fairly common in plants and increases the rate of allopolyploidy in this group.

- Speciation by polyploidy is virtually instantaneous as compared to gradual speciation by geographic isolation or disruptive selection in sympatry (**Table 26.3**).

(f) Polyploid mutations are an important cause of speciation in plants but not in animals. Why?

_____

_____

_____

26.4 What Happens When Isolated Populations Come into Contact?

- Populations that have been isolated may evolve differences in mating mechanics, time, place, or behavior (prezygotic isolation) that continue to prevent mating and reduce gene flow even if the populations regain contact. These populations may continue to diverge after contact is regained.

- When prezygotic isolation does not occur, populations may successfully interbreed and cause gene flow that might eventually eliminate differences that evolved during their separation. Other possible outcomes are reinforcement, hybrid zones, and speciation by hybridization.

Reinforcement

- Sometimes individuals from populations that had been separated can mate with each other, but the hybrid offspring are aborted, die young, or survive but are sterile. Generally speaking, the hybrid offspring may have lower fitness. This is known as postzygotic isolation.

- Postzygotic isolation causes natural selection against interbreeding, because individuals mating within their own population will be more successful at passing their genes on to future generations. Selection for traits that reproductively isolate populations is called **reinforcement**.

- Researchers placed flies from different *Drosophila* species together in the lab and found that fruit flies from sympatric species typically will not mate with one another, whereas flies from allopatric species often will.

- This pattern suggests reinforcement selection occurred among sympatric *Drosophila* species. Reinforcement occurs *only* when species live in the same geographic area; if two species never actually interbreed in the wild, natural selection will not act to reduce interbreeding.

- Researchers continue to debate whether reinforcement selection has been important in other groups of organisms.

Hybrid Zones

- Sometimes hybrid offspring are healthy, with traits that are intermediate between the two parental populations. A geographic area where interbreeding between the two populations is common and there are lots of hybrid offspring is called a **hybrid zone**.

- Researchers studied a hybrid zone between Townsend's warblers and hermit warblers in North America's Pacific Northwest. The species' ranges overlap in Washington State, and hybrid offspring with intermediate characteristics are common (**Figure 26.11**).

- Gene sequencing of the maternally inherited mitochondrial DNA (mtDNA) showed that most hybrids are the offspring of Townsend's males and hermit females.

- Townsend's males attack hermit males, but hermit males do not challenge Townsend's males. Thus Townsend's males may invade hermit territories, chase off the males, and mate with the females.

- Further investigation showed that many other northern Townsend's warblers have hermit mtDNA. On some islands all the warblers had hermit mtDNA, even though they looked like typical Townsend's warblers.

- The Townsend's warblers may have gradually expanded their range into hermit territory, hybridizing with female hermit warblers along the way. As each new generation mates with Townsend's males, the offspring look more and more like Townsend's warblers but retain the original hermit mtDNA.

- Continued extension of the Townsend's warbler range could lead to the extinction of hermit warblers; but hybridization can sometimes lead to the creation of a new species.

New Species through Hybridization

- The sunflower, *Helianthus anomalus*, resembles hybrids of *H. annus* and *H. petiolaris*. *H. anomalus* also contains some gene sequences similar to those in *H. annus* and others similar to those in *H. petiolaris*. However, *H. anomalus* has a distinct appearance and grows in drier habitats.

- Biologists experimentally hybridized *H. annus* with *H. petiolaris* for several generations. These hybrids looked like *H. anomalus*, and genetic maps using genetic markers showed that the experimental

CHECK YOUR UNDERSTANDING

(g) Why isn't there any Townsend's warbler mtDNA on some of the northern islands within the current Townsend's warbler range?

_____

_____

_____

_____

(b) Hybrids inherit species-specific mtDNA sequences from their mothers.

N

All individuals have Townsend's mtDNA

Present range of Townsend's warblers (in red)

Individuals that look like Townsend's warblers but have hermit mtDNA

Some individuals have Townsend's mtDNA, others have hermit mtDNA

Pacific Ocean

Present hybrid zones (where two ranges meet)

Present range of hermit warblers (in orange)

All individuals have hermit mtDNA

hybrids had the same combination of *H. annus* and *H. petiolaris* genes found in *H. anomalus* (**Figure 26.13**).

• These results support the hypothesis that *H. anomalus* originated from hybridization of *H. annus* and *H. petiolaris*.

• To summarize, when two populations regain contact, many outcomes are possible (**Table 26.4**).

B. INTEGRATING YOUR KNOWLEDGE

*Looking Back—Key Concepts
from Earlier Chapters*

Evolution—Chapters 1, 4, 24, and 25

Earlier chapters have focused on processes that cause evolution or change within a population. Chapter 26 focuses on how different evolution in different populations can cause speciation.

Mitochondrial DNA—Chapter 20

Maternally inherited mitochondrial DNA, studied to investigate the nature of warbler hybrid zones, was first mentioned in **Chapter 20**.

Meiosis—Chapter 12

If the description about why triploid offspring would be unable to make healthy gametes is not clear to you, consider reviewing the process of meiosis. What would happen to triploid chromosomes during synapsis?

*Looking Forward—Key Concepts
in Later Chapters*

Speciation Patterns—Chapter 27

Now that you understand how speciation occurs, you are ready to consider large-scale speciation patterns (e.g., adaptive radiations), discussed in the next chapter.

Conservation—Chapter 55

Understanding of speciation is critical to conservation efforts, as demonstrated by the dusky seaside sparrow example from this chapter. Biodiversity and conservation are further discussed in the final chapter of the text.

C. MASTERING CHALLENGING TOPICS

In describing the process of speciation, this chapter shows that speciation is usually a gradual process. Because it is gradual, there is no obvious signpost announcing to researchers the point at which two populations have sufficiently diverged to be considered different species. (This determination is similar to the judgment call that is sometimes necessary when deciding whether a cell is in a particular phase of mitosis). Worse, questions that are relevant to these decisions are often unanswerable; for example, if two species are allopatric now, how long will they remain isolated? What will happen if contact is regained?

This ambiguity can be frustrating for beginning biology students, but it is an important reality in many areas of research. Biologists work with extremely complex systems, and sometimes there are no hard-and-fast rules for solving certain scientific questions. Researchers must then just do the best they can with the information they have.

D.   ASSESSING WHAT YOU'VE LEARNED

1. The two key factors responsible for speciation among populations are:
   a. lack of gene flow and mutation
   b. mutation and genetic drift
   c. genetic isolation and genetic divergence
   d. postzygotic isolation and morphological change
   e. hybrid vigor and heterozygote advantage

2. Which of the following are examples of postzygotic reproductive isolation?
   a. Two sympatric bird species are similar in plumage, but engage in dramatically different courtship dances.
   b. Two species of yucca plants can pollinate each others' seeds, but the seeds are then aborted.
   c. Two plant species have wind-dispersed pollen that lands on the styles and grows a pollen tube through the ovary of either species; however, in hybrid matings, the sperm cannot fertilize the ovum.
   d. Two frog species meet and can mate with each other, but the hybrid offspring are infertile.
   e. Two beetle species are superficially similar in appearance; but the structure of the male penis and the female genitalia prevent males of one species from copulating with females of the other.

3. A monophyletic group would be best described as:
   a. a grouping of all species descended from a common ancestor, including that ancestor
   b. a grouping of all species descended from a common ancestor, excluding that ancestor
   c. a grouping of all species that share a similar set of traits
   d. a grouping of species descended from two or more closely related species

4. Paleontologists studying fossilized therapsids (mammal-like reptiles that are now extinct) would probably be using which of the following species concepts?
   a. the biological species concept
   b. the morphospecies concept
   c. the phylogenetic species concept
   d. None of the above; fossil species cannot be classified.

5. To which of the following situations would it be difficult to apply the biological species concept? (You may wish to choose more than one situation.)
   a. two populations of morphologically similar organisms that differ in breeding habits and physiology
   b. two populations of asexual clonal organisms that are morphologically similar
   c. two morphologically similar populations, one in Africa and one in South America, that exhibit no apparent prezygotic or postzygotic isolation
   d. lions and tigers do not interbreed in nature, but can sometimes hybridize in zoos
   e. two bird populations that live next to each other and look similar, but have different songs that attract only females that grew up in a nest with a father that sang that song

6. Why shouldn't the hybrid offspring of the male *A. m. nigrescens* and female *A. m. peninsulae* be released to replenish the dusky seaside sparrow's genetic diversity?
   a. There is already enough genetic diversity in the *A. m. nigrescens* population.
   b. Releasing these offspring would artificially cause gene flow between two genetically distinct populations.
   c. Due to hybrid sterility, these hybrids cannot reproduce.
   d. The morphospecies concept does not recognize subspecies.

7. Vicariance refers to:
   a. a pattern of speciation in which a population is subdivided by a geographic barrier
   b. a kind of speciation where a small propagule founds a population on a habitat island
   c. a pattern of speciation whereby new species evolve reproductive isolation in sympatry
   d. a technique for constructing phylogenetic trees based on vicariant traits shared by all members of a taxon

8. Peter and Rosemary Grant observed a small group of the large ground finch colonizing Daphne Major in the Galápagos Islands. In a few years, the descendants of the colonists had evolved beaks that were much larger than those in the original source population. What factors likely influenced this evolution? (Choose all that apply.)
   a. artificial selection due to the measurements researchers made
   b. continued gene flow with the source population
   c. the original small size of the colonist pool resulting in divergence due to genetic drift
   d. reinforcement due to postzygotic isolation on secondary contact
   e. natural selection imposed by larger seeds found on Daphne Major than on the ancestral island

9. Which of the following is the best example of isolation and divergence in allopatry?
   a. Two similar ground-squirrel species are separated by the Grand Canyon.
   b. Different species of tapeworms infect humans and cats.
   c. Horticulturalists have created tetraploid flower species.
   d. Soapberry bug populations appear to be diverging into a long-beaked and a short-beaked subspecies.

10. Which of the following would best be described as a case of speciation in sympatry?
    a. A population of lizards is subdivided by a natural barrier and subsequently diverges to form two species that cannot interbreed.
    b. A new, isolated population of fruit flies is founded by a small group of colonists, which then diverges from the ancestral source population.
    c. An individual hermaphroditic plant undergoes meiotic failure, producing diploid pollen and ovules; these self-fertilize, germinate, and grow into several fully fertile tetraploid plants.
    d. Speciation cannot take place in sympatry, only in allopatry, where geography poses a barrier to gene flow.

11. Why have soapberry bugs feeding on nonnative hosts evolved shorter beaks than those feeding on their native host plant?
    a. Beak size is phenotypically plastic; the changes seen are not genetic and do not reflect genetic differentiation.
    b. A small population of soapberry bugs with short beaks colonized the nonnative hosts; the differentiation is due to genetic drift.
    c. This difference is due to a change in ploidy, resulting in a genetically differentiated population on nonnative hosts.
    d. The nonnative host species have smaller fruits; natural selection favors bugs with a shorter beak length on these hosts.

12. Parents from two different species mate and produce a fertile offspring with a polyploid chromosome number. This is called:
    a. triploidy
    b. allopatric speciation
    c. autopolyploidy
    d. mutation selection
    e. allopolyploidy

13. Reinforcement of speciation refers to:
    a. selection for increased prezygotic isolation stemming from the reduced fitness of hybrid offspring
    b. natural selection for increased postzygotic isolation due to differences in courtship rituals or pollination strategies
    c. genetic drift that causes two subsets of a population to become more differentiated
    d. the evolution of compatibility between hybrids, causing the fusion of two diverging populations

14. Researchers used mitochondrial DNA to study the hybrid zone between Townsend's warblers and hermit warblers. Why was mitochondrial DNA so useful in this study?
    a. The mtDNA is maternally inherited, so researchers were able to discover that most hybrids resulted from Townsend's males mating with hermit females.
    b. The mtDNA is easy to extract, so the researchers could obtain lots of genetic information about the hybrids.
    c. The mtDNA shows that reinforcement has been occurring in this hybrid zone.
    d. The mtDNA is biparentally inherited, like most of the nuclear genetic material. Any gene could have been used in this study.

15. Biologists recently tested the hypothesis that the sunflower *Helianthus anomalus* originated as a hybrid between *H. annuus* and *H. petiolaris*. Which of the following findings would *not* support this hypothesis?
    a. *H. anomalus* phenotypically resembles hybrids between *H. annuus* and *H. petiolaris*.
    b. Most of the gene sequences in *H. anomalus* are different from either *H. annuus* or *H. petiolaris*.
    c. Experimental crosses between *H. annuus* and *H. petiolaris* produced hybrids with species-specific patterns of DNA sequences similar to those found in *H. anomalus*.
    d. *H. annuus* and *H. petiolaris* hybridize in the geographic region where *H. annuus* is found.

CHAPTER 26—ANSWER KEY

Checking Your Understanding

(a) You would probably have to use the morphospecies concept for a new fossil species because phylogenetic trees with the species would be unavailable, and there is no way to determine with which other populations a fossil was able to mate. You would probably use the phylogenetic species concept for a well-studied bacterial group, although you might be able to use the morphospecies concept. Bacteria are asexual, so the biological species concept could not be used in this case. Finally, you could really use any of the species definitions for the well-studied mammal species; ideally, you would use all concepts and compare your conclusions using each method.

(b) Greater gene flow within the two monophyletic groups (perhaps by occasional dispersal of the young to a different population) would tend to eliminate any genetic differences.

(c) Movement of an aquatic organism from one lake to another (for example, during an unusual wet period that temporarily joins two lakes) also causes physical isolation and may involve a small colonist population. Similarly, mountaintops in the Southwest United States are separated by desert areas that are difficult for many mountaintop organisms to cross (these mountains are sometimes referred to as the sky islands). Colonization by a small population that then experiences genetic drift and natural selection can also occur in these habitats.

(d) Colonization events are more likely to produce small population sizes (at least initially).

(e) These two populations are still in the process of possibly separating into two species. Whether or not this process becomes complete depends on future events. For example, the two soapbug populations do not interbreed, because these bugs mate on their food plants. However, if the Asian sapindaceous plants were to suddenly undergo a population crash, the soapbugs on them would presumably switch back to the native plants, even though they feed less efficiently on them. Interbreeding would then resume, eliminating differences between the two populations.

(f) Many plants can self-fertilize; thus, polyploid individuals can often mate with themselves to produce healthy offspring. However, animals typically cannot self-fertilize, so that any surviving polyploid mutants would be unable to reproduce.

(g) Presumably there is no Townsend's warbler mtDNA on those islands because no female Townsend's have made it to the islands (perhaps the male Townsend's warblers disperse farther than do the females). Thus all the island birds are descended from the original female hermit warblers that first mated with the Townsend's males.

Assessing What You've Learned

1. c; 2. b, d; 3. a; 4. b; 5. b, c; 6. b; 7. a; 8. c, e; 9. a; 10. c; 11. d; 12. e; 13. a; 14. a; 15. b

27

Phylogenies and the History of Life

A. KEY CONCEPTS

- This chapter examines large-scale patterns in the family tree of life: What has been the history of life since life evolved on Earth? How do we study this history? How do evolutionary innovations occur, leading to adaptive radiation? Why do mass extinctions occur?

27.1 Tools for Studying History: Phylogenetic Trees

- The evolutionary history of an organismal group is called a **phylogeny**. A **phylogenetic tree** shows ancestor-descendant relationships among populations or species and clarifies who is related to whom. In a phylogenetic tree, a tree **branch** represent populations through time, the point where two branches diverge, called a **node** (or fork) represents the point in time when an ancestral group split into two or more descendant groups; and a tip (or terminal node), the endpoint of a branch, represents a group (a species or larger taxon) living today or one that ended in extinction.

How Do Researchers Estimate Phylogenies?

- Both morphological and genetic characteristics are used to estimate phylogenetic relationships among species. Closely related species should share many traits, while distantly related species should share fewer traits.

- The **phenetic approach** uses all traits to evaluate the relatedness of species. For example, a researcher using this approach might assess similarity using *genetic distance*, a measure of the average percentage of difference in DNA bases between two populations or species.

- The **cladistic approach**, on the other hand, claims that some traits are more informative than other traits in determining relatedness.

- Researchers using this approach focus on **synapomorphies**—shared derived characters found in the species of interest (**Figure 27.1**). Descendant groups share synapomorphies because they are inherited from their common ancestor (e.g., lactation and fur in mammals).

- As evolution occurs in each lineage, new synapomorphies appear. Computer programs help researchers identify these unique traits that define monophyletic groups.

How Can Biologists Distinguish Homology from Homoplasy?

- Difficulties determining the true phylogeny arise when similar traits evolve independently in two species.

- **Homology**, in biology, refers to similarities due to shared ancestry. **Homoplasy** refers to similar traits that are independently derived.

- For example, icthyosaurs and dolphins show morphological similarities (e.g., streamlined bodies) not found in their reptilian and mammalian ancestors. These traits were selected for in both these groups because they are advantageous in fish-eating marine animals (**Figure 27.2**).

- Species inhabiting similar environments may have similar traits due to **convergent evolution** when natural selection favors similar solutions to the challenges of a particular habitat. These convergent traits are not found in the species' common ancestor.

- Thus, the key to distinguishing homology from homoplasy is determining whether the similar structures or genes derive from a common ancestor.

- For example, the *HOM* genes of insects and the *Hox* genes of vertebrates *are* thought to derive from a common ancestor because they (1) show similar organizations, (2) share a *homeobox* sequence, and (3) code for products with similar functions and patterns of expression. Further, many animals from lineages that branch off between these two groups have similar genes.

- Sometimes it is difficult to distinguish homology from homoplasy. Researchers often use **parsimony** in trying to identify the phylogenetic tree that minimizes the overall number of convergent evolution events (**Figure 27.3**). This approach assumes that convergent evolution should occur less frequently than similarity due to shared descent.

CHECK YOUR UNDERSTANDING

(a) Are bat wings and pterosaur wings an example of homology or homoplasy? Why?

_____

_____

_____

Whale Evolution: A Case History

- Traditionally, phylogenetic trees based on morphological data show whales outside of the artiodactyls (mammals with hooves and an even number of toes) and identify the evolution of a pulley-shaped astralagus as a synapomorphy, a shared derived trait, for the artiodactyls group (**Figure 27.4a**).

- However, DNA sequence data suggest a close relationship between whales and hippos. A phylogenetic tree showing closely related whales and hippos is less parsimonious for morphological data because it requires evolution and then loss of the pulley-shaped astralagus in whales (**Figure 27.2b**).

- Data on short interspersed nuclear elements (SINEs) show that whales and hippos share several SINE genes that are absent in other artiodactyl groups (**Figure 27.2c**). These SINEs are synapomorphies supporting the hypothesis that whales and hippos are closely related.

- Recent fossil finds of whale-like artiodactyls with pulley-shaped astralagus bones lend further support to the hypothesis that whales and hippos are closely related and that whales are, in fact, artiodactyls.

27.2 Tools for Studying History: The Fossil Record

- **Fossils** are physical evidence left by organisms from the past. They provide the only direct evidence of what ancient organisms looked like and when they existed. The **fossil record** includes all fossils that have been found and recorded.

How Do Fossils Form?

- Most fossils form when an organism is buried in sediment before decomposition occurs (**Figure 27.5**).

- Four types of fossils are (**Figure 27.6**): (1) organic remains that resist decomposition and are preserved intact; (2) sediments above the organism that cement into rock and compress the organic material into a thin film; (3) remains that decay after burial, leaving a hole that fills with dissolved minerals, creating a **cast**; (4) *permineralized* fossils that form if the dissolved minerals enter cells.

- Centuries later, fossils may be exposed due to erosion or human action. Once found, fossils must be carefully picked from the surrounding rock. Fossils are then named, described in a scientific publication, dated, and added to a collection so they are available for study as part of the fossil record.

- Most organisms decompose rapidly and are buried slowly, if at all. Thus, fossilization is an extremely rare event.

Limitations of the Fossil Record

- Habitat bias: Organisms living where sedimentation occurs (beaches, swamps, etc.) are more likely to become buried in sediments and be fossilized than are organisms living in other habitats. Burrowing organisms are more likely to fossilize because they are pre-buried at death.

- Taxonomic bias: Organisms with hard parts (bones, shells) are more likely to resist decay long enough to be fossilized than are organisms with only soft parts. Similarly, within an organism the hard parts (pollen, teeth) are more likely to fossilize than are the soft parts.

- Temporal bias: Rocks in Earth's crust may be worn down by erosion or melted as one tectonic plate is pushed underneath another one. Because these processes destroy rocks and the fossils in them, many ancient fossils have been destroyed; recent fossils are thus more common than ancient ones.

- Abundance bias: Because fossilization of any organism is a rare event, organisms that are common, widespread, and extant as a species for long periods are more likely to become fossilized and be found by researchers.

- The fossil record is highly biased; but it gives the *only* direct evidence about the morphology and ecology of extinct organisms. Analysis of the fossil record (by **paleontologists**) is thus central to our understanding of the history of life.

(b) There are no known fossils of the very first living organisms. Give at least two reasons to explain why this lack of fossil evidence for these first organisms is not surprising.

_____

_____

_____

_____

Life's Timeline

- Major events in the history of life are marked on the timeline shown in **Figure 27.8**. Dates are primarily calculated using radiometric dating (but see **Box 27.1** for a molecular approach). Figure 27.8 has been broken into four segments.

- The **Precambrian** (almost 2 billion years!) includes the **Hadean, Archaean,** and **Proterozoic eons**. During most of this time period, almost all life was unicellular and hardly any oxygen was present in Earth's oceans or atmosphere.

- The **Phanerozoic eon** includes the interval between 543 million years ago (mya) and the present and is divided into three eras.

- The appearance of many animal groups marks the beginning of the **Paleozoic era**. Land animals, land plants, and fungi appear and diversify during this period. This era ends with a mass extinction at the end of the Permian.

- The end-Permian extinction is followed by the **Mesozoic era**, also known as the Age of Reptiles. The Mesozoic ends with the extinction event that killed the dinosaurs. Gymnosperms were the dominant plants.

- The Mesozoic is followed by the **Cenozoic era**, also known as the Age of Mammals because mammals diversified after the dinosaurs disappeared. Angiosperms became the dominant plants.

27.3 The Cambrian Explosion

- Animals first originated around 565 million years ago. Soon after that, at the start of the Cambrian period, animals diversified into almost all the major groups extant today. This is known as the **Cambrian explosion**.

- Life had been predominantly unicellular for almost 3 billion years when this spectacular period of evolutionary change occurred.

Cambrian Fossils: An Overview

- Three major fossil beds record this explosion of animal life. In China, the Doushantuo fossils record the first animal life about 570 Ma; the Ediacaran fossils in Australia date from 565 to 544 Ma; and the Burgess Shale fossils in Canada show animal life from 525 to 515 Ma (**Figure 27.9a**).

- Each of these fossil beds is extraordinary because each of them includes more than one habitat and the fossils of soft-bodied as well as hard-shelled animals, greatly increasing number of **fauna** (animal species) represented.

The Doushantuo Microfossils

- Researchers identified several submillimeter-sized sponges in samples dated to 580 mya. Other biologists found what seem to be animal embryos in the one-, two-, four-, and eight-cell stages. As in modern embryos, during these early divisions the cell number increases and the volume of the embryo stays the same (**Figure 27.9b**)

- Cyanobacteria and multicellular algae were also present in the samples, suggesting a shallow-water marine habitat. The sponges and other tiny creatures probably filtered organic debris from the water for food.

The Ediacaran Faunas

- Sponges, jellyfish, comb jellies, and traces of other animals are found in these Australian deposits (**Figure 27.9c**). However, no animals with shells, heads, mouths, or feeding appendages are present-either at this site or at other fossil sites from this time period (565–544 mya).

- This shallow-water marine habitat had lots of filter feeders, but no animals that actively hunted for or captured food. Ediacaran animals burrowed in sediments, floated in water, or sat immobile on the seafloor.

The Burgess Shale Faunas

- Just about every major animal group can be found among the Burgess Shale fossils (525–515 mya): arthropods, mollusks, echinoderms, worms, even a chordate (**Figure 27.9d**). These animals show incredible morphological and ecological diversity.

- Besides filter feeders, there are predators, scavengers, and grazers. These animals had eyes, mouths, limbs, and shells; they swam, burrowed, walked, ran, slithered, clung, and/or floated. How did this incredible diversification happen?

Did Gene Duplications Trigger the Cambrian Explosion?

- Homeotic genes code for proteins that determine, in early development, the three-dimensional layout of multicellular organisms. Because these genes have such a major effect on morphology, researchers have examined correlations between the evolutionary history, morphology, and number of homeotic genes in different animal groups.

- For example, the *Hox* genes help specify the location and shape of limbs, antenna, and other key structures in animals. These genes are found in clusters, suggesting origination via gene duplication.

- The "new genes, new bodies" hypothesis proposes that gene duplications produced new copies of existing homeotic genes. These new copies were then able to take on new roles in development, making possible the diversity of new body plans and appendages that appear during the Cambrian explosion.

- The number and type of *Hox* loci in different animal groups are shown along with the evolutionary relationships among animal groups in **Figure 27.10**. Generally speaking, these data support the hypothesis that homeotic genes are responsible for the evolution of animal morphological diversity.

- The number and type of *Hox* genes in animals is variable, and, in general, animal groups that branched off early in evolution have smaller, simpler bodies and fewer *Hox* genes than do groups that branched off later. These patterns support the new genes, new bodies hypothesis.

- Structure and base sequence similarity supports the idea that gene duplications were responsible for the origin of many *Hox* loci. Gene duplication provides a mechanism for the new genes, new bodies hypothesis, and the order in which different genes appear can be estimated from their phylogenetic pattern.

- Vertebrates, some of the largest and most complex animals, have several copies of the entire *Hox* cluster. This suggests that the complete set of 13 genes was duplicated at least twice. A third duplication occurred among the species-rich and morphologically diverse ray-finned fish.

- On average, greater complexity is associated with more *Hox* genes among the groups shown in **Figure 27.10**. However, this correlation does not always hold up within these groups, especially among the arthropods where, for example, crabs have only five *Hox* genes.

- Duplication of *Hox* genes is an important factor in the diversification of animal body plans. However, changes in gene expression and function have also been important in creating animal morphological diversity.

CHECK YOUR UNDERSTANDING

(c) Crustaceans and centipedes have fewer *Hox* genes than velvet worms or fruit flies do. Based on the phylogeny shown in **Figure 27.10**, does it look like the *Hox* genes found in fruit flies were secondarily lost in crustaceans, or were these genes never present in crustacean ancestors? Explain your answer.

_____

_____

_____

_____

27.4 Adaptive Radiations

- One broad pattern that can be observed in the tree of life is that dense groups of branches are scattered throughout the tree. These dense branching patterns, known as star phylogenies (**Figure 27.11**), represent major diversification of organism groups over a relatively short period of time.

- When a single lineage produces many descendant groups in a short period of time, an **adaptive radiation** has occurred. Adaptive radiations are an important pattern in the history of life.

- Examples of adaptive radiations include the diversification of Hawaiian honeycreeper birds and Hawaiian silverswords following colonization of the Hawaiian Islands, the Cambrian explosion, and the diversification of mammals from a common ancestor between 65 and 60 mya.

- In all these examples, the descendant groups inhabit myriad habitats and have many different forms and lifestyles. What causes these adaptive radiations?

Ecological Opportunity as a Trigger

- Adaptive radiations follow periods of ecological opportunity when new or novel resources become available. Often this occurs when habitats are unoccupied by competitors. For example, mammals evolved to occupy habitats left vacant after extinction of the dinosaurs.

- Researchers have been able to document the details of an adaptive radiation that occurred following the colonization of unoccupied island habitats.

- One study focused on *Anolis* lizards of the Caribbean islands. This group includes 150 species, and the size and shape of a species closely correlates with its habitat (e.g., species living on the ground or on tree trunks have longer legs than do those that spend their time clinging to twigs).

- Researchers estimated *Anolis* phylogeny using DNA sequence data. They then compared habitat (and associated morphology) to the evolutionary relationships among species (**Figure 27.12**).

- For the islands of Hispaniola and Jamaica, the founding lizard colonists belonged to different ecological types; but over time, the lizards evolved to occupy all island habitats, forming four different species with four different morphologies. Each island had its own adaptive radiation.

- Several of these miniature adaptive radiations, each on a different island, were possible following the colonization of each new available habitat that lacked competitors.

Morphological Innovation as a Trigger

- Morphological innovation can provide opportunities for exploiting new habitats. Multicellularity, shells, exoskeletons, and limbs were new traits that produced the adaptive radiation known as the Cambrian explosion.

- Other examples of traits that allowed exploitation of new habitats, causing adaptive radiations, are (1) the evolution of wings, three pairs of legs, and a protective external skeleton in insects; (2) the evolution of flowers in angiosperms; (3) the evolution of a second set of jaws in cichlid fish; and (4) the evolution of feathers and wings in birds (**Figure 27.13**).

CHECK YOUR UNDERSTANDING

(d) Why isn't the "new genes, new bodies" hypothesis sufficient to explain all animal morphological diversity?

_____

_____

_____

_____

27.5 Mass Extinctions

- **Mass extinctions** are the rapid extinction of many groups throughout the tree of life (loss of at least 60 percent of all species within 1 million years). These extinction events are caused by catastrophic episodes.

- Traditionally, five mass extinction events are distinguished from a lower **background extinction** rate that represents the normal loss of some species that always occurs (**Figure 27.14**).

How Do Background and Mass Extinctions Differ?

- Background extinctions typically occur when normal environmental change or competition reduces a population to the point that it dies out. Mass extinctions occur when unusual large-scale environmental change causes the extinction of many normally well-adapted species.

- Natural selection causes most background extinctions, whereas random chance plays a large role in mass extinctions.

What Killed the Dinosaurs?

- At the end of the Cretaceous, 65 mya, an asteroid about 10 km across hit Earth and splashed material over much of North America. Destruction caused by the asteroid impact led to dinosaur extinction, paving the way for the mammalian radiation. Several lines of evidence support this **impact hypothesis**.

  1. High levels of iridium, common in asteroids but rare on Earth, are found worldwide in sedimentary rocks from the Cretaceous-Tertiary (K-P) boundary.
  2. Minerals found only at meteorite impact sites (e.g., shocked quartz and microtektites) are common in Haitian and other rocks dated to 65 Ma.
  3. There is a huge crater off the northwest coast of the Yucatán peninsula (see **Figure 27.16**).

- Computer models and geologic data suggest that the asteroid impact caused a fireball of hot gas that spread out from the impact site, causing worldwide wildfires (evidenced by soot and ash deposits dated to 65 mya). The dust, soot, and ash then blocked the Sun, causing rapid global cooling and decreased plant productivity. Additionally, SO_4^{2-} that was released from rock at the impact site formed sulfuric acid, causing acid rain.

- The large-scale environmental change triggered by the asteroid impact caused the extinction of 60 to 80 percent of all species. For another example of a mass extinction, see **Box 27.2**.

Selectivity

- Some evolutionary lineages withstood the environmental change better than others did. Thus dinosaurs, pterosaurs, and large marine reptiles went extinct; but most mammals, crocodiles, amphibians, and turtles survived.

- One hypothesis is that extended darkness and cold affected large organisms more than small ones because they require more food. However, size selectivity is not found among marine clams and snails. Other hypotheses are still being investigated.

Recovery

- Ferns appear to have replaced diverse woody and flowering plants in many habitats following the K-P extinction. Mammals diversified to fill the niches left empty following the dinosaur extinction. Succeeding flora and fauna of the Tertiary were very different from those of the Cretaceous.

CHECK YOUR UNDERSTANDING

(e) What might happen if Earth were to be hit by another asteroid the size of the one that struck in 65 mya?

_____

_____

_____

_____

B. INTEGRATING YOUR KNOWLEDGE

Looking Back—Key Concepts from Earlier Chapters

Homeotic Genes—Chapter 21

Clearly, homeotic genes (e.g., *Hox* loci) have played a major role in evolution because they affect the overall morphology of organisms.

Early Embryonic Development in Animals—Chapter 22

You learned about embryo development in **Chapter 22**. Here you learn additional details about how gene expression of homeotic genes controls limb development. Changes in expression of these genes may have been responsible for the evolution of tetrapod limbs.

Evolution—Chapters 1, 4, 24–26

Because evolution is one of the unifying themes of biology, many chapters have addressed this topic. Chapter 27, however, is the last to concentrate specifically on evolution. Nonetheless, watch out for continued references to evolution in future chapters.

Phylogenies—Chapters 1 and 26

Chapter 1 depicted a SSU-RNA phylogeny for the tree of life, and **Chapter 26** went into more detail on phylogenies and monophyletic groups. Now you are learning about large-scale patterns in the tree of life, and how phylogenetic information can be compared with information on specific genes to infer how different groups and morphologies evolved.

Looking Forward—Key Concepts in Later Chapters

Diversity of Life—Chapters 28–35

Now that you know a bit about how evolution occurs and about evolutionary patterns in the tree of life, you are ready to learn specifics about the major groups of organisms comprising the tree of life.

Competition—Chapter 53

Chapter 27 mentioned how adaptive radiations tend to occur when niches are unoccupied by competitors. **Chapter 53** discusses what happens when two or more species do compete within the same habitat.

C. MASTERING CHALLENGING TOPICS

To thoroughly understand evolution, you must integrate the concepts of how mutations and natural selection at the level of the individual are related to large-scale evolutionary patterns.

For example, consider a mutation in a fish that causes altered expression of *Hoxd-11*. Suppose that the altered expression causes the fish to have limb-like fins. If the fish can still swim effectively but can also more easily draw itself onto land to escape predators, it may survive better and leave more offspring (the altered *Hoxd-11* expression has been selected for).

Over time, the altered expression may increase in frequency throughout the fish species' population. Over time, other traits that favor the ability to exist on land may also be selected for. If eventually some of the population takes up living partly on land, they have colonized a new habitat and may experience sympatric speciation (different habitats in same location leading to decreased gene flow—just like the soapberry bugs in **Chapter 26**). Now the first vertebrate land dweller has lots of new habitats available without much competition. This situation causes an adaptive radiation of tetrapods on land.

In this way, individual and population-level evolutionary processes can lead to the large-scale patterns discussed in this chapter.

D.   ASSESSING WHAT YOU'VE LEARNED

1. A monophyletic group is:
   a. a taxonomic group known to have diverged prior to the rest of the taxa in the study
   b. an ancestor and all its descendants
   c. the branch tips of a phylogenetic tree
   d. any two taxa that show convergent evolution

2. Which of the following is *not* true of the **cladistic** approach for inferring phylogenetic trees?
   a. Synapomorphies are used to infer evolutionary relationships.
   b. When convergent evolution occurs, biologists use parsimony to choose the best tree.
   c. Computer programs are used to identify traits that are unique to each monophyletic group.
   d. Genetic distance can be used to create phylogenetic trees.

3. Bird feathers in warblers and ostriches are:
   a. an example of homology
   b. an example of homoplasy
   c. an example of convergent evolution
   d. an example of parsimony
   e. all of the above

4. Researchers conclude that whales are most closely related to hippos because (choose all that apply):
   a. Whales and hippos share several SINEs that are not present in other artiodactyls.
   b. Modern whales have tiny pulley-shaped astralagus bones.
   c. Non-SINE whale and hippo DNA sequences are also similar.
   d. Whales look more like hippos than they look like cows.
   e. all of the above

5. The fossil record is a biased sample of organisms that are now extinct. Which of the following is *not* a reason for that bias?
   a. Organisms with hard parts, such as teeth, shells, and skeletons, are more likely to be preserved.
   b. Organisms living in certain habitats were more likely to be buried in sediments and preserved than were organisms in other habitats.
   c. Organisms with broad geographic distributions and large populations are more likely to be preserved than are rare endemic species.
   d. Exposed fossils are often subject to weathering, whereas those beneath the surface are protected.

6. Which of the following correctly lists the geological eras from most ancient to most recent?
   a. Paleozoic, Mesozoic, Precambrian, Cenozoic
   b. Paleozoic, Mesozoic, Cenozoic, Precambrian
   c. Precambrian, Paleozoic, Mesozoic, Cenozoic
   d. Precambrian, Cenozoic, Paleozoic, Mesozoic

7. Why are the Doushantuo, Ediacaran, and Burgess Shale fossil deposits so unusual?
   a. because they preserve a diverse array of microorganisms that we believe gave rise to multicellular life at the start of the Cambrian period
   b. because they capture a wide range of soft-bodied organisms that are rarely preserved in the fossil record at a time when the diversity of metazoans was greatly increasing
   c. because these assemblages show precisely how animals began as single-celled organisms, then became two-celled, then four-celled, and so on to become the multicellular forms we know today
   d. because these fossils record the origin of animals from seaweeds and other photosynthetic organisms

8. Fossils of organisms from the Precambrian are extraordinarily rare, but many species are found in the Burgess Shale deposits. Which of the following statements best describes the Burgess Shale organisms?
   a. mostly bacteria and simple one-celled plants and animals
   b. primarily sponges, algae, and seaweeds
   c. arthropods, mollusks, chordates, and a variety of soft-bodied organisms
   d. jellyfish and sponges, along with some tracks of various other organisms

9. What is one genetic hypothesis proposed for the rapid diversification of body plans during the Cambrian period?
   a. Many mutations accumulated due to high exposure to UV radiation, creating many new kinds of organisms.
   b. Reproductive isolation was nearly nonexistent; many "species" interbred to form hybrids that went on to become new species.
   c. New genes appeared in many different organisms through a mechanism that has not been identified; these are associated with new body plans.
   d. Several duplications of the homeotic genes that control major body plans apparently arose during the Cambrian, creating new body plans and appendage configurations.

10. Which of the following observations is *not* consistent with the hypothesis that increasing metazoan complexity is associated with an increase in number of *Hox* loci?

    a. Within phyla, there is no correspondence between the number of *Hox* genes and complexity.

    b. Phyla that branch off early, such as sponges and cnidarians, have simple body plans with relatively few *Hox* loci.

    c. *Hox* loci are similar in structure and DNA sequence and are grouped in clusters, hence they are assumed to have arisen through duplication; when a new *Hox* gene appears within a lineage, most of the descendent taxa have a homologous *Hox* locus.

    d. In vertebrates, the *Hox* clusters themselves appear to be duplicated several times.

11. Which of the following cases best illustrates an adaptive radiation?

    a. A fishless lake is colonized by a single fish species, which over a few thousand years gives rise to several species, each with a series of unique feeding adaptations.

    b. A small group of tree-dwelling lizards of a single species migrates to an uninhabited, treeless island and adapts to use the open grassland habitat.

    c. A population of a cricket species takes up residence in a cave. Over many generations, the cave crickets eventually lose their eyes, as have many other cave-dwelling animals.

    d. A colonizing species of fruit fly takes up residence on a new continent and displaces two closely related native species.

12. Why are morphological innovations often associated with adaptive radiations?

    a. Morphological innovations often allow a species to replace competitors.

    b. Morphological innovations can open up new adaptive options that may allow a species to colonize an underutilized resource.

    c. Morphological innovations disrupt natural selection, forcing species to find new adaptations.

    d. Morphological innovations prevent adaptive radiations most of the time.

13. In the study of *Anolis* lizards colonizing various Caribbean Islands, researchers found similar habitats and similar ecological types on Hispaniola and Jamaica. They conducted a phylogenetic analysis of the species on each island. What observation suggested that adaptive radiations occur in response to habitat availability and the absence of competitors?

    a. Different islands were initially colonized by different species that differed in habitat preference, but in both cases subsequent speciation produced a range of ecological specialists occupying similar habitat niches. Phylogeny of each island was unique, but ended up with species having similar sets of adaptations for each habitat.

    b. The first colonists on every island were always the same species, a twig-dwelling specialist. But in every case, that species gave rise to the same set of other species that specialized on different habitats. The phylogeny for each island had the same root and the same pattern of speciation.

    c. Each island was initially colonized by a different species that occupied its preferred habitat. Subsequent colonizing species were successful only if they could use an unoccupied habitat type. The phylogeny for each island was the same throughout the region because no evolution occurred, only colonization.

    d. Each island was colonized by a different species initially, and that species underwent a radiation, giving rise to new species that occupied different habitat types. But there was no similarity in the adaptations from one island to another. The phylogeny for each island was completely unique and totally unlike that of any other island.

14. Which of the following statements represents an important contrast between background extinctions and mass extinctions?

    a. Background extinctions occur only sporadically; mass extinctions happen on a regular, periodic basis.

    b. Background extinctions are caused by rapid changes in the environment; mass extinctions occur primarily in response to biological competition.

    c. During mass extinctions, adaptations for survival and competition make little difference in the likelihood of extinction; the opposite is true of background extinctions.

    d. Mass extinctions, such as the K-P event, tend to take out large-bodied organisms; smaller-bodied organisms tend to be removed by background extinction.

15. Which of the following statements best represents current thinking about the rise of the mammals after the K-P extinction?
    a. Mammalian diversity increased dramatically prior to the impact, but the population sizes of every species remained small until after the dinosaurs were extinct.
    b. Mammals were inherently better competitors than the dinosaurs; the K-P impact simply speeded up the replacement, which was already well under way.
    c. By chance, the K-P impact caused extinction of the dinosaurs but did not have as great an effect on small, nocturnal, scavenging mammals. The release from competition allowed the mammals to diversify.
    d. Both mammals and dinosaurs survived the asteroid, but because mammals were more advanced, they rapidly underwent an adaptive radiation and outcompeted the dinosaurs.

CHAPTER 27—ANSWER KEY

(a) Bat wings and pterosaur wings are an example of homoplasy, because the early mammal ancestor of bats did not have wings and neither did the reptile ancestor of pterosaurs. This is a case of convergent evolution.

(b) It is no surprise that there are no known fossils from the very first living organisms, because (1) most of the sedimentary rocks that were formed then have since been eroded or otherwise destroyed; and (2) those organisms would have had no hard parts, and they would thus have been less likely to be fossilized in the first place.

(c) Gene deletions must have occurred in the Crustacean and Centipede groups because the pattern of *Hox* genes presence shows that the common ancestor of the arthropods must have had (according to parsimony) the genes that are missing from the Crustacean and centipede groups.

(d) The "new genes, new bodies" hypothesis cannot explain *all* animal diversity in morphology, because within both the arthropod group and the vertebrates group, there is no correlation between the number of *Hox* loci and organismal complexity.

(e) The impact could cause worldwide wildfires and global cooling from all the dust and ash in the air. However, acid rain would occur again only if the impact site contained the sulfate-rich anhydrite.

Assessing What You've Learned

1. b; 2. d; 3. b; 4. a, c; 5. d; 6. c; 7. b; 8. c; 9. d; 10. a; 11. a; 12. b; 13. a; 14. c; 15. c

28

Bacteria and Archaea

A. KEY CONCEPTS

28.1 Why Do Biologists Study Bacteria and Archaea?

- Bacteria are fascinating—some bacteria live over 1 km underground or in 95°C hot springs.
- To preserve diversity, we must understand it.

Some Bacteria Cause Disease

- Bacteria that cause disease are said to be **pathogenic**.
- **Key Research**—Robert Koch (1880) established that bacteria can cause disease. He tested his hypothesis using **anthrax**.
- Koch's Postulates:
  1. The microbe must be present in individuals suffering from the disease and absent from healthy individuals.
  2. The organism must be isolated and grown in a pure culture away from the host organism.
  3. If organisms from the pure culture are injected into a healthy experimental animal, the disease symptoms should appear.
  4. The organism should be isolated from the diseased experimental animal, again grown in pure culture, and demonstrated by its size, shape, and color to be the same as the original organism.
- Koch's experiments also became the basis for the **germ theory of disease**.
- **Antibiotics** are molecules that kill bacteria. Widespread use of antibiotics has led to drug-resistant strains of bacteria.

Bacteria Can Clean Up Pollution

- Bioremediation is the use of bacteria and archaea to degrade pollutants.
  1. Fertilize contaminated sites to encourage the growth of existing bacteria that degrade toxic compounds.
  2. "Seed" or add specific species of bacteria to contaminated sites.

Extremophiles

- Bacteria or archaea that live in high-salt, high- or low-temperature, or high-pressure habitats are **extremophiles**.
- Astrobiologists are using extremophiles as their model organisms in the search for extraterrestrial life.

How Do Small Cells Affect Global Change?

The Oxygen Revolution

- No free molecular oxygen existed for the first 2.5 billion years because:
  1. There was no plausible source of oxygen at the time the planet cooled to a solid.
  2. The oldest Earth rocks indicate any oxygen that formed immediately reacted with iron atoms to produce iron oxides.
- Even in the ocean, oxygen was produced by **cyanobacteria** through **oxygenic photosynthesis**.
- Quantities of oxygen did not begin to build up until 2.0 billion years ago. At 1.8 billion years ago, iron oxides in terrestrial environments appeared.
- Once oxygen was common in the oceans, it could be used as an electron acceptor; thus, **aerobic respiration** was now a possibility.
- The rate of energy production and metabolism could now rise dramatically. Coincidentally, 2.1 billion years ago is when the first macroscopic algae first appear in the fossil record.

The Nitrogen Cycle

- Plant growth and overall productivity are often limited by the availability of nitrogen.
- Redox reactions that result in the production of ammonia (NH_3) from nitrogen (N_2) are referred to as **nitrogen fixation**.
- Certain species of cyanobacteria are capable of fixing nitrogen.
- On land, nitrogen-fixing bacteria live in close association with plants—often taking up residence in **nodules**.

- Plants need this nitrogen to make up large quantities of proteins.

Nitrate Pollution

- Widespread use of ammonia fertilizers (NH_3) on many crops not associated with N_2-fixing bacteria had an unforeseen consequence.
- Nitrate (NO_3^-) produced as a by-product of bacterial ammonia metabolism dissolves readily in water—eventually it enters aquatic ecosystems.

CHECK YOUR UNDERSTANDING

(a) How does high nitrate affect aquatic ecosystems?

_____

_____

_____

28.2 How Do Biologists Study Bacteria and Archaea?

Using Enrichment Cultures

- Biologists rely heavily on the ability to culture organisms in the lab.
- **Enrichment cultures** are based on establishing a specific set of growing conditions.

Using Direct Sequencing

- **Direct sequencing** is a strategy for documenting the presence of bacteria and archaea that cannot be grown in culture and studied in the laboratory.
- This technique is vitally important in understanding diversity since only a tiny fraction of the bacteria and archaea that exist can be grown in the lab.
- Direct sequencing has been used to discover new lineages of Archaea tentatively called the **Korarchaeota** and **Nanoarchaeota**.

Evaluating Molecular Phylogenies

- **Ribosome complexes** are found in all organisms. Ribosomes are the site of protein synthesis and contain a **small subunit (SSU)** RNA.
- **Key Research**—In 1960 Carl Woese described base sequences from SSU in many bacterial species. He drew a diagram now known as the **universal tree**, or **tree of life** (**Figure 28.1**).

CHECK YOUR UNDERSTANDING

(b) What did Woese's follow-up work determine?

_____

_____

28.3 What Themes Occur in the Diversification of Bacteria and Archaea?

Morphological Diversity

- Bacteria and archaea range from the smallest of all free-living cells—mycoplasmas have a volume of 0.03 μm^3—to the largest bacterium known, *Thiomargarita namibiensis*, with volumes as large as $200 \times 10^6 \, \mu m^3$.
- Bacterial shapes include **filaments**, **spheres**, **rods**, **chains**, and **spirals** (**Figure 28.13b**). Some bacteria are **sedentary** and some are **motile**—using **flagella** or gliding.
- Within bacteria, two types of cell walls exist and are distinguished by a **Gram stain**:
  1. **Gram-positive** cells have a plasma membrane surrounded by a cell wall with extensive peptidoglycan.
  2. **Gram-negative** cells have a thin gelatinous layer containing peptidoglycan and an outer phospholipid bilayer.

Metabolic Diversity

- Bacteria produce ATP in three ways:
  1. **Phototrophs**—use light energy to promote electrons to the top of electron transport chains. ATP is produced by cellular respiration.
  2. **Chemoorganotrophs**—oxidize reduced organic molecules with high potential energy. ATP is produced by cellular respiration or fermentation.
  3. **Chemolithotrophs**—oxidize inorganic molecules with high potential energy. ATP is produced by cellular respiration with the inorganic compound serving as the electron donor.
- **Autotrophs** manufacture their own carbon-containing compounds, and **heterotrophs** must acquire them.

Producing ATP via Photosynthesis: Variation in Electron Sources and Pigments

- Bacteria can use kinetic energy in light to raise electrons to higher energy states.

- As electrons are stepped down to lower energy states, the energy released is used to make ATP.
- This process requires a source of electrons—cyanobacteria and plants use water. Splitting water to obtain electrons produces oxygen.
- Other bacteria split hydrogen sulfide or ferrous iron.
- It is important to recall the light-absorbing properties of chlorophylls *a* and *b*. Cyanobacteria have the same two pigments.

Producing ATP via Cellular Respiration: Variation in Electron Donors and Acceptors

- **Cellular respiration**—A molecule with high potential energy serves as an electron donor and is oxidized, while a molecule with low potential energy serves as a final electron acceptor and becomes reduced. The potential energy difference is converted into ATP.
- A common **electron donor** is glucose, and a **common electron acceptor** is oxygen.
- When cellular respiration is complete, glucose is completely oxidized to CO_2.
- When oxygen acts as the final electron acceptor, the by-product is water.
- Electron donors range from hydrogen molecules and hydrogen sulfide to ammonia and methane.
- Other electron acceptors are sulfate, nitrate, carbon dioxide, or ferric ions.

Producing ATP via Fermentation: Variation in Substrates

- **Fermentation** is a strategy for making ATP without using electron transport chains.
- No electron acceptor is used—redox reactions are internally balanced.
- Fermentation is less efficient compared to respiration and is often used as an alternative.
- Glucose is fermented into lactic acid or ethanol.
- *Clostridium aceticum* can ferment ethanol, fatty acids, and acetate as well.
- Species that ferment amino acids produce end-products like cadaverine and putrescine—responsible for odors of rotting flesh.

Obtaining Building-Block Compounds: Variation in Pathways for Fixing Carbon

- Autotrophs make their own building-block compounds; heterotrophs do not.
- In cyanobacteria and plants, enzymes from the **Calvin cycle** transform CO_2 into organic molecules. Other bacteria, animals, and fungi must obtain carbon by eating plants or animals or by absorbing organic compounds released in dead tissues.
- **Methanotrophs** use methane as their primary electron donor and carbon source.
- Some bacteria use carbon monoxide or methanol as starting material.

CHECK YOUR UNDERSTANDING

(c) Explain the difference between a chemoorganotroph and a chemolithotroph.

_____

_____

_____

28.4 Key Lineages of Bacteria and Archaea

Bacteria

Firmicutes

- They are also called "low-GC Gram positives" and are rod shaped or spherical.
- Large metabolic diversity; include anthrax, botulism, tetanus, and strep throat.

Spirochaeles (Spirochetes)

- They are distinguished by their corkscrew shape and unusual flagella.
- Syphilis is caused by a spirochete, and so is Lyme disease.

Actinobacteria

- They are called "high-GC Gram positives" and vary in shape from rods to filaments.
- Over 500 distinct antibiotics have been isolated within the genus

Streptomyces.

- Many species are abundant in the soil and vital as decomposers. Tuberculosis and leprosy are caused by member of this group.

Cyanobacteria

- They produce much of the oxygen and nitrogen and many of the organic compounds that feed other organisms in freshwater and marine environments.

Chlamydiales

- They are spherical and very tiny.
- All known species live as parasites inside host cells (**endosymbionts**).
- *Chlamydia trachomatis* is a very common sexually transmitted disease.

Proteobacteria

- Species in this group cause Legionnaire's disease, cholera, dysentery, and gonorrhea. Certain species can produce vinegars. *Rhizobium* can fix nitrogen.

Archaea

Crenarchaeota

- None of these species have yet to cause disease in humans. Can inhabit very extreme environments.

Euryarchaeota

- None of these species have yet to cause disease in humans. *Ferroplasma* live in piles of waste rock near abandoned mines. Methanogens live in the soils of swamps and guts of animals.

 CHECK YOUR UNDERSTANDING

(d) From which genus have researchers obtained over 500 distinct antibiotics? How do you think antibiotics are made from bacteria?

_____

_____

_____

_____

B. INTEGRATING YOUR KNOWLEDGE

Looking Back—Key Concepts from Earlier Chapters

Respiration—Chapter 9

The diverse world of bacteria and archaea depends on the presence of electron donors and acceptors. There are many variations on the basic theme in bacteria and archaea; they can employ not only glucose, oxygen, and standard fermentation but also an entire suite of new substrates and reactions.

Photosynthesis—Chapter 10

Plants use water as their source of electrons, and bacteria can use many other molecules. Chlorophylls are also used in cyanobacteria in the same manner as in plants.

Carbon Fixing—Chapter 10

Plants use the Calvin cycle to get carbon from CO_2 and while some bacteria do this, others may fix carbon from other inorganic sources of carbon.

Looking Forward—Key Concepts in Later Chapters

Biogeochemical Cycling—Chapter 54

Chapter 28 just touches on the importance of the nitrogen cycle in the global ecosystem, which is discussed in greater detail in **Chapter 54**. Many other interrelated cycles exist and function to maintain the Earth's balance.

Animal Behavior—Chapter 51

Animal behavior is intimately related to the process of disease transmission by bacteria. Many diseases travel to several animal species before ultimately settling on a host.

C. MASTERING CHALLENGING TOPICS

In your first-year biology course, you are being exposed to many new terms in a short time. Often, these terms are introduced with a new branch of biology. In this chapter, the terms describing the metabolic diversity of bacteria can be confusing (phototroph, chemoorganotroph, chemolithotroph, autotroph, and heterotroph). Take special care in learning what each term means, and be able to verbally explain the differences between them to a classmate, laboratory partner, or friend. "Compare and contrast" questions are often used on tests, so be able to create a list of similarities and differences between all of these terms. To simplify, categorize the groups into terms explaining how bacteria make ATP (phototroph, chemoorganotroph, or chemolithotroph) and how the bacteria obtain carbon (autotroph or heterotroph). After that, the prefixes help a lot in defining the terms.

D. ASSESSING WHAT YOU'VE LEARNED

1. The key substance making up bacterial cell walls is:
   a. polysaccharides
   b. peptidoglycan
   c. ribosomes
   d. polymerases

2. Any bacteria that can cause disease are said to be:
   a. pathogenic
   b. contagious
   c. resistant
   d. prolific

3. The use of bacteria and archaea to degrade pollutants is called:
   a. biocontamination
   b. seeding
   c. biodetoxification
   d. bioremediation

4. The oxygen revolution was so important because available oxygen in the oceans could now be used as:
   a. an electron donor
   b. an electron acceptor
   c. for oxygen fixation
   d. a reducer

5. Plant growth and overall productivity is often limited by the availability of:
   a. oxygen
   b. nitrogen
   c. carbon
   d. xenon

6. Direct sequencing is used to document the presence of bacteria and archaea that .
   a. cannot be found in animals
   b. cannot be detected
   c. cannot be grown in culture
   d. cannot infect humans

7. Researchers generally evaluate bacterial molecular phylogenies by using sequences from:
   a. DNA
   b. cDNA
   c. membrane proteins
   d. ribosomes

8. Bacteria that use light to promote electrons to the top of electron transport chains and produce ATP using cellular respiration are called:
   a. phototrophs
   b. autotrophs
   c. heterotrophs
   d. organotrophs

9. The strategy for making ATP without using electron transport chains is called:
   a. aerobic respiration
   b. fermentation
   c. electron transport
   d. carbon fixation

10. The lineage of bacteria characterized by causing diseases such as tuberculosis and leprosy are the _____.
    a. high-GC gram positives
    b. low-GC gram positives
    c. cyanobacteria
    d. spirochaeles

CHAPTER 28—ANSWER KEY

Check Your Understanding

(a) Cyanobacteria and algae use nitrates and grow and die in huge numbers. Decomposers like heterotrophic bacteria and archaea then grow in huge numbers. These organisms use up all the available oxygen and produce "dead zones."

(b) Follow-up work documented that Bacteria were the first of the three lineages to diverge from the common ancestor of all living organisms.

(c) Chemoorganotrophs oxidize reduced organic molecules with high potential energy, whereas chemolithotrophs oxidize inorganic molecules with high potential energy.(

(d) Over 500 distinct antibiotics have been isolated within the genus *Streptomyces*. Originally, the metabolic products each species generated were used as antibiotics. Now, these products can be synthesized without the use of the bacteria.

Assessing What You've Learned

1. b; 2. a; 3. d; 4. b; 5. b; 6. c; 7. d; 8. a; 9. b; 10. a

29

Protists

A. KEY CONCEPTS

29.1 Why Do Biologists Study Protists?

Impacts on Human Health and Welfare

- The most spectacular crop failure in history, the Irish potato famine, was caused by a protist.

Malaria

- ***Plasmodium*** is the causative agent of malaria and is transmitted to humans by mosquitoes. At least 300 million people worldwide are sickened by it each year, and over 1 million people die from the disease annually.

- Because each *Plasmodium* species spends part of its life cycle inside mosquitoes (**Figure 29.3**), most antimalaria campaigns have focused on controlling these insects.

Harmful Algal Blooms

- **Red tides** occur when toxin-producing protists called dinoflagellates reach high densities in a particular area. Shellfish eat these protists, and people can become sick eating the contaminated shellfish.

Ecological Importance of Protists

- Although protists represent 10 percent of the total number of named species, a single teaspoon of pond water can contain over 1000 flagellated protists. Dinoflagellates can reach 60 million cells per liter.

- Photosynthetic species are major players in the global carbon cycle. Primary productivity by the world's oceans, in turn, is responsible for almost half of the total carbon that is fixed on the planet.

Protists Play a Key Role in Aquatic Food Chains

- Photosynthetic protists (**plankton**) are the basis of marine and freshwater **food chains**.

- Without protists, most food chains in freshwater and marine habitats would collapse.

Could Protists Help Reduce Global Warming?

- Protists could help reduce global warming by becoming carbon sinks.

- Several lineages of protists have shells made of calcium carbonate. When these protists die, they sink to the ocean floor and form sedimentary rock.

- Petroleum formation begins with accumulations of dead bacteria, archaea, and protists at the bottom of the ocean.

CHECK YOUR UNDERSTANDING

(a) How is stable carbon deposited in the ocean?

_____

_____

29.2 How Do Biologists Study Protists?

Microscopy: Studying Cell Structure

- Detailed studies of cell structure revealed that many protists have a characteristic overall form.

- Researches interpret these distinct morphological features as **synapomorphies**—shared, derived traits that distinguish major monophyletic groups (**Table 29.2**).

- Eight major groups have been identified on the basis of structures that protect or support the cell or that influence the organism's ability to move or feed.

Evaluating Molecular Phylogenies

- Using the gene that codes for the RNA molecule in the small subunit of ribosomes, eight monophyletic groups from the Eukarya were found.

- The current tree (**Figure 29.8**) shows that Amoebozoa and Opisthokonta form a monophyletic group recently named the Unikonta and the Alveolata and Stramenopila form a monophyletic group celled the Chromalveolata.

Discovering New Lineages via Direct Sequencing

- Direct sequencing has found a wide array of distinct ribosomal RNA sequences and thus a vast amount of new species.

- Eukaryotic cells are much more variable in size than previously imagined; many very tiny species were just discovered.

CHECK YOUR UNDERSTANDING

(b) What do the synapomorphies listed in Table 29.2 represent?

_____

_____

_____

29.3 What Themes Occur in the Diversification of Protists?

What Morphological Innovations Evolved in Protists?

- The earliest eukaryotes were probably single-celled organisms with a nucleus and endomembrane system, mitochondria, and a cytoskeleton, but no cell wall.

- It is also likely that these cells swam using a novel type of flagellum.

The Nuclear Envelope

- Although researchers are still gathering evidence on how the nuclear envelope originated, the leading hypothesis is that it derived from infoldings of the plasma membrane.

- The nuclear envelope is advantageous because it separates RNA transcription inside the nucleus and RNA translation outside. This structure allows for further control of gene expression.

The Mitochondrion

- Mitochondria are organelles that generate ATP using pyruvate as an electron donor and oxygen as the ultimate electron acceptor.

- The endosymbiosis theory proposes that mitochondria originated when a bacterial cell took up residence inside a eukaryote about 2 billion years ago.

- The **host** supplied the bacterial **symbiont** with protection and reduced carbon compounds from its other prey, while the symbiont supplied the host with ATP.

- **Key Research**—Lynn Margulis expanded on the original hypothesis of the origin of mitochondria.

- Observations consistent with the theory:

  1. Mitochondria are about the size of an average bacterium and replicate by fission, as do bacterial cells.

  2. Mitochondria have their own ribosomes and manufacture their own proteins. The ribosomes are similar to bacterial ribosomes.

  3. Mitochondria have double membranes, consistent with the engulfing mechanism (**Figure 29.11**).

  4. Mitochondria have their own genomes, which are organized as circular molecules, like bacteria.

- Researchers compared gene sequences isolated from the nuclear DNA of eukaryotes, mitochondrial DNA from the same eukaryote, and DNA from several species of bacteria. Mitochondrial gene sequences were found to be much more closely related to the sequences from the bacteria.

- Once the mitochondrion was present in the ancestor of today's protists, it underwent diversification.

Structures for Support and Protection

- Many protists have a rigid internal skeleton or a hard external structure called a **test** or a **shell** (**Figure 29.13**).

- Diatoms are surrounded by a glass-like, silicone-oxide shell.

- Dinoflagellates have a cell wall made up of cellulose plates.

- Foraminifera can use calcium carbonate or cover themselves with tiny pebbles.

- Parabasalids use an internal support of cross-linked microtubules.

- Euglenids have a collection of protein strips just under the plasma membrane.

Multicellularity

- Differentiation of cell types is a crucial criterion for defining multicellularity.

- In simple multicellular species, some cells are for producing or obtaining food, while others are specialized for reproduction.

- Not all cells express the same genes.

- Multicellularity is a synapomorphy shared by all of the brown algae and all of the plasmodial and cellular slime molds. It also arose in some lineages of red algae.

How Do Protists Find Food?

Ingestive Feeding

- Predation and scavenging—Protists can eat bacteria, archaea, and other protists—dead or alive.
- The engulfing process is possible because protists lack a cell wall, and because their flexible membrane and cytoskeleton move as in **Figure 29.14a**.

Absorptive Feeding

- Some absorptive groups are decomposers, like the oomycetes.
- Many others live inside organisms as parasites, although in many cases the relationships are beneficial or mutualistic.

Photosynthesis

- The endosymbiosis theory contends that the organelle where photosynthesis takes place in eukaryotes originated when a protest engulfed a cyanobacterium (**primary endosymbiosis**).
- **Secondary endosymbiosis** occurs when an organism engulfs an already photosynthetic organism and retains the acquired chloroplasts as intracellular symbionts.

Diversity in Lifestyles

- All three lifestyles just described occur in many different eukaryotic lineages, and all three can occur in the same group of species.

How Do Protists Move?

- **Amoeboid motion** is a gliding motion. Fingerlike projections called pseudopodia stream forward over a substrate. This motion involves interactions between actin, myosin, and ATP.
- The other major mode of locomotion involves flagella or cilia. Flagella and cilia have identical structures and are actually extensions of the cell. Both consist of nine sets of doublet microtubules around two central, single microtubules.

How Do Protists Reproduce?

Sexual versus Asexual Reproduction

- **Asexual reproduction** produces offspring that are genetically identical to the parent.
- **Sexual reproduction** produces offspring that are genetically different from their parents.
- **Meiosis** is the reduction division that produces a set of haploid daughter cells from a diploid parent cell. Genetic differences between parents and offspring are increased when **crossing over** occurs during meiosis.

- Importance:
  1. Production of better-quality offspring.
  2. Genetically variable offspring have a better chance of surviving during environmental changes.

Variation in Life Cycles

- Every aspect of a life cycle is variable among protists.
- **Alternation of generations** is a phenomenon in which the haploid and diploid phases of the life cycle are multicellular.
- Multicellular haploid and diploid forms of the same individual can be identical in morphology or radically different.

CHECK YOUR UNDERSTANDING

(c) Compare and contrast asexual and sexual reproduction.

_____

_____

_____

_____

29.4 Key Lineages of Protists

Excavata

Diplomonadida

- Each cell has two nuclei, each associated with four flagella, and all lack a cell wall (e.g., *Giardia*).

Parabasalida

- Cells lack both a cell wall and mitochondria, and have a single nucleus. A distinctive rod of cross-linked microtubules runs the length of the cell. They feed by engulfing (e.g., *Trichomonas*).

Discicristata

Euglenida

- Cells lack a cell wall but have a system of interlocking protein strips.
- One-third of species perform photosynthesis; euglenids are important components of freshwater plankton and food chains.

Alveolata

Ciliata

- Cells have a micronucleus and a macronucleus, can reproduce by conjugation, and use cilia for locomotion.

Dinoflagellata

- Cells from sexual reproduction may form cysts to remain dormant. Second only to diatoms in amount of carbon fixed.

Apicomplexa

- System of organelles at one end called the apical complex.
- All are parasitic. Malaria is caused by *Plasmodium*.

Stramenopila (Heterokonta)

Oomycota

- Fungus-like and extremely important decomposers in aquatic ecosystems. An oomycete caused the Irish potato famine.

Diatoms

- Cells are supported by silicon-rich, glassy shells. They are photosynthetic and therefore the most important producer of carbon compounds in the water.

Phaeophyta (Brown Algae)

- Cell walls contain cellulose, and all species are multicellular.
- They are photosynthetic and sessile, though their reproductive cells may be motile. Form forests that are important habitats.

Rhizaria

Foraminifera

- Cells have multiple nuclei, and their tests are usually made of organic material. Dead forams often form extensive deposits.

Plantae

Rhodophyta (Red Algae)

- Cell walls composed of cellulose and other polymers and have no flagella. Species contribute to reef building.

CHECK YOUR UNDERSTANDING

(d) What is alternation of generations, and which of the groups above is likely to demonstrate this phenomenon?

_____

_____

_____

_____

Amoebozoa

Myxogastrida (Plasmodial Slime Molds)

- Have a huge supercell with many nuclei.
- Move by amoeboid motion and are important decomposers in forest ecosystems.

B. INTEGRATING YOUR KNOWLEDGE

*Looking Back—Key Concepts
from Earlier Chapters*

Cell Organelles—Chapter 7

All of the organelles of the cell, their structure, and function are described in detail in **Chapter 7**. Chapter 29 explains some of the diversity in function of the same organelles.

Diversity—Chapter 28

The variation in photosynthetic pigments observed in bacteria echoes the theme in diversification of protists. Photosynthetic species harvest different wavelengths of light to avoid competition.

Genetic Recombination—Chapter 18

Chapter 18 introduces the mechanisms that bacteria use to transfer genes and undergo genetic recombination. These cells exchange short stretches of DNA between individuals, usually via plasmids. In protists, the entire genome is involved.

Meiosis—Chapter 11

The reduction division called meiosis is explored in **Chapter 11**. The process is now applied and explained in terms of the first organisms to adopt this process.

Looking Forward—Key Concepts in Later Chapters

Muscle Contraction—Chapter 47

The mechanism of amoeboid movement is closely related to muscle movement in animals, which is detailed in **Chapter 47**.

C. MASTERING CHALLENGING TOPICS

In this chapter, the theory of endosymbiosis is used to explain the origins of the first eukaryotes. While the theory itself may be easy to understand, explaining it and the techniques used to support it is a different matter. Try starting with the basics. Outline the theory and note what it tries to explain. Then outline each experiment in the chapter that supports the theory and note why it supports the theory. As the text now has separated mitochondrion endosymbiosis from chloroplast endosymbiosis, so should you. If you do not understand the techniques used in a particular experiment, go back in the text to previous chapters that outline those techniques in more detail. Theories are modified constantly as new evidence (for or against) is published.

D. ASSESSING WHAT YOU'VE LEARNED

1. The following ranks as one of the most devastating diseases and is caused by a protist.
   a. red tides
   b. AIDS
   c. malaria
   d. plasmodium

2. The theory of endosymbiosis proposes that mitochondria originated _____.
   a. when a virus invaded and stayed in a bacterial cell
   b. when a bacterial cell took up residence inside a eukaryote
   c. when a virus took up residence inside a eukaryote
   d. when a bacterial cell took up residence inside a prokaryote

3. Which one of the following statements *does not* support endosymbiosis?
   a. Cyanobacteria look nothing like chloroplasts.
   b. Mitochondria and chloroplasts are about the same size as an average bacterium.
   c. Both mitochondria and chloroplasts have a double membrane.
   d. Both mitochondria and chloroplasts reproduce by fission.

4. The crucial criterion for defining multicellularity is:
   a. differentiation
   b. colonial growth
   c. lysosomes
   d. cell walls

5. Which of the following is *not* a mode of locomotion in protists?
   a. ameboid movement
   b. using flagella
   c. using cilia
   d. using mitochondria

6. The phenomenon in which the haploid and diploid phases of a life cycle are multicellular is called:
   a. double diploid
   b. alternation of generations
   c. haplodiplo generation
   d. multigenerational

7. Which type of reproduction produces offspring that are genetically different from their parents?
   a. sexual reproduction
   b. asexual reproduction
   c. multisexual reproduction
   d. unisexual reproduction

8. This group of protists includes the species responsible for malaria.
   a. diatoms
   b. red algae
   c. apicomplexans
   d. dinoflagellates

9. This group was responsible for the Irish potato famine.
   a. diatoms
   b. apicomplexans
   c. ciliates
   d. oomycetes

10. The key characteristic of the plasmodial slime molds is:
    a. their single nuclei per cell
    b. their ability to parasitize
    c. their huge supercell with many nuclei
    d. their carbon fixing in aquatic habitats

CHAPTER 29—ANSWER KEY

Check Your Understanding

(a) Carbon turnover in the ocean is significantly more rapid than in the terrestrial environment. However, limestone is a stable repository for carbon in the oceans. The sugars that protists produce are the basis for food chains in both freshwater and marine environments.

(b) The synapomorphies listed in the table represent changes in structures that protect or support the cell or that influence the organism's ability to move or feed.

(c) Asexual reproduction produces offspring that are genetically identical to the parent. Sexual reproduction produces offspring that are genetically different from their parents. Meiosis is the reduction division that produces a set of haploid daughter cells from a diploid parent cell. Genetic differences between parents and offspring are increased when crossing over occurs during meiosis.

(d) Alternation of generations is a phenomenon in which the haploid and diploid phases of the life cycle are multicellular. Multicellular haploid and diploid forms of the same individual can be identical in morphology or radically different. This phenomenon evolved independently in the brown algae and the red algae.

Assessing What You've Learned

1. c; 2. b; 3. a; 4. a; 5. d; 6. b; 7. a; 8. c; 9. d; 10. c

Green Plants

30

A. KEY CONCEPTS

30.1 Why Do Biologists Study the Green Plants?

Plants Provide Ecosystem Services

* Plants produce oxygen via oxygenic photosynthesis.
* Plants build soil by providing food for decomposers.
* Plants hold soil and prevent nutrients from being washed or blown away.
* Plants hold water and increase water-holding capacity of an area.
* Plants moderate the local climate.

Plants Provide Humans with Food, Fuel, Fiber, Building Materials, and Medicines

* The grains that form the basis of our current food supply were derived from wild species about 10,000 to 2,000 years ago. Humans were actively selecting seeds to plant the next generation of crops—a process called **artificial selection**.
* Sugars synthesized during the Carboniferous period laid the groundwork for the industrial revolution. There will probably not be another generation of plant-based fuels.
* Cotton and other plant fibers are still important sources of raw materials for clothing, rope, and other household articles.
* Primary interest in woody plants is for building materials and fibers used in papermaking. Relative to its density, wood is a stiffer and stronger building material than concrete, cast iron, aluminum alloys, or steel.
* In both traditional and modern medicine, plants are a key source of drugs.

CHECK YOUR UNDERSTANDING

(a) Describe the process of artificial selection in producing corn with more oil.

_____

_____

_____

_____

_____

30.2 How Do Biologists Study Green Plants?

Analyzing Morphological Traits

* The 12 most important Phyla are grouped into three categories:
    1. **Nonvascular plants** (Hepaticophyta—liverworts, Anthocerophyta—hornworts, and Bryophyta—mosses)
    2. **Seedless vascular plants** (Lycophyta—club mosses, Psilophyta—whisk ferns, Sphenophyta—horsetails, and Pterophyta—ferns)
    3. **Seed plants** (Cycadophyta—cycads, Ginkophyta—ginkoes, Coniferophyta—conifers, Gnetophyta—gnetophytes, Pinophyta—pines, and Anthophyta—angiosperms)

Using the Fossil Record

* The fossil record for land plants is massive; it began 476 million years ago (**Figure 30.7**) and is divided into five segments:

1. **475 million years ago**—duration 30 million years

Most fossils are microscopic and consist of reproductive cells called **spores** and sheets of a waxy coating called **cuticle**.

2. **"Silurian-Devonian Explosion"**—445 to 359 million years ago

Macroscopic fossils from most major plant lineages are found. Virtually all adaptations that allow plants to occupy dry, terrestrial habitats are present, including water-conducting cells, tissues, and roots.

3. **Carboniferous Period**—359 to 299 million years ago

Extensive deposits of coal are found; coal is a carbon-rich rock packed with fossil spores, branches, leaves, and tree trunks. Usually derived from lycophytes, horsetails, and ferns and in the presence of water, the makeup of coal indicates the presence of extensive forested swamps.

4. **Gymnosperms**—299 to 145 million years ago

These lineages include the cycads, conifers and gingkoes, gnetophytes and pines. Because gymnosperms grow readily in dry habitats, biologists infer that both wet and dry environments on the continents became blanketed with green plants for the first time during this interval.

5. **Angiosperms**—145 million years ago to present

The woody plants that produced the first flowers are the ancestors of today's grasses, orchids, daisies, oaks, maples, and roses.

Evaluating Molecular Phylogenies

- The phylogenetic tree in **Figure 30.9** is a recent version from laboratories studying plant phylogenies. Note the following:
  1. Land plants most likely evolved from green algae.
  2. Charales is the sister group to land plants.
  3. Green plants are monophyletic, and "green algae" is paraphyletic.
  4. The land plants are monophyletic.
  5. The nonvascular plants are the most basal groups among land plants.
  6. The earliest land plants lacked water-conducting tissues.
  7. Lycophytes are the sister groups to all other seedless vascular plants.
  8. Morphological simplicity of the whisk ferns is likely a derived trait.
  9. Seeds and flowers evolved only once.
  10. The gymnosperms are monophyletic.

(b) What are the key innovations that allowed plants to tolerate dry conditions?

_____

_____

_____

_____

30.3 What Themes Occur in the Diversification of Green Plants?

Transition to Land, I: How Did Plants Adapt to Dry Conditions?

- Plants had to adapt to situations where only a portion of their tissues were bathed in fluid.
- Solving the water problem was a breakthrough in evolution by natural selection, achieved in two steps:
  1. preventing water loss from cells
  2. transporting water from tissues with access to water to tissues without access

Preventing Water Loss: Cuticle and Stomata

- **Cuticle** is a waxy, watertight sealant that gives plants the ability to survive in dry environments. Gas exchange is accomplished on wax-covered plants by the presence of **stoma**, which consists of an opening covered by **guard cells**.
- The **pore** opens or closes as the guard cells change shape. The presence of guard cells in later groups implies that many of the early land plants had the ability to regulate gas exchange and control water loss by opening and closing their pores.

Transporting Water: Vascular Tissue and Upright Growth

- **Key Research**—Paul Kenrick and Peter Crane examined extraordinary fossils found in Scotland in a formation called the Rhynie Chert. This formation contains many plants fossilized in an upright position. Cells in these plants were able to move water to different areas of the plant using water-conducting cells. Some cells had thickened rings containing **lignin**, a polymer built from six-carbon rings and extraordinarily strong for its weight.

- 380 million years ago, the fossil record contains **tracheids** (**Figure 30.10c**). Tracheids are elongated cells, die after maturing, do not contain cytoplasm, and are the first true vascular tissues that made possible the efficient transport of water to aboveground tissues.
- 250–270 million years ago, plants have true **vessel elements**, which are the most advanced type of water-conducting tissue. Tracheids and/or vessel elements can form **wood**.

Mapping Evolutionary Changes on the Phylogenetic Tree

- Cuticles, pores, stomata, vascular tissue, and tracheids evolved just once.
- Vessels evolved independently in gnetophytes and angiosperms.

Transition to Land, II: How Do Plants Reproduce in Dry Conditions?

- Spores resisted drying because they were encased in a tough coat of sporopollenin.
- Gametes were produced in complex multicellular structures.
- Embryo was retained on the parent and nourished by it.

Retaining and Nourishing Offspring: Land Plants as Embryophytes

- The gametophytes of early land plants contained reproductive organs called **gametangia**. The sperm-producing structure is the **antheridium**, and the egg-producing structure is the **archegonium**.
- In land plants, the zygote is retained on the gametophyte after fertilization; therefore, these plants do not have to make their own food early in life.

Alternation of Generations

- Individuals have a multicellular haploid phase (**gametophyte**) and a multicellular diploid phase (**sporophyte**).
- The two phases are connected by the reproductive cells: gametes and spores.
- Development always follows the same basic events:
  1. Gametophytes produce gametes by meiosis.
  2. Two gametes unite during fertilization to form a zygote.
  3. The zygote divides by mitosis and develops into a multicellular sporophyte.
  4. The sporophyte produces spores by meiosis.
  5. Spores divide by mitosis and develop into a haploid gametophyte.

The Gametophyte-Dominant to Sporophyte-Dominant Trend in Life Cycles

- In nonvascular plants, the sporophyte is small, short lived, and dependent on the gametophyte for nutrition.
- In ferns and other vascular plants, the sporophyte is much larger and longer lived.
- The trend to sporophyte-dominated life cycles is one of the most striking trends in land plant evolution.

Heterospory

- **Heterospory** is the production of two distinct types of spore-producing structures and thus of spores.
- **Microsporangia** produce **microspores**, which develop into the male gametophyte.
- **Megasporangia** produce **megaspores**, which develop into the female gametophyte.

Pollen

- Nonvascular plants and ferns need water for gametes to swim and perform fertilization.
- In heterosporous seed plants, the microspore germinates to form a tiny male gametophyte surrounded by a tough, dessication-resistant coat of sporopollenin—a **pollen grain**.

Seeds

- A **seed** is a structure that includes a developing embryo and a food supply surrounded by a tough coat. Seeds are often attached to a structure that aids in dispersal by wind, water, or animals.

Flowers

- Flowers contain stamens (where microsporangia develop) and carpels (where ovaries are found).
- When a pollen grain lands on a carpel and produces sperm, fertilization takes place.
- **Double fertilization** involves one sperm fusing with the egg to form the zygote and the other sperm fusing with two nuclei to form triploid nutritive tissue called **endosperm**.
- Flowers are diverse in size, shape, and coloration. They also produce a wide range of scents. Scientists have hypothesized that flowers are adaptations that increase the probability that **pollination** will occur. Flowers provide a food reward to the pollinator in the form of sugar-rich nectar or protein-rich pollen grains—a mutually beneficial relationship.

Fruits

- A fruit is a structure that is derived from the ovary and encloses one or more seeds.

- Tissues derived from the ovary are often nutritious and brightly colored. Animals often eat fruits and disperse seeds.

The Angiosperm Radiation

- Angiosperms represent one of the great **adaptive radiations** in the history of life.
- The diversification of angiosperms is associated with three key adaptations: (1) vessels, (2) flowers, and (3) fruits.

Monocots are monophyletic, but **dicots** are paraphyletic. Thus, a new term for the monophyletic group of dicots is the **eudicots**.

CHECK YOUR UNDERSTANDING

(c) Why is a flower a key innovation in plants?

_____

_____

_____

30.4 Key Lineages of Green Plants

Green Algae

Ulvophyceae (Ulvophytes)

- Many large green algae such as *Ulva*, the sea lettuce, belong to this group.
- Unicellular and small multicellular species inhabit the plankton of freshwater systems.

Coleochaetophyceae (Coleochaetes)

- Most grow as flat sheets of cells (e.g., water lilies), and the multicellular individuals are haploid.

Charaphyceae (Stoneworts)

- They commonly accumulate crusts of calcium carbonate over their surfaces.
- They can form extensive beds on lake bottoms.

Nonvascular Plants ("Bryophytes")

Hepaticophyta (Liverworts)

- They have liver-shaped leaves and can grow on bare rock or tree bark, which helps in soil formation.

Anthocerophyta (Hornworts)

- Sporophytes have a horn-like appearance and have stomata.

Bryophyta (Mosses)

- Can be abundant in extreme environments and can become dormant.
- *Sphagnum* species are among the most abundant.

Seedless Vascular Plants

Lycophyta (Lycophytes or Club Mosses)

- The most ancient plant lineage with roots.
- Tree-sized lycophytes dominated the coal-forming forests of the Carboniferous.

Psilotophyta (Whisk Ferns)

- They are restricted to tropical regions and have no fossil record.

Sphenophyta (or Equisetophyta) (Horsetails)

- They can flourish in waterlogged soils by allowing oxygen to diffuse down their hollow stem.

Pteridophyta (Ferns)

- The only seedless vascular plants to have large, well-developed leaves.

Seed Plants

Gymnosperms > Cycadophyta (Cycads)

- They harbor large numbers of symbiotic, nitrogen-fixing cyanobacteria, which are important sources of nutrients.

Gymnosperms > Ginkgophyta (Ginkoes)

- One species is alive today. It is deciduous, and individual trees are either male or female.

Gymnosperms > Other Conifers (Redwoods, Junipers, Yews)

- This group is named for its reproductive structure, the cone. Dominate all high-latitude and high-altitude forests.

Gymnosperms > Gnetophyta (Gnetophytes)

- Three living genera; the drug ephedrine has been isolated from this group.

Gymnosperms > Pinophyta (Pines, Spruces, Firs)

- Pines are a mainstay in the forest products industry—wood, paper and turpentine.

Anthophyta (Angiosperms)

- The defining adaptation is the flower. They supply the food that supports virtually every other species.

▰▰▰ CHECK YOUR UNDERSTANDING ▰▰▰

(d) What do the gymnosperms and angiosperms produce during reproduction that is different from all other groups?

_____

_____

B.   INTEGRATING YOUR KNOWLEDGE

Looking Back—Key Concepts
from Earlier Chapters

Evolutionary Relationships—Chapter 24

Chapter 24 introduces us to how molecular and morphological data are used to infer evolutionary relationships. This investigation of evolutionary relationships is continued in Chapter 30 with the land plants.

Alternation of Generations—Chapter 29

The phenomenon of alternation of generations was introduced in **Chapter 29** in the protists. This phenomenon persists with further variation in the land plants in Chapter 30.

Genetic Engineering—Chapter 19

The mechanisms and procedures for genetic engineering are explained in **Chapter 19** and applied to the production of land plants in Chapter 30.

Looking Forward—Key Concepts in Later Chapters

Water Conduction—Chapter 37

Water conduction is mentioned as only one explanation for the massive aboveground growth of land plants. The mechanisms for water conduction are explored in detail in **Chapter 37**.

The Seed—Chapter 40

The structure of the seed is introduced in Chapter 30, but how it protects the developing embryo and the changes that occur during growth and development are discussed in detail in **Chapter 40**.

C.   MASTERING CHALLENGING TOPICS

By far, understanding how alternation of generations works, and then changes during the evolution of land plants, is the toughest concept to master in this chapter. First, think about the definition of alternation of generations and look at the differences between sporophytes and gametophytes. Next, slowly go through Figures 30.16–30.19 and really take in the general ideas first, like how much time the groups spend in haploid versus diploid stages. Lastly, try to describe to your classmates these major changes in the different plant groups.

D.   ASSESSING WHAT YOU'VE LEARNED

1. The process of actively choosing seeds to plant the next generation of crops is called:
   a. natural selection
   b. artificial selection
   c. general selection
   d. hydroponics

2. The Carboniferous period is most noted for:
   a. its extensive deposits of coal
   b. its extensive angiosperm radiation
   c. its extensive radiation of aquatic plants
   d. an explosion of macroscopic fossils

3. Which of the following is *not* a major grouping of land plants?
   a. seedless vascular plants
   b. nonvascular plants
   c. cone-bearing plants
   d. seed plants

4. The latest and most impressive plant radiation was that of the   .
   a. algae
   b. angiosperms
   c. gymnosperms
   d. club mosses

5. Which of these statements about plant molecular phylogenies is *true*?
   a. The group "green algae" is paraphyletic.
   b. The Charales is not the sister group to green algae.
   c. The land plants are paraphyletic.
   d. The nonvascular plants are the most recent group.

6. Plants can prevent water loss from cells by:
   a. secreting a waxy cuticle
   b. exchanging respiratory gases through a pore using guard cells
   c. regulating the opening and closing of these pores
   d. all of the above

7. Upright growth in nonwoody plants is attained by:
   a. extensive root systems for support
   b. the production of thicker leaves
   c. the absence of leaves
   d. pumping water into cells to increase their rigidity

8. The structure that encloses and protects the developing plant embryo is:
   a. the dicot
   b. the seed
   c. the pollen grain
   d. the ovule

9. Diversification of the angiosperms is associated with the following adaptation:
   a. vessels
   b. flowers
   c. fruits
   d. all of the above

10. The only major group of seed plants with only one species alive today is:
    a. Ginkophyta
    b. Cycadophyta
    c. Coniferophyta
    d. Gnetophyta

CHAPTER 30—ANSWER KEY

Check Your Understanding

(a) By selecting and planting seeds coming from corn plants with the highest oil content, eventually corn plants will yield more oil per plant.

(b) Solving the water problem was a breakthrough in evolution by natural selection achieved in two steps:

   1. preventing water loss from cells

   2. transporting water from tissues with access to water to tissues without access

(c) Flowers are diverse in size, shape, and coloration. They also produce a wide range of scents. Scientists have hypothesized that flowers are adaptations to increase the probability that pollination will occur. Flowers provide a food reward to the pollinator in the form of sugar-rich nectar or protein-rich pollen grains—a mutually beneficial relationship.

(d) Angiosperms and gymnosperms produce seeds instead of spores.

Assessing What You've Learned

1. b; 2. a; 3. c; 4. b; 5. a; 6. d; 7. d; 8. b; 9. d; 10. a

31

Fungi

A. KEY CONCEPTS

31.1 Why Do Biologists Study Fungi?

Fungi Provide Nutrients for Land Plants

- **Mycorrhizal** (fungal-root) associations between fungi and the roots of land plants are extremely common.

- Tree species can grow 3 to 4 times faster with these fungal associations.

Fungi Speed the Carbon Cycle on Land

- **Saprophytes** are fungi that make their living by digesting dead plant material. They did not grow in the Carboniferous period because the pH was too low.

- At the end of the great Permian extinction, researchers documented a huge fungal spike.

- Fungi make the carbon cycle (the fixation of carbon by plants and the release of CO_2 via respiration) turn more rapidly.

Fungi Have Important Economic Impacts

- Incidence of fungal infections in humans is low. The major destructive impact is on crops.

- Fungi have important positive impacts on the human food supply.

- In nature, fungi have killed 4 billion chestnut trees and tens of millions of American elm trees. The fungi were accidentally imported.

Fungi Are Key Model Organisms in Eukaryotic Genetics

- The yeast *Saccharomyces cerevisiae* has been incredibly important in basic research on cell biology and molecular genetics.

- It has become the species of choice for studies on control of the cell cycle and regulation of gene expression in eukaryotes.

(a) Describe some positive impacts fungi have on the human food supply.

_____

_____

_____

_____

31.2 How Do Biologists Study Fungi?

Analyzing Morphological Traits

- Only two growth forms exist: (1) single-celled forms called **yeasts** and (2) multicellular filamentous forms called **mycelia**.

The Nature of the Fungal Mycelium

- Mycelia constantly grow in the direction of food and die back in areas that lack food.

- The filaments that make up a mycelium are called **hyphae**, and each filament is separated by cross-walls called **septa**.

- Most hyphae are haploid or **heterokaryotic**, meaning each cell contains two haploid nuclei—one from each parent.

- Mycelia are incredibly thin and have an extraordinarily high surface area.

- Fungi are adapted to an absorptive lifestyle. The only thick, fleshy structures are the reproductive structures.

Reproductive Structures

- By analyzing reproductive structures, biologists have identified four major groups of fungi:

1. Chytridiomycota (chytrids)

- The chytrids are the only group of fungi that live in water.

- Chytrids reproduce sexually and asexually. They are the only fungi that have motile cells.

2. Zygomycota

- Most Zygomycota live in soil, and many are key partners in the mutualistic interactions with plants.
- The hyphae of Zygomycota are haploid and come in several different mating types or "sexes," If chemicals released by two hyphae indicate that they are from different mating types, the individuals become yoked together—**zygosporangium**.

3. Basidiomycota (club-fungi)

- Mushrooms, bracket fungi, and puffballs are reproductive structures produced by members of the Basidiomycota. Their size, shape, and color vary enormously; but they all originate from hyphae of mated individuals.
- These fungi form specialized structures called **basidia**, which produce spores.

4. Ascomycota (sac-fungi)

- They produce complex reproductive structures that are cup-shaped. The tips of these hyphae have distinctive sac-like cells called **asci**.

Evaluating Molecular Phylogenies

- To understand the relationships among the groups of fungi, biologists have sequenced a series of genes from an array of fungal species and used the data to estimate the phylogeny of the group. The data support several important conclusions:
  1. Chytrids are the most basal group of fungi, supporting the idea that fungi evolved from aquatic ancestors and made an early transition to land.
  2. The Chytridiomycota and Zygomycota are paraphyletic. This means that either swimming gametes or yolked hyphae evolved more than once, or both were present in a common ancestor but then were lost in certain lineages.
  3. Microsporidians are actually within the monophyletic group called fungi.
  4. The Glomeromycota are monophyletic.
  5. The Basidiomycota and Ascomycota are monophyletic and are the most highly derived groups of fungi.

Experimental Studies of Mutualism

- Fungi and land plants often have a **symbiotic** relationship.
- **Mutualism** is a relationship that provides benefits to the host in return for the food the fungi absorbs.

- Researchers have begun to categorize these relationships as mutualistic, **parasitic**, or **commensal** (only the fungi benefit).
- By using radioactive isotopes of CO_2, phosphorus, and nitrogen, researchers have documented that up to 20 percent of sugars produced by a plant are transferred to fungi, and fungi facilitate the transfer of phosphorus and nitrogen to the plant.

CHECK YOUR UNDERSTANDING

(b) What is the likely ancestor of fungi, and what two things are similar between fungi and animals?

_____

_____

_____

31.3 What Themes Occur in the Diversification of Fungi?

Fungi Participate in Several Types of Mutualisms

Ectomycorrhizal Fungi (EMF)

- EMF form a dense network of hyphae around roots and are found on virtually all of the tree species that grow in temperate and boreal forests.
- EMF hyphae penetrate between cells in the outer layer of the root, but they do not enter root cells. Most are basidiomycetes, and a few are ascomycetes.
- Where the growing season is short, the decomposition of needles, leaves, twigs, and trunks is often sluggish. As a result, nitrogen tends to remain tied up in dead tissues.
- EMF however, release peptidases that cleave the peptide bonds between amino acids. The nitrogen released by this cleavage is absorbed by the hyphae and transported close to the tree roots, where it can be absorbed by the plant. In return, the fungi receive sugars and other reduced-carbon compounds from the plant.

Arbuscular Mycorrhizal Fungi (AMF)

- AMF grow *into* the cells of root tissue, and most are members of the Glomeromycota. They are found in 80 percent of all land plant species and are especially common in grasslands and in the forests of warm tropical habitats.
- In these habitats, phosphorus is the limiting nutrient, so fungi supply plants with phosphorus in exchange for carbon.

Are Endophytes Mutualists?

- Endophytes do not cause disease in the tissues they invade. Most are mutualistic, and some are commensal.

What Adaptations Make Fungi Such Effective Decomposers?

Extracellular Digestion

- Fungi secrete enzymes outside their hyphae into their food. Digestion takes place outside the fungus, or **extracellularly**.
- Only a handful of organisms on Earth can digest **cellulose**, and members of the basidiomycetes are the only organisms that can degrade lignin completely to CO_2.

Lignin Degradation

- **Lignin peroxidase** is an enzyme that catalyzes the removal of a single electron from an atom in the ring structures of lignin. This oxidation step creates a free radical and leads to a series of uncontrolled and unpredictable reactions that end up splitting the polymer into smaller units.
- The uncontrolled oxidation reactions triggered by lignin peroxidase are analogous to the uncontrolled oxidation reactions that occur when wood burns. The uncontrolled nature of the reactions means that the oxidation of lignin cannot be harnessed to drive the production of ATP. However, by oxidizing lignin, the hyphae gain access to huge supplies of energy-rich cellulose within.

Cellulose Digestion

- **Cellulases** are enzymes secreted into the extracellular environment by fungi. Some cleave long strands of cellulose into the disaccharide called cellobiose. Other cellulases are equally specific, and together they eventually convert cellulose into glucose.
- Biologists have purified seven different cellulases from the fungus *Trichoderma reesei*. These enzymes could be used to make paper production more efficient.

Variation in Life Cycles

- The **spore** is the most fundamental reproductive cell in fungi. Spores are the dispersal stage of the fungal life cycle and are produced during both asexual and sexual reproduction.

Unique Aspects of Fungal Life Cycles

- The process of sexual reproduction begins when hyphae from two individuals fuse to form. When cytoplasm fuses, **plasmogamy** is said to occur. When nuclei of a hybrid hypha. The two individuals stay independent, the mycelium is **heterokaryotic**. When the nuclei fuse, it is called **karyogamy**.

- The four major types of life cycles in fungi are as follows:

  1. The Chytridiomycota are the only fungi to exhibit alternation of generations. Swimming gametes fuse and form a diploid zygote.

  2. The haploid hyphae of Zygomycota come into contact with a different individual, and their nuclei fuse and develop a tough, resistant coat. Meiosis occurs, and the products grow into a structure that produces haploid spores. When spores are released and germinate, they grow into new mycelia.

  3. In Basidiomycota, hyphae of mated individuals fuse and then the cytoplasms fuse; but the nuclei remain independent. The diploid nucleus undergoes meiosis, and haploid spores mature from the meiotic products. Sexual reproduction concludes when the spores are expelled and dispersed by the wind.

  4. A dikaryotic mycelium, containing one nucleus from each parent, emerges and eventually grows into a complex reproductive structure with a distinctive cell called the **ascus** (sac) at its tip. The nuclei fuse inside the sac, and spores are produced.

CHECK YOUR UNDERSTANDING

(c) How is cellulose degraded by fungi?

_____

_____

_____

31.4 Key Lineages of Fungi

Chytridomycota (Chytrids)

- Largely aquatic, members of this group are the only fungi that can produce cells that are motile.
- An epidemic of chytrid infections may be responsible for catastrophic declines that recently occurred in frog populations all over the world.

Microsporidia

- All of the estimated 1200 species are single celled and parasitic.
- They are distinguished by their polar tube, which allows them to enter the interior of cells they parasitize.

Zygomycota

- Many members of this group are saprophytic, but some are important parasites of insects and spiders.
- The black bread mold *Rhizus stolonifer* is a common household pest, and some species of *Mucor* are used in the production of steroids.

Glomeromycota

- All AMF are members of this group.
- Because grasslands and tropical forests are among the most productive habitats on Earth, the AMF are enormously important to both human and natural economies.

Basidiomycota (Club Fungi)

- Basidiomycetes are important saprophytes and are the only lineage capable of digesting lignin.
- EMF are enormously important in forest management. Mushrooms are also cultivated for food. Some of the toxins found in poisonous mushrooms are used in medicine.

Ascomycota (Sac Fungi)

Lichen-Formers

- The fungal hyphae form a dense protective layer that shields the photosynthetic species; in return, the cyanobacterium or alga provides carbohydrates to the fungus.
- Lichens dominate the Arctic and Antarctic tundras. They are a major source of food, and they break down minerals from rocks—the first step in soil formation.

Non-Lichen-Formers

- Most are saprophytic, but parasitic forms exist.
- Some species are capable of growing on jet fuel or paint, and they can be used along with bacteria in efforts to clean up contaminated sites.
- *Penicillium* is an important source of antibiotics.

CHECK YOUR UNDERSTANDING

(d) What is the significance of the microsporidians belonging within the fungi?

_____

_____

_____

B. INTEGRATING YOUR KNOWLEDGE

Looking Back—Key Concepts from Earlier Chapters

Monophyletic Groupings—Chapter 29

This term was first used in describing protists since they are not monophyletic. Similarly, fungi are not monophyletic because their phylogenetic groupings do not include all descendents of the same common ancestor.

Direct Sequencing—Chapter 19

This approach to analyzing genes was introduced in **Chapter 19**. This chapter analyzes the gene that codes for the RNA molecule in the small subunit of fungal ribosomes.

Cell Structure—Chapter 7

Fungi synthesize a tough structural material called chitin. Chitin is a prominent component of the cell walls of fungi.

Looking Forward—Key Concepts in Later Chapters

Emergence—Chapter 35

Like the emerging viruses in **Chapter 35**, fungi that recently switched host organisms show the same outbreak trend.

C. MASTERING CHALLENGING TOPICS

It is usually difficult to understand, initially, the different associations between fungi and other organisms. Ectomycorrhizal fungi (EMF) and arbuscular mycorrhizal fungi (AMF) are often confused in the student's mind, and it is even harder to explain examples once the terms have been confused. If you focus on the word *ecto* as meaning "outside," this should at least help you to get the two terms organized properly. Even though both may appear to penetrate the root tissue, EMF do not penetrate the root cell wall; AMF actually do penetrate the root cell. Once you have learned this concept, pick a detailed example of each type and work through it. By understanding an example of each fungi and being able to explain it to a classmate, lab partner or friend, you will begin to master the material.

D. ASSESSING WHAT YOU'VE LEARNED

1. The associations between fungi and the roots of land plants are called:
   a. saprophytic associations
   b. parasitic associations
   c. mycorrhizal associations
   d. mycelial associations

2. Saprophytes are fungi that make their living by:
   a. digesting dead plant material
   b. digesting live plant material
   c. living inside the roots of land plants
   d. living on the leaves of vascular plants

3. The filaments that make up the multicellular fungi are called:
   a. yeasts
   b. hyphae
   c. septa
   d. gametes

4. Which of the following is *not* one of the four major groups of fungi?
   a. Zygomycota
   b. Chitinomycota
   c. Basidiomycota
   d. Ascomycota

5. The microsporidians are:
   a. a monophyletic group within the fungi
   b. a paraphyletic group within the fungi
   c. the closest living relatives to the fungi
   d. the closest extinct relatives to the fungi

6. Which of the following symbiotic relationships offers benefits to *both* the host and the invader?
   a. parasitism
   b. commensalism
   c. mutualism
   d. all of the above

7. Which of these items best describes ectomycorrhizal fungi?
   a. a mutualistic relationship where the fungi grow into the cells of plant roots
   b. a mutualistic relationship where the fungi grow around the plant roots
   c. a commensal relationship where the fungi grow into the cells of plant roots
   d. a commensal relationship where the fungi grow around the plant roots

8. How do fungi digest their food?
   a. They ingest particles and break them down.
   b. They absorb complex molecules directly.
   c. They have a rudimentary gut.
   d. They secrete enzymes extracellularly onto their food and then absorb the broken-down products.

9. Fungi have the ability to break down:
   a. carbohydrates
   b. cellulose
   c. lignin
   d. all of the above

10. When nuclei of two different individual fungi fuse, this is called:
    a. karyogamy
    b. plasmogamy
    c. nuclear fusion
    d. homokaryotism

CHAPTER 31—ANSWER KEY

Check Your Understanding

(a) Mushrooms are consumed in many cultures. The yeast *S. cerevisiae* is essential in the manufacturing of bread, soy sauce, cheese, beer, wine, whiskey, and other products. Enzymes derived from fungi are being used to improve the characteristics of foods ranging from fruit juice and candy to meat.

(b) Fungi likely evolved from an aquatic ancestor. Fungi are more closely related to animals than to plants. Most animals and fungi produce chitin. Also, the flagella that develop in chytrid spores and gametes are very similar to those observed in animals.

(c) Cellulases are secreted into the extracellular environment by fungi. Some cleave long strands of cellulose into the disaccharide called cellobiose. Other cellulases are equally specific and together, they eventually convert cellulose into glucose.

(d) This was an important result because microsporidians cause serious disease in bee colonies, silkworm colonies, and people with AIDS. With this evidence, perhaps fungicides may prove helpful in combating these microsporidian infections.

Assessing What You've Learned

1. c; 2. a; 3. b; 4. b; 5. a; 6. c; 7. b; 8. d; 9. d; 10. a

32

An Introduction to Animals

A. KEY CONCEPTS

32.1 Why Do Biologists Study Animals?

- Animals are heterotrophs and are the dominant consumers and predators in both aquatic and terrestrial habitats.

- Animals are a particularly species-rich and morphologically diverse lineage of multicellular organisms on the tree of life.

- Humans depend on wild and domesticated animals for food.

- Efforts to understand human biology depend on advances in animal biology.

32.2 How Do Biologists Study Animals?

Analyzing Comparative Morphology

The Evolution of Tissues

- The number of tissue layers that exist in an embryo helps to identify body plans. A **tissue** is a highly organized and functionally integrated group of cells.

- Although sponges lack many types of tissues, they do have an **epithelium**—a layer of tightly joined cells that covers the surface.

- **Diploblasts** are animals whose embryos have two types of tissues. Only two groups of diploblastic animals are alive today: the cnidarians and ctenophores.

- **Triploblasts** are animals whose embryos have three types of tissues. In triploblasts, the **ectoderm** gives rise to the skin and nervous system, the **endoderm** gives rise to the digestive tract, and the **mesoderm** gives rise to the circulatory system, muscle, and internal structures.

Symmetry and Cephalization

- A basic feature of a multicellular body is whether it has a plane of symmetry. All animals except sponges exhibit either **radial** or **bilateral** symmetry. Animals with radial symmetry have at least two planes of symmetry. Bilaterally symmetric organisms face their environment in one direction.

- This structure allowed **cephalization**—the evolution of a head, where structures for feeding, sensing the environment, and processing information are concentrated.

Why Was the Evolution of a Body Cavity Important?

- A body cavity creates a medium for circulation, along with space for internal organs. Fluid-filled chambers are central to the operation of a **hydrostatic skeleton** (**Figure 32.7**). This gave bilaterally symmetrical organisms the ability to move efficiently.

- Diploblasts do not have a body cavity, and neither does the triploblast phylum of flatworms.

- Roundworms and rotifers have a **pseudocoelom** that forms between the endoderm and mesoderm.

- In all other triploblasts, the body cavity forms from within the mesoderm itself and is lined with cells from the mesoderm. Muscle and blood vessels can then form on either side of the body cavity or **coelom**.

What Are the Protostome and Deuterostome Patterns of Development?

- Except for the echinoderms, all true coelomates are bilaterally symmetric and have three embryonic tissue layers. This huge group can be divided into **protostomes** (arthropods, mollusks, and segmented worms) and **deuterostomes** (vertebrates and echinoderms).

- In protostomes, the early cell division pattern is spiral cleavage; deuterostomes show radial cleavage.

- **Gastrulation** is a series of cell movements that result in the formation of the three embryonic layers. This invagination of cells creates a pore that opens to the outside (**Figure 32.8b**). In **protostomes** this pore becomes the mouth, and the anus forms later in development. In **deuterostomes** this initial pore becomes the anus, and the mouth forms later in development.

- Protostomes have two blocks of mesoderm beside the gut; deuterostomes have layers of mesodermal cells located on either side of the gut.

The Tube-within-a-Tube Design

- Over 99 percent of animal species alive are bilaterally symmetric triploblasts that have coeloms.

- The outer tube forms the body wall, and the inner tube is the animal's gut. The mesoderm in between forms the muscles and organs.

CHECK YOUR UNDERSTANDING

(a) Describe the two main patterns of organization of embryonic tissues.

_____

_____

Evaluating Molecular Phylogenies

- The phylogenetic tree is based on genes for ribosomal RNA and several proteins (**Figure 32.10**). Several important observations emerge from the data:

  1. A group of protists called the choanoflagellates are the closest living relatives of the animals, and the Porifera are the most basal animal phylum.

  2. Sponges are paraphyletic, meaning they represent several independent lineages.

  3. The radially symmetric diploblastic phyla split off from animals slightly later than sponges.

  4. The most ancient group of bilaterally symmetric triploblasts lacked a coelom entirely.

  5. The major event in the evolution of the Bilateria was the split between protostomes and deuterostomes. Also, within the protostomes, another fundamental split occurred: (1) the Ecdysozoa and (2) the Lophotrochozoa.

  6. Flatworms are acoelomate; but this trait is derived, meaning they evolved from a coelomate ancestor.

  7. Pseudocoeloms arose from coeloms twice in evolutionary history: (1) nematodes and (2) rotifers.

  8. Segmentation evolved independently in annelids and arthropods, as well as in vertebrates.

32.3 What Themes Occur in the Diversification of Animals?

Feeding

How Animals Feed: Four General Tactics

- **Suspension feeders** feed by filtering out food particles suspended in water. Cilia on gills pump water in one siphon and out the other. Food particles are trapped by the gills and swept toward the mouth by cilia.

- Krill can suspension-feed as they swim. Projections on their legs trap food particles as they flow past (**Figure 32.12b**).

- **Deposit feeders** eat their way through a substrate. For example, earthworms swallow soil as they move through it and leave mineral material behind as feces.

- Food for deposit feeders consists of soil-dwelling bacteria, protists, fungi, and archaea, along with **detritus**—the dead and partially decomposed remains of organisms.

- Deposit feeders are similar in appearance. They usually have simple mouthparts and a wormlike body shape (**Figure 32.13**).

- **Fluid feeders** suck or mop up liquids like nectar, plant sap, blood, or fruit juice.

- They often have mouth parts that allow them to pierce a structure so they can withdraw the fluids inside.

- **Mass feeders** take chunks of food into their mouths.

- The structure of the mouthparts correlates with the type of food pieces that are harvested or ingested.

What Animals Eat: Three General Sources

- **Herbivores** are animals that digest algae or plant tissues. **Carnivores** are animals that feed on animals. **Detritivores** are animals that feed on dead organic matter.

- Herbivores and carnivores can be subclassified as predators or parasites.

- **Predators** kill other organisms for food by using an array of mouthparts and hunting strategies.

- **Parasites** are much smaller than their victims and often harvest nutrients without causing death.

- **Endoparasites** live inside their hosts. They are often wormlike in shape and can be extremely simple morphologically. The tapeworms found inside humans have no digestive system; however, most endoparasites ingest their food and have a digestive tract.

CHECK YOUR UNDERSTANDING

(b) What is the difference between radial and bilateral symmetry?

_____

_____

_____

- **Ectoparasites** live outside their hosts. They usually have grasping mouthparts that allow them to pierce their host's exterior and suck the nutrient-rich fluids inside. A louse is an example of an insect ectoparasite (**Figure 32.17b**).

Movement

Types of Limbs: Unjointed and Jointed

- Unjointed limbs (such as in velvet worms) are sac-like.
- Jointed limbs move when muscles that are attached to a skeleton contract or relax.
- Ecdysozoans have an exoskeleton, and vertebrates have an endoskeleton.

Are All Animal Appendages Homologous?

- Biologists have concluded that at least a few of the same genes are involved in the development of *all* appendages observed in animals.
- The current hypothesis is that all animal appendages have some degree of genetic homology, and that they are all derived from appendages that were present in a common ancestor.

Reproduction and Life Cycles

- Animal reproduction is extremely diverse.
- Animals can reproduce sexually or asexually.
- Sexual reproduction can be with internal or external fertilization.
- Eggs (**ovoviviparous**) or embryos (**viviparous**) can be retained in the female's body during development, or eggs may be laid outside the body (**oviparous**).
- **Metamorphosis** is the change from juvenile to adult body type. This change can either be subtle or spectacular.
- **Hemimetabolous** development is a subtle change from metamorphosis; it refers to the limited morphological difference between juvenile and adult. It is a one-step process of sexual maturation. Grasshoppers undergo hemimetabolism as a series of molts gradually change it from a wingless immature juvenile (**nymph**) to a sexually mature adult capable of flight.
- **Homometabolous** development is a two-step process, from larvae to pupae to adult, involving dramatic changes in morphology and habitat use. This process is found in 10 times as many species as the hemimetabolous process.

32.4 Key Lineages of Animals: Basal Groups

Porifera (Sponges)

- All are suspension feeders, and although the adults are sessile, larvae can swim with the aid of flagella.

(c) What are the various types of metamorphosis in insects, and how do they differ from each other?

_____

_____

_____

_____

- Reproduction can be asexual or sexual.
- Sponge cells are totipotent, meaning that an isolated cell has the capacity to develop into a complete adult.

Cnidaria (Jellyfish, Corals, Anemones, Hydroids)

- Cnidarians are radially symmetric diploblasts consisting of ectoderm and endoderm layers that sandwich gelatinous material known as mesoglea.
- The gut has only one opening. A specialized cell, a **cnidocyte**, is used for prey capture.
- Most life cycles have a sessile polyp form and a mobile medusa.
- Polyps can reproduce asexually by budding, fission, or fragmentation. During sexual reproduction, fertilization takes place in open water.

Ctenophora (Comb Jellies)

- Ctenophores are transparent, ciliated, gelatinous diploblasts that live in marine habitats.
- They are predators, and adults move via beating of cilia.
- Most species have male and female organs and routinely self-fertilize, though fertilization is external.

Acoelomorpha

- These animals lack a coelom. They are bilaterally symmetric worms that have distinct anterior and posterior ends and are triploblastic.
- They feed on detritus and prey on small animals.
- They can reproduce asexually by fission or budding. Sexual reproduction is also possible; fertilization is internal, and fertilized eggs are laid outside the body.

CHECK YOUR UNDERSTANDING

(d) Organize the basal lineages of animals into their tissue types and symmetry types.

_____

_____

_____

_____

B. INTEGRATING YOUR KNOWLEDGE

Looking Back—Key Concepts
from Earlier Chapters

Embryonic Tissue Layers—Chapter 22

The basic layers of embryonic tissues and their different paths of development are explained in **Chapter 22** and utilized in Chapter 32 to characterize the evolution of animals.

Homologous Structures—Chapter 24

These structures have been traced to a common ancestor and used to identify points of origin on an evolutionary tree.

Natural Selection—Chapter 24

Natural selection is explained in detail in **Chapter 24** and is applied to the many evolutionary processes of the animal kingdom in Chapter 32. The prevalence of this term in this section should indicate its importance.

Looking Forward—Key Concepts
in Later Chapters

Mechanisms of Movement—Chapter 41

Movement is used to explain the diversity of feeding strategies used in animals. **Chapter 41** focuses on the mechanisms of animal movement in detail.

Reproduction—Chapter 48

Chapter 32 begins to describe many reproductive strategies of animals. **Chapter 48** goes into further detail and focuses on vertebrate reproduction.

C. MASTERING CHALLENGING TOPICS

In this chapter, the terminology referring to body tissue layers and body cavity organizations can be confusing, especially when attempting to match these classifications with different animal groups. If these terms are not mastered during the study of this chapter, it will make the next two chapters on animals even more confusing. Try to study the tissue layers first (endoderm, ectoderm, and mesoderm); next, list the animal groups that are diploblastic and triploblastic. Then define coelom, acoelomate, and pseudocoelom and try to list animal groups that match these terms. Finally, try to combine both sets and groups to get an overall summary, perhaps in a chart or other diagram. Explaining the trend to a classmate will help you organize your thoughts and prepare you for incorporating these terms into the next two chapters.

D. ASSESSING WHAT YOU'VE LEARNED

1. Diploblasts and triploblasts are terms that describe the number of:
   a. invaginations during development
   b. head regions during development
   c. tissue layers during development
   d. cell types during development

2. In triploblasts, the body cavity or coelom forms:
   a. from within the mesoderm
   b. between the endoderm and mesoderm
   c. between the ectoderm and mesoderm
   d. from within the endoderm

3. During gastrulation, the initial pore eventually forms the mouth, this a developmental process that defines the lineage as:
   a. the coelomates
   b. the deuterostomes
   c. the protostomes
   d. the gastrosomes

4. What is the earliest-branching lineage of animals?
   a. the insects
   b. the sponges
   c. the echinoderms
   d. the chordates

5. Animals that eat and digest algae or plant materials are called:
   a. herbivores
   b. carnivores
   c. detritivores
   d. parasites

6. Which parasites live inside their hosts?
   a. endoparasites
   b. ectoparasites
   c. mesoparasites
   d. none of the above

7. The current hypothesis regarding animal limbs is that _____.
   a. animal limbs are not homologous structures
   b. all animal limbs have some degree of genetic homology
   c. animal limbs all perform the same function
   d. animal limbs all arose at the same time

8. Which specific term describes the change from larvae to pupa to adult?
   a. hemimetabolous
   b. metabolism
   c. heterometabolous
   d. homometabolous

9. A cell such as a sponge cell can be isolated and form a complete adult. The term used to describe this cell is:
   a. omnipotent
   b. totipotent
   c. heteropotent
   d. allipotent

10. The motile form of the Cnidaria is called the.
    a. medusa
    b. polyp
    c. cnidocyte
    d. mesoglea

CHAPTER 32—ANSWER KEY

Check Your Understanding

Diploblasts are animals whose embryos have two types of tissues. Only two groups of diploblastic animals are alive today: the cnidarians and ctenophores. Triploblasts are animals whose embryos have three types of tissues. In triploblasts, the ectoderm gives rise to the skin and nervous system, the endoderm gives rise to the digestive tract, and the mesoderm gives rise to the circulatory system, muscle, and internal structures.

(a) A basic feature of a multicellular body is whether it has a plane of symmetry. All animals exhibit either radial or bilateral symmetry. Animals with radial symmetry have at least two planes of symmetry. Bilaterally symmetric organisms face their environment in one direction.

(b) A basic feature of a multicellular body is whether it has a plane of symmetry. All animals exhibit either radial or bilateral symmetry. Animals with radial symmetry have at least two planes of symmetry. Bilaterally symmetric organisms face their environment in one direction.

(c) Metamorphosis is the change from juvenile to adult body type. This change can either be subtle or spectacular. Hemimetabolous development is a subtle change from metamorphosis; its name refers to the limited morphological difference between juvenile and adult. It is a one-step process of sexual maturation. Grasshoppers undergo this process as a series of molts gradually change it from a wingless, immature juvenile to a sexually mature adult capable of flight. Homometabolous development is a two-step process, from larvae to pupae to adult, involving dramatic changes in morphology and habitat use. This is found in 10 times as many species as the hemimetabolous process.

(d) The Porifera are mostly asymmetrical and have only epithelial tissues; therefore, they lack the tissue organization of other animals. The Cnidaria are radially symmetric diploblasts. The Ctenophora are bilaterally symmetric diploblasts. The Acoelomorpha are bilaterally symmetric triploblasts.

Assessing What You've Learned
1. c; 2. a; 3. c; 4. b; 5. a; 6. a; 7. b; 8. d; 9. b; 10. a

33

Protostome Animals

A. KEY CONCEPTS

33.1 An Overview of Protostome Evolution

What Is a Lophotrochozoan?

- The 14 phyla of lophotrochozoans include the mollusks, annelids, and flatworms.

- The name came from two structures found in some but not all of the phyla in the lineage.

- A **lophophore** is a specialized structure that rings the mouth and specializes in suspension feeding. They consist of tentacles that have ciliated cells on their surface.

- **Trochophores** are a type of larvae common to marine mollusks, annelids that live in the ocean, and several other oceanic phyla. The larvae have a ring of cilia around their middle that aids in locomotion.

What Is an Ecdysozoan?

- Ecdysozoans have an exoskeleton and grow by molting.

- The most prominent phyla in this group are the roundworms and the arthropods.

CHECK YOUR UNDERSTANDING

(a) Compare and contrast lophotrochozoans with ecdysozoans.

_____

_____

_____

_____

33.2 Themes in the Diversification of Protostomes

How Do Body Plans Vary among Phyla?

- All protostomes are triploblastic and bilaterally symmetric, and all of them undergo embryonic development in a similar way.

- Radical changes occurred in coelom formation as protostomes diversified.

The Arthropod Body Plan

- Arthropods have segmented bodies that are organized into prominent regions called **tagma**.

- Arthropod locomotion is based on muscles that apply force against an exoskeleton.

- Arthropods have a spacious body cavity called the **hemocoel** that provides space for internal organs and circulation of fluids.

The Molluscan Body Plan

- The **foot** is a large muscle that is located at the base of the animal and is usually used in movement.

- The **visceral mass** is the region that contains the main internal organs and external gill.

- The **mantle** is a tissue layer that covers the visceral mass and that in many species secretes a shell made of calcium carbonate.

Variation Among Body Plans of the Wormlike Phyla

- The body plan is extremely similar in these phyla, and they are distinguished by specialized mouthparts.

- **Echiurans** (spoon worms) burrow into marine mud and feed using an extended structure called a **proboscis**.

- **Priapulids** (penis worms) burrow in substrate but act as sit-and-wait predators. This worm everts its toothed, cuticle-lined throat to grab prey.

- **Nemerteans** (ribbon worms) are active predators; they have a spear-like proboscis that can extend or retract.

The Water-to-Land Transition

- It is clear that water-to-land transitions occurred multiple times as protostomes diversified.

- It was much easier for protostomes to move from water to land than it was for plants.

- They already had hydrostatic skeletons, exoskeletons, appendages, or other adaptations for support and locomotion.

How Do Protostomes Feed, Move, and Reproduce?

Adaptations for Feeding

- A wide diversity of feeding strategies are reflected in the wide diversity of mouthparts.

- Some protostomes have jaws and mouth; others have a proboscis. Arthropods show the most diversity—they can pierce, suck, grind, bite, mop, chew, engulf, cut, and mash (**Figure 33.9**).

Adaptations for Moving

- Variation in movement depends on (1) the presence or absence of limbs and (2) the type of skeleton that is present.

- The evolution of the **jointed limb** allowed movement in these unique ways, along with others:

  1. The insect wing is one of the most important adaptations in the history of life.

  2. In mollusks, waves of muscle contractions sweep down the length of the large, muscular foot, allowing individuals to glide.

  3. Squid fill a cavity surrounded by their mantle and expel water through a siphon to propel them through the water.

Adaptations in Reproduction

- Regarding variation in protostome life cycles, the most basic issue is whether a juvenile's body is rearranged during metamorphosis to form an adult.

- Reproduction is highly variable, from sexual to asexual. Many crustacean species can reproduce asexually via **parthenogenesis**.

CHECK YOUR UNDERSTANDING

(b) Describe a few unique ways of protostome movement.

_____

_____

_____

33.3 Key Lineages: Lophotrochozoans

Rotifera (Rotifers)

- Rotifers have a cluster of cilia at their anterior called a **corona**, which they use for suspension feeding.

- Females produce unfertilized eggs by mitosis; the eggs then hatch into new, asexually produced individuals (parthenogenesis).

Platyhelminthes (Flatworms)

- This phylum includes (1) the free-living flatworms, (2) the endoparasitic tapeworms, and (3) the endo- or ectoparasitic flukes.

- Flatworms are unsegmented, do not have a coelom, and have no circulatory system. Their high surface-to-volume ratio allows for efficient absorption of nutrients from their host.

- In many cases, the life cycles of tremadotes and cestodes involve 2 or 3 distinct host species.

Annelida (Segmented Worms)

- Annelids have a segmented body plan and a coelom that functions as a hydrostatic skeleton.

- They are divided into two major subgroups:

  1. The Polychaeta are named for their numerous, bristle-like extensions called chaetae, which extend from appendages called parapodia.

  2. The Clitellata are comprised of the oligochaetes (which include earthworms) and leeches.

- Annelids can crawl, burrow, or swim using their hydrostatic skeletons.

Mollusca (Mollusks)

- This is by far the most species-rich and morphologically diverse group in the Lophotrochozoa.

- They have a specialized body plan based on a muscular foot, a visceral mass, and a mantle that may or may not secrete a calcium carbonate shell.

Bivalvia (Clams, Mussels, Scallops, Oysters)

- Bivalves have two separate shells, which are hinged.

- They are suspension feeders and use gills for respiration.

- Most are sessile, but some can move using their muscular foot.

- Only sexual reproduction occurs. Externally fertilized eggs develop into trochophore larvae. They metamorphose into a **veliger**, which continues to feed before metamorphosing into the adult form.

Gastropoda (Snails, Slugs, Nudibranchs)

- Gastropods are named for the large, muscular foot on their ventral side.

- In many species the **radula** functions like a rasp to scrape away algae.

- Some species can reproduce via parthenogenesis, but most reproduction is asexual.

Polyplacophora (Chitons)

- These animals are named for the eight calcium carbonate plates along their dorsal side.

- They are marine and feed using a radula. They move using a muscular foot, just as in the gastropods.

- Sexes are separate and fertilization is external.

Cephalopoda (Nautilus, Cuttlefish, Squid, Octopuses)

- Cephalopods have a well-developed head and a foot that is modified to form long, muscular tentacles.

- Except for the nautilus, most have highly reduced shells or none at all.

- They are highly intelligent predators that hunt by sight. They have a radula and a beak for biting.

- They have separate sexes and highly elaborate courtship rituals. Males deposit sperm encased in a spermatophore.

CHECK YOUR UNDERSTANDING

(c) Describe one unique structure within the group Mollusca.

_____

_____

33.4 Key Lineages: Ecdysozoans

Nematoda (Roundworms)

- Nematodes are unsegmented worms with a pseudocoelom.

- Several species are common parasites of humans. Pinworms infect about 40 million people in the United States.

- Sexes are separate, and asexual reproduction is rare or unknown. Internal fertilization is used with egg laying and direct development.

Arthropoda (Arthropods)

- Arthropods are easily the most successful lineage of eukaryotes.

- The body is organized into a distinct head and trunk. A **compound eye** contains many light-sensing structures, each with its own lens.

- Most arthropods have a pair of antennae on their head.

Myriapods (Millipedes, Centipedes)

- Myriapods have relatively simple bodies with a series of short segments.

- They have mouthparts that can bite and chew. Millipedes are detritivores, and centipedes are predators.

- Sexes are separate, and fertilization is internal. Males deposit sperm packets that are picked up by the female.

Chelicerata (Spiders, Ticks, Mites, Horseshoe Crabs, Daddy Longlegs, Scorpions)

- This group is named for appendages called **chelicerae** found near the mouth.

- The most prominent group is the arachnids (spiders, scorpions, mites, and ticks).

- Spiders, scorpions, and daddy longlegs capture and sting their prey. Mites and ticks are ectoparasitic.

- Sexual reproduction via internal fertilization and direct development is the most common in this group.

Insecta (Insects)

- Insects are distinguished by having a **head**, **thorax**, and **abdomen**. In most species, two pairs of wings are mounted on the dorsal side of the thorax.

- Mating usually takes place through direct copulation.

Crustaceans (Shrimp, Lobster, Crabs, Barnacles, Isopods, Copepods)

- The segmented body of the crustacean is divided into the cephalothorax (which combines the head and thorax) and the abdomen.

- Many crustaceans have a **carapace**—a platelike section of their exoskeleton that covers and protects the cephalothorax.

- Crabs and lobsters have **mandibles** that can bite or chew.

- Sexual reproduction via internal fertilization is the norm.

- Many species include a larval stage called a **nauplius**, which is usually planktonic.

CHECK YOUR UNDERSTANDING

(d) Compare and contrast the body plan of insects and crustaceans.

_____

_____

_____

_____

_____

B. INTEGRATING YOUR KNOWLEDGE

*Looking Back—Key Concepts
from Earlier Chapters*

Genetic Defects during Early Development—Chapter 22

The use of flies in developmental studies has shown that flies with different genetic defects in early development could show how various gene products influence the development of eukaryotes.

Genomics—Chapter 20

The entire genome of *Caenorhabditis elegans*, as well as that of *Drosophila melanogaster*, has now been sequenced.

Phylogenies—Chapter 27

Recall from **Chapter 27** that careful descriptions of events in embryonic development of the anatomy of juveniles (often larvae) and adults, analyses of the fossil record, and phylogenies based on comparisons of DNA sequence data allowed researchers to make conclusions about the evolutionary history of protostomes.

*Looking Forward—Key Concepts
in Later Chapters*

Deuterostome Development—Chapter 34

The protostomes develop in a dramatically different way than the other major lineage of bilaterians, the deuterostomes, do.

Mechanisms of Movement—Chapter 41

Movement is used to explain the diversity of feeding strategies used in animals. **Chapter 41** focuses on the mechanisms of animal movement in detail.

Gas Exchange—Chapter 44

In the transition to land, animals had to solve the problem of exchanging gas and prevent drying out. As **Chapter 44** will detail, land animals exchange gases with the atmosphere readily as long as a moist body surface is exposed to air.

C. MASTERING CHALLENGING TOPICS

The most difficult topic in this chapter is the organization of all the key lineages of protostomes. The sheer number of phyla and their characteristics present a large body of information that can be incredibly confusing to remember and recall on a test. Although strict memorization can get the job done, it can often yield less than desirable results on certain types of tests. The preferred way to learn this material is to take the time to outline

the major divisions of the lineages first, starting with what makes a protostome a protostome. Then, distinguish the Lophotrochozoans from the Ecdysozoans, even if you can name only a few things that are different. Lastly, begin to fill in the key phyla under each group, starting with the easiest to remember. Looking at pictures and explaining characteristics to classmates will also help you in learning all of these key lineages.

D. ASSESSING WHAT YOU'VE LEARNED

1. Rotifers have a cluster of cilia at their anterior called:
   a. a pore
   b. a radula
   c. a corona
   d. a lophophore

2. The group Lophotrochozoans is so named because:
   a. They all have a feeding structure called a lophophore.
   b. They all have a type of larvae called a trochophore.
   c. They are all wormlike.
   d. Answers a and b are both correct.

3. Platyhelminthes include:
   a. free-living flatworms
   b. endoparasitic tapeworms
   c. endoparasitic and ectoparasitic flukes
   d. all of the above

4. The Annelida are divided into two major subgroups called:
   a. Polyplacophora and Monoplacophora
   b. Polychaeta and Polyplacophora
   c. Polychaeta and Clitellata
   d. Polychaeta and Mollusca

5. The most species-rich and morphologically diverse group of Lophotrochozoa are:
   a. Mollusca
   b. Annelida
   c. Nematoda
   d. Polyplacophora

6. Which of the following is *not* a Bivalvia?
   a. clam
   b. mussel
   c. snail
   d. scallop

7. The most intelligent of the mollusks are the:
   a. Bivalvia
   b. Gastropoda
   c. Cephalopoda
   d. Polyplacophora

8. Which of the following are unsegmented worms with a pseudocoelom?
   a. nematodes
   b. annelids
   c. cephalopods
   d. platyhelminthes

9. Spiders belong to the major group called:
   a. Myriapods
   b. Chelicerata
   c. Insecta
   d. Crustaceans

10. The anterior region of the Crustaceans is called the:
    a. thorax
    b. head
    c. cephalum
    d. cephalothorax

CHAPTER 33—ANSWER KEY

Check Your Understanding

(a) The two monophyletic groups of protostomes are the Lophotrochozoa and the Ecdysozoa. Lophotrochozoa are characterized by two structures found in some but not all of the phyla in the lineage. A **lophophore** is a specialized structure, consisting of tentacles with ciliated cells on their surface, that rings the mouth and specializes in suspension feeding. **Trochophores** are a type of larvae common to marine mollusks, annelids that live in the ocean, and several other oceanic phyla. The larvae have a ring of cilia around their middle that aids in locomotion. In contrast, ecdysozoans have an exoskeleton and grow by molting.

(b) Difference in body movement mechanisms depends on (1) the presence or absence of limbs and (2) the type of skeleton that is present. The evolution of the joined limb allowed movement in these unique ways, along with others:
   • The insect wing is one of the most important adaptations in the history of life.
   • In mollusks, waves of muscle contractions sweep down the length of the large, muscular foot, allowing individuals to glide.
   • Squid fill a cavity surrounded by their mantle and expel water through a siphon to propel themselves through the water.

(c) Bivalves have two separate shells and metamorphose into a veliger stage. Gastropods, polyplacophorans, and cephalopods all have a radula for feeding.

(d) Insects and crustaceans both have an exoskeleton. Insects have a head, thorax, and abdomen while crustaceans have a fused cephalothorax and an abdomen.

Assessing What You've Learned

1. c; 2. d; 3. d; 4. c; 5. a; 6. c; 7. c; 8. a; 9. b; 10. d

34

Deuterostome Animals

A. KEY CONCEPTS

34.1 An Overview of Deuterostome Evolution

What Is an Echinoderm?

- Although all deuterostomes are bilaterally symmetric, echinoderms reverted to a type of radial symmetry called pentaradial symmetry. Recall that echinoderm larvae are bilaterally symmetric.

- Another unique feature of the enchinoderms is their **water vascular system** and associated **tube feet** and **podia**.

- Echinoderms also have an **endoskeleton**.

- Two other phyla form a monophyletic group with echinoderms—Xenoturbellida ("strange-flatworms") and Hemichordata ("half-chordates").

What Is a Chordate?

- This phylum is defined by the presence of these traits:

  1. Pharyngeal gill slits

  2. A stiff and supportive but flexible rod called a **notochord**

  3. A bundle of nerve cells that run the length of the body and form a **dorsal hollow nerve cord**

  4. A muscular, **post-anal tail**

- This phylum is also subdivided into three major lineages:

  1. **Urochordates** are also called tunicates. The pharyngeal gill slits function in suspension feeding, while the notochord functions as a simple endoskeleton.

  2. **Cephalochordates** are also called lancelets or amphioxus. They are small, mobile suspension feeders that look like a little fish. Their notocords stiffen their bodies to aid in swimming.

  3. **Vertebrates** include sharks, several lineages of fishes, amphibians, reptiles (including birds), and mammals. The dorsal hollow nerve cord is elaborated into the spinal cord. The notochord helps organize the body plan during embryonic development.

What Is a Vertebrate?

- Vertebrates are a monophyletic group distinguished by two traits: **vertebrae** and a **cranium**.

- The vertebrae protect the spinal column, and the cranium protects the brain and sensory organs.

CHECK YOUR UNDERSTANDING

(a) What are the major similarities and differences between chordates and vertebrates?

_____

_____

_____

_____

34.2 An Overview of Vertebrate Evolution

The Vertebrate Fossil Record

- About 530 million years ago, the earliest vertebrates had streamlined bodies and appeared to have skulls made of **cartilage**.

- In the Ordovician period (480 mya), the first fossils to contain **bone** appear. They looked like plates (exoskeleton) and were presumably used as armor.

- First vertebrates with **jaws** appear about 430 mya. This gave vertebrates the ability to bite and chew.

- The transition to land is dated at about 357 mya. These were the first **tetrapods**—animals with four limbs.

- About 20 mya, the first amniotes were present. An **amniotic egg** is one with a watertight shell or case enclosing a membrane-bound food supply, a water supply, and a waste repository.

Evaluating Molecular Phylogenies

- Figure 34.7 provides a phylogenetic tree based on morphology and DNA.
- The closest living relatives of the vertebrates are the cephalochordates.
- The most basal groups of chordates lack the skull and vertebral column that define the vertebrates, and the most ancient lineage of vertebrates lack jaws and bony skeletons.

CHECK YOUR UNDERSTANDING

(b) What did bone first look like in the fossil record of vertebrates?

_____

_____

_____

34.3 What Themes Occur in the Diversification of Deuterostomes?

Feeding

How Do Echinoderms Feed?

- Echinoderms suspension feed, deposit feed, or harvest algae or other animals. In most cases, the podia play a key role in obtaining food.

The Vertebrate Jaw

- The leading hypothesis is that natural selection acted on the gill arches by increasing its size and modifying its orientation.
- Three lines of evidence:
  1. Both gill arches and jaws consist of flattened bars of bony or cartilaginous tissue that hinges and bends forward.
  2. During development, the same population of cells gives rise to the muscles that move the jaws and the muscles that move gill arches.
  3. Both jaws and gill arches derive from specialized cells in the embryo—neural crest cells.

Movement

- The hypothesis that tetrapod limbs evolved from fish fins is supported by fossil evidence as well as molecular genetic evidence.
- Once the limb evolved, natural selection elaborated it into structures used for running, gliding, crawling, burrowing, or swimming.

Reproduction

What Is an Amniotic Egg?

- These eggs have shells that minimize water loss because the embryo develops inside.
- The egg contains a membrane-bound supply of water in a protein-rich solution called **albumen** (**Figure 34.16**). The embryo is enveloped in a membrane called the **amnion**. The **yolk sac** is the membranous pouch that contains the nutrients, and the **allantois** is the pouch that holds waste material. The **chorion** is the membrane that provides the surface for gas exchange.

The Placenta

- The mother retains the egg in her body. The placenta is an organ that is rich in blood vessels, and it facilitates the flow of oxygen and nutrients from mother to offspring (**Figure 34.17**).

Parental Care

- Parental care is any action by a parent that improves the ability of its offspring to survive—feeding, keeping young and/or eggs warm, and protection.
- The most extensive parental care seen is by mammals and birds.

CHECK YOUR UNDERSTANDING

(c) Describe the main characteristics of the amniotic egg.

_____

_____

_____

34.4 Key Lineages: Echinodermata

Asteroidea (Sea Stars)

- They have star-shaped bodies with five or more long arms radiating from a central region that contains the mouth, stomach, and anus (**Figure 34.20**).
- Sea stars are predators or scavengers. Some species can pull apart bivalves.

Echinoidea (Sea Urchins and Sand Dollars)

- Sea urchins have globular bodies and long spines, and they crawl along substrates (**Figure 34.21a**). Most are herbivores of algae.
- Sand dollars are flattened and disk-shaped, have short spines, and burrow in soft sediments (**Figure 34.21b**). They use mucus-covered podia to collect food particles in the sand.

(d) Describe the structure and function of the water vascular system in echinoderms.

_____

_____

_____

_____

34.5 Key Lineages: Chordata

Urochordata (Tunicates)

- Tunicates have an exoskeleton-like coat of polysaccharide, called a tunic, a U-shaped gut, and two openings called siphons.
- They use their pharyngeal gill slits to suspension feed.

Cephalochordata (Lancelets)

- The notochord is retained in adult lancelets, where it functions as an endoskeleton.
- They take water in through their mouth and trap food particles in their pharyngeal gill slits.

Craniata > Myxinoidea (Hagfish) and Petromyzontoidea (Lampreys)

- Hagfish and lampreys belong to independent lineages, although both groups lack jaws. Hagfish lack any sort of vertebral column, but lampreys have small pieces of cartilage along their length.
- Hagfish are scavengers and predators (**Figure 34.25a**) that deposit feed on the carcasses of dead fish and whales.
- Lampreys are ectoparasites—they attach to fish and suck blood and other body fluids. They are also anadromous—they spend their adult life in the ocean and swim up streams to breed.

Vertebrata > Condrichthyes (Sharks, Rays, Skates)

- These species are distinguished by their cartilaginous skeleton. Most species are marine, and sharks are top predators of marine ecosystems.

Vertebrata > Actinopterygii (Ray-Finned Fishes)

- These fish have fins supported by long, bony rods arranged in a ray pattern. They are the most ancient group of vertebrates that have a skeleton made of bone.
- They avoid sinking in the water with the aid of a gas-filled **swim bladder**.

- The most important subgroup of ray-finned fishes is the Teleosti, with 20,000 species.

Vertebrata > Sarcopterygii: Actinistia (Coelocanths) and Dipnoi (Lungfish)

- Although these are independent lineages, they are often grouped together and called **lobe-finned fishes**.
- They represent a crucial evolutionary link between the ray-finned fishes and the tetrapods.

Vertebrata > Amphibia (Frogs, Salamanders, Caecilians)

- Adults of most species feed on land, but many species have to lay their eggs in water.
- The larvae undergo metamorphosis into the adult form, which is mostly carnivorous.

Vertebrata: Mammalia (Mammals)

- Mammals have hair or fur and mammary glands for lactation.

Monotremata (Platypuses, Echidnas)

- Most ancient group of mammals living. They lay eggs and have lower metabolic rates than other mammals.

Marsupiala (Marsupials)

- Although females have a placenta, the young are born poorly developed. They crawl to the female's nipples, where they suck milk. They stay there until they become independent.

Eutheria (Placental Mammals)

- The six (of the 18) most species-rich orders are rodents, bats, insectivores, artiodactyls, carnivores, and primates.
- At birth, the young are much better developed than marsupials, although there is still a prolonged period of parental care.

Reptilia > Turtles, Snakes and Lizards, Crocodiles, Bird)

Testudinia (Turtles)

- Distinguished by a shell comprised of bony plates that fuse to the vertebrae and ribs.
- Their skulls are more similar to amphibians than to other reptiles. Although they lack teeth, their jawbone and lower skull form a bony beak.

Lepidosauria (Lizards, Snakes)

- Most lizards have well-developed jointed legs, but snakes are limbless, evolving from limbed ancestors.
- Most lay eggs, although a few give birth to live young that have hatched from the egg and begun development.

Crocodilia (Crocodiles, Alligators)

- Only 21 species known. They have eyes and nostrils located on the top of their skulls and can sit underwater for extended periods.
- They are top predators and, although they are oviparous, parental care is extensive.

Aves (Birds)

- They are the only **endotherms** within this group. This means their metabolic rate and insulation rate are so high, they retain a high constant body temperature.
- Most birds are omnivores, and almost all species can fly.
- Birds are oviparous, but parental care is extensive.

CHECK YOUR UNDERSTANDING

(e) What distinguishes the three orders of mammals?

_____

_____

_____

_____

34.6 Key Lineages: The Hominin Radiation

The Primates

- The **prosimians** are composed of lemurs, tarsiers, pottos, and lorises. Most of these animals live in trees and are active at night.
- The **anthropoids** include the New World monkeys found in Central and South Americas, the Old World monkeys that live in Africa and tropical regions of Asia, and the great apes—orangutans, gorillas, chimpanzees, and humans.
- Primates are distinguished by having their eyes on the front of their face.
- The lineage of the great apes is called the **hominids**. Humans are the only great ape that is bipedal.

Fossil Humans

- The fossil record of humanlike hominids is not nearly complete, but it is rapidly improving. These species can be organized into four general groups (**Figure 34.41a**):
  1. Three species of small apes called **gracile australopithecines** lived from 4.12 to 2.4 million years ago and were bipedal.

2. Three distinct species of **robust australopithecines** lived from 2.7 to 1.0 million years ago and were also bipedal. The robust forms had more massive cheek teeth and jaws, very large cheekbones, and a sagittal crest.
3. The earliest species in the genus *Homo* are called **humans** and date from 2.4 to 1.5 million years ago. *Homo* species have flatter, narrower faces and smaller jaws, and their braincases are as much as three times larger than those of australopithecines.
4. More recent species of *Homo* date from 1.2 million years ago to the present. This group includes **Cro-Magnons** and **Neanderthals**.

- Humans did not evolve through a simple, steady progression from a chimpanzee-like ancestor. Instead, a complex radiation of bipedal hominids occurred in Africa during the past 4–5 million years.

The Out-of-Africa Hypothesis

- Because the first lineage to branch off leads to descendent populations that live in Africa today, it is logical to infer that the ancestral population also lived in Africa (**Figure 34.43**).
- The lineages branched off to form three monophyletic groups. These lineages are thought to descend from three major waves of migration that occurred as *Homo sapiens* population dispersed from east Africa to:
  1. north Africa, Europe, and central Asia
  2. northeast Asia and the Americas, and
  3. southeast Asia and the South Pacific.
- This hypothesis also contends that *Homo sapiens*, Neanderthals, *H. erectus*, and *H. floriensis* never interbred.

CHECK YOUR UNDERSTANDING

(f) Explain the Out-of-Africa hypothesis.

_____

_____

_____

_____

_____

B.   INTEGRATING YOUR KNOWLEDGE

*Looking Back—Key Concepts
from Earlier Chapters*

Protostome Development—Chapter 33

The deuterostomes develop in a dramatically different way than the other major lineage of bilaterians, the protostomes, do.

Genetic Defects during Early Development—Chapter 22

The use of zebra fish in developmental studies has shown that zebra fish with different genetic defects in early development could show how various gene products influence the development of eukaryotes.

Genomics—Chapter 20

The entire genome of zebra fish, as well as that of humans, has now been sequenced.

Phylogenies—Chapter 27

Recall from **Chapter 27** that careful descriptions of events in embryonic development of the anatomy of juveniles (often larvae) and adults, analyses of the fossil record, and phylogenies based on comparisons of DNA sequence data allowed researchers to make conclusions about the evolutionary history of deuterostomes.

*Looking Forward—Key Concepts
in Later Chapters*

Mechanisms of Movement—Chapter 41

Movement is used to explain the diversity of feeding strategies used in animals. **Chapter 41** focuses on the mechanisms of animal movement in detail.

Gas Exchange—Chapter 44

In the transition to land, animals had to solve the problem of exchanging gas and prevent drying out. As **Chapter 44** will detail, land animals exchange gases with the atmosphere readily as long as a moist body surface is exposed to air.

C.   MASTERING CHALLENGING TOPICS

The particulars of hominid evolution are confusing since they consist of hypotheses that are constantly being tested and updated. This is the inherent problem in studying the particulars of a branch of science that is unfolding as textbooks are being printed and classes are being taught. The best place to start is **Figure 34.41** and then move to **Figure 34.42**. Draw the lineages presented in this figure, and try to work out the timeline of presence and extinction of the different

hominid species. Next, work through the hypotheses presented in this chapter and the experimental evidence for each.

D.   ASSESSING WHAT YOU'VE LEARNED

1. The sections of echinoderm tube feet that project outside the body are called:
   a. podia
   b. pseudopods
   c. postiums
   d. endoskeleton

2. Which of the following statements regarding the Chordates is *false*?
   a. Chordates have pharyngeal gill slits.
   b. Chordates do not have a notochord.
   c. Chordates have a dorsal hollow nerve cord.
   d. Chordates have a muscular tail.

3. The leading hypothesis regarding the evolution of the vertebrate jaw is that it arose from natural selection acting on the _____.
   a. gills
   b. gill arches
   c. opercula
   d. fins

4. The hypothesis that tetrapod limbs evolved from fish fins is:
   a. rubbish
   b. only slightly supported
   c. supported only by fossil evidence
   d. supported by fossil and molecular genetic evidence

5. The organ rich in blood vessels and facilitating the flow of oxygen and nutrients from the mother to offspring is called:
   a. the allantois
   b. the yolk
   c. the placebo
   d. the placenta

6. The key difference between lampreys and sharks is:
   a. Lampreys lack jaws.
   b. Lampreys lack cartilage.
   c. Lampreys do not have tails.
   d. Lampreys cannot swim.

7. The key difference between Condrichthyes and Actinopterygii is:
   a. Actinopterygii have fins.
   b. Actinopterygii have jaws.
   c. Actinopterygii have true bone.
   d. Actinopterygii have no tail.

8. Endothermy is found among the Reptilia only within the group:
   a. Testudinia
   b. Lepidosauria
   c. Crocodilia
   d. Aves

9. The only mammalian group to lay eggs are the _____.
   a. Monotremata
   b. Marsupiala
   c. Eutheria
   d. Primates

10. The prosimians include all of the following *except*:
    a. lemurs
    b. gibbons
    c. tarsiers
    d. lorises

CHAPTER 34—ANSWER KEY

Check Your Understanding

(a) Chordates are a phylum of animals defined by the following characteristics:
   1. pharyngeal gill slits
   2. a stiff and supportive but flexible rod called a notochord
   3. a bundle of nerve cells that run the length of the body and form a dorsal hollow nerve cord, and
   4. a muscular post-anal tail.

   Vertebrates, however, are a subphylum within the chordates. Vertebrates include sharks, several lineages of fishes, amphibians, reptiles (including birds), and mammals. The dorsal hollow nerve cord is elaborated into the spinal cord. The notochord helps organize the body plan during embryonic development. Vertebrates are a monophyletic group distinguished by two traits: vertebrae and a cranium.

(b) Bone first appeared as an exoskeleton rather than an endoskeleton. They looked like plates and presumably looked like armor. Animals in this lineage were the size and shape of large fish.

(c) Amniotic eggs have shells that minimize water loss as the embryo develops inside. The egg contains a membrane-bound supply of water in a protein-rich solution called albumin. The embryo is enveloped in a membrane called the amnion. The yolk sac is the membranous pouch that contains the nutrients, and the allantois is the pouch that holds waste material. The chorion is the membrane that provides the surface for gas exchange.

(d) The echinoderm body contains a series of fluid-filled tubes and chambers called the water vascular system. Tube feet are elongated, fluid-filled structures; podia are sections of the tube feet that project outside the body. Echinoderms have an endoskeleton—a hard supportive structure inside the body.

(e) The three orders of mammals are the monotremes, the marsupials, and the eutherians. All of these orders have fur and mammary glands. The monotremes lay eggs and have lower metabolic rates than the other two orders. The marsupials give birth to a very underdeveloped offspring, which continues to develop in a pouch. The eutherians use a placenta to exchange nutrients and waste between mother and embryo.

(f) Because the first lineage to branch off leads to descendent populations that live in Africa today, it is logical to infer that the ancestral population also lived in Africa. The lineages branched off to form three monophyletic groups. These lineages are thought to descend from three major waves of migration that occurred as *Homo sapiens* population dispersed from east Africa to:

   • north Africa, Europe, and central Asia
   • northeast Asia and the Americas, and
   • southeast Asia and the South Pacific.

   This hypothesis also contends that *Homo sapiens*, Neanderthals, *H. erectus*, and *H. floriensis* never interbred.

Assessing What You've Learned

1. a; 2. b; 3. b; 4. d; 5. d; 6. a; 7. c; 8. d; 9. a; 10. b

35

Viruses

A. KEY CONCEPTS

35.1 Why Do Biologists Study Viruses?

Recent Viral Epidemics in Humans

- Viruses have caused the most devastating **epidemics** in recent human history. The Spanish flu of 1918–1919 was the most devastating epidemic—20–50 million people died.

Current Viral Epidemics in Humans: HIV

- Human immunodeficiency virus will surpass influenza as the most deadly of diseases. This virus is the cause of **acquired immune deficiency syndrome (AIDS)**.

How Does HIV Cause Disease?

- HIV parasitizes immune cells called T cells. The number produced by the body does not keep pace with the number of T cells destroyed by HIV.
- When the T cell drops, the body becomes less able to fight incoming pathogens.

What Is the Scope of the AIDS Epidemic?

- AIDS has already killed 28 million people, and it is estimated that 39.5 million people worldwide are infected with HIV.

CHECK YOUR UNDERSTANDING

(a) Are viruses living organisms? Justify your answer.

_____

_____

_____

_____

35.2 How Do Biologists Study Viruses?

Analyzing Morphological Traits

- Viruses can either be (1) enclosed by just a shell of protein called a **capsid** or (2) enclosed by both a capsid and a membrane-like **envelope**.
- An important distinction is therefore whether they are **enveloped** or **nonenveloped**.
- A vaccine is a preparation that primes a host's immune system to respond to a specific type of virus.

Analyzing Variation in Growth Cycles: Replicative and Latent Growth

- Viruses all infect their host cells in one of two ways: via replicative growth or in a dormant form referred to as latency in animal viruses or lysogeny in bacteriophages.
- All viruses undergo replicative growth, but some can halt this cycle.
- A mature virus particle is called a **virion**; the life cycle is complete when the virions exit the cell, usually killing the host cell in the process.
- **Figure 35.10** diagrams a **lysogenic replication cycle**, or **lysogeny**, in a bacteriophage.
- It is not possible to treat a latent infection with **antivirals**, because the viruses are quiescent.

Analyzing the Phases of the Replicative Cycle

- Six phases are common to replicative growth in virtually all viruses: (1) entry into a host cell, (2) transcription of the viral genome and production and processing of viral proteins, (3) replication of the viral genome, (4) assembly of a new generation of virions, (5) exit from the infected cell, and (6) transmission to a new host.

How Do Viruses Enter a Cell?

- Many plant viruses are transmitted through mouthparts of sucking insects.
- In bacterial or animal cells, viruses gain entry by binding to specific molecules on cell walls or plasma membranes.

Producing Viral Proteins

- A virus must exploit the host cell's biosynthetic machinery in order to make the ribosomes and tRNAs necessary for translating its own mRNAs into proteins.
- Production of viral proteins begins soon after a virus enters a cell and continues after the viral genome is replicated.
- RNAs that code for a virus's envelope proteins follow a route through the cell identical to the RNAs of the cell's transmembrane proteins.

How Do Viruses Copy Their Genomes?

- Most DNA viruses copy their genomes by using their own DNA polymerase enzyme. This protein synthesizes copies of the viral genome using nucleotides provided by the host.
- Most RNA viruses use a viral enzyme called **RNA replicase**. RNA replicase is an RNA polymerase that synthesizes RNA from an RNA template, using ribonucleotides provided by the host cell.
- In other RNA viruses, the genome is transcribed from RNA to DNA by a viral enzyme called **reverse transcriptase**. Reverse transcriptase is a DNA polymerase that makes a double-stranded DNA from a single-stranded RNA template.
- **Retroviruses** are viruses that reverse-transcribe their genome. Enzymes then insert this copied DNA (cDNA) into a stretch of host-cell chromosome. The viral genes are then transcribed to mRNA, which is translated into proteins by the host cell's ribosomes.

Assembly of New Virions

- Once the viral genome has been replicated and the viral proteins produced, they are transported to locations where a new generation of virions assembles inside the infected host cell.
- During assembly, the capsid forms around the viral genome.

Exiting an Infected Cell

- Viruses leave the host cell in one of two ways: by budding from cellular membranes or by bursting out of the cell.

How Are Viruses Transmitted to New Hosts?

- If the host cell is part of a multicellular organism, the new generation of particles travels through the blood or lymph. Antibodies or macrophages can then attempt to destroy them before they infect another host cell.
- Natural selection favors alleles that allow viruses to replicate within a host and be transmitted to a new host.

- Transmission of HIV has taken place through direct exchange of blood or semen between an uninfected person and an infected person.
- HIV is quickly destroyed if it is heated or dried; therefore, it does not survive long outside the host tissues.

CHECK YOUR UNDERSTANDING

(b) What are retroviruses, and what specific viral enzyme do they use?

_____

_____

_____

35.3 What Themes Occur in the Diversification of Viruses?

The Nature of the Viral Genetic Material

- Researchers were able to separate the protein and nucleic acid components of the tobacco mosaic virus (TMV). They showed in this virus that RNA—*not* DNA—functions as the genetic material.
- In some groups of viruses, like measles and flu, the genome consists of RNA. In others, like herpes and smallpox, the genome is composed of DNA. RNA and DNA genomes of viruses can either be **positive sense** (single stranded) or **negative sense** (double stranded). In **positive-sense** viruses, the genome contains the same sequences as the mRNA required to produce viral proteins. In **negative-sense** viruses, the base sequences in the genome are complementary to those in viral mRNAs. **Ambisense** viruses contain both positive- and negative-sense sections.
- Viruses can have as few as three loci (tymoviruses) or up to hundreds, as in the genome of smallpox.

Where Did Viruses Come From?

- Many biologists suggest that viruses may be derived from plasmids and transposable elements. Viruses are distinguished from plasmids because they encode proteins that form a capsid and allow the genes to exist outside the cell.
- The "escaped-gene sets" hypothesis states that these elements descended from clusters of genes that physically escaped from bacterial or eukaryotic chromosomes long ago.
- The degeneration hypothesis states that organisms gradually degenerated into viruses by gradually

losing the genes required to synthesize ATP and other compounds.

- To date, neither of these hypotheses has been tested rigorously.

Emerging Viruses, Emerging Diseases

- Hantavirus and Ebola are examples of emerging diseases—illnesses that suddenly affect significant numbers of individuals in a host population.

- Hantavirus and Ebola are considered emerging viruses because they had switched from their traditional host species to a new host—humans.

- In an outbreak, physicians need to (1) identify the agent causing the new illness and (2) determine how it is being transmitted.

CHECK YOUR UNDERSTANDING

(c) Compare and contrast negative-sense and positive-sense DNA. How do these DNA types relate to the virus types?

_____

_____

_____

_____

_____

35.4 Key Lineages of Viruses

Double-Stranded DNA (dsDNA) Viruses

- This is a large group of 21 families and 65 genera, of which smallpox and herpes are the most familiar.

- These viruses parasitize hosts throughout the tree of life, with the exception of land plants.

- The viral genes have to enter the nucleus to be replicated. So they can infect only cells that are actively dividing, such as those that line the respiratory tract or urogenital tract.

RNA Reverse-Transcribing Viruses (Retroviruses)

- There is only one family called the retroviruses, which includes HIV.

- Species are known to parasitize only vertebrates—specifically birds, fish, or mammals.

- When the virus's RNA genome and reverse transcriptase enter a host cell's cytoplasm, the enzyme

catalyzes the synthesis of a single-stranded cDNA from the original RNA. Reverse transcriptase makes this cDNA double stranded. The DNA then integrates with the host chromosome.

Double-Stranded RNA (dsRNA) Viruses

- There are 7 families and a total of 22 genera.

- Species in this group infect a wide variety of hosts, including fungi, land plants, insects, vertebrates, and bacteria.

- For humans, the most important viruses in this group cause disease in major crops.

- Once in the host cell, the double-stranded DNA synthesizes RNA, which are then translated into viral proteins.

Negative-Sense Single-Stranded RNA ([−]ssRNA) Viruses

- There are 7 families and 30 genera in this group.

- The single-stranded virus genome is complementary to the viral mRNA.

- A wide variety of plants and animals are parasitized by these viruses. Flu, mumps, measles, Ebola, hantavirus, and rabies all belong to this group.

- Once in a host cell, a viral RNA polymerase uses the negative-sense template to make viral mRNA. The viral mRNAs are then translated to form viral proteins and new negative-sense single-stranded RNA copies.

Positive-Sense Single-Stranded RNA ([+]ssRNA) Viruses

- This is the largest group known, with 81 genera organized into 21 families.

- Since the sequence of bases is the same as mRNA, it does not need to be transcribed before proteins are produced.

- Most of the commercially important plant viruses belong to this group. This group also includes colds, polio, and hepatitis A, C, and E.

- Single-stranded RNA is immediately translated into viral proteins.

CHECK YOUR UNDERSTANDING

(d) Which of the main viruses discussed in this section is officially extinct, and to which group does it belong?

_____

_____

B. INTEGRATING YOUR KNOWLEDGE

*Looking Back—Key Concepts
from Earlier Chapters*

Translation—Chapter 16

This process of creating proteins from codons attaching to ribosomes is applied in Chapter 35 to virus replication. An extensive review of DNA replication and protein synthesis would be beneficial.

Natural Selection—Chapter 24

Just as HIV has become resistant to drugs that were once effective, natural selection has recently favored strains of *Mycobacterium tuberculosis* that are resistant to the drug rifampin, and these strains are responsible for the resurgence of tuberculosis in the industrialized world.

DNA Polymerase—Chapter 14

Normal DNA polymerase is analyzed in **Chapter 14**, but the DNA polymerase of HIV in Chapter 35 has no 3 exonuclease activity. Comparing the two will help you understand the overall functions of DNA polymerase.

*Looking Forward—Key Concepts
in Later Chapters*

The Immune System—Chapter 49

In Chapter 35, we just touch on the role of T cells and HIV. **Chapter 49** explains just how crucial these cells are to the immune system's response to invading bacteria and viruses. Chapter 49 also explores epidemic control through vaccination and explains why researchers haven't been able to design an effective vaccine against HIV.

C. MASTERING CHALLENGING TOPICS

In your first-year biology course, you are being exposed to many new terms in a short time. Often, these terms are introduced with each new branch of biology. In this chapter, the HIV virus is used as an example in describing a viral replication cycle and the process of researchers in combating a deadly virus. To make matters more complex, HIV is only one virus, and thus only one type of life cycle is explained. To better understand the life cycle of HIV and how it differs from the cell cycle, it would first be beneficial to review the processes of DNA replication and the processes of transcription and translation. Then, by consulting the diagrams in Chapter 35, make a list of differences and similarities between the viral life cycle and the cell cycle. Finally, try to apply this to the HIV virus so you can easily explain to a classmate, lab partner, or friend why we haven't yet found a successful vaccine for HIV.

D. ASSESSING WHAT YOU'VE LEARNED

1. The basic viral replication cycle that usually results in the death of the host cell is called:
   a. lysogeny
   b. the lysogenic cycle
   c. the lytic cycle
   d. the translation cycle

2. Viruses can enter a cell by:
   a. transmission through mouthparts of sucking insects
   b. binding to specific proteins on cell membranes
   c. diffusion
   d. only a and b

3. The RNA polymerase that synthesizes RNA from an RNA template by using ribonucleotides provided by the host cell is called:
   a. RNA replicase
   b. reverse transcriptase
   c. retrovirus transcriptase
   d. RNA transcriptase

4. Viruses have to do which of the following in order to make viral proteins?
   a. ramp up their metabolism
   b. exploit the host cell's biosynthetic machinery
   c. enter a dormant phase
   d. create their own biosynthetic machinery

5. Viruses leave a host cell by:
   a. budding from the cell membrane
   b. bursting out of the cell
   c. forming spores
   d. only a and b

6. The escaped-gene hypothesis states that viral elements descended from clusters of genes that
   _____ .
   a. physically escaped from bacterial or eukaryotic organisms
   b. degenerated over evolutionary time
   c. merged with other viral particles to become active
   d. none of the above

7. The main difference between emerging viruses and emerging diseases is:
   a. Emerging diseases involve only one isolated case.
   b. Emerging diseases involve a significant number of infected individuals.
   c. Emerging viruses involve a significant number of infected individuals.
   d. none of the above

8. The group of viruses to which smallpox belongs is:
   a. double-stranded DNA (dsDNA) viruses
   b. RNA reverse-transcribing viruses
   c. double-stranded RNA (dsRNA) viruses
   d. negative-sense single-stranded RNA ([−]ssRNA) viruses

9. The group of viruses to which HIV belongs is:
   a. double-stranded DNA (dsDNA) viruses
   b. RNA reverse-transcribing viruses
   c. double-stranded RNA (dsRNA) viruses
   d. negative-sense single-stranded RNA ([−]ssRNA) viruses

10. Negative-sense single-stranded RNA must do the following in order to replicate successfully:
    a. use the negative-sense template to make viral proteins
    b. use the negative-sense template to make viral RNA
    c. use the negative-sense template to make viral DNA
    d. use the negative-sense template to make replicase

CHAPTER 35—ANSWER KEY

Check Your Understanding

(a) Viruses are an obligate intracellular parasite. They are not made up of cells; therefore, they are not organisms. Viruses are referred to as particles or agents.

(b) Retroviruses are viruses that reverse-transcribe their genome. Enzymes then insert this copied DNA (cDNA) into a stretch of host-cell chromosome. The viral genes are then transcribed to mRNA, which is translated into proteins by the host cell's ribosomes. In other RNA viruses, the genome is transcribed from RNA to DNA by a viral enzyme called reverse transcriptase. Reverse transcriptase is a DNA polymerase that makes a double-stranded DNA from a single-stranded RNA template.

(c) RNA and DNA genomes of viruses can either be positive sense (single-stranded) or negative sense (double stranded). In positive-sense viruses, the genome contains the same sequences as the mRNA required to produce viral proteins. In negative-sense viruses, the base sequences in the genome are complementary to those in viral mRNAs.

(d) Smallpox is currently extinct in the wild, and the only remaining samples of the virus are stored in research labs. Smallpox is a double-stranded DNA (dsDNA) virus.

Assessing What You've Learned

1. b; 2. d; 3. a; 4. b; 5. d; 6. a; 7. b; 8. a; 9. b; 10. b

36

Plant Form and Function

A. KEY CONCEPTS

36.1 Plant Form: Themes with Many Variations

The Root System

- Many root systems have a vertical section called a **taproot**, as well as numerous **lateral roots**.
- The root system functions to anchor the plant in the soil, absorb water and ions from the soil, conduct water and selected ions to the shoot, and store material produced in the shoot for later use.

Morphological Diversity

- Root systems are incredibly diverse.
- As an example, the prairie plants of North America have a vast array of root systems. Some have a **fibrous** root system, while others have a taproot. Others contain very large reservoirs of food stored in the form of starch.
- This diversity allows prairie plants to coexist with less competition for soil resources, and most individuals can therefore survive intense water stress.

Phenotypic Plasticity

- Morphological diversity in roots occurs within species as well as among species.
- Roots show a great deal of **phenotypic plasticity**, meaning that they are changeable, depending on environmental conditions.

Modified Roots

- In **Figure 36.6**, you can begin to appreciate the huge diversity of modified roots.

The Shoot System

- The shoot system consists of one or more **stems** with each stem consisting of **nodes** and **internodes**. A **leaf** is an appendage attached to a node, and the leaf functions as a photosynthetic organism.
- Sometimes, an **axillary bud** may develop a branch. At the tip of each stem and branch is an **apical bud**.

Morphological Diversity

- The shoot systems of land plants range in size from mosses and liverworts (a few millimeters tall) to redwood trees that reach heights of over 100 m.
- Their enormous variation allows plants to harvest light at different locations and thus minimize competition.

Phenotypic Plasticity

- Shoots respond to variation in light ability by altering their shape and size to maximize the number of photons absorbed in a given environment.

Modified Shoots

- Cactus stems are often modified into water-storage organs, and the stems contain the plant's photosynthetic tissue.
- **Stolons** are modified stems that run over the soil surface, producing roots and leaves at each node.
- **Rhizomes** are similar to stolons—they are stems that grow horizontally, producing new plants at nodes. However, these stems grow *below* ground.
- **Tubers** are rhizomes that are modified as carbohydrate storage organs.
- **Thorns** are modified stems that help protect the plant from attacks by large herbivores.

The Leaf

- The leaf is composed of the expanded portion called the **blade** and a stalk called the **petiole**.

Morphological Diversity

- The arrangement of leaves can vary as much as overall shape.
- Needlelike leaves are interpreted as adaptations that minimize **transpiration** in water-short habitats.

Phenotypic Plasticity

- On the same tree, shade leaves are relatively large and broad while sun leaves are smaller.

Modified Leaves

- It is important to recognize that not all leaves function primarily in photosynthesis.

- Cactus spines are modified leaves that protect the stem.
- The leaves of the flowerpot plant are used as homes by ant colonies.

CHECK YOUR UNDERSTANDING

(a) Describe four main types of modified shoots.

_____

_____

_____

_____

36.2 Primary Growth

- Plants grow because they have **meristems**. When a **meristematic cell** divides, one of the daughter cells remains in the meristem, allowing the meristem to persist, while the other cell specializes.

How Do Apical Meristems Produce the Primary Plant Body?

- A tissue is a group of cells that functions as a **tissue**.
- There are three primary meristematic cell types:
  1. **Protoderm** gives rise to the dermal tissue system or epidermis.
  2. **Ground meristem** gives rise to the ground tissue system, which makes up the bulk of the plant body.
  3. **Procambium** gives rise to the vascular tissue system, which provides support and transports water, nutrients, and photosynthetic products.

How Is the Primary Root System Organized?

- The growing region of the root is covered by protective cells—**the root cap**. These cells also secrete **mucigel** for lubrication.
- Three distinct cell populations exist behind the root cap:
  1. The **zone of cellular division** contains the apical meristem.
  2. The **zone of cellular elongation** is where cells increase in length.
  3. The **zone of cellular maturation** is where older cells complete their differentiation. In this region, **root hairs** increase surface area and actively absorb water and nutrients.

How Is the Primary Shoot System Organized?

- On the top of the stem is the shoot apical meristem.
- The primary meristematic cells give rise to the dermal, ground, and vascular tissues.
- Vascular tissues are found around in groups called **vascular bundles**.
- The ground tissue inside the vascular bundles is called the **pith**, while the ground tissue outside the vascular bundles is called the **cortex**.

A Closer Look at the Cells and Tissues of the Primary Plant Body

The Dermal Tissue System

- Epidermal cells secrete the **cuticle**—a waxy layer that protects the leaves and reduces water loss.
- In addition to water loss, the cuticle forms a barrier to protect the plant from **pathogens**.
- To allow CO_2 to enter photosynthetically active tissues, most land plant leaves have **stomata**—a hole produced by two bean-shaped **guard cells**.
- **Trichomes** are appendages from epidermal cells which function to minimize water loss and attacks by herbivores.

The Ground Tissue System

- Ground tissue is made up of cells below the epidermis and is made up of parenchyma stiffened by collenchyma and sclerenchyma.
- This makes up the bulk of the plant body and is the primary location of photosynthesis and carbohydrate storage.
- **Parenchyma cells** serve as storage cells for starch deposits in roots, stems, and leaves. They are filled with chloroplasts and are the primary site of photosynthesis.
- **Collenchyma cells** function mainly in support. They have thickened primary walls and serve to stiffen leaves and stems.
- **Sclerenchyma cells** also stiffen stems and other structures, but they are distinguished by thickened **secondary cell walls** that are strengthened by tough **lignin** molecules.
- There are two types of sclerenchyma cells:
  1. **Fibers** are usually associated with vascular tissue and are extremely elongated.
  2. **Sclereids** are relatively short, have variable shapes, and often function in protection.

The Vascular Tissue System

- **Vascular tissue** is made up of two complex conducting tissues called xylem and phloem.

- **Xylem** is made up primarily of sclerenchyma cells and conducts water and dissolved ions from the root system to the shoot system.
- Xylem contains two types of conducting cells:
  1. **Tracheids** are dead cells that are long and slender with tapered ends. The sides and ends of the tracheids have **pits** that reduce resistance to water flow.
  2. **Vessel elements** are shorter and wider and have perforations that produce little resistance to water flow.
- **Phloem** is made up primarily of parenchyma cells and conducts sugar, amino acids, chemical agents, and other substances throughout the plant body.
- The cells that make up phloem are alive at maturity.
  1. **Sieve-tube members** are long, thin cells that have perforated ends called sieve plates. They lack major organelles.
  2. **Companion cells** are attached to sieve-tube members by plasmodesmata. They provide materials to the sieve-tube members to maintain the cytoplasm and membrane.

CHECK YOUR UNDERSTANDING

(b) What is the difference between collenchyma cells and sclerenchyma cells?

_____

_____

_____

36.3 Secondary Growth

- Whereas primary growth increases the length of roots and shoots, **secondary growth** increases width.
- Without the structural support provided by secondary growth, roots could not anchor and stems would fall over.
- **Lateral meristems** are also called **secondary meristems** or **cambium**. Lateral meristems differ from apical meristems in two ways:
  1. Apical meristems are localized at root and shoot tips. Lateral meristems form cylinders that run the length of a root or stem.
  2. Cells in apical meristems divide in a plane parallel to the long axis of the root or stem. In the lateral meristem, the cells divide perpendicularly to the long axis of the root or stem, which increases width.

- **Cork cambium** is a ring of meristematic cells located near the perimeter of the stem.
- **Vascular cambium** is a ring of meristematic cells just inside the cork cambium.

What Does Vascular Cambium Produce?

- The vascular cambium produces both phloem and xylem, which, as new layers form, develop into **secondary phloem** and **secondary xylem**.
- Secondary phloem contributes to bark, and secondary xylem forms the material we call **wood**.
- Much more secondary xylem is produced than secondary phloem. As a result, mature woody stems are dominated by wood.

What Does Cork Cambium Produce?

- The cork cambium produces **cork cells** to the outside and a smaller layer of cells called the **phelloderm**. Together, these cells make up the **periderm**.
- These cells eventually die, forming a protective layer for the mature root or shoot.
- This is an important component for **bark**. Bark actually refers to all cells outside the vascular cambium.

The Structure of a Tree Trunk

- Trees are perennial—they live for many years.
- A period of **dormancy** occurs in winter or dry seasons. When growth is rapid during the spring or wet season, large thin-walled cells are produced in the secondary xylem. When dormancy is approached, the secondary xylem cells are smaller and have thick walls.
- When growth is seasonal, **annual growth rings** can be observed in cross sections of wood.
- The inner xylem region is called **heartwood**, while the outer xylem is called **sapwood**.

CHECK YOUR UNDERSTANDING

(c) Why is secondary growth important to many plants?

_____

_____

_____

B. INTEGRATING YOUR KNOWLEDGE

Looking Back—Key Concepts
from Earlier Chapters

Photosynthesis—Chapter 10

This chapter details how plants, along with algae, cyanobacteria, and a variety of protists, obtain the energy and carbon they need to grow and reproduce.

The Plant Cell—Chapter 7

Chapter 7 introduced a generalized version of the plant cell. This information should be reviewed before tackling the specialized cells seen in Chapter 36.

Looking Forward—Key Concepts
in Later Chapters

Water—Chapter 37

Plants must have water as a source of electrons to run photosynthesis. In **Chapter 37**, you will see that plants also need water to keep their cells in proper working order.

Plant Nutrition—Chapter 38

Plants must obtain nitrogen, phosphorus, potassium, magnesium, and a host of other nutrients to synthesize nucleic acids, enzymes, phospholipids, and the other macromolecules needed to build and run cells. **Chapter 38** explores how plants acquire these key elements.

Sensory Response—Chapter 39

A shoot's growth is controlled by sophisticated sensory and response systems. **Chapter 39** explores how plants sense certain wavelengths of light and respond by bending and growing in that direction.

Reproductive Structures—Chapter 40

Twigs and stems are structures that arrange leaves in space. **Chapter 40** explores the reproductive structures housed in twigs and stems.

C. MASTERING CHALLENGING TOPICS

The main aspect of this chapter that creates problems is the overwhelming amount of anatomical terms. Before you can understand the function of anatomy, you need to understand the anatomy itself. Drawing diagrams and labeling them is the best way to learn anatomical terms. Adding the functions on top of the anatomy will help to solidify all the terms and organize them. To completely prove you have mastered the anatomy and function of angiosperms, go out and collect various samples of plants and test yourself. This is a great way to see firsthand the variation of form and function.

D. ASSESSING WHAT YOU'VE LEARNED

1. The term for an individual changing in response to the environment is:
   a. natural selection
   b. allele frequency modulation
   c. phenotypic plasticity
   d. artificial selection

2. Which of the following is a modified stem that runs over the soil surface?
   a. stolon
   b. rhizome
   c. tuber
   d. thorn

3. Which of the following is a modified stem used for carbohydrate storage?
   a. stolon
   b. rhizome
   c. tuber
   d. thorn

4. Plants in desert or northern habitats tend to have leaves that are needlelike because:
   a. They reduce herbivory.
   b. They reduce sun damage.
   c. They increase water absorption.
   d. They reduce water loss.

5. The undifferentiated cells responsible for plant growth are called:
   a. meristematic cells
   b. parenchyma cells
   c. mesophyll cells
   d. apical cells

6. Which of the following cells serve as storage cells for starch deposits in roots, stems, and leaves?
   a. parenchyma cells
   b. meristematic cells
   c. collenchyma cells
   d. sclerenchyma cells

7. Which of the following are water-conducting cells that are dead at maturity?
   a. sclerenchyma cells
   b. tracheids
   c. sieve-tube members
   d. sclereids

8. The waxy layer that is secreted by the epidermis is called:
   a. the dermis
   b. the stoma
   c. the cuticle
   d. the sterol

9. Which of the following is *not* a group of cells behind the root cap?
   a. the zone of cellular division
   b. the zone of cellular elongation
   c. the zone of cellular communication
   d. the zone of cellular maturation

10. Wood is:
   a. produced by the cork cambium
   b. secondary xylem
   c. secondary phloem
   d. produced by the apical meristem

CHAPTER 36—ANSWER KEY

Check Your Understanding

(a) **Stolons** are modified stems that run over the soil surface, producing roots and leaves at each node. **Rhizomes** are similar to stolons—they are stems that grow horizontally, producing new plants at nodes. However, these stems grow *below* ground. **Tubers** are rhizomes that are modified as carbohydrate storage organs. **Thorns** are modified stems that help protect the plant from attacks by large herbivores.

(b) Collenchyma cells have thickened primary walls and serve to stiffen leaves and stems. Sclerenchyma cells also stiffen stems and other structures, but they are distinguished by thickened secondary cell walls that are strengthened by tough lignin molecules.

(c) Whereas primary growth increases the length of roots and shoots, **secondary growth** increases width. Without the structural support provided by secondary growth, roots could not anchor and stems would fall over.

Assessing What You've Learned

1. c; 2. a; 3. c; 4. d; 5. a; 6. a; 7. b; 8. c; 9. c; 10. b

37

Water and Sugar Transport in Plants

A. KEY CONCEPTS

37.1 Water Potential and Water Movement

- Loss of water via evaporation from the aerial parts of a plant is called **transpiration**.
- Water potential indicates the potential energy that water has in a particular environment compared to control conditions.
- Water always flows from areas of high water potential to areas of low water potential.
- **Water potential gradient** is the overall movement of water when a series of water potential differences are contrasted. Plants tend to gain water from the soil and lose it to the atmosphere.

What Factors Affect Water Potential?

- A **solution** is a homogeneous liquid mixture containing **solutes**.
- **Osmosis** is the movement of water across membranes in response to differences in water potential.
- The **solute potential** or **osmotic potential** is the difference in solute concentrations between two cells.
- **Turgor pressure** is the force of the cell membrane swelling and pushing against the cell wall. The cell wall exerts an equal and opposite force called **wall pressure**.
- **Pressure potential** is the sum of all the types of pressure on water.

Calculating Water Potential

- The water potential is the pressure potential plus the solute potential, calculated as follows:

$$\psi = \psi_p + \psi_s$$

- The unit of measurement is a pressure unit called the **megapascal** (**MPa**).

Water Potentials in Soils, Plants, and the Atmosphere

- The water contained within a leaf or root system or plant has a pressure potential and a solute potential, just as the water inside a cell does. The water potentials of plant tissues, air, and soil are changing all the time in response to heat, cold, rain, etc.
- If the water potential in the space that surrounds a cell drops, water moves out of a cell in response, causing it to shrink. If the cells do not quickly regain turgor pressure, dehydration and death result.
- Because water moves along the water potential gradient, plants tend to gain water from the soil and lose it to the atmosphere.

CHECK YOUR UNDERSTANDING

(a) Explain the difference between water potential and water gradient. What is the significance of this difference?

_____

_____

_____

_____

37.2 How Does Water Move from Roots to Shoots?

Root Pressure and Short-Distance Transport

- Look at **Figure 37.7** and note the different tissues. Starting at the outside of the root and working inward, they are the **epidermis** and **root hairs**, the **cortex**, the **endodermis**, the **pericycle**, and the **vascular tissues**.

- When water enters a root along a water potential gradient, it does so through the root hairs. Before it reaches the vascular tissue, it can be transported through two pathways:
  1. The **apoplast** lies within cell walls, which are porous.
  2. The **symplast** consists of the continuous connection through cells that exist via plasmodesmata.
- **Endodermal** cells are tightly packed and secrete a waxy layer called the **Casparian strip**. It prevents water from creeping through the walls of the endodermal cells (i.e., blocks the apoplastic pathway).
- When water is forced to travel through the plasma membrane of endodermal cells, it can be filtered. The plasma membrane can also prevent water from flowing out of the vascular tissue.
- The endodermis is also responsible for **root pressure**—active transport of ions brings more water into the root that is lost via transpiration.
- Water droplets can even be forced out of leaves—**guttation**.

Water Movement via Capillary Action

- Capillary movement occurs in response to these forces:
  1. **Surface tension** is a downward pull that exists on water molecules at an air–water interface.
  2. **Cohesion** is the mutual attraction among like molecules, such as the hydrogen bonding among water molecules.
  3. **Adhesion** is the attraction of unlike molecules.

The Cohesion-Tension Theory

- The leading hypothesis to explain water movement in vascular plants is called the **cohesion-tension theory** (**Figure 37.11**).
- The leaf area just below the stoma is filled with moist air such that, when the stoma opens, the humid air is exposed to the atmosphere. Water exits through the stomatal pore as a result.
- The water potential of the atmosphere is extremely low compared to the water potential of the space inside the leaf, meaning that there is a steep water potential gradient between the leaf interior and the air.
- The water inside the cell walls, at the interface with the air, forms a concave boundary layer called a **meniscus**. If the water potential gradient is especially steep, then water molecules leave the surface rapidly and the menisci in the cell walls become more concave.

The Role of Surface Tension in Water Transport

- The key idea of the cohesion-tension theory is that the negative force or pull (tension) generated at the air–water interface is transmitted through the water present in the leaf cells to the water present in the xylem, on to the water in the vascular tissue of the roots, and finally on to the water in the soil.
- This continuous transmission of pulling force is possible because water is present throughout the plant and because all of the water molecules present hydrogen-bond to one another.

What Evidence Do Biologists Have for the Cohesion-Tension Theory?

- Researchers also confirmed that water tension in trees is great enough to actually make tree trunks shrink. The dendrograph is the instrument used to measure trunk diameter.
- If you take a leaf and cut its petiole, the watery fluid in the xylem withdraws from the edge. This confirms that the xylem sap is under tension.
- By placing a leaf or branch in an airtight container and then applying a steadily increasing external pressure, the pressure required to push the **xylem sap** back to the cut surface can be measured. This pressure is equal to the tension exerted on the xylem sap (**Figure 37.12**).
- Researchers collected data on water potential on Sitka spruce leaves. They harvested branches from the tops of 10-m-tall spruce trees several times a day for several days, used a pressure bomb to measure the water potential of the tissues, and produced a data set.
- Water potential of these tissues dropped during the day and rose at night. This information correlates with stomata opening during the day and closing at night.
- **Key Research**—Chunfang Wei, Melvin Tyree, and Ernst Steudle set out to study the cohesion-tension theory using a root bomb and a xylem pressure probe. This technique allowed the investigators to record changes in xylem pressure instantly and directly (**Figure 37.15**).
- First, they added pressure to the root systems of their experimental plants using a root bomb. Then, they released pressure on the root system and began to alter light levels.
- Pulses of pressure applied to through the root bomb produced sharp decreases in the tension exerted on the xylem sap. The tension on the xylem sap increased each time that light intensity was increased.

(b) Explain what surface tension means and why it is important to plant physiology.

_____

_____

_____

_____

37.3 Water Absorption and Water Loss

- One of the most important features of the cohesion-tension theory is that it does **not require the expenditure of energy** by plants. The Sun furnishes the energy that breaks hydrogen bonds between water molecules at the air–water interface inside leaves and causes transpiration.

- The balance between conserving water and maximizing photosynthesis is termed the **photosynthesis-transpiration compromise**.

Limiting Water Loss

- To cope with water loss, the oleander has (**Figure 37.16**):
  1. a thick cuticle on the upper surface of the leaves
  2. several layers of epidermis
  3. stomata located in deep pits on the undersurface of leaves

Obtaining Carbon Dioxide under Water Stress

- **Crassulacean acid metabolism** (CAM) plants are able to continue photosynthesizing even though their stomata are closed during the day. CAM plants open their stomata at night and store the CO_2 that diffuses into their tissues by adding organic acids.

- When photosynthesis begins during the day, CO_2 is released from organic acids to **rubisco**—the enzyme that initiates the Calvin cycle.

- C_4 plants use CO_2 so efficiently that they are able to keep their stomata closed more than competing plants can. CO_2 is transferred to **bundle-sheath cells**, where rubisco is abundant.

37.4 Translocation

- **Translocation** is the movement of sugars through a plant. Sugars move from sources to sinks. In

(c) Why must a plant compromise between photosynthesis and transpiration?

_____

_____

_____

vascular plants, a **source** is defined as a tissue where sugar enters the phloem. A **sink** is a tissue where sugar exits the phloem.

- During the growing season, leaves and stems that are actively photosynthesizing and producing sugar in excess of their own needs act as sources. Early in the growing season, though, the situation is reversed.

- [14]C experiments have made two important generalizations possible:
  1. Sugars are translocated very rapidly.
  2. There is a strong correspondence between the physical location of certain sources and certain sinks.

The Anatomy of Phloem

- Phloem is made up of **sieve-tube elements** and **companion cells**. Both cells are alive at maturity. In most plants, sieve-tube elements lack nuclei and many major organelles; they are connected to one another, end to end, by perforated structures called **sieve plates** (**Figure 37.19**).

- Sieve plates create a direct connection between the cytoplasm of adjacent cells. Companion cells, in contrast, have nuclei and a large number of ribosomes, mitochondria, and chloroplasts.

The Pressure-Flow Hypothesis

- **Key Research**—Ernst Münch proposed a mechanism for translocation in 1926. The **pressure-flow hypothesis** states that events at source tissues and at sink tissues create a steep pressure potential gradient in phloem. The water in phloem moves down this gradient, and sugar molecules are carried along by **bulk flow**—a mass movement of molecules along a pressure gradient.

- The net result of these events in **Figure 37.20** is high turgor pressure near the source and low turgor pressure near the sink. Because the pores of the sieve plate have no membranes, the difference in pressure potential drives phloem sap from source to sink. The pressure contrast is responsible for a one-way flow of sucrose molecules.

Phloem Loading

- At source cells, sugar has to be loaded into sieve-tube elements against a concentration gradient. Therefore, loading requires the use of ATP and a membrane transport mechanism. Unloading also requires ATP and a second membrane transport system.

How Are Sugars Concentrated in Sieve-Tube Members at Sources?

- The observation of strong pH differences between the interior and exterior of phloem cells led researchers to suspect that phloem loading depends on an **H⁺-ATPase**.

Where Are H⁺-ATPases Located?

- Proton pumps are found in a variety of organisms, including bacteria, fungi, and animals. Researchers found that the *Arabidopsis* (a widely studied plant model from the mustard family) genome actually codes for 10 different pump proteins. One of these genes, called *AHA3*, appeared to be expressed primarily in vascular tissues.

- **Key Research**—Natalie DeWitt and Michael Sussman used this information and hypothesized that *AHA3* encoded the proton pump responsible for phloem loading. Their goal was to treat *Arabidopsis* leaves with the AHA3 antibody, examine treated leaves under the electron microscope, and determine exactly where the pump proteins are located.

- The proton pumps responsible for phloem loading are found almost exclusively in the membranes of companion cells.

- This result supported the following model for phloem loading:

  1. Proton pumps in the membranes of companion cells create a strong gradient that favors flow of protons into companion cells.

  2. A cotransporter protein in the membrane of companion cells uses the proton gradient to bring in sucrose.

  3. Once inside companion cells, sucrose travels into sieve-tube members via direct cytoplasmic connection.

Phloem Unloading

- Sugar transport is an energy-demanding process, but the membrane proteins involved and mechanism of movement vary among different types of sinks within the same plant, as well as among different species.

(d) Compare and contrast sieve-tube elements and companion cells.

_____

_____

_____

_____

B. INTEGRATING YOUR KNOWLEDGE

Looking Back—Key Concepts from Earlier Chapters

Leaf Structure—Chapter 36

Recall that the surfaces of leaves are dotted with structures called stomata. Chapter 37 begins to discuss the functions of the stomata and apply the anatomy learned in **Chapter 36**.

Osmosis—Chapter 8

Water will begin moving into or out of a cell via osmosis—water moves from regions of low solute concentrations to regions of high solute concentrations. Chapter 37 discusses osmosis in relation to the plant and its transport of fluids.

Hydrogen Bonding—Chapter 2

Water molecules are polar and interact with one another through hydrogen bonding. Under an air–water interface, in the body of a solution, all of the water molecules present are surrounded by other water molecules and form hydrogen bonds in all directions. The water molecules on the surface, however, can form hydrogen bonds in only one direction—with the water molecules below them.

CAM and C₄ Photosynthesis—Chapter 10

Chapter 10 introduced two novel biochemical pathways that are found in species native to deserts and other hot, dry habitats. These pathways are called crassulacean acid metabolism (CAM) and C_4 photosynthesis. CAM plants are able to continue photosynthesizing even though their stomata close during the heat of the day.

Proton Pumps—Chapter 9 and 10

As **Chapters 9 and 10** indicated, proteins like these are called proton pumps, or more formally H⁺-ATPases. Their activity establishes a large difference in the charge and hydrogen ion concentration on either side of the membrane.

Looking Forward—Key Concepts in Later Chapters

Creating Antibodies—Chapter 49

Researchers raised antibodies to the AHA3 protein using the types of techniques introduced in **Chapter 49**. As that chapter points out, an antibody is a polypeptide that binds to a specific protein.

C.   MASTERING CHALLENGING TOPICS

Now that you have mastered the anatomy of vascular plants from Chapter 36, you begin to learn the physiology of fluid transport in Chapter 37. The terminology can be quite confusing, especially on a test. Try to explain, using plain language, how water and sugars move through a plant. Then apply the appropriate terms in your explanation. List the following terms, and make sure you are able to tell the difference between them:

- solute potential
- turgor pressure
- wall pressure
- pressure potential
- water potential
- water potential gradient

Try using each of these terms in a sentence that explains their meaning, and then try to compare and contrast the terms. If you can master the meaning of these terms, you are well on your way to understanding the physiology of fluid transport in vascular plants.

D.   ASSESSING WHAT YOU'VE LEARNED

1. The force of the cell membrane swelling and pushing against the cell wall is called the _____.
   a. turgor pressure
   b. wall pressure
   c. pressure gradient
   d. solute potential

2. When the solute potential and the pressure potential of a cell are added together, the resulting quantity is called the cell's _____.
   a. turgor pressure
   b. wall pressure
   c. pressure gradient
   d. water potential

3. Water potential gradient is:
   a. the overall movement of solutes
   b. the evaporation of water from the leaves
   c. the overall movement of water
   d. the pressure gradient minus the water potential

4. The barrier inside the root that prevents water from leaking out of the vascular tissue is the _____.
   a. epidermis
   b. Casparian strip
   c. apoplast
   d. root hair

5. The pathway for water that lies within the cell walls is:
   a. the apoplastic pathway
   b. the symplastic pathway
   c. the Casparian pathway
   d. the endodermal pathway

6. Cohesion is:
   a. the mutual attraction among like molecules
   b. the mutual attraction among unlike molecules
   c. the repelling of like molecules
   d. the repelling of unlike molecules

7. A movement in response to pressure can also be called:
   a. solute flow
   b. bulk flow
   c. transpiration
   d. CAM metabolism

8. Which type of plants open their stomata at night to store CO_2?
   a. C_4 plants
   b. rubisco plants
   c. CAM plants
   d. transpiring plants

9. Translocation is:
   a. the movement of water through a plant
   b. the movement of ions through a plant
   c. the movement of sugars through a plant
   d. the movement of xylem though the plant

10. Where are the proton pumps responsible for phloem loading located?
    a. on the membranes of companion cells
    b. on the membranes of sieve-tube members
    c. on the membranes of root cells
    d. on the membranes of root hairs

CHAPTER 37—ANSWER KEY

Check Your Understanding

(a) When the solute potential and the pressure potential of a cell are added together, the resulting quantity is called the cell's water potential. It can be thought of as the tendency of water to move from one location to another. Water potential gradient is the overall movement of water when a series of water potential differences are contrasted. Plants tend to gain water from the soil and lose it to the atmosphere.

(b) Surface tension is the pull that occurs on water molecules at the air–water interface.

(c) Photosynthesis requires that the stomata on the leaves be open. The greater the requirement for photosynthesis, the greater the leaf surface area and the more stomata per unit area. More surface area and more stomata mean greater transpiration rates. There is a compromise because transpiration can damage a plant if there is not enough water.

(d) Phloem is made up of sieve-tube elements and companion cells. Both cells are alive at maturity. In most plants, sieve-tube elements lack nuclei and many major organelles; they are connected to one another, end to end, by open pores. These sieve-like pores create a direct connection between the cytoplasm of adjacent cells. Companion cells, in contrast, have nuclei and a large number of ribosomes, mitochondria, and chloroplasts.

Assessing What You've Learned

1. a; 2. d; 3. c; 4. b; 5. a; 6. a; 7. b; 8. c; 9. c; 10. a

38

Plant Nutrition

A. KEY CONCEPTS

38.1 Nutritional Requirements

Which Nutrients Are Essential?

- An **essential nutrient** should fulfill the following criteria:
  1. It is required for normal growth and reproduction—the plant cannot complete its life cycle without this nutrient.
  2. It is required for a specific structure or metabolic function.
- **Macronutrients** are the building blocks of nucleic acids, proteins, carbohydrates, phospholipids, and other key molecules required in relatively large quantities.
- N, P, and K are particularly important because they are often **limiting nutrients**.
- **Micronutrients** (e.g., selenium) are required in relatively small quantities and usually function as cofactors for specific enzymes.

What Happens When Key Nutrients Are in Short Supply?

- Researchers used **hydroponic growth** to explore the effect of copper deficiency on tomatoes. They found that the copper-deprived individuals had stunted shoots and roots, dark foliage, curled leaves, and no flowers.
- Copper has since been revealed as a cofactor or component in several enzymes involved in redox reactions.

38.2 Soil

- The process of soil building begins with solid rock. Water, wind, and organisms continually break tiny pieces off large rocks—this process is called **weathering**.
- As organisms occupy the substrate, they add their waste products and carcasses—this organic matter is **humus**.

(a) What are the criteria for an essential nutrient in plants?

_____

_____

_____

- **Texture** refers to the proportions of different-sized particles present in soil and is important because:
  1. texture affects the ability of roots to penetrate
  2. texture affects the soil's ability to hold water
  3. texture and water content dictate availability of oxygen

Soil Conservation

- **Soil erosion** occurs when soil is carried away from a site by wind or water.
- When plant cover is removed for forestry, farming, or suburbanization, soil erosion is accelerated.
- Techniques that maintain long-term soil quality and productivity are collectively called **sustainable agriculture**.

What Factors Affect Nutrient Availability?

- The elements required for plant growth exist in the soil as **ions**. Ions with negative charge (anions) usually dissolve in water because they interact with water molecules via hydrogen bonding. Ions with positive charges (cations) often interact with the negative charges found in organic matter and on the surfaces of the tiny, sheetlike particles called clay—they are less readily available.
- As solutes, negatively charged ions are available to plants for absorption; however, they are also easily washed out of the soil by rain. The loss of nutrients via washing is called **leaching**.

- **Cation exchange** occurs when protons or other cations bind to negative charges on soil particles and cause bound cations such as magnesium or calcium to be released to nearby roots.

CHECK YOUR UNDERSTANDING

(b) Describe the process of leaching and its effects. How can leaching be prevented?

_____

_____

_____

_____

_____

_____

_____

38.3 Nutrient Uptake

Mechanisms of Nutrient Uptake

- Some of the proteins found in certain cells span the lipid bilayer of a cell membrane and act as channels that allow the transit of specific ion.

- The large surface area of **root hairs** holds large numbers of membrane proteins, which contact the soil and selectively facilitate the passage of ions into the cell.

Establishing and Using a Proton Gradient

- **Proton pumps** use ATP and pump an excess of H^+ ions on the exterior of the plasma membrane of the root-hair cells.

- This separation of charge is measured as a **voltage** across the membrane, and these pumps maintain a **membrane potential**.

- The effect of concentration and electrical charge on an ion is called the **electrochemical gradient**. When an electrochemical gradient favors the movement of an ion, no energy expenditure is required and the movement is described as passive.

- Researchers found that K^+ transport was passive; they found that the plant could import K^+ even when the outside concentrations of K^+ were extremely low.

Nutrient Transfer via Mycorrhizal Fungi

- Fungi that live in close association with roots are called **mycorrhizae**, and they are **symbiotic**.

- Many plants living in northern forests receive large quantities of nitrogen from fungi that wrap themselves around the epidermal cells of roots and radiate out into the surrounding soil. Plants that live in grasslands and in tropical forests receive much of the phosphorus they need from species of fungi whose bodies actually penetrate into the plant root interior.

Mechanisms of Ion Exclusion

- Many of the metal ions found in soils are poisonous to plants; even essential nutrients can become toxic if they are present in high levels.

Passive Exclusion

- The Casparian strip prevents some ions from entering the symplast and reaching the xylem.

- Passive exclusion also occurs in root hairs.

- Researchers explored the molecular mechanism responsible for variation in salt tolerance among corn varieties. They grew seedlings from salt-tolerant and salt-intolerant populations in hydroponic cultures containing a high NaCl concentration. They found that the salt-intolerant plants had taken up almost twice as much Na^+ as had the salt-tolerant individuals.

- Researchers suggest that individuals from salt-tolerant populations have fewer sodium channels in their root hairs than do salt-intolerant individuals.

Active Exclusion

- Plants that grow on waste rock and soils from copper-mining operations experience large concentration gradients that favor an influx of this nutrient. How do plants neutralize excess nutrients?

- **Metallothioneins** are small proteins that bind to metal ions and prevent them from acting as a poison.

CHECK YOUR UNDERSTANDING

(c) What is a membrane potential?

_____

_____

_____

_____

_____

- The membrane that surrounds the vacuole—the **tonoplast**—contains transport proteins that move sodium ions from the cytosol into the vacuoles, where they cannot poison the enzymes.

38.4 Nitrogen Fixation

- Only selected bacteria are able to take up N_2, convert it to ammonia, and use it to fuel growth. This conversion process is called **nitrogen fixation**. Nitrogen fixation requires a series of specialized enzymes and cofactors, including an enzyme called nitrogenase.

- Bacteria in the genus *Rhizobium* are located in the roots of pea plants in structures called **nodules**. The bacteria provide the plant with ammonia, while the legume provides the bacteria with sugar and protection.

How Do Nitrogen-Fixing Bacteria Colonize Plant Roots?

- Roots do not contain a population of rhizobia from the moment of germination. The rhizobia must colonize the plant.

- Colonization is a complex process involving a series of specific interactions between the rhizobia and the legume.

- Recognition is possible because the root hairs of the pea family plants contain compounds called flavonoids. When rhizobia contact the flavonoids, they produce **Nod factors**, which in turn bind to the proteins on the membrane surface of the root hairs.

How Do Host Plants Respond to Contact from a Symbiotic Bacterium?

- Once Nod factors bind to the root-hair surface, they set off a chain of events that leads to a dramatic morphological change in the legume.

- As **Figure 38.15** shows, rhizobia proliferate around the tip of the root hair and then enter the epidermal cell through an invagination in the root hair membrane called an **infection thread**.

![pencil icon] **CHECK YOUR UNDERSTANDING**

(d) Why would certain plants allow fungi to inhabit their roots? Are there any disadvantages to the plant?

_____

_____

_____

_____

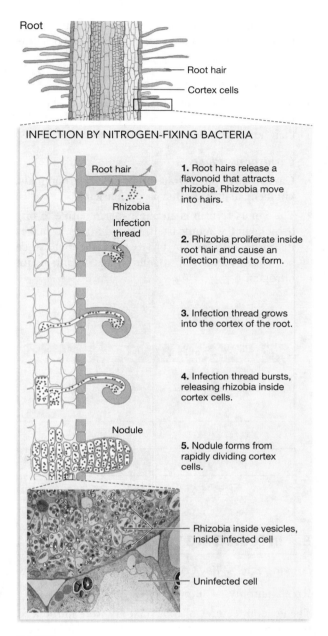

FIGURE 38.15

- The infected cortex cells begin to divide rapidly, forming root nodules.

- Genetic changes also occur; for example, the production of **leghemoglobin** allows oxygen to bind to it rather than binding to and poisoning nitrogenase.

38.5 Nutritional Adaptations of Plants

Epiphytic Plants

- **Epiphytes** are plants that never make contact with the soil and usually grow in the leaves or branches of trees. They absorb most of the nutrients they need from rainwater that collects in their tissues or in the crevices in bark.

Parasitic Plants

- In most cases, parasitic plants use photosynthesis to make their own sugars and tap the root systems of their hosts for water and essential nutrients.

Carnivorous Plants

- **Carnivorous** plants trap insects and other animals, kill them, and absorb the prey's nutrients.
- The Venus flytrap is an angiosperm native to bog habitats in the southeastern United States. It makes its carbohydrates via photosynthesis; but, since bogs have very low nitrogen, it traps organisms for nitrogen.

CHECK YOUR UNDERSTANDING

(e) What is an epiphyte?

_____

_____

_____

B. INTEGRATING YOUR KNOWLEDGE

Looking Back—Key Concepts from Earlier Chapters

Root Anatomy—Chapter 36

The general features of root anatomy were introduced in **Chapter 36**. Chapter 38 investigates further how nutrition is obtained through the roots.

Root Function—Chapter 37

The specifics of root function in water uptake were analyzed in **Chapter 37**. Again, this is broadened in Chapter 38 to include roots and plant nutrition.

Phospholipid Bilayer—Chapter 6

The interior of the phospholipid bilayer is uncharged; therefore, it resists the passage of ions. Some of the proteins found in certain cells span the bilayer and act as channels that allow the transit of specific ions. The ions are discussed in terms of plant nutrition in Chapter 38.

Mycorrhizae—Chapter 31

The fungi that live in close association with roots are called mycorrhizae. **Chapter 31** introduced two major types of mycorrhizae and presented evidence that these fungi transfer nitrogen and phosphorus, respectively, from soil to plant roots. This is further discussed in Chapter 38 in terms of nutrition.

Looking Forward—Key Concepts in Later Chapters

Mutualism—Chapter 50

A mutually beneficial relationship was shown between mycorrhizae and plants in Chapter 38. This topic is further explored in **Chapter 50**.

Bog Habitats—Chapter 50

Bog habitats are notoriously poor in nutrients—particularly in nitrogen. **Chapter 50** explains why this is so, while Chapter 38 gives an example of a plant adaptation to this nutrient-poor environment.

Plants and Stimuli—Chapter 39

Chapter 39 focuses on how a signal travels from the hairs on a leaf to the rest of the plant, and it examines how plants respond to light.

Ammonia Fertilizers—Chapter 55

The use of ammonia fertilizers has been causing serious pollution problems, as discussed in **Chapter 55**. For this reason, there has been intense interest in the phenomenon of nitrogen fixation.

C. MASTERING CHALLENGING TOPICS

You will find the concept of gradients to be the most confusing topic in this chapter. Understanding the movement of ions based on concentration and based on charge is actually quite straightforward, if they are separate entities. Once both concentration and charge are incorporated in one group of ions, predicting their movement begins to get quite complicated. Adding a selectively permeable membrane—like the plasma membrane of a cell—and then considering different ion types can confuse even the well-versed physiologist. Luckily, you only need to understand the basics, so make sure that you do. As with any other concept, it is best to start simple and gradually add complexity. Make sure you understand the concepts of concentration gradients and electrical gradients before you attempt to study the electrochemical gradient. Stick with one type of ion, and then try to set up gradients with other ion types and predict their movement. Save the study of the movement of mixed ions for a physiology class.

D.   ASSESSING WHAT YOU'VE LEARNED

1. Which of the following is *not* a criterion for an essential nutrient?
   a. It is required for growth and reproduction.
   b. A cofactor is needed for absorption.
   c. No other element can substitute for it.
   d. It is required for a specific structure or metabolic function.

2. Researchers can use one of the following techniques to explore plant nutrient deficiencies:
   a. hydroponics
   b. sun exposure
   c. crop rotation
   d. hyperbaric chambers

3. The process of water and wind breaking off tiny pieces of rock is called:
   a. soil erosion
   b. humus
   c. blasting
   d. weathering

4. Soil texture is important because:
   a. Texture affects the ability of roots to penetrate.
   b. Texture affects the soil's ability to hold water.
   c. Texture and water content dictate oxygen availability.
   d. All of the above answers are correct.

5. Which of the following statements regarding soil erosion is *false*?
   a. Soil erosion is a very slow, natural process.
   b. Removal of trees can accelerate soil erosion.
   c. Suburbanization does not affect soil erosion.
   d. Farming can accelerate soil erosion.

6. The loss of nutrients via washing by rain is called:
   a. erosion
   b. filtering
   c. leaching
   d. binding

7. Fungi that live in close association with roots are called:
   a. rhizomes
   b. mycorrhizae
   c. rhizobium
   d. rice

8. Passive exclusion of ions occurs mainly by:
   a. the Casparian strip
   b. the root hairs
   c. both a and b
   d. sodium channels

9. Nitrogen fixation is:
   a. the ability to convert N_2 to ammonia
   b. the ability to convert ammonia to N_2
   c. the ability to absorb N_2
   d. the ability to metabolize nitriles

10. When rhizobia contact the flavonoids of pea root hairs:
   a. The rhizobia produce Nod factors.
   b. The rhizobia produce flavonoids.
   c. The pea root hairs produce Nod factors.
   d. Nothing happens.

CHAPTER 38—ANSWER KEY

Check Your Understanding

(a) An **essential nutrient** should fulfill the following criteria:
   1. It is required for normal growth and reproduction—the plant cannot complete its life cycle without this nutrient.
   2. It is required for a specific structure or metabolic function.

(b) As solutes, negatively charged ions are available to plants for absorption; but they are also easily washed out of the soil by rain. The loss of nutrients by means of washing is called leaching. Ions with positive charges, in contrast, often interact with negative charges found in organic matter and on the surfaces of the tiny, sheetlike particles called clay. As a result, the presence of organic matter and clay in soil slows leaching.

(c) Proton pumps use ATP and pump an excess of H^+ ions on the exterior of the plasma membrane of the root-hair cells. This separation of charge is measured as a voltage across the membrane, and these pumps maintain a membrane potential.

(d) Certain plants allow fungi to inhabit their roots so that they have a mutualistic or symbiotic relationship. The plant can gain nitrogen from the fungi and, in return, the fungi receive sugars and protection.

(e) An epiphyte is a plant that never makes contact with the soil.

Assessing What You've Learned

1. b; 2. a; 3. d; 4. d; 5. c; 6. c; 7. b; 8. c; 9. a; 10. a

39

Plant Sensory Systems, Signals, and Responses

A. KEY CONCEPTS

39.1 An Overview of Information Processing

How Do Cells Receive and Transduce an External Signal?

- When a receptor changes shape in response to a stimulus, the information changes from an external signal to an intracellular signal—**signal transduction**.

- **Phosphorylation cascades** are triggered when the change in the receptor protein's shape leads to the addition of a phosphate group from ATP to the receptor or a nearby protein.

- **Second messengers** are produced when hormone binding results in the release of an intracellular signal from storage areas.

- A **hormone** is an organic compound that is produced in small amounts in one part of the plant and transported to target cells in another region of the plant, where it causes a physiological response.

How Are Cell–Cell Signals Transmitted?

- The molecules that function as hormones are wildly diverse in structure.

- Plant hormones have several features in common, but they (1) can elicit a response only if a cell has an appropriate receptor, and (2) they are active at extremely small concentrations.

How Do Cells Respond to Cell–Cell Signals?

- The signal in the cell's periphery is rapidly transduced to increased activity inside the cell via a signal transduction pathway.

- This results in the production of many phosphorylated proteins or the release of many second messengers

- Therefore, the original signal can be amplified many times over.

(a) What is transduction, and why is it important to plants?

_____

_____

_____

_____

_____

39.2 Blue Light: The Phototropic Response

- Charles Darwin and his son (1881) performed the first experiments showing that plants respond to light. They also concluded that the blue part of the spectrum was involved.

- A **coleoptile** is a modified leaf that forms a sheath protecting the stems and leaves of young grasses.

- **Phototropism** is any type of directed movement in response to light. Plants are positively phototrophic.

Phototropins as Blue-Light Receptors

- A **pigment** is a molecule that absorbs certain wavelengths of light.

- Researchers found a membrane protein that is abundant in the tips of emerging shoots and that becomes phosphorylated in response to blue light. But was this protein the membrane receptor itself?

- Researchers began analyzing *A. thaliana* mutants that do not show a phototrophic response to blue light. The modified gene, named *PHOT1*, which encodes the sunlight detector in plants. A phototropic response is triggered when the protein expressed by this gene is phosphorylated.

- Most recent research indicates that there are multiple blue-light receptors related to PHOT1—called **phototropins**.

Auxin as the Phototropic Hormone

- Although the Darwins published their hypothesis about chemical signals in 1881, Peter Boysen-Jensen confirmed it in 1913. He cut the tips off young oat shoots and placed either a porous block or nonporous block between the shoot and tip. The stems with the porous block showed normal phototropism, so he concluded that the phototropic signal was indeed chemical.

- Fritz Went (1925) collected the phototrophic hormone in gelatinous blocks of agar. Stems responded by bending away from the source of the hormone, without a source of light present.

- The physical basis of the phototrophic response is cell elongation. Cells on the side of the shoot opposite to the source of light elongate in response to the phototrophic hormone. Because it promotes cell elongation, Went named the hormone **auxin** (from the Greek *auxein*, to increase).

Isolating and Characterizing Auxin

- Researchers succeeded in isolating and characterizing auxin. It is called IAA or indole acetic acid.

The Cholodny-Went Hypothesis

- This hypothesis contends that auxin produced in the tips of coleoptiles is shunted from one side of the tip to the other, in response to light; then it is transported down the shoot.

- Others proposed an alternative hypothesis: that auxin is broken down by blue light, which produces an asymmetric distribution.

- Winslow Briggs grew corn seedlings in the dark, cut off their tips, and placed the tips on agar blocks. He either kept these in the dark or exposed them to light from one side. Later, he put agar blocks on one side of the decapitated shoots. In response, the shoots from each treatment bent the same amount.

- This result is inconsistent with the auxin destruction hypothesis. If light destroys auxin, then the block exposed to light should be much less effective in inducing bending.

- Briggs generated additional evidence in support of the Cholodny-Went hypothesis by dividing tips fully or partially with mica. He showed that auxin had been transported from one side to the other, as in **Figure 39.10**.

The Auxin Receptor

- Researchers set out to attach a radioactive label to the hormone, treat cells with the labeled hormone,

and purify the protein or proteins to which it binds.

- Researchers succeeded in isolating auxin-binding protein 1, or ABP1, from corn plants in 1985.

- Experiments have confirmed that ABP1 is an auxin receptor that is located in the plasma membrane of cells in the stem.

- When it binds auxin, it triggers cell elongation.

The Cell-Elongation Response

- Researchers proposed that once ABP1 is bound to auxin, the signal transduction cascade leads to an increase in the number of membrane H^+-ATPases.

- Because the pH of the cell wall goes down when H^+-ATPases are active, the **acid-growth hypothesis** resulted.

- Two things have to happen for a plant cell to get larger:

  1. Water has to enter the cell. Proton pumps could lead to water entry because potassium and other positively charged ions often enter a cell after protons are pumped out. Water follows K^+ via osmosis.

  2. The cell wall must also expand. Cell wall expansion occurs in an **acidic** environment. Researchers believe that proteins called **expansins** are somehow responsible.

CHECK YOUR UNDERSTANDING

(b) Is there an auxin receptor? What proof can you provide?

_____

_____

_____

_____

39.3 Red and Far-Red Light: Germination, Flowering, and Stem Elongation

Phytochromes as Red/Far-Red Receptors

- Photoreversibility occurs when light causes a photoreceptor pigment to change shape. Each conformation would be responsible for a different response.

- **Phytochromes** from young corn shoots were found to be photoreversible. When they were placed in a solution and exposed to red or far-red light, phytochrome color switched from blue to blue-green and black.

- **Photoperiodism** is any response by an organism that is based on the photoperiod—the relative lengths of day and night.

- Plants fall into three categories (**Figure 39.13**):

  1. **Long-day plants** bloom in midsummer when days are at their longest.

  2. **Short-day plants** bloom in spring, late summer, or fall.

  3. **Day-neutral plants** flower without regard to photoperiod.

How Were Phytochromes Isolated?

- The genome of *Arabidopsis thaliana* is a small, weedy mustard plant that serves as a model organism in the molecular studies of plants. It actually has five distinct loci that encode phytochrome proteins—*PHYA*, *PHYB*, etc.

- Each of these proteins holds a small pigment molecule that absorbs light in the red and far-red parts of the spectrum.

CHECK YOUR UNDERSTANDING

(c) What is the difference between a pigment and a phytochrome?

_____

_____

_____

39.4 Gravity: The Gravitropic Response

- **Gravitropism** is the ability of plants to move in response to gravity. Charles and Francis Darwin (1881) found that roots stop responding to gravity if their **root caps** are removed.

- Researchers recently demonstrated that by killing tiny blocks of cells in *Arabidopsis* with a laser beam, they were able to determine that the cells directly under the epidermal cells at the tip of the root are the most important for initiating the gravitropic response.

The Statolith Hypothesis

- The **statolith hypothesis** holds that **amyloplasts** (starch-storing organelles) are the primary gravity sensors in plants. The idea is that gravity pulls the amyloplasts to the bottom of cells, where the force of the amyloplasts on the cell membrane activates pressure or stretch receptors that initiate the gravitropic response.

- Although recent experiments strongly support the statolith hypothesis, the receptor itself has yet to be found.

Auxin as the Gravitropic Signal

- Root cap cells that sense changes in the direction of gravitational pull respond by changing the distribution of auxin in the root tip.

- Auxin flows down the middle of the root and then toward the perimeter and away from the root cap.

- The differences in auxin concentrations results in differences in cell elongation (**Figure 39.19**).

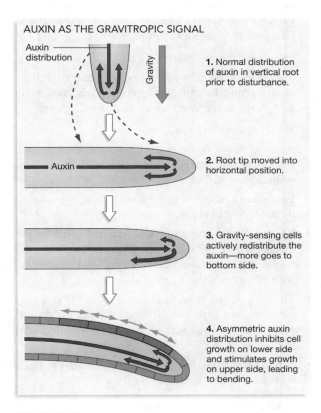

AUXIN AS THE GRAVITROPIC SIGNAL

Auxin distribution

Gravity

1. Normal distribution of auxin in vertical root prior to disturbance.

Auxin

2. Root tip moved into horizontal position.

3. Gravity-sensing cells actively redistribute the auxin—more goes to bottom side.

4. Asymmetric auxin distribution inhibits cell growth on lower side and stimulates growth on upper side, leading to bending.

FIGURE 39.19

CHECK YOUR UNDERSTANDING

(d) Explain the statolith hypothesis.

_____

_____

_____

_____

39.5 Youth, Maturity, and Aging: The Growth Responses

Auxin and Apical Dominance

- **Apical dominance** is a growth in which the majority of stem elongation occurs at the apical meristem of the main shoot. The presence of this topmost meristem inhibits growth by apical meristems that are present lower down on the plant, in nodes.

- Apical dominance occurs because a continuous flow of auxin from the tips of growing shoots to the tissues below signals the direction of growth. If the signal stops, it means that growth has been interrupted. In response, lateral branches sprout and begin to take over for the main shoot.

Polar Transport of Auxin

- Transport of auxin is **polar**, or unidirectional. It travels in an "inverted umbrella" pattern. At the root, auxin is redistributed before it travels back up the plant.

What Is Auxin's Overall Role?

- Auxin clearly plays a key role in controlling growth via apical dominance, phototropism, and gravitropism. Auxin has other important effects.

- Fruit development is influenced by auxin produced by seeds within the fruit.

- Falling auxin concentrations are involved in the **abscission**, or the shedding of leaves and fruits, associated with **senescence**, or aging. Auxin interacts with ethylene in these processes.

- The presence of auxin in growing roots and shoots is essential for the proper differentiation of xylem and phloem cells in vascular tissue and the development of vascular cambium.

- Auxin stimulates the development of adventitious roots in tissue cultures and cuttings.

Cytokinins and Cell Division

- **Cytokinins** are a group of plant hormones that promote cell division.

- Cytokinins are synthesized in root tips, young fruits, seeds, growing buds, and other developing organs.

- The cytokinin that has been found in the most species is derived from adenine and is called zeatin.

- Cytokinins act as growth hormones by activating the genes that keep the cell cycle going. In the absence of cytokinins, cells arrest at the G_1 checkpoint in the cell cycle and cease growth.

Gibberellins and ABA: Growth and Dormancy

- **Dormancy** is a temporary state of reduced or no metabolic activity.

- Growth responses and dormancy are mediated by two main hormones, **abscisic acid** (**ABA**) and the **gibberellic acids** (**GAs**).

Gibberellins Stimulate Shoot Elongation

- In stems, gibberellins appear to promote both cell elongation and rates of cell division. In seeds, GAs activate the transcription of digestive enzymes that support germination and growth.

- In these tissues, GA action must be coordinated with the effects of cytokinins and auxin.

- The strong association between shoot elongation and gibberellin dosage gave researchers an important tool for dissecting how GAs work.

- By using forward genetics, researchers began with mutant phenotypes and attempted to characterize the loci responsible for the defect.

- The locus responsible for the stem-length differences came to be known as *Le*. Early work on *Le* mutants showed that they attain normal height if they were treated with the gibberellin called GA_1.

- Researchers confirmed that the *Le* locus encodes an enzyme involved in GA synthesis by finding a locus in the pea genome that encodes an enzyme called **3β-hydroxylase**. This enzyme adds a hydroxyl group to GA_{20} to produce GA_1. Mutant enzymes were unable to convert GA_{20} to GA_1.

Gibberellins and ABA Interact during Seed Dormancy and Germination

- By applying hormones to seeds, researchers learned that in many plants, ABA is the signal that inhibits seed germination and gibberellins are the signal that triggers embryonic development.

- During the germination of a barley seedling, **α-amylase** is released from a tissue called the **aleurone layer**. α-amylase is significant because it acts as a digestive enzyme that breaks the bonds between sugar subunits of starch. The enzyme diffuses into the carbohydrate-rich endosperm tissue and releases sugars that can be transported and used by the growing embryo.

- Adding GA to the aleurone layer increases the production of α-amylase; adding ABA decreases α-amylase levels.

- Summary of hormone action:
  1. A cell's response to a hormone often occurs because specific genes are turned on or off.
  2. Hormones rarely act on DNA directly. Instead, a receptor on the surface of a cell usually receives the message and responds by initiating a chain of events that leads to gene activation or repression.

3. Different hormones interact at the molecular level because they induce different transcription activators and repressors.

- Researchers noticed that the promoter sequence near the α-amylase gene resembles the DNA sequences targeted by a class of transcription factors called the **Myb proteins**. Mybs are DNA-binding proteins that turn gene transcription on or off.

- It was proposed that a Myb might turn α-amylase on in response to a signal from GA or off in response to a signal from ABA (more specifically, see **Figure 39.29a**).

- Researchers began by isolating all of the mRNAs produced in the aleurone layer of a germinating barley seed. After they used reverse transcriptase to make DNA copies of the mRNAs, they tested each DNA sequence to see if it would react with the DNA-binding sequence that characterizes Mybs. One DNA sequence did, and they confirmed that a Myb protein exists in activated aleurone tissue.

- Was this transcription factor produced specifically in response to GA? Follow-up experiments confirmed that this protein, which they named GAMyb, binds to the α-amylase promoter and acts as a transcription activator that stimulates α-amylase production.

ABA Closes Guard Cells in Stomata

- Early work on stomatal opening and closing suggested that ABA is involved in communicating from roots to leaves about water conditions. Applying ABA to the exterior of guard cells causes them to close.

- Researchers were able to confirm two important predictions:

1. ABA concentrations in roots on the dry side of the pot were extraordinarily high.

2. ABA concentrations in the leaves of experimental plants were much higher than in the controls.

- These results suggested that ABA from roots is transported to leaves, and that it actually does serve as an early warning system of drought stress.

- Changes in cell shape are related to changes in the activity of H⁺-ATPases in the plasma membrane. See above section on stem elongation and the acid-growth hypothesis.

Ethylene and Senescence

- **Senescence** is the regulated process of aging, decline, and eventual death of an entire organism or particular organs, such as plant fruits and leaves.

- **Ethylene** is strongly associated in three aspects of senescence in plants:

1. Fruit ripening, which eventually leads to rotting

2. The fading of flowers

3. Abscission, or the detachment of leaves

- The **abscission zone** is a region of the leaf petiole that is more sensitive to ethylene in the tissue. As a result, it degrades first and the leaf breaks off at this point.

CHECK YOUR UNDERSTANDING

(e) What is a Myb protein, and how does it relate to plant growth?

_____

_____

_____

39.6 Pathogens and Herbivores: The Defense Responses

How Do Plants Sense and Respond to Pathogens?

- The ability to cause disease is called **virulence**, and disease-causing agents are called **pathogens**.

- Responses to attacks are called **induced defenses**.

- The rapid and localized death of one or a few infected cells is called the **hypersensitive response (HR)**.

The Gene-for-Gene Hypothesis

- Plants have disease-resistant genes known as **resistance (R) loci**. The genes associated with virulence or antivirulence in pathogens are known as **avirulence (avr) loci**.

- The idea that R and avr gene products interact in a specific way is called the **gene-for-gene hypothesis**. R genes produce receptors, and avr genes produce **ligands**—molecules that bind to receptors.

Why Is the Existence of So Many Resistance Genes and Alleles Significant?

- R genes are similar in structure, and within a population of plants there are usually many different alleles at each R locus.

- Different alleles allow plants to recognize different proteins from the same pathogen.

The HR Response

- The binding of an R gene product to an avr protein triggers the production of NO and **reactive oxygen intermediates (ROIs)**.

- ROIs disrupt enzymes in cells at the point of infection and thus kill those cells.

Salicylic Acid Extends the HR

- Direct application of salicylic acid to tissues triggers **systemic acquired resistance (SAR)**—a slower, widespread set of events.

- Breaking down **salicylic acid (SA)** in tobacco plants made them susceptible to a wide variety of pathogens.

- Radiolabeled SA has been shown to be transported from infected to uninfected tissues.

How Do Plants Sense and Respond to Herbivore Attack?

- Protease inhibitors found in plants makes herbivores sick. They can then detect this compound by taste and avoid the plant.

- **Systemin** is a peptide that triggers a series of reactions that eventually result in the synthesis of jasmonic acid, which initiates the production of defense compounds.

Pheromones Released from Plant Wounds Recruit Help from Wasps

- An organism that is free living as an adult but parasitic as a larvae is called a **parasitoid**.

- Plants produce **pheromones**—chemical messengers that are synthesized by an individual and released into the environment—to attract wasps in response to attack by caterpillars.

CHECK YOUR UNDERSTANDING

(f) Describe a wound response hormone and how it is activated.

_____

_____

_____

_____

B. INTEGRATING YOUR KNOWLEDGE

Looking Back—Key Concepts from Earlier Chapters

Absorbing Light—Chapter 10

Recall from **Chapter 10** that land plants absorb many of the wavelengths in the red and blue portions of the spectrum, but they do not absorb wavelengths in the far red.

Phosphorylation—Chapter 9

Many proteins switch from inactive to active states, or vice versa, when a phosphate group (PO_4^{2-}) from ATP is added to them.

Potential Energy—Chapters 2 and 37

Chapter 2 explained that the chemical potential electron describes its tendency to move closer to a nucleus. **Chapter 37** pointed out that the water potential of a tissue or substance expresses the tendency of water to move toward it or away from it. In the same way, electrical potential is the tendency of charged particles to move toward an area of opposite charge.

Gene Hunting—Chapter 19

Gene hunting and other strategies for finding the ABP1 locus were discussed in detail in **Chapter 19**.

Looking Forward—Key Concepts in Later Chapters

Fruit Ripening—Chapter 40

As **Chapter 40** will show, fruit ripening is interpreted as an adaptation that enhances the attractiveness of fruits to birds, mammals, and other animals that disperse seeds to new locations.

The Action Potential—Chapter 45

In **Chapter 45** we will examine the molecular mechanisms responsible for the action potential in much more detail.

C. MASTERING CHALLENGING TOPICS

The world of plant hormones is a complicated one, and this chapter just begins to scratch the surface of complexity. The key concepts you must understand are that a hormone can have (1) a multitude of actions, depending on the location in the organism; and (2) a multitude of interactions, depending on both the location and the nature of the interactive chemical. You will encounter more actions and interactions of ABA and the Gas, so it is important to learn their present functions as discussed in this chapter. Organize both hormones in a chart and list their actions, locations of those actions, and possible interactions with each other. Continue to describe their actions in a broad sense and then proceed to describe these at the molecular level, citing the research used to determine each hormone. This process will be of great benefit as you move to the chapters dealing with hormones in animals.

D. ASSESSING WHAT YOU'VE LEARNED

1. Any directed movement in response to light is called:
   a. photoperiodism
   b. tropism
   c. phototropism
   d. gravitropism

2. An organic substance produced in small quantities in one part of the plant and transported to target cells, causing a physiological response, is:
   a. sodium
   b. a protein
   c. a hormone
   d. nucleotides

3. Which plants would be likely to bloom in midsummer?
   a. day-neutral plants
   b. short-day plants
   c. long-day plants
   d. none of the above

4. Which molecules hydrolyze ATP and catalyze the addition of a phosphate group?
   a. pigments
   b. phototropins
   c. phosphate genes
   d. protein kinases

5. The technical name for blue-light receptors is:
   a. phytochromes
   b. phototropins
   c. pigments
   d. transducins

6. The process that translates a signal from the environment into an energy form that the plant decode is:
   a. signal transduction
   b. signal deciding
   c. signal transmission
   d. neural transmission

7. The hypothesis stating that amyloplasts are the primary gravity sensors in plants is the _____.
   a. statolith hypothesis
   b. gravitational pressure hypothesis
   c. amyloplast hypothesis
   d. none of the above

8. What is the best explanation for how cell elongation occurs?
   a. active pumping of water
   b. active pumping of ions
   c. active pumping of ions, passive movement of water
   d. active pumping of ions, active movement of water

9. Which of the following is *true* about auxin transport?
   a. Auxin transport is random.
   b. Auxin transport is bidirectional.
   c. Nothing is known about the specifics of auxin transport.
   d. Auxin transport is unidirectional.

10. The two main hormones that mediate growth responses and dormancy are:
    a. auxin and cytokinins
    b. gibberellins and cytokinins
    c. gibberellins and auxin
    d. gibberellins and abscisic acid

CHAPTER 39—ANSWER KEY

Check Your Understanding

(a) Transduction is the changing of energy from one form into another. In an animal, a signal from the environment, such as light, can be transduced into chemical or electrical energy.

(b) Yes. Lobler and Klambt succeeded in isolating auxin-binding protein 1, or ABP1, from corn plants in 1985.

(c) Plants contain a wide variety of pigments that absorb light. These molecules include the chlorophylls and carotenoids involved in photosynthesis. Phytochrome is a specific type of pigment that is known for being photoreversible.

(d) The statolith hypothesis holds that amyloplasts (starch-storing organelles) are the primary gravity sensors in plants. The idea is that gravity pulls the amyloplasts to the bottom of cells, where the force of the amyloplasts on the cell membrane activates pressure or stretch receptors that initiate the gravitropic response.

(e) Gubler and associates noticed that the promoter sequence near the α-amylase gene resembles the DNA sequences targeted by a class of transcription factors called the Myb proteins. Mybs are DNA-binding proteins that turn gene transcription on or off. Gubler's team proposed that a Myb might turn α-amylase on in response to a signal from GA or off in response to a signal from ABA.

(f) A wound response hormone is called systemin. It has been confirmed to move from damaged sites to undamaged sites. Systemin initiates a series of reactions that eventually produce jasmonic acid, which in turn initiates the production of many insecticide-type compounds including protease inhibitors.

Assessing What You've Learned

1. c; 2. c; 3. c; 4. d; 5. b; 6. a; 7. a; 8. c; 9. d; 10. d

40

Plant Reproduction

A. KEY CONCEPTS

40.1 An Introduction to Plant Reproduction

Sexual Reproduction

- Most plants reproduce sexually. **Sexual reproduction** is based on the reduction division known as **meiosis** and on **fertilization**, the fusion of haploid cells called **gametes**.

- **Sperm** are small cells from the male that contribute genetic information in the form of DNA but few or no nutrients to the offspring. Female gametes are called **eggs**, which contribute a store of nutrients to the offspring.

- **Outcrossing** occurs when male and female gametes are exchanged between individuals of the same species. When **self-fertilization** occurs, a sperm and an egg from the same individual unite to form a progeny.

- The primary advantage of **selfing** is that successful **pollination** is virtually assured.

The Land Plant Life Cycle

- Plants are the only organisms that have both a multicellular form that is diploid and a multicellular form that is haploid. This life cycle is called **alternation of generations**.

- The diploid phase is called the **sporophyte**, and the haploid phase is called the **gametophyte**.

- A **spore** is a cell that grows directly into an adult individual. A **gamete** is a reproductive cell that must fuse with another gamete before growing into a new individual.

- Meiosis occurs in sporophytes and results in the production of haploid spores, which are produced in structures called **sporangia**.

- Spores divide by mitosis to form multicellular, haploid gametophytes.

- Fertilization occurs when two gametes fuse to form a diploid zygote. The zygote grows by mitosis to form the sporophyte.

- Study the following diagram (**Figure 4.20**) to make sure you understand these life-cycle intricacies.

Asexual Reproduction

- **Asexual reproduction** does not involve meiosis or fertilization. It leads to offspring that are genetically identical to the parent plant. The major advantage of asexual reproduction is its high efficiency.

- However, if a fungus or other disease-causing agent infects an individual, it will probably succeed in infecting the plant's asexual offspring as well.

CHECK YOUR UNDERSTANDING

(a) What is the difference between outcrossing and self-fertilization?

_____

_____

_____

40.2 Reproductive Structures

When Does Flowering Occur?

- A flower is a modified shoot that develops from a compressed stem and highly modified leaves.

- Flower formation can be stimulated by external or internal cues, or both.

The General Structure of the Flower

- **Sepals** are leaflike structures that comprise the outermost part of the flower.

- **Petals** are arranged in a stem in a whorl. They are often brightly colored and advertise the flower to visually oriented animals such as bees, wasps, and hummingbirds.

- The **nectary** produces sugar-rich fluid called **nectar**, which is harvested by many of the animals that visit.

(a) Liverworts: Gametophytes are large and long lived; the sporophyte is small and short lived.

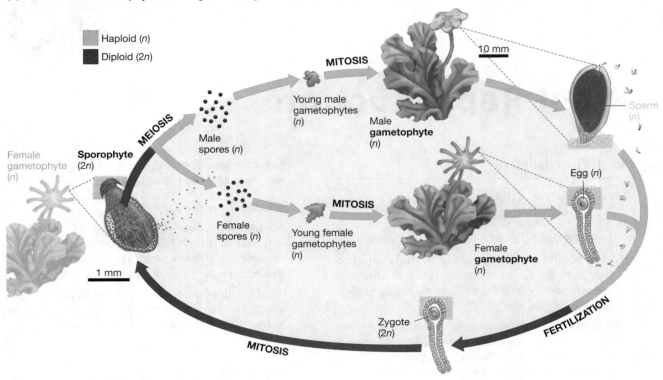

(b) Angiosperms: Sporophyte is large and long lived; gametophytes are small (microscopic) and short lived.

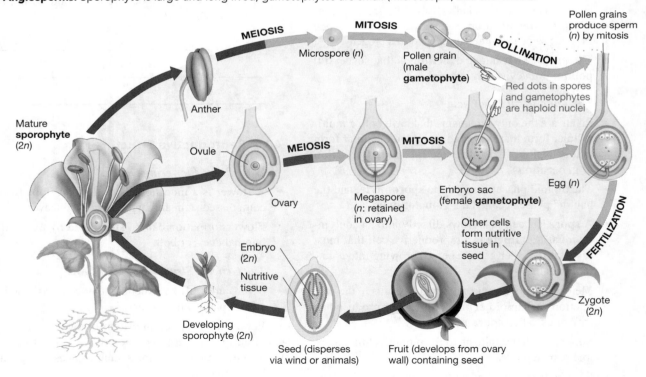

- An entire group of petals is called a **corolla**. The male reproductive structure of angiosperms is called a **stamen**, while the female reproductive structure is called the **carpel**.

- Flowers that contain both stamens and carpels are referred to as **perfect**. If they have either stamens or carpels, they are called **imperfect**.

- When imperfect flowers have both stamens and carpels on the same plant (but not the same flower), they are called **monoecious**.

- If an individual plant produces either stamens or carpels (but not both), the species is **dioecious**.

How Are Female Gametophytes Produced?

- The carpel consists of three parts: the **stigma**, the **style**, and the **ovary**. The **ovule**, inside the ovary, is where meiosis takes place and where the female gametophyte is produced. Four nuclei result from meiosis, but three of theme degenerate. The remaining haploid nucleus is the **megaspore**.

- The megaspore divides by mitosis to produce eight **nuclei** (the number varies among species), which segregate to different positions and form seven cells, comprising the **embryo sac**. The most important elements of the embryo sac are the **egg** and the **polar nuclei**. The egg is located near the base of the structure by an opening called the **micropyle**.

How Are Male Gametophytes Produced?

- The **stamen** consists of the **anther** and a **filament**. Cells inside the anther undergo meiosis. Each of the haploid cells that results is called a **microspore**.

- Each microspore becomes a **pollen grain**. When mature, a pollen grain consists of a small **generative cell** enclosed within a large *vegetative cell*. The vegetative cell develops a *hard coat*, which protects the cell contents and generative cell when pollen grains are shed.

- Pollen grains *do not* contain sperm. Instead, they are tiny **gametophytes**. The haploid generative cell will produce sperm cells via mitosis.

CHECK YOUR UNDERSTANDING

(b) Compare and contrast a spore and a gamete.

_____

_____

_____

40.3 Pollination and Fertilization

Pollination

- **Pollination** is the transfer of pollen from an anther to a stigma.

- In many angiosperms, pollen is carried from the anther of one individual to the stigma of a different individual (**cross-pollination**) on the body of an insect, bird, or bat that moves from flower to flower. Animal pollination is an example of **mutualism**.

What Is the Adaptive Significance of Pollination?

- The more recently evolved plant groups do not need water in order for sexual reproduction to occur. As a result, the evolution of pollen made these species less dependent on wet habitats and allowed colonization of dryer habitats.

- Pollination evolved into a much more precise process when animals began to act as pollinators. Insect pollination is an important adaptation because it makes sexual reproduction much more efficient.

Does Pollination by Animals Encourage Speciation?

- Insect pollination appears to be closely associated with the formation of new flower and insect species.

- A biologist documented differences in alpine skypilot flowers. In the lower timberline regions, flowers are small and have a skunk-like odor, while in the tundra regions, flowers are large and smell sweet.

- Large bumblebees pollinate the tundra flowers, while small flies pollinate the timberline flowers. Flower morphology is in tune to attract the type of insect that is abundant in that area. These two types of flowers are on their way to evolving into two distinct species.

Fertilization

- The male gametophyte produces a long projection called a **pollen tube** that grows down the length of the style.

- This growth is directed by chemical attractants released by **synergids**—cells in the female gametophyte that lie close to the egg.

- When the pollen tube reaches the micropyle, the sperm are discharged into the ovule.

- **Fertilization** occurs when a sperm and an egg actually unite to form a diploid zygote.

- In angiosperms, an event called **double fertilization** takes place.

- One sperm unites with the egg nucleus to form the zygote. The other sperm nucleus moves through the embryo sac and fuses with the two polar nuclei to form a single triploid ($3n$) cell.

- The triploid cell resulting from the second fertilization begins a series of mitotic divisions that form tissue called **endosperm**, which stores nutrients.

CHECK YOUR UNDERSTANDING

(c) What is "double fertilization," and what makes it different from the typical fertilization process?

_____

_____

_____

_____

40.4 The Seed

Embryogenesis

- As a seed matures, the embryo and endosperm develop inside the ovule and become surrounded by a covering called the **seed coat**. At the same time, the ovary around the ovule develops into a fruit.
- **Embryogenesis** is the process by which a single-celled zygote becomes a multicellular embryo.
- When a zygote divides, it produces two **daughter cells**: the **basal cell** divides to form a row of single cells, and the **terminal cell** is the parent cell of all the cells in the embryo.
- The terminal cell and its progeny divide and sort into three groups:
  1. The exterior layer forms the **protoderm**.
  2. The cells just inside the exterior layer form the **ground meristem**.
  3. A group of cells (**procambium**) in the core of the embryo become the vascular tissue.
- **Cotyledons** are the seed leaves that take up the nutrients in the endosperm and store them. The **hypocotyl** is the initial stem, and the **radicle** is the first root structure.
- By the time the embryo matures, the three tissue types have differentiated and the structures mentioned above have formed. The seed tissues then dry, and the embryo becomes **quiescent**.

The Role of Drying in Seed Maturation

- Water loss is an adaptation that prevents seeds from germinating on the parent plant.

- Researchers have shown that as water leaves the seed during drying, **sugars** begin to interact with the hydrophilic parts of cell components. These interactions stabilize the proteins and membranes before damage occurs.

Fruit Development and Seed Dispersal

- After fertilization occurs in angiosperms, the cells that make up the ovary develop a structure called the **pericarp**. This protects the seeds. The mature structure is called a **fruit**.
- **Simple fruits** like the apricot develop from a single flower that contains a single carpel. **Aggregate fruits** like the raspberry develop from a single flower containing many separate carpels. **Multiple fruits** like the pineapple develop from many flowers with many carpels.
- Most dry fruits are either dispersed by wind or simply fall to the ground. Animals are the most common dispersal agent of fleshy fruits.

Seed Dormancy

- Even though water and oxygen are available, seeds may not germinate; this process is called **seed dormancy**.
- Researchers have studied lotus plant seeds from an old dried-up lakebed in China. The oldest seed that germinated was 1300 years old.
- Dormancy is usually a feature of seeds from species that inhabit seasonal environments.
- Dormancy is usually interpreted as an adaptation that allows seeds to remain viable until conditions improve.

What Role Does ABA Play in Dormancy?

- ABA levels have not been universally correlated with dormancy. Researchers have concluded that there is no single universal mechanism for initiating and maintaining seed dormancy.

How Is Dormancy Broken?

- For some seeds, the seed coat must be disrupted, or **scarified**, for germination to occur. For others, exposure to water, light, oxygen, or fire will initiate germination.
- The cue that triggers germination is a reliable signal that conditions for seedling growth are favorable for a particular species in a particular environment.

Seed Germination

- Seeds do not germinate without water, because water uptake is the first event in germination.

- During the first phase of water uptake, oxygen consumption and protein synthesis in the seed increase dramatically, but no new messenger RNAs are transcribed. Based on these observations, biologists have concluded that the earliest events in germination are driven by mRNAs that are stored in the seed.

- During the second phase, when water uptake stops, newly transcribed mRNAs appear and are translated into protein products. Mitochondria also begin to multiply. In effect, seeds take up enough water to hydrate their existing proteins and membranes and then begin to manufacture the proteins and mitochondria needed to support growth. Water uptake resumes as growth begins. This second bout of water uptake enables cells to enlarge and the embryo to burst from the seed coat.

CHECK YOUR UNDERSTANDING

(d) What protection mechanisms do seeds have to protect them during dormancy?

_____

_____

_____

B.   INTEGRATING YOUR KNOWLEDGE

*Looking Back—Key Concepts
from Earlier Chapters*

Inbreeding—Chapter 25

Self-fertilization is the most extreme form of inbreeding. Inbreeding is defined as reproduction among relatives. **Chapter 25** explored the consequences of inbreeding in detail and presented data showing that inbred offspring have poor fitness.

Diploid and Haploid Cells—Chapters 11 and 12

Except for certain red, brown, and green algae, plants are the only organisms that have both a multicellular form that is diploid and a multicellular form that is haploid. Recall from **Chapters 11 and 12** that diploid cells have two copies of each chromosome, while haploid cells have one.

Life Cycles of Plants—Chapter 30

In flowering plants, the gametophyte generation is tiny, short lived, and dependent on the sporophyte for nutrition. **Chapter 30** explored the diversity of life cycles found in plants in more depth.

Tissue Layers—Chapter 32

The three tissue types formed during embryogenesis were discussed in detail in **Chapter 32**. Chapter 40 shows the relationship between these tissue types and the originating cells.

*Looking Forward—Key Concepts
in Later Chapters*

Disease—Chapter 50

If a fungus or other disease-causing agent infects a big bluestem individual, it will probably succeed in infecting the plant's asexual offspring as well. As **Chapter 50** will show, plants fight disease with a wide variety of molecules.

C.   MASTERING CHALLENGING TOPICS

The concepts in this chapter are fairly straightforward, but there are many new anatomical terms and physiological processes to learn. The best way to begin studying the anatomy is by drawing pictures and labeling them. Once you are comfortable with the anatomy, move on to the processes of pollination, fertilization, and embryogenesis. Using the figures in your textbook, map out each process while drawing panels and describing each step. Use the proper terms as you have learned them for each structure. Lastly, try describing each process with only words, no pictures. This will help you to understand the concepts and not just simply memorize facts.

D.   ASSESSING WHAT YOU'VE LEARNED

1. Which kind of flower contains both male and female structures?
   a. perfect flower
   b. imperfect flower
   c. haploid flower
   d. diploid flower

2. Which kind of reproduction increases the genetic variation of the offspring?
   a. asexual reproduction
   b. alternation of generations
   c. sexual reproduction
   d. sporophyte production

3. What are the leaflike structures that comprise the outermost part of the flower?
   a. the petals
   b. the sepals
   c. the nectary
   d. the leaves

4. The sugar-rich fluid in the flower that is harvested by many animals is called:
   a. juice
   b. nectar
   c. pollen
   d. fruit

5. Which of these statements about reproductive structures is *false*?
   a. The stamen is the female reproductive structure of angiosperms.
   b. The carpel consists of the stigma, the style, and the ovary.
   c. The stamen consists of the anther and filament.
   d. The pollen grains are tiny gametophytes.

6. When pollen is carried from the anther of one individual to the stigma of a different individual, it is called:
   a. self-pollination
   b. self-fertilization
   c. cross-pollination
   d. cross-fertilization

7. Animal pollination is an example of:
   a. parasitism
   b. competition
   c. commensalism
   d. mutualism

8. "One sperm unites with an egg nucleus, and another sperm unites with two polar nuclei." This scenario best describes:
   a. fertilization in gymnosperms
   b. double fertilization in angiosperms
   c. pollination in angiosperms
   d. fertilization in angiosperms

9. Which of these statements about seed dormancy is *false*?
   a. ABA is the universal hormone initiating seed dormancy.
   b. Seeds can lie dormant for hundreds of years.
   c. Dormancy is usually a feature of seeds from species that inhabit seasonal environments.
   d. Dormancy is usually interpreted as an adaptation that allows seeds to remain viable until conditions improve.

10. Which of the following is absolutely required for seed germination?
    a. fire
    b. light
    c. water
    d. soil

CHAPTER 40—ANSWER KEY

Check Your Understanding

(a) Outcrossing occurs when male and female gametes are exchanged between individuals of the same species. When self-fertilization occurs, a sperm and an egg from the same individual unite to form a progeny.

(b) A spore is a reproductive cell that grows into a new individual directly. A gamete is a reproductive cell that must fuse with another gamete before growing into a new individual.

(c) Fertilization occurs when a sperm and an egg actually unite to form a diploid zygote. In angiosperms, an event called double fertilization takes place. One sperm unites with the egg nucleus to form the zygote. The other sperm nucleus moves through the embryo sac and fuses with the two polar nuclei to form a single triploid ($3n$) cell.

(d) Researchers have shown that as water leaves the seed during drying, sugars begin to interact with the hydrophilic parts of cell components. These interactions stabilize the proteins and membranes before damage occurs.

Assessing What You've Learned

1. a; 2. c; 3. b; 4. b; 5. a; 6. c; 7. d; 8. b; 9. a; 10. c

Animal Form and Function

41

A. KEY BIOLOGICAL CONCEPTS

41.1 Form, Function, and Adaptation

- Adaptive evolution shapes animal populations.

- Adaptive evolution occurs when the frequency of alleles subject to natural selection increases from one generation to the next *and* the resulting changes in the characteristics of the population lead to higher average fitness in a particular environment.

- Natural selection is not the only process that leads to changes in allele frequencies over time. Evolution occurs via the random process called genetic drift, the movement of alleles into and out of populations by migration (gene flow), and the constant introduction of new alleles by mutation.

CHECK YOUR UNDERSTANDING

(a) What processes lead to changes in allele frequency in a population?

_____

_____

_____

The Role of Fitness Trade-Offs

- One of the most important constraints on adaptation involves **trade-offs**. This may involve expenditures of time or energy.

- Comparative physiologists investigated the predicted trade-off in egg size and egg number by manipulating these parameters in side-blotched lizards (**Figure 41.1**). They were able to induce the production of large clutches of small eggs by catching females and surgically removing yolk from their eggs. By removing eggs early on, they found that the remaining eggs grew larger.

Adaptation and Acclimatization

- **Adaptation** occurs when a population changes in response to natural selection. **Acclimatization** occurs when an individual changes in response to a change in environmental conditions.

CHECK YOUR UNDERSTANDING

(b) Using examples, explain the difference between adaptation and acclimatization.

_____

_____

_____

41.2 Tissues, Organs, and Systems: How Does Structure Correlate with Function?

- Researchers studying medium ground finches on the Galápagos island of Daphne Major found that beak size and shape vary among individuals, and that this trait is heritable. During a major drought, much of the population disappeared. The researchers found that survivors tended to have much deeper beaks than did the birds that died. What was the likely cause of this enduring morphological trait?

- Natural selection is reflected in a strong correlation between structure and function. This relationship is true at multiple levels: structure correlates with function at molecular, cellular, and tissue levels as well as at the gross anatomical level (**Figure 41.3**).

Tissues

- A **tissue** is a group of cells with the same structure and function.

- There are four basic types of tissues (**Figure 41.3**):
    1. The **epithelial tissues** cover the outside of the body and line the surfaces of organs. This tissue consists of layers of tightly packed cells.

(c) Why aren't all traits adaptive? Give an example.

_____

_____

_____

2. **Connective tissue** is made up of cells that are loosely arranged in a liquid, jellylike, or solid extracellular matrix. **Loose connective tissue** serves as a packing material between organs. **Cartilage** and **bone** support the body, and **blood** transports material throughout the body.

3. **Muscle tissue** functions in movement. Muscle cells are packed with specialized proteins that move in response to **phosphorylation**.

4. Nerve cells or neurons make up **nervous tissue**. Although their shapes vary widely, all neurons have connections to other cells and deliver signals in the form of electrical impulses.

Organs and Systems

- An **organ** is a structure that serves a specialized function and consists of several tissues (**Figure 41.ba**).

- A **system** consists of tissues and organs that work in conjunction to perform a function (**Figure 41.7b**).

(d) Explain the difference between a tissue and an organ.

_____

_____

_____

41.3 How Does Body Size Affect Animal Physiology?

Surface Area/Volume Relationships

- As a cell gets larger, its volume increases much faster than its surface area. Examine **Figure 41.9** and convince yourself that this is true.

- Given this relationship, the transport processes that occur across the cell membrane would have to support disproportionately more and more cell volume as the cell increased.

Comparing Mice and Elephants

- **Metabolic rate** is defined as the overall rate of energy consumption by an individual. Because it is usually based on aerobic respiration, it is measured in units of ml O_2 consumed per hour.

- Even though an elephant consumes a great deal more oxygen per hour than a mouse does, if you looked at 1 gram of mouse tissue, it would consume about 12 times more oxygen per hour than 1 gram of elephant tissue would (**Figure 41.10**).

- The leading hypothesis to explain this pattern is based on surface area/volume ratios. Many aspects of metabolism, digestion, the delivery of nutrients, and the removal of wastes depend on the exchange of materials across surfaces.

Adaptations That Increase Surface Area

- Because gills have flattened, sheetlike structures called lamellae, the organ has an extremely high surface area relative to its volume (**Figure 41.13**). Diffusion of gases from water into blood can take place rapidly enough to keep up with the growth in volume of the developing fish.

- If the function of a cell or tissue depends on diffusion, its structure most likely has a shape that increases its surface area relative to its volume. Can you explain how the structure of the digestive system and circulatory systems are consistent with these observations?

(e) Why does surface area increase more slowly than volume in a growing organism?

_____

_____

_____

_____

41.4 Homeostasis

- **Homeostasis** is the maintenance of relatively constant chemical and physical conditions in an animal's cells, tissues, and organs.

- Epithelium plays a vital role in creating an internal environment that is dramatically different from the external environment, and in maintaining physical and chemical conditions inside an animal that are relatively constant. Its most basic function is to control the exchange of materials across surfaces.

The Role of Regulation and Feedback

- Achieving homeostasis requires a regulatory system that has a specific **set point**, or normal value. The temperature that you set on the dial of a room's thermostat is an example of a set point.

- Regulatory systems typically consist of three components: a sensor, an integrator, and an effector (**Figure 41.14**).

- A **sensor** is a structure that senses some aspect of the external or internal environment.

- An **integrator** is a component of the nervous system that evaluates the incoming sensory information and "decides" if a response is necessary to achieve homeostasis.

- An **effector** is any structure that helps to restore the desired internal condition.

- Regulatory systems depend on **negative feedback** to maintain constant conditions. Negative feedback occurs when the regulatory system makes a change in the opposite direction to a change in internal conditions (equivalent to a heater turning on and raising temperature when a room has cooled off).

41.5 How Do Animals Regulate Body Temperature?

- Animals exchange heat with the environment by varying combinations of conduction, convection, radiation, and evaporation (**Figure 41.15**).

- **Conduction** is the direct transfer of heat between two physical bodies that are in contact with each other. The rate at which conduction occurs depends on the surface area, the steepness of temperature difference, and how well each body conducts heat.

- **Convection** occurs when air or water moves over the body surface. As the speed of water or air flow increases, so does the rate of heat transfer.

- **Radiation** is the transfer of heat between two bodies that are not in direct physical contact. The major source of radiant energy is the Sun.

- **Evaporation** is the phase change that occurs when liquid water becomes a gas. It leads to heat loss only. As a result, water is an efficient coolant on a hot day.

Body Temperature and Homeostasis

- Many animals can control their body temperature through the process of **thermoregulation** (**Figure 41.16**). In mammals, this process is coordinated by neurons in the hypothalamus. Neural signals from temperature receptors in the skin (the "sensors") are interpreted by the hypothalamus (the "integrator"), which then causes varying responses, ranging from metabolic to behavioral (the "effectors").

Gaining and Losing Heat

- An **endotherm** produces heat in its own tissue, while an **ectotherm** relies on heat gained from the environment. Birds and mammals are endotherms; most other animals are ectotherms.

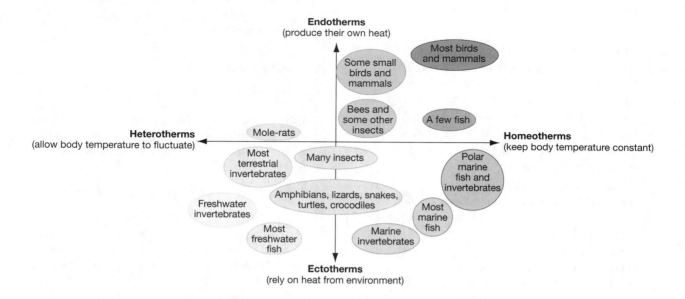

- **Homeotherms** keep their body temperature constant, whereas **heterotherms** experience changes in body temperature.

- There are numerous exceptions to these classifications—ectothermic pythons can generate body heat when they need to warm their eggs; tuna, mackerel, and certain other species of ectothermic fish generate heat to warm certain sections of their bodies.

- Compared to endotherms, ectotherms have low metabolic rates and lower mitochondrial enzyme activity.

Where Does Body Heat Come From?

- Heat is produced by muscle activity during movement and sometimes by the involuntary muscle contractions known as shivering.

- **Brown adipose tissue** (BAT) is a collection of specialized heat-generating cells (**Figure 41.19**). BAT has a high density of mitochondria and stored fats. When fats are oxidized by the mitochondria in BAT, no ATP is produced. Instead, all of the stored energy is released as heat.

CHECK YOUR UNDERSTANDING

(g) Would you expect to find brown fat in a reptile? Explain your answer.

_____

_____

_____

How Do Animals Conserve Heat?

- Air conducts heat poorly; thus, it is a good insulator. Endothermic animals have elaborate external structures that trap air, slow the rate of heat exchange, and conserve body heat.

- Water is a good conductor; thus, the body temperature of most aquatic invertebrates and fish is the same as the water they inhabit.

- To conserve heat in water, otters have dense, **water-repellent fur** that maintains a layer of trapped air against the fur. Thick layers of **fatty blubber** insulate seals and whales.

- Some aquatic mammals and birds have **countercurrent heat exchangers**, where arteries and veins lie beside each other and freely exchange heat (**Figure 41.20**). Explain how this anatomical arrangement conserves heat.

Ectothermy versus Endothermy

- Endotherms are able to maintain enzymes at optimal temperatures at all times, so mammals and birds can remain active in winter and at night. Due to their high metabolic rates and insulation, endotherms are also able to sustain very high levels of aerobic activities like running or flying.

- To fuel their high metabolic rates, endotherms have to obtain large quantities of energy-rich food.

- Ectotherms are able to thrive with much lower intakes of food. They can also use a greater proportion of their energy intake to support reproduction.

- Chemical reactions are temperature dependent, so muscular activity and digestion slow down dramatically as body temperature drops. As a result, ectotherms are more vulnerable to predation in cold weather than endotherms are.

B. INTEGRATING YOUR KNOWLEDGE

Looking Back—Concepts from Earlier Chapters

Natural Selection and Adaptation—Chapter 24

As **Chapter 24** explained, an adaptation is a trait that allows individuals to survive and reproduce better than individuals that lack this trait. This chapter also introduced some of the techniques that biologists use to establish that traits have a genetic basis.

Genetic Drift—Chapter 25

Natural selection is not the only process that leads to changes in allele frequencies over time. Evolution occurs via the random process called genetic drift—the movement of alleles into and out of populations by migration (gene flow) and the constant introduction of new alleles by mutation.

Form and Function—Chapter 7

The shape of proteins often relates to their role as enzymes or structural components of the cell. **Chapter 7** pointed out that strong correlations exist between the structure and function of the rough ER, Golgi apparatus, mitochondrion, chloroplast, and other organelles.

Looking Forward—Concepts in Later Chapters

The Vertebrate Eye—Chapter 46

The importance of historical constraint will surface again in the discussion of the vertebrate eye in **Chapter 46**. Although nonbiologists sometimes use the vertebrate eye as an example of a "perfect" adaptation, optometrists can attest that the organ is often defective.

Membrane Transport and Surface Area— Chapter 42

Nutrients such as glucose must diffuse into the cell, and waste products such as urea and carbon dioxide must diffuse out. As **Chapter 42** explains, the rate at which these and other molecules and ions diffuse depends in part on the amount of surface area available. In contrast, the rate at which nutrients are used and waste products are produced depends on the volume of the cell. **Chapter 42** will also explore how the gill functions as an adaptation for increased surface area.

C. MASTERING CHALLENGING TOPICS

When studying the section on body size and scaling, you may find that some of the concepts are rather abstract, and you will need concrete examples for each concept described. Surface area and volume relationships cannot simply be written down and studied. You should draw the examples listed in your text to learn what it means for the volume of a cell or organism to outgrow its surface area.

The study of allometry is very complex, and this chapter only touches on the basics. It is intuitive to understand that an elephant consumes more oxygen than a mouse does, but do you really understand why the mouse consumes more oxygen per gram of tissue than an elephant does? As you imagine these comparisons between animals, also place the animals in their natural environment and try to apply what you find about their anatomy and physiology to how they interact with their environment. Maintaining a "big picture" of these concepts in addition to learning everything in little chunks will also help you in understanding this chapter.

D. ASSESSING WHAT YOU'VE LEARNED

1. Which of the following descriptions is an example of *acclimatization*?
   a. Over many generations, the average length of giraffe necks has increased.
   b. The optimal temperature at which goldfish can swim at maximum speed will decrease if the goldfish is maintained in cold water.
   c. The average brain size of the ancestors of modern humans has increased dramatically over time.
   d. The emergence of the seed represents an important event in the evolution of land plants.

2. Consider several objects of different size that are heated and then allowed to cool. Which of the following would cool at the fastest rate?
   a. an object with a surface area of 20 and a volume of 10

   b. an object with a surface area of 40 and a volume of 30
   c. an object with a surface area of 8 and a volume of 2
   d. an object with a surface area of 10 and a volume of 30

3. What happens to the mass-specific metabolic rate of an organism as it gets bigger?
   a. Mass-specific metabolic rate decreases.
   b. Mass-specific metabolic rate remains constant.
   c. Mass-specific metabolic rate increases.

4. Patrick Wells and Alan Pinder found that the percentage of oxygen uptake by the gills increases as an Atlantic salmon grows from a small larva to an adult fish. Which of the following could explain this result?
   a. The gills provide a much lower surface-to-volume ratio, so that gas transfer becomes more efficient as the organism gets larger.
   b. The surface area of skin is much larger than the surface area of the gills.
   c. As an individual grows, the skin surface area decreases in relation to its volume (the surface-to-volume ratio drops). To keep the individual from suffocating, gills must take over most of the gas exchange activity.
   d. As an individual grows, the skin surface area increases in relation to its volume (the surface-to-volume ratio increases). To keep the individual from suffocating, gills must take over most of the gas exchange activity.

5. Which of the following describes an allometric relationship?
   a. a relationship in which two quantities change at the same rate
   b. a relationship in which one quantity increases while one quantity decreases
   c. a relationship in which two quantities both decrease
   d. a relationship in which two quantities change at different rates

6. The relationship between heart mass and body mass differs for dogs and cats. Which of the following statements explains these differences?
   a. Cat hearts are smaller because they have smaller bodies. The heart does not have to work as hard in a smaller animal.
   b. There is no difference in the size of dog and cat hearts when you compare animals of similar size.
   c. Cats are much more active than dogs. This increased activity is reflected in the data when comparing heart mass to body mass for dogs and cats.

d. The hunting style of dogs involves long-distance chases. Cats use short, quick sprints to ambush their prey. The increased activity in the hunting style of dogs requires a larger heart.

7. Which of the following terms implies the maintenance of relatively constant physical conditions?
   a. homeostasis
   b. acclimatization
   c. adaptation
   d. isometry

8. Which of the following terms describes an animal that maintains body temperature by producing heat in its own tissues?
   a. isotherm
   b. ectotherm
   c. endotherm

9. Which of the following describes an ectotherm?
   a. a person shivering in the cold
   b. a turtle basking in the sun on a log
   c. a small animal increasing its rate of oxidation of fat in brown adipose tissue to produce heat
   d. a hormonally induced increase in basal metabolism for the purpose of generating heat

10. The direct transfer of heat between two physical bodies that are in contact with each other would be:
    a. evaporation
    b. radiation
    c. convection
    d. conduction

11. Which of the following would explain the observation that marine mammals maintain very thick layers of insulating fat?
    a. The fat helps prevent heat loss through the process of evaporation.
    b. The fat helps prevent heat loss through the process of conduction.
    c. The fat prevents heat loss through the process of convection.

12. When researchers compare the cells in tissues collected from endotherms and ectotherms of similar size, what can they expect to find?
    a. Endotherm cells will have a much greater density of mitochondria.
    b. Ectotherm cells will have a much greater density of mitochondria.
    c. Endotherm cells will have a much greater density of ribosomes.
    d. Ectotherm cells will have a much greater density of ribosomes.

CHAPTER 41—ANSWER KEY

Checking Your Understanding

(a) Changes in allele frequency in a population can occur by natural selection, genetic drift, gene flow, and mutation.

(b) Adaptation occurs when a population changes in response to natural selection. For example, the heavy fur of a mammal living at high altitudes has evolved over time to allow the animal to survive winter climates. Acclimatization occurs when an individual changes in response to a change in environmental conditions. For example, a mammal migrating in spring from a low-altitude valley to high mountain slopes will develop more red blood cells to cope with the lower oxygen at higher altitude.

(c) Animals possess a wide variety of structures that were present in ancestral populations but are not currently adaptive. Vestigial traits like this are widespread. Some traits exist as holdovers from structures that appear early in development. Human males have rudimentary mammary glands only because nipples form in the early embryo before sex hormones begin influencing the development of organs.

(d) A tissue is a group of cells with the same structure and function. An organ is a structure that serves a specialized function and consists of several tissues.

(e) In a cell, the surface area grows as a power of 2; the volume increases as a power of 3. As a result, its volume increases much faster than its surface area. Given this, the transport processes that occur across the cell membrane would have to support disproportionately more and more cell volume as the cell increased.

(f) Allometry occurs when changes in body size are accompanied by disproportionate changes in anatomical structures or physiological processes. In skeletal size and body size, the volume involved is the mass that must be supported by the skeleton. The area involved is the cross-sectional area of the bones supporting the mass. As mass increases, the amount of bone required for support has to increase disproportionately.

(g) Reptiles are ectothermic, so you would not expect to see brown fat in them because this would mean they would be producing heat from within—a requirement for endothermy.

Assessing What You've Learned

1. b; 2. c; 3. a; 4. c; 5. d; 6. d; 7. a; 8. c; 9. b; 10. d; 11. b; 12. a

42

Water and Electrolyte Balance in Animals

A. KEY CONCEPTS

42.1 Osmoregulation and Osmotic Stress

- Substances move from regions of higher concentrations to lower concentrations by the process of **diffusion** (**Figure 42.1a**).

- The diffusion of water from areas of higher to lower concentration is called **osmosis** (**Figure 42.1b**). The concentration of dissolved substances in a solution, measured in moles per liter, is referred to as its **osmolarity**.

- Cells and tissues that are gaining too much water are under **osmotic stress**. Fish must rid these cells of incoming water, or the cells will burst and the individual will die.

- The ability to maintain homeostasis with respect to water and electrolyte balance in the face of osmotic stress is called **osmoregulation**.

- In freshwater fish, gill epithelium is **hypertonic** in relation to the surrounding water (**Figure 42.2b**). Gill epithelial cells can gain water via osmosis. Marine fish are **hypotonic** to seawater, so water tends to flow out of the gill epithelium. Marine fish drink large quantities of seawater and excrete salt so that their cells do not shrivel and die.

- Freshwater animals must replace electrolytes that are lost by obtaining them in food or from the surrounding water. Marine animals must rid the body of excess ions they gain from drinking seawater (**Figure 42.2a**).

- Depending on conditions, terrestrial animals may need to conserve or to excrete electrolytes to maintain homeostasis. Water is lost in urine and feces, and it is gained by drinking and eating, which are also sources of electrolytes.

CHECK YOUR UNDERSTANDING

(a) How do freshwater fish and saltwater fish differ in their osmotic stresses?

_____

_____

_____

42.2 Water and Electrolyte Balance in Aquatic Environments

How Do Sharks Osmoregulate?

- The shark rectal gland secretes a concentrated salt solution. Ions can be concentrated in this way only if they are actively transported against a concentration gradient.

- Solutes move across membranes by passive or active transport. **Passive transport** often occurs via channels or by transmembrane proteins that act as carriers.

- **Active transport** requires membrane proteins that act as a pump, requiring energy in the form of **ATP**.

The Role of Na^+/K^+-ATPase

- Epithelial cells along the inner surface, or lumen, of the shark rectal gland contain **Na^+/K^+-ATPase**. The location of the pumps was paradoxical because they were situated along the **basolateral membrane** of the cells and not on the **apical membrane**. This meant that sodium would be pumped opposite to the direction it is secreted.

A Molecular Model for Salt Excretion

- In 1977, Silva and colleagues developed the following molecular model for salt excretion:

  1. Na^+/K^+-ATPase pumps sodium ions out of epithelial cells across the basolateral surface and into the surrounding extracellular fluid (**Figure 42.5**). The pump creates a large electrochemical gradient favoring the diffusion of Na^+ into the cell.

  2. A Na^+/Cl^- cotransporter, powered by the gradient favoring Na^+ diffusion, brings these two ions from the extracellular fluid into epithelial cells across their basolateral surfaces.

  3. Although sodium ions are pumped back out by Na^+/K^+-ATPase, chloride ion concentrations build up inside the cell as a result of the cotransport process. A chloride channel located in the apical membrane of the epithelial cells allows Cl^- to diffuse out down its concentration gradient.

  4. Sodium ions also diffuse into the lumen of the gland, following their charge and concentration gradient. But instead of passing through the epithelial cells as Cl^- does, Na^+ diffuses out through spaces between the cells.

- Using the steps provided here, can you draw the model for salt excretion?

CHECK YOUR UNDERSTANDING

(b) How did the study of the shark rectal gland yield practical applications for human medicine?

_____

_____

_____

A Common Molecular Mechanism Underlies Many Instances of Salt Excretion

- Marine birds and reptiles drink salt water and must excrete NaCl. In their nostrils, they have **salt-excreting glands** that function much like the shark rectal gland.

- Because marine teleost (bony) fish are hypertonic to seawater, salt constantly diffuses in through their **gills**. Their gills contain specialized cells that are configured precisely like the cells lining the rectal gland.

- Cells with the same configuration of **pumps**, **cotransporters**, and **channels** are responsible for transporting salt in the **kidneys** of mammals.

How Do Salmon Osmoregulate?

- Young chum salmon living in salt water have a large number of chloride cells at the base of their gill filaments. In contrast, most of the chloride cells observed in freshwater chum are located in the sheetlike lamellae that extend from the base of gill filaments.

- Chloride cells in the gill lamellae import electrolytes, while chloride cells at the base of the gills secrete salt.

CHECK YOUR UNDERSTANDING

(c) Why is a migrating salmon such an interesting animal with respect to osmotic stress?

_____

_____

_____

_____

_____

42.3 Water and Electrolyte Balance in Terrestrial Insects

How Do Insects Minimize Water Loss from the Body?

- Water loss is an inevitable by-product of respiration. Insects have a relatively large surface area with which to lose water and a small volume in which to retain it.

- An insect's **trachea** connects with the atmosphere through openings called **spiracles**. Muscles just inside each spiracle close or open the pore. This is an important adaptation for minimizing water loss (**Figure 42.9**).

- The insect **exoskeleton** consists of **chitin** and layers of protein. A layer of **wax** covers the surface, which is highly impermeable to water.

CHECK YOUR UNDERSTANDING

(d) What structures and processes are used by terrestrial insects to minimize water loss from the body?

_____

_____

_____

_____

Types of Nitrogenous Wastes: Impact on Water Balance

- **Ammonia** is a by-product of **catabolic** reactions. Those animals that excrete ammonia directly usually lose a lot of water. Fish detoxify ammonia by diluting to a low concentration in watery urine.

- Humans convert ammonia to nontoxic **urea** and excrete it. Urea requires water for excretion, however, but not as much as excreting ammonia directly.

- Birds, reptiles, and terrestrial arthropods convert ammonia to **uric acid**. Because uric acid is not soluble in water, it can be excreted as a dry paste.

- Does the type of nitrogenous waste produced by an animal correlate with the amount of osmotic stress it endures?

CHECK YOUR UNDERSTANDING

(e) Why is uric acid such an efficient waste product?

_____

_____

_____

_____

Maintaining Homeostasis: The Excretory System

- Insects must carefully regulate the composition of a bloodlike fluid called **hemolymph**.

Pre-Urine Formation in the Malpighian Tubules

- Pre-urine formation occurs in the **Malpighian tubules,** which have a large surface area and are in direct contact with the hemolymph.

- Because the urine that is finally expelled from the anus is often strongly hyperosmotic to pre-urine, a large amount of water must be reabsorbed. Selective reabsorption of electrolytes and water occurs in the hindgut. The epithelial cells transport ions out of the hindgut, and water follows by osmosis.

- The formation of pre-urine is not particularly selective. Reabsorption, in contrast, is highly selective. Waste products do not pass through the rectal membrane, instead remaining in the urine and feces.

The Hindgut: Selective Reabsorption of Electrolytes and Water

- Water and electrolyte recovery in the hindgut is achieved using membrane pumps and channels, which are precisely controlled to maintain homeostasis.

Regulating Water and Electrolyte Balance: An Overview

- There are no mechanisms for actively pumping water. Water moves between cells or body compartments via osmotic gradients.

CHECK YOUR UNDERSTANDING

(f) What is a Malpighian tubule, and which animals have them?

_____

_____

_____

_____

- The formation of the pre-urine is not selective.

- In contrast to pre-urine formation, reabsorption is highly selective. Only valuable ions and molecules are reabsorbed.

- Reabsorption is tightly regulated by the membrane pumps and channels involved in reabsorption.

- Reabsorption of water is based on osmotic gradients created by ion pumps.

42.4 Water and Electrolyte Balance in Terrestrial Vertebrates

The Structure of the Kidney

- The paired **kidneys** receive considerable blood flow via the **renal artery**. Urine formed in each kidney is carried to the **bladder** via a long, thin tube called the **ureter**. Urine leaves the bladder and exits the body via the **urethra** (**Figure 42.10**).

Filtration: The Renal Corpuscle

- **Nephrons** are the functional units of the kidney—they perform the work in maintaining water and electrolyte balance (**Figure 42.11**).

- Urine formation begins in the **renal corpuscle** (**Figure 42.12**). Water and small solutes are pushed out of the capillary's pores, through the slits in the surrounding cells, and into the fluid-filled space inside **Bowman's capsule**. Because proteins, cells, and other large components of blood would not fit through the pores, they would not enter the nephron.

- Pressure is much higher inside the capillaries than it is in the surrounding capsule. This pressure differential forces water and solutes out of the blood and into the capsule space.

- Can you predict how glomerular filtration would be affected by high blood pressure?

Reabsorption: The Proximal Tubule

- Fluid leaves the Bowman's capsule and enters a convoluted structure called the **proximal tubule**. The fluid entering this area has the composition of plasma minus the blood cells and large proteins.

- The epithelial cells of the tubule have a prominent series of small projections called **microvilli** facing the lumen (**Figure 42.13**).

- Na^+/K^+-ATPase in the basolateral membranes removes intracellular Na^+ and creates a gradient for Na^+ entry. In the apical membrane, Na^+-dependent cotransporters simultaneously bind Na^+ and another solute such as glucose, an amino acid, or Cl^-.

CHECK YOUR UNDERSTANDING

(g) What are aquaporins, and how do they relate to water and ion balance?

_____

_____

_____

_____

- Water follows the movement of these solutes by osmosis. It leaves the proximal tubule via membrane proteins called **aquaporins**.

Creating an Osmotic Gradient: The Loop of Henle

- In 1942 Werner Kuhn hypothesized that the loop of Henle functions as a countercurrent multiplier, setting up an osmotic gradient. He proposed that an exchange of water and solutes occurs between each of the segments in the loop and cells outside.

Testing Kuhn's Hypothesis

- A strong gradient in osmolarity existed from the cortex to the medulla, as predicted in Kuhn's hypothesis (**Figure 42.14**).

How Is the Osmotic Gradient Established?

- In the ascending loop of Henle, Na^+ and Cl^- constitute at least 60 percent of the solutes; urea is about 10 percent. In contrast, in the collecting duct both ions were rare, and the major solute was urea (**Figure 42.15**).

- Na^+ and Cl^- are actively pumped out of the ascending limb.

- The descending limb is highly permeable to water but almost completely impermeable to solutes. The thick ascending limb, in contrast, is highly permeable to Na^+ and Cl^-, moderately permeable to urea, and almost completely impermeable to water.

(a) Three regions of the nephron

(b) Permeability

Descending limb is highly permeable to water but impermeable to solutes

Ascending limb is impermeable to water but highly permeable to Na^+ and Cl^-

A Comprehensive View of the Loop of Henle

- As water flows down the descending limb, the fluid inside the loop loses water to the tissue surrounding the nephron.

- The fluid inside the nephron begins to lose Na^+ and Cl^-, and as it ascends it encounters surrounding tissues with progressively lower osmolarity.

- In the thick ascending limb, additional Na^+ and Cl^- ions are actively transported out of the nephron.

Regulating Water and Electrolyte Balance: The Distal Tubule and the Collecting Duct

- The amount of Na^+, Cl^-, and water reabsorbed in the distal tubule and collecting duct varies with the animal's condition and is under **hormonal control**.

- If an individual is dehydrated, a molecule called **antidiuretic hormone (ADH)** is released from the brain. When ADH interacts with cells lining the distal tubule and collecting duct, aquaporin channels are inserted into the membrane. The cells thus become highly permeable to water, and large amounts of water are reabsorbed (**Figure 42.17**).

- What happens to water reabsorption in the distal tubule and collecting duct in the absence of ADH?

CHECK YOUR UNDERSTANDING

(h) How does ADH affect water balance in the human kidney?

_____

_____

_____

_____

_____

B. INTEGRATING YOUR KNOWLEDGE

Looking Back—Concepts from Earlier Chapters

Osmosis and Diffusion—Chapter 6

Chapter 6 introduced the processes called diffusion and osmosis. Diffusion describes the movement of substances from regions of higher concentration to regions of lower concentration. The movement of water from regions of higher to lower concentration is called osmosis.

Gas Exchange in Plants—Chapter 36

Water balance in land animals mirrors the situation in land plants. Land plants also lose water as an inevitable by-product of gas exchange. As **Chapter 36** pointed out, land plants lose huge quantities of water in the course of obtaining CO_2 through their stomata.

Surface-to-Volume Ratio—Chapter 41

Small organisms have a high surface area/volume ratio. Insects have a relatively large surface area with which to lose water, and a small volume in which to retain it.

Looking Forward—Concepts in Later Chapters

Membrane Transport and Surface Area—Chapter 44

Nutrients such as glucose must diffuse into the cell, and waste products such as urea and carbon dioxide must diffuse out. As explained in **Chapter 44**, the rate at which these and other molecules and ions diffuse depends partly on the available surface area. In contrast, the rate at which nutrients are used and waste products are produced depends on the volume of the cell. This chapter will also explore how the gill functions as an adaptation for increased surface area, and how animals exchange gases in detail.

The Sodium-Potassium Pump—Chapter 45

The best-characterized membrane protein pump involving ions is the sodium-potassium pump, also known as Na^+/K^+-ATPase. **Chapter 45** discusses how this enzyme was discovered and explores its function in the transmission of electrical signals in animals. That discussion also introduces a plant defense compound called ouabain. Ouabain is toxic to animals because it binds to Na^+/K^+-ATPase, prevents it from functioning, and poisons transmission of nerve impulses.

C. MASTERING CHALLENGING TOPICS

The entire chapter has a central theme of sodium-potassium pumps. They are always involved in some way with water and ion balance. Once you have learned this basic tenet, the difficult part is understanding the most complex process of water and ion balance—which takes place in the vertebrate kidney. By working through this chapter and learning how more primitive vertebrates and invertebrates maintain water and ion homeostasis, you can focus on the vertebrate kidney. Just like the chapter and study guide organize the functions of the kidney, so should you. Study the filtration process first, then reabsorption, and then secretion, as you trace the path of urine formation beginning at the renal corpuscle and ending at the collecting duct. Since the formation of urine is such a dynamic process, make sure to note changes in urine formation under stressful conditions—dehydration, high blood pressure, and response to the various hormones discussed.

D. ASSESSING WHAT YOU'VE LEARNED

1. The major challenges to water and ion balance for most marine vertebrates are:
   a. gaining water from and losing ions to the environment
   b. gaining water and ions from the environment
   c. losing water to and gaining ions from the environment
   d. losing water and ions to the environment

2. Marine elasmobranchs, like the shark, avoid water loss to the environment by:
   a. drinking seawater and using their rectal gland to excrete excess ions
   b. maintaining the osmolarity of their body fluids as hypotonic relative to seawater
   c. gaining water through their gills and rectal gland
   d. gaining water through their gills

3. Which of the following statements accurately describes the role of Na^+K^+-ATPase in the shark rectal gland?
   a. Na^+K^+-ATPase, localized in the apical membrane of the lumen cells, pumps Na^+ into the lumen of the rectal gland.
   b. Na^+K^+-ATPase, localized in the basolateral membrane of the lumen cells, pumps Na^+ into the blood.
   c. Na^+K^+-ATPase, localized in the apical membrane, pumps Na^+ into the blood.
   d. Na^+K^+-ATPase, localized in the basolateral membrane of the lumen cells, pumps Na^+ out of the blood and into the lumen cells.

4. The major site for osmoregulation in teleosts (bony fishes) is:
   a. the rectal gland
   b. the gill
   c. the kidney
   d. the integument (skin)

5. An animal that lives in a dry environment needs to minimize water loss and will thus most likely secrete nitrogenous waste in the form of:
   a. uric acid
   b. urea
   c. creatinine
   d. ammonia

6. Which of the following statements correctly summarizes the primary role of the Malpighian tubule system of insects in formation of pre-urine?
   a. K^+ ions are actively secreted into the lumen of the Malpighian tubules; other electrolytes, water, and waste products follow passively.
   b. Na^+ and Cl^- ions are actively transported into the lumen of the Malpighian tubules; water and waste products follow.

   c. Cl^- is actively transported into the lumen by the actions of chloride cells; Na^+, waste products, and water follow.
   d. Waste products such as uric acid are actively secreted into the lumen, and water follows.

7. Which of the following statements concerning the composition of the filtrate formed by the renal corpuscle of the human kidney is correct?
   a. Blood and filtrate are identical in composition.
   b. Filtrate has a higher concentration of the waste product urea than blood does.
   c. The filtrate is similar to blood but without proteins and blood cells.
   d. The filtrate has a lower concentration of glucose than blood does.

8. Which of the following represents the primary driving force for the filtration of blood within the renal corpuscle?
   a. Pressure within the glomerular capillaries is lower than pressure in Bowman's capsule.
   b. Pressure within the glomerular capillaries is greater than pressure in Bowman's capsule.
   c. The collecting duct creates a suction that draws fluid into Bowman's capsule.
   d. The filtration slits are present in the glomerular capillaries.

9. Which statement most accurately describes the primary role of the loop of Henle in urine formation?
   a. The loop of Henle deposits Na^+ and Cl^- in the medullary region of the kidney, increasing its osmolarity.
   b. The hormone ADH acts on the loop of Henle to increase water reabsorption.
   c. The ascending limb of the loop of Henle contributes to the high osmolarity of the medullary region by depositing urea.
   d. The loop of Henle is responsible for most of the Na^+ reabsorbed by the nephron tubule.

10. The amount of Na^+ excreted in the urine is primarily determined by:
   a. the amount of Na^+ secreted into the proximal convoluted tubule
   b. the amount of Na^+ reabsorbed by the proximal convoluted tubule
   c. the amount of ADH released by the posterior pituitary gland
   d. hormonal regulation of Na^+ reabsorption in the distal convoluted tubule

11. Which of the following would result in the production of hypotonic urine?
   a. dehydration in an individual
   b. an increase in the permeability of the collecting duct to water

c. a decrease in ADH release by the brain

d. a decrease in water reabsorption by the loop of Henle

12. From an evolutionary perspective, which of the following adaptations makes the most sense?
    a. long loops of Henle in the kidneys of a river otter
    b. long loops of Henle in the kidneys of a desert fox
    c. an extensive Malpighian tubule system in a freshwater beetle
    d. renal corpuscles that are few in number and small in size in the kidney of a freshwater fish

CHAPTER 42—ANSWER KEY

Check Your Understanding

(a) Freshwater animals must replace electrolytes that are lost by obtaining them in food or from the surrounding water. Marine animals must rid the body of excess ions they gain from drinking seawater.

(b) Investigators realized that the amino acid sequence of CFTR (cystic fibrosis trans-membrane regulator) was 80 percent identical to the shark chloride channel. It was their first hint that CFTR involved chloride transport.

(c) Young chum salmon living in salt water have a large number of chloride cells at the base of their gill filaments. In contrast, most of the chloride cells observed in freshwater chum are located in the sheetlike lamellae that extend from the base of gill filaments. This allows researchers to study the distinctive challenges of osmoregulating in fresh and in salt water.

(d) Water loss is an inevitable by-product of respiration. Insects have a relatively large surface area with which to lose water, and a small volume in which to retain it. The tracheal system connects with the atmosphere at openings called spiracles. Muscles just inside each spiracle close or open the pore. This is an important adaptation for minimizing water loss. The insect exoskeleton consists of chitin and layers of protein. A layer of wax covers the surface, which is highly impermeable to water.

(e) Birds, reptiles, and terrestrial arthropods convert ammonia to uric acid. Because uric acid is not soluble in water, it can be excreted as a dry paste. In this way, the maximum amount of water can be conserved compared to excreting other types of nitrogenous wastes.

(f) Insects must carefully regulate the composition of a blood-like fluid called hemolymph. Malpighian tubules have a large surface area and are in direct contact with the hemolymph. They are the water and ion balance organs in insects. J. A. Ramsay and colleagues found that K^+ accumulated inside the tubules against a concentration gradient. They hypothesized that cells in the membranes of Malpighian tubules contain a pump that transports K^+ into the organ.

(g) Na^+/K^+-ATPase in the basolateral membranes removes intracellular Na^+ and creates a gradient for Na^+ entry. In the apical membrane, Na^+-dependent cotransporters simultaneously bind Na^+ and another solute such as glucose, an amino acid, or Cl^-. Water follows the movement of these solutes via osmosis. It leaves the proximal tubule through membrane proteins called aquaporins.

(h) The amount of Na^+, Cl^-, and water that is reabsorbed in the distal tubule and collecting duct varies with the animal's condition and is under hormonal control. If an individual is dehydrated, a molecule called antidiuretic hormone (ADH) is released from the brain. When ADH interacts with the cells lining the distal tubule and collecting duct, aquaporin channels are inserted into the membrane. As a result, the cells become highly permeable to water and large amounts of water are reabsorbed.

Assessing What You've Learned

1. c; 2. d; 3. b; 4. b; 5. a; 6. a; 7. c; 8. b; 9. a; 10. d; 11. c; 12. b

43

Animal Nutrition

A. KEY CONCEPTS

43.1 Nutritional Requirements

Meeting Basic Needs

- **Nutrients**, substances needed by an organism to remain alive, are contained within **food**. An organism needs chemical energy in the form of reduced carbon compounds and the elements and molecules needed to synthesize body components and sustain cells.

- Recommended Daily Allowances (RDAs) are placed on packaging in the USA to specify amounts of each essential nutrient that must be ingested to meet the needs of practically all healthy people. Specifically:

  1. **Essential amino acids** are the 8 (out of a total of 20) amino acids that are required to manufacture proteins, but that cannot be synthesized by humans. We can synthesize the other 12 amino acids.

  2. **Vitamins** are carbon-containing molecules that function as coenzymes in critical reactions (**Table 43.1**).

  3. **Essential elements** cannot be synthesized by the body, but they serve a wide variety of functions (**Table 43.2**).

  4. **Electrolytes** form ions in solution. They influence osmotic balance and are required for normal membrane function.

CHECK YOUR UNDERSTANDING

(a) What are the four areas of emphasis of the Recommended Daily Allowances?

_____

_____

_____

_____

_____

Can Nutrition Enhance Athletic Performance?

- RDAs identify nutrients that healthy people need to stay healthy, but they target people with normal needs and activities.

- The dietary regime known as **carbohydrate loading** is now a routine part of race preparations. For endurance athletes, a high-carbohydrate meal is the breakfast of champions (**Figure 43.2**).

43.2 Obtaining Food: The Structure and Function of Mouthparts

- **Heterotrophs** are organisms that obtain energy and nutrients from other organisms. Animals are heterotrophs.

- Given the diversity of food sources that animals exploit, they have a wide variety of structures in and around the mouth to obtain food.

- A mosquito's proboscis allows it to pierce the hide of an animal and suck up blood, and the mandibles of some ants are modified to cut leaves and other plant material.

- The mouthparts of mammals are adapted to different diets. Human teeth include sharp canines for tearing meat and flattened molars for crushing seeds, roots, and other sources of carbohydrates.

CHECK YOUR UNDERSTANDING

(b) Give an example of an animal with a specialized mouth for feeding.

_____

_____

_____

43.3 How Are Nutrients Digested and Absorbed?

- Digestion takes place in the **alimentary canal**, also known as the **gastrointestinal tract** (**Figure 43.5**).

- Mechanical breakdown occurs primarily in the mouth and stomach, but chemical breakdown begins with enzymes in the saliva.

The Mouth and Esophagus

- **Amylase**, contained in saliva, cleaves the bonds that link glucose monomers in starch, glycogen, and other glucose polymers.

- **Salivary glands** also release water and glycoproteins called **mucins**. The combination of water and mucus makes food soft and slippery enough to be swallowed.

- Food then enters the esophagus and is propelled by a wave of contractions called **peristalsis** (**Figure 43.7**). Food then enters the stomach.

CHECK YOUR UNDERSTANDING

(c) What is peristalsis, and where in the body does it occur?

_____

_____

_____

The Stomach

- When the stomach is filled by a meal, muscular contractions result in churning that mixes the contents (**Figure 43.8**).

- The predominant acid in the stomach is hydrochloric acid. There is also the enzyme pepsin, which breaks down proteins. Pepsinogen is converted to active pepsin by contact with the acidic environment of the stomach.

Which Cells Produce Stomach Acid?

- **Parietal cells** are the source of the HCl in gastric juice, whose pH can be as low as 1.5. **Goblet cells** secrete **mucus**, which lines the gastric epithelium and protects the stomach from damage by HCl (**Figure 43.9**).

How Do Parietal Cells Secrete HCl?

- The enzyme **carbonic anhydrase** is present in high concentration in parietal cells. In solution, the carbonic acid that is formed immediately dissociates to form the bicarbonate ion and a proton.

- Protons formed by dissociation of carbonic acid are actively pumped into the lumen of the stomach. Chloride ions enter parietal cells in exchange for bicarbonate ions and then move into the lumen via a chloride channel.

The Ruminant Stomach

- Cattle, sheep, deer, and other ruminants have complex, four-chambered stomachs (**Figure 43.10**). The largest chamber, the **rumen**, serves as a vat for fermenting food before food is regurgitated and re-chewed as **cud.** It then passes to the **reticulum** for bacterial breakdown of cellulose, to the **omasum** for water removal, and to the **abomasum** (true stomach) for further chemical digestion.

CHECK YOUR UNDERSTANDING

(d) Which stomach cells secrete acid?

_____

_____

The Small Intestine

- In the small intestine, partially digested material mixes with secretions from the pancreas and liver and begins a journey of 6 meters.

Protein Processing by Pancreatic Enzymes

- The activating enzyme in the intestinal juice is **enterokinase,** which activates a pancreatic enzyme called **trypsinogen**, yielding **trypsin**. Trypsin triggers the activation of other protein-digesting enzymes secreted by the pancreas.

- **Secretin** is produced in the small intestine in response to the arrival of food to the stomach. This hormone induces a flow of bicarbonate ions (HCO_3^-) from the pancreas to the small intestine, where they neutralize stomach acids. The discovery of secretin was important because it confirmed that digestion is under both neural and hormonal control.

CHECK YOUR UNDERSTANDING

(e) How is the hormone secretin involved in the digestion process?

_____

_____

_____

_____

Digesting Lipids: Bile and Transport

- The pancreatic secretions include digestive enzymes that act on fats and carbohydrates as well as proteins. **Lipase** breaks certain bonds present in complex fats.

- Can you outline the steps in digestion of lipids in the small intestine? (See **Figure 43.12**.)

- **Bile salts** are synthesized in the liver and secreted in a complex solution called **bile** that is stored in the **gallbladder**. When bile enters the small intestine, it raises the pH and emulsifies fats so that they can be digested.

CHECK YOUR UNDERSTANDING

(f) How and where is fat digested?

_____

_____

_____

_____

_____

How Are Carbohydrates Digested and Transported?

- Cells lining the small intestine absorb the sugars released by amylase digestion. The enormous surface area of the small intestine (because of **villi** and **microvilli**) increases the efficiency of nutrient absorption.

- Nutrient absorption depends on the presence of an electrochemical gradient favoring an influx of sodium ions into the epithelium. This process depends on a series of Na^+ cotransporters, like the **Na^+/glucose cotransporter** (**Figure 43.14**).

- Water follows the solutes into the epithelium passively, via osmosis. This movement of water is the mechanistic basis of an extremely important medical strategy called **oral rehydration therapy**.

The Large Intestine

- The primary function of the large intestine is to compact wastes that remain and absorb enough water to form **feces**.

- To date, four distinct transmembrane water channels called **aquaporins** have been found in the large intestine of rats, mice, or humans. AQP3 and AQP4 are located in the basolateral membrane of cells in the epithelium of the large intestine of the rat.

- Mechanisms of water absorption in the large intestine are not well understood, but aquaporins are hypothesized to have a major role.

43.4 Nutritional Homeostasis— Glucose as a Case Study

- People with **diabetes mellitus** experience abnormally high glucose levels in their blood.

- Banting and Best (1921) removed a pancreas from a dog, injected an extract made from it, and injected the extract into a diabetic dog; its blood glucose stabilized. They had discovered **insulin**.

Insulin's Role in Homeostasis

- **Insulin** is a hormone produced in the pancreas when blood glucose levels are high. It travels through the bloodstream and binds to receptors on cells throughout the body. In response, the cells increase their rate of glucose uptake and respiration.

- If blood glucose levels fall, cells in the pancreas secrete a hormone called **glucagon**. Cells in the liver and skeletal muscle catabolize glycogen, and cells that store lipids catabolize fatty acids. The result is that glucose levels in the blood rise.

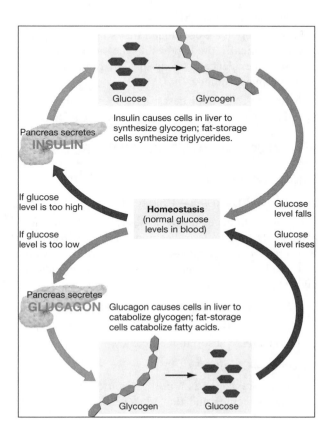

- **Diabetes mellitus** develops in people who do not synthesize insulin or who have defective versions of the insulin receptor. Type I diabetes mellitus is treated with insulin injections; type II diabetes is managed mainly via diet.

CHECK YOUR UNDERSTANDING

(h) What is the difference between insulin and glucagon?

_____

_____

_____

_____

_____

The Type II Diabetes Mellitus Epidemic

- In the United States, about 6.6 percent of people ages 20–74 have type II diabetes. Among the Pima, almost 50 percent of the adults over 35 years old have type II diabetes (**Figure 43.17**).

- Researchers have concluded that some individuals are genetically predisposed to developing the disease, but evidence is strong that environmental conditions have an impact.

- Nutrition-related diseases like obesity are being linked to diabetes, especially in young people.

B. INTEGRATING YOUR KNOWLEDGE

Looking Back—Concepts from Earlier Chapters

Osmosis and Diffusion—Chapter 6

Chapter 6 introduced the processes called diffusion and osmosis. Diffusion describes movement of substances from regions of higher concentration to regions of lower concentration. Movement of water from regions of higher to lower concentration is called osmosis.

Surface-to-Volume Ratio—Chapter 41

Organs like the small intestine increase surface area in order to increase absorption efficiency.

Glycogen—Chapter 5

Recall that glycogen is a polysaccharide made up of glucose molecules and that glucose is the preferred

starting compound for the production of ATP via cellular respiration.

Adaptive Radiation—Chapter 25

Cichlids inhabiting the Rift Lakes of East Africa are a spectacular example of adaptive radiation. Recall from **Chapter 25** that adaptive radiation refers to diversification of a single ancestral population into many species, each of which lives in a different habitat or utilizes a distinct feeding method.

Embryonic Tissue Layers—Chapter 22

The interior tube of the GI tract communicates directly with the environment. Embryologically, it derives from the hollow tube that forms as sheets of cells invaginate during gastrulation.

Aquaporins—Chapter 42

In an attempt to identify the mechanism of water absorption in the large intestine, researchers focused on aquaporins. Recall that aquaporins are water channels and are common in the descending loop of Henle of the kidney.

Looking Forward—Concepts in Later Chapters

Membrane Transport and Surface Area— Chapter 44

Nutrients such as glucose must diffuse into the cell, and waste products such as urea and carbon dioxide must diffuse out. As **Chapter 44** explains, the rate at which these and other molecules and ions diffuse depends partly on the amount of surface area available. In contrast, the rate at which nutrients are used and waste products are produced depends on the volume of the cell. This chapter also explores how the gill functions as an adaptation for increased surface area, and how animals exchange gases in detail.

The Sodium-Potassium Pump—Chapter 45

The best-characterized membrane protein pump involving ions is the sodium-potassium pump, also known as Na^+/K^+-ATPase. **Chapter 45** discusses how this enzyme was discovered and explores its function in the transmission of electrical signals in animals. That discussion also introduces a plant defense compound called ouabain. Ouabain is toxic to animals since it binds to Na^+/K^+-ATPase, prevents it from functioning, and poisons transmission of nerve impulses.

Muscle Types—Chapter 46

The action of peristalsis and churning is unconscious. **Chapter 46** explains the differences between the structure and function of different types of muscle and related nerve actions.

C. MASTERING CHALLENGING TOPICS

Following food through the digestive system is a fairly simple process; but the chemical reactions that occur during this process, coupled with the neural and hormonal signals, can quickly get confusing. Make sure you have mastered the anatomy of the digestive system before you tackle the physiology. It will be difficult to study the functions of the liver if you can't picture where it's located in relation to the digestive system and the body.

Begin by following a piece of food through the digestive system, starting at the mouth. Note the chemical reactions occurring at each stage, and any neural or hormonal signals involved. The most confusing section will be the small intestine because secretions from the liver and pancreas are involved with digesting food entering from the stomach. Draw arrows mapping the hormonal and neural signals so you are clear on which way the signal is traveling.

D. ASSESSING WHAT YOU'VE LEARNED

1. Which of the following represent examples of electrolytes?
   a. K$^+$ and Cl
   b. valine and glycine
   c. niacin and folate
   d. glucose and fructose

2. As you've probably noticed, recommendations for basic nutritional requirements for humans change fairly frequently. Which of the following statements provides the most likely explanation?
   a. As humans evolve, their nutritional requirements change.
   b. The nutritional information provided on packaged food labels contains many errors.
   c. It is difficult to perform controlled experiments on humans in order to make reliable nutritional recommendations.

3. Which of the following experimental results led Jonas Bergstrom and co-workers to challenge the prevailing hypothesis concerning the nutritional requirements of endurance athletes?
   a. Athletes eating a diet rich in complex carbohydrates had higher blood glucose levels while exercising than did those who ate a diet rich in fats and proteins.
   b. Athletes eating a diet rich in complex carbohydrates experienced a decrease in the length of time it took to reach exhaustion when compared to those who ate a diet rich in fats and protein.
   c. Athletes eating a diet rich in fat and protein experienced an increase in levels of stored glycogen in their muscles when compared to those who ate a diet rich in complex carbohydrates.
   d. Athletes who ate a diet rich in simple carbohydrates (such as glucose) experienced an increase in performance level due to enhanced glycogen storage.

4. Which of these adaptations to the pharyngeal jaw system of cichlids does *not* represent an example of how mouthparts are adaptations to allow for efficient feeding?
   a. Because the pharyngeal jaws articulate with the braincase, they can be used for biting as well as for moving food to the back of the throat.
   b. Because the lower pharyngeal jaw exists as two separate pieces, the pharyngeal jaws are better able to grasp prey.
   c. The pharyngeal jaws of cichlids have more muscles attached, improving their use for biting prey.
   d. The toothlike projections of the cichlid pharyngeal jaws vary in shape and size depending on the primary food source of each species.

5. Chemical digestion begins in the mouth. Which of the following best describes the type of digestion that occurs in the mouth?
   a. Lipase begins to break down proteins.
   b. Amylase begins to break down starch.
   c. Pepsin begins to break down proteins.
   d. Amylase begins to break down fats.

6. Which of these statements concerning acid production by the stomach is *not* correct?
   a. The enzyme carbonic anhydrase, located in the parietal cells, catalyzes the formation of carbonic acid from carbon dioxide and water.
   b. Carbonic acid is actively secreted by parietal cells into the stomach lumen.
   c. Protons are actively pumped into the stomach lumen by the parietal cells.
   d. Chloride ions move into the lumen of the stomach through chloride channels located in the parietal cells.

7. Based on what you've learned about the role of the parietal cells in HCl production in the stomach, predict the relative acidities of these three compartments: A = the stomach lumen; B = blood supply arriving to the stomach; C = blood supply leaving the stomach.
   a. pH A > pH C > pH B
   b. pH B > pH C > pH A
   c. pH A > pH B > pH C
   d. pH C > pH B > pH A

8. Which of the following does *not* represent a function of HCl in the stomach?
   a. HCl cleaves peptide bonds.
   b. HCl activates digestive enzymes like pepsin.
   c. HCl acts as a barrier to pathogens present in food.
   d. HCl disrupts the tertiary and secondary structure of proteins, enhancing their enzymatic degradation.

9. Which of the following represents the correct sequence for protein digestion and absorption?
   a. In the stomach, pepsin degrades large polypeptides to smaller polypeptides. In the small intestine, the smaller polypeptides are further broken down into amino acids by the actions of enzymes like trypsin, secreted from the pancreas. Amino acids are absorbed in the small intestine by a Na^+-amino acid cotransport mechanism.
   b. In the stomach, enterokinase degrades large polypeptides to smaller polypeptides. In the small intestine, the smaller polypeptides are degraded to amino acids by the actions of pancreatic proteases, like trypsin, and brush border enzymes. Amino acids are absorbed in the small intestine by a Na^+-amino acid cotransport mechanism.
   c. In the stomach, pepsin degrades large polypeptides to small polypeptides. In the small intestine, the smaller polypeptides are broken down into amino acids by the actions of pancreatic enzymes, like trypsin, and brush border enzymes. Amino acids are absorbed in the large intestine by a Na^+-amino acid cotransport mechanism.
   d. In the mouth, large polypeptides are degraded to smaller polypeptides by the actions of amylase. In the small intestine, the smaller polypeptides are further broken down into amino acids by the actions of pancreatic enzymes, like trypsin, and brush border enzymes. Amino acids are absorbed in the small intestine by a Na^+-amino acid cotransport mechanism.

10. Which of the following represents the correct sequence for fat digestion?
    a. The bulk of fat digestion takes place in the small intestine with the release of lipases from the gallbladder. Lipases cleave fats into fatty acids and other small lipids that are absorbed by the small intestine.
    b. The bulk of fat digestion takes place in the small intestine. Bile released from the gallbladder helps to break up large fat globules to smaller ones, while lipases released from the pancreas cleave the fats to fatty acids and other small lipids. Absorption of these small lipids occurs via a Na^+-fatty acid cotransport mechanism in the small intestine.
    c. The bulk of fat digestion takes place in the small intestine. Bile released from the gallbladder helps to break up large fat globules to smaller ones, while amylase released from the pancreas cleaves the fats to fatty acids and other small lipids. Absorption of these small lipids occurs via a Na^+-fatty acid cotransport mechanism in the small intestine.
    d. The bulk of fat digestion takes place in the small intestine. Bile released from the gallbladder helps to break up large fat globules to smaller ones, while lipases released from the pancreas cleave the fats to fatty acids and other small lipids. Absorption of these small lipids occurs using fatty-acid-binding proteins located in the small intestine.

11. Which of these statements concerning pancreatic secretion is *false*?
    a. The introduction of HCl into the upper regions of the small intestine stimulates pancreatic secretion.
    b. Nerves innervating the upper region of the small intestine sense the arrival of food from the stomach and signal the pancreas to secrete enzymes.
    c. The arrival of food into the small intestine from the stomach stimulates release of the hormone secretin, which triggers pancreatic secretion.
    d. Digestive enzymes secreted by the pancreas are activated by an enzyme, called enterokinase, secreted by the small intestine.

12. All of the following occur with the condition known as diabetes *except*:
    a. high blood levels of insulin
    b. high blood levels of glucose
    c. reduced rates of glucose uptake by cells
    d. presence of glucose in the urine

CHAPTER 43—ANSWER KEY

Check Your Understanding

(a) In 1943, the Recommended Daily Allowances (RDAs) were published to specify the amount of each essential nutrient that must be ingested to meet the needs of practically all healthy people. The following were their focus:

1. Of the 20 amino acids required to manufacture proteins, humans can synthesize 12. The other 8 must be obtained from food and are called essential amino acids.
2. Vitamins are carbon-containing molecules that function as coenzymes in critical reactions.

3. Essential elements serve a wide variety of functions.

4. Electrolytes form ions in solution; they influence osmotic balance and are required for normal membrane function.

(b) Human teeth include sharp canines for tearing meat and flattened molars for crushing seeds, roots, and other sources of carbohydrates.

(c) Peristalsis is a wave of contraction that propels food through the digestive system. It occurs primarily in the esophagus and the intestines.

(d) Parietal cells in the lining of the stomach produce hydrochloric acid.

(e) Secretin is a hormone produced in the small intestine in response to the arrival of food to the stomach. This hormone induces a flow of bicarbonate ions (HCO_3^-) from the pancreas to the small intestine, where they neutralize stomach acids.

(f) The pancreatic secretions include digestive enzymes that act on fats and carbohydrates, as well as proteins. Lipase breaks certain bonds present in complex fats. Bile salts are synthesized in the liver and secreted in a complex solution called bile, which is stored in the gallbladder. When bile enters the small intestine, it raises the pH and emulsifies fats so that they can be digested.

(g) The primary function of the large intestine is to compact wastes that remain and absorb enough water to form feces.

(h) Insulin is a hormone produced in the pancreas when blood glucose levels are high. It travels through the bloodstream and binds to receptors on cells throughout the body. In response to insulin, the cells increase their rate of glucose uptake and respiration. If blood glucose levels fall, cells in the pancreas secrete a hormone called glucagon. Cells in the liver and skeletal muscle catabolize glycogen and cells that store lipids catabolize fatty acids. The result is that glucose levels in the blood rise.

Assessing What You've Learned

1. a; 2. c; 3. a; 4. b; 5. b; 6. b; 7. d; 8. a; 9. a; 10. d; 11. b; 12. a

44

Gas Exchange and Circulation

A. KEY BIOLOGICAL CONCEPTS

44.1 The Respiratory and Circulatory Systems

- The transport of O_2 from environment to cells, and of CO_2 from cells to environment, requires both the respiratory and circulatory systems (**Figure 44.1**). Four stages are involved:
    1. **Ventilation**, where air or water moves through a specialized gas exchange organ (e.g., lungs or gills)
    2. **Gas exchange**, where O_2 and CO_2 diffuse between air or water and blood at the ventilatory surface
    3. **Circulation**, where dissolved O_2 and CO_2 are transported throughout the body
    4. **Cellular respiration**, where O_2 and CO_2 diffuse between blood and cells

44.2 Air and Water as Respiratory Media

How Do Oxygen and Carbon Dioxide Behave in Air?

- **Partial pressure** is the pressure of a particular gas in a mixture. To calculate this, multiply the percent composition of that gas by the total pressure exerted by the entire mixture.

CHECK YOUR UNDERSTANDING

(a) The composition of air at 760 mm Hg is 20.95% oxygen, 78.89% nitrogen, and 0.03% carbon dioxide. What are the partial pressures for each of these gases? As you climb Mt. Everest, what happens to these gas percentages and these gas partial pressures?

_____

_____

_____

_____

How Do Oxygen and Carbon Dioxide Behave in Water?

- The amount of gas that dissolves in a liquid depends on several factors:
    1. Solubility of the gas in that liquid
    2. Temperature of the liquid
    3. Presence of other solutes
- CO_2 is almost 30 times more soluble in water than O_2 is. As a result, fish rid themselves of CO_2 more easily than they can obtain oxygen.
- As the temperature of the water increases, the amount of gas that dissolves decreases. Cold-water habitats have much more oxygen available than warm-water habitats do.
- Because seawater has a much higher concentration of solutes than freshwater, it can hold much less dissolved gas.
- Mixing improves the oxygenation of water. Rapids and waterfalls tend to be the most highly oxygenated of all aquatic environments.
- A liter of air can contain up to 209 ml of O_2 while a liter of water may contain up to 7 ml of O_2. To extract the same amount of oxygen, water breathers must breathe 30 times more water than air.
- Water is also about 1000 times denser than air. As a result, water breathers expend considerably more energy to ventilate their respiratory surfaces.

CHECK YOUR UNDERSTANDING

(b) What are the three factors that determine the amount of gas that will dissolve in water? Which factor contributes the most in the difference between O_2 and CO_2 dissolving in water?

_____

_____

_____

_____

44.3 Organs of Gas Exchange

Design Parameters: The Law of Diffusion

- Gas diffusion rate, defined by Fick's law of diffusion (**Figure 44.5**), depends on five factors:
  1. Solubility of the gas
  2. Temperature
  3. Surface area available for diffusion
  4. Difference in partial pressures of the gas across the respiratory medium
  5. Thickness of the barrier to diffusion

CHECK YOUR UNDERSTANDING

(c) Which of the factors affecting rate of gas diffusion has the potential to be quickly altered by an animal?

_____

_____

_____

_____

_____

How Do Gills Work?

- **Gills** are outgrowths of the body surface that are used for gas exchange in aquatic animals.
- Among invertebrates, the structure of gills is extremely diverse whereas the gills of bony fish are similar in structure.
- To move water through these structures, fish open and close their mouths and the stiff flap of tissue that covers the gills. Fish can also force water through their gills by swimming with their mouths open; this is called **ram ventilation**.
- Movement of water over the gills is **unidirectional**, which sets up a **countercurrent exchange system** (**Figure 44.8**). This ensures that the difference in the amount of O_2 and CO_2 in water versus blood is large over the entire respiratory surface.

How Do Trachea Work?

- **The tracheal system** transports air close enough to cells for gas exchange to take place directly. As a result, insects do not require a circulatory system to transport gases from a respiratory structure to the tissues.

- Trachea connect to the exterior via an opening called a **spiracle** that can be closed. Spiracles are interpreted as adaptations to minimize water loss.

CHECK YOUR UNDERSTANDING

(d) Do insects have a "respiratory system" by definition? If not, what do they have instead?

_____

_____

_____

_____

How Do Lungs Work?

- Mammalian lungs are divided into tiny sacs called **alveoli** (**Figure 44.11**). Each human lung contains about 150 million of these structures. It has about 100 m² of surface area, and the barrier of diffusion is only 0.2 μm thick.
- In the simple lungs of snails and spiders, air movement takes place by diffusion only.
- Vertebrates actively ventilate their lungs by pumping air via muscular contractions.

Homeostatic Control of Ventilation

- The rate of breathing is established by the **medullary respiratory center**, which stimulates the rib and diaphragm muscles to contract about 12–14 times per minute in humans.
- The medullary respiratory center receives neural input from sensory receptors sensitive to O_2 levels in arterial blood. The cells of the respiratory center are also sensitive to pH changes that result from changes in CO_2.
- Signals received by the medullary center stimulate breathing when O_2 levels fall and CO_2 levels rise. Thus, through a negative-feedback system, O_2 and CO_2 delivery are maintained and blood pH stabilized.

44.4 How Are Oxygen and Carbon Dioxide Transported in Blood?

- Blood is a **connective tissue** that consists of cells in a fluid extracellular matrix called **plasma**. **Formed elements** in the blood include red blood cells, platelets, and several types of white blood cells.

- The functions of blood include
  1. Carrying O_2 and CO_2 between mitochondria and lungs
  2. Transporting nutrients from the digestive tract to other tissues in the body
  3. Moving waste products to the kidney and liver for processing
  4. Conveying hormones from glands to target tissues
  5. Delivering immune system cells to sites of infection
  6. Distributing heat from deeper organs to the surface
- **White blood cells** are part of the immune system and fight infections. **Platelets** are involved in clot formation.
- **Red blood cells** contain an oxygen-carrying molecule called **hemoglobin**. It consists of four polypeptide chains, each of which binds to a nonprotein group called **heme**. Each heme contains an iron molecule that can bind to an oxygen molecule.

CHECK YOUR UNDERSTANDING

(e) List the functions of the blood.

_____

_____

_____

_____

_____

Structure and Function of Hemoglobin

- Hemoglobin is a tetramer consisting of four polypeptide chains, each bound to an iron-containing nonprotein group called **heme**. Each heme molecule contains an iron ion that can bind an oxygen molecule.
- Blood leaving the lungs has a P_{O_2} of 100 mm Hg, while muscle and other tissues have P_{O_2} levels of about 40 mm Hg at rest. This partial pressure difference drives the unloading of O_2 from the hemoglobin to the tissues via diffusion.

(a) With cooperative binding, large amounts of O_2 are delivered to resting and exercising tissues.

(b) Without cooperative binding, smaller amounts of O_2 would be delivered to resting and exercising tissues.

- Because of the sigmoidal relationship, hemoglobin responds quickly and effectively to small changes in oxygen demand.
- Hemoglobin is also sensitive to changes in pH and temperature. During exercise, the temperature and the P_{CO_2} increases in the tissue. This results in a drop in pH.
- Decreases in pH and increases in temperature alter hemoglobin's conformation such that it is more likely to release O_2 at all values of P_{O_2}. This is known as the **Bohr shift**.
- Biologists wanted to determine how the oxygen transport system in fish responded to sustained exercise. Arterial O_2 levels remained fairly constant as swimming speed increased, but O_2 levels in the venous blood dropped steadily as swimming speed increased. In hard-working tissues, the combination of increased temperature, lower pH, and lower P_{O_2} caused hemoglobin to become almost completely deoxygenated.

(f) What is the Bohr shift?

_____

_____

_____

CO₂ Transport and the Buffering of Blood pH

- **Carbonic anhydrase** in red blood cells catalyzes the formation of carbonic acid from carbon dioxide and water. As a result, CO_2 that diffuses into red blood cells is quickly converted into bicarbonate ions and protons (**Figure 44.20**). The same reaction occurs much more slowly in the plasma surrounding the red blood cells.

- When Hb is not carrying O_2, it has a high affinity for protons. As a result, Hb takes up much of the H^+ that is produced by the dissociation of carbonic acid. In this way, Hb acts as a **buffer**—a compound that minimizes changes in pH.

- At the lungs, CO_2 diffuses from the blood into the alveoli and the P_{CO_2} in the blood declines. Bicarbonate is converted back to CO_2, which then diffuses into the alveoli and is exhaled from the lungs. Hb picks up O_2 during inhalation and the cycle begins again.

(g) Describe the significance of carbonic anhydrase in carbon dioxide elemination.

_____

_____

_____

_____

_____

44.5 The Circulatory System

What Is an Open Circulatory System?

- One or more hearts pump hemolymph into an artery that empties into many small, open, fluid-filled spaces. Hemolymph is returned to the heart when it relaxes and creates suction.

- Hemolymph is not tightly constrained inside vessels, so the overall pressure in the system is low.

What Is a Closed Circulatory System?

- Blood is completely contained within blood vessels and flows in a continuous circuit through the body under pressure generated by the heart.

- In a closed circulatory system, the heart ejects blood into a large artery, usually called the **aorta**, which carries blood into the distribution vessels, or **arteries**.

- Small arteries lead into arterioles, which are surrounded by sphincters that help regulate blood flow.

- **Capillaries** are the smallest blood vessels. With walls only a single cell layer thick, the capillaries allow gases, nutrients, and wastes to pass between blood and cells.

- **Interstitial fluid** is constantly augmented by water and proteins from blood that leak out between the cells forming the walls of the capillaries (**Figure 44.23**).

- The lymphatic system is a collection of vessels that branches throughout the body. The fluid inside the lymph vessels is called lymph and is eventually returned to the body.

How Does the Heart Work?

- The number of distinct chambers in the heart has increased as vertebrates diversified. The circulatory system in fish forms a single loop; in all other vertebrates, there are two separate circuits—to the lungs and the body. (**Figure 44.24**).

- In the human, the atrium receives blood returning from circulation and the ventricle generates force to propel the blood through the system.

- The right ventricle powers the blood to the lungs and back to the heart in a section called the **pulmonary circulation**. Blood flow from the atrium to the ventricle moves only one way due to the tissue flaps called **valves**. The left ventricle pumps blood to the entire body and back, or to the **systemic circulation**.

- A cardiac cycle consists of one complete contraction phase, called **systole**, and one complete relaxation phase, called **diastole**.

- **Blood pressure** is the force that blood exerts on the walls of arteries, capillaries, and veins.

- Peak pressure is due to the contraction of the ventricle and is called **systolic pressure**. The lowest pressure occurs at the end of ventricular relaxation and is called **diastolic pressure**.

- A typical blood pressure might be 120/80 (systolic/diastolic). People with pressures of 150/90 and higher are considered to have high blood pressure or **hypertension**.

(h) What is the difference between systole and diastole, and how does this correlate with the pulsatile pressure in arterial blood?

_____

_____

_____

_____

_____

- The "pacemaker cells" that initiate contraction are located in a region called the **SA (sinoatrial) node**.
- The pace at which the SA node initiates electrical signals determines heart rate. The pace, in turn, is regulated by electrical signals from the brain.
- Heart rate may be altered by electrical signals from emotional centers in the brain or by signals from the chemical messengers called epinephrine and norepinephrine.
- Heart rate can also vary with exposure to drugs.

(i) What is the electrical event that regulates heart rate?

_____

_____

_____

Patterns in Blood Pressure and Blood Flow

- As the same volume of blood travels through a greater cross-sectional area, the pressure in the fluid drops. Overall, blood pressure is highest in the artery that leads away from the left ventricle (100 mm Hg) and lowest in the veins that return blood from the body to the right atrium (10 mm Hg; **Figure 44.30**).
- Sensory cells are located next to an artery in the neck, an artery near the heart, and the wall of the heart. These cells act as **baroreceptors**, or pressure receptors. When they detect a change in blood pressure, they trigger electrical signals that change the heart's output and vessel diameter.

- Changes in blood P_{O_2}, P_{CO_2}, and pH are detected by specialized nerve cells in the neck, near the heart, and in the brain region just above the spinal cord. These cells act as **chemoreceptors** and send signals to the lungs and heart to maintain homeostasis.

B.   INTEGRATING YOUR KNOWLEDGE

Looking Back—Concepts from Earlier Chapters

Osmosis and Diffusion—Chapter 6

Chapter 6 introduced the processes called diffusion and osmosis. Diffusion describes the movement of substances from regions of higher concentration to regions of lower concentration. The movement of water from regions of higher to lower concentration is called osmosis.

Surface-to-Volume Ratio—Chapter 41

Organs such as the small intestine have a large internal surface area to increase absorption efficiency.

Water Loss—Chapter 42

As **Chapter 42** explained, many marine mammals lose water across their respiratory surface via osmosis; animals that live in freshwater lose ions and other solutes by diffusion. Chapter 42 also introduced the trachea of insects. Biologists found that when spiracles were kept open experimentally, insects were likely to die of dehydration.

Carbonic Anhydrase—Chapter 43

Recall from **Chapter 43** that carbonic anhydrase catalyzes the formation of carbonic acid from carbon dioxide and water. As a result, CO_2 that diffuses into red blood cells is quickly converted to bicarbonate ions and protons.

Looking Forward—Concepts in Later Chapters

Electrical Signals—Chapter 45

Because intercalated discs of cardiac muscle contain numerous gaps, electrical signals pass directly from one cell to the next. **Chapter 45** explores how electrical signals in animals are initiated and propagated.

Fight-or-Flight—Chapter 47

As part of the fight-or-flight response analyzed in **Chapter 47**, blood is directed away from the skin and digestive system and toward the heart, brain, and muscles.

The Lymphatic System—Chapter 49

The lymphatic system acts as a type of circulatory system. Because its primary function is in defense against disease-causing agents, however, the lymphatic system is analyzed in more detail in **Chapter 49**.

C. MASTERING CHALLENGING TOPICS

The two most difficult topics in this chapter have to do with the respiratory properties of blood. The first topic deals with the partial pressures of respiratory gases. This concept is central to respiratory physiology, so you must feel comfortable using partial pressures to describe elements of respiration and know the difference between partial pressures and percent composition in respiratory gases. In air, think of percent composition as constant but partial pressures as changing with altitude. Run through a few calculations, and follow the partial pressures of O_2 and CO_2 in the blood going from the lungs to the tissues and back. Have classmates ask you questions and try to stump each other. Mastering this concept will make future lessons in respiratory physiology more enjoyable.

The second topic that most students have difficulty with is the sigmoidal curve of hemoglobin and O_2 loading/unloading. Always think of the loading and unloading of O_2 as a dynamic process. Think of the binding of one oxygen molecule like a person getting into a car. Once this person is inside, she can open the doors of the car to allow three other people easier access to the car. In this way, when one oxygen binds to Hb, it changes the conformation such that it is very easy for the next three O_2 molecules to bind. This is the primary reason for the sigmoidal curve in Hb-O_2 loading.

D. ASSESSING WHAT YOU'VE LEARNED

1. Which of the following most accurately describes the gas composition of the atmosphere?
   a. nitrogen > oxygen > carbon dioxide > inert gases (such as argon)
   b. oxygen > nitrogen > carbon dioxide > inert gases (such as argon)
   c. nitrogen > oxygen > inert gases (such as argon) > carbon dioxide
   d. oxygen > carbon dioxide > nitrogen > inert gases (such as argon)

2. Of these gases found in the atmosphere—nitrogen, oxygen, argon, and carbon dioxide—which are considered to be physiologically active?
   a. carbon dioxide and oxygen
   b. nitrogen, carbon dioxide, and oxygen
   c. nitrogen and oxygen
   d. oxygen only

3. All of the following represent examples of how the design of respiratory organs conforms to the dictates of Fick's law of diffusion *except*:
   a. The respiratory surface of the human lung is 70 square meters.
   b. The lung epithelium and blood vessels are extremely thin.
   c. The epithelium of both internal and external gills is in direct contact with oxygen-bearing water.
   d. The flow of blood through the gills is in the same direction as the flow of water over the gill surface.

4. Which of the following represents a disadvantage of breathing air (at lower elevations) versus breathing water?
   a. A liter of water holds less O_2 than a liter of air.
   b. Water is denser than air.
   c. Carbon dioxide is quite soluble in liquids like water.
   d. Air-breathing animals are better able to retain heat generated by their bodies.

5. The insect tracheal system differs from the mammalian respiratory system in all of the following ways *except*:
   a. O_2 is delivered by hemolymph, not blood, to cells.
   b. The insect tracheal system uses spiracles to reduce evaporative water loss.
   c. There is no respiratory pigment for gas transport.
   d. Muscles of locomotion help to ventilate the respiratory system.

6. Which of the following statements most accurately describes the structure of hemoglobin?
   a. Hemoglobin consists of a single polypeptide chain with a heme center containing a copper ion to which a molecule of O_2 binds.
   b. Hemoglobin consists of a single polypeptide chain with a heme center containing an iron ion to which a molecule of O_2 binds.
   c. Hemoglobin consists of four polypeptide chains, each of which have a heme group containing a copper ion. One hemoglobin molecule can bind a total of four O_2 molecules.
   d. Hemoglobin consists of four polypeptide chains, each of which have a heme group containing an iron ion. One hemoglobin molecule can bind a total of four O_2 molecules.

7. Which of the following best describes the significance of the sigmoidal relationship between P_{O_2} levels and hemoglobin saturation?
   a. As tissue P_{O_2} levels decline, the amount of O_2 released increases proportionately.
   b. As it circulates, hemoglobin releases the same amount of oxygen to all tissues.
   c. As tissue P_{O_2} levels decline with an increase in activity, the oxygen reserve increases.
   d. Once tissue P_{O_2} levels drop below 40 mm Hg, the amount of oxygen released by hemoglobin increases greatly.

8. Which of the following best describes the phenomenon known as the Bohr shift?
   a. As tissue CO_2 levels increase, pH decreases, resulting in an increase in O_2 delivery to the tissues.
   b. As tissue CO_2 levels decrease, pH decreases, resulting in an increase in O_2 delivery to the tissues.
   c. As tissue CO_2 levels increase, pH increases, resulting in a decrease in O_2 delivery to the tissues.
   d. As tissue CO_2 levels decrease, pH increases, resulting in an increase in O_2 delivery to the tissues.

9. Which of the following situations would result in the greatest degree of O_2 saturation for hemoglobin, assuming P_{O_2} remains constant?
   a. increased CO_2 levels, decreased temperature
   b. increased CO_2 levels, increased temperature
   c. decreased CO_2 levels, decreased temperature
   d. decreased CO_2 levels, increased temperature

10. The transport of CO_2 by the blood is primarily dependent on:
    a. the solubility of CO_2 in blood
    b. the presence of carbonic anhydrase in red blood cells
    c. the ability of hemoglobin to bind and transport CO_2
    d. the ability of other blood proteins to bind CO_2

11. Hemoglobin acts as a buffer by:
    a. binding O_2 and CO_2
    b. binding CO_2
    c. binding H^+
    d. binding bicarbonate ions

12. One of the main advantages of a closed over an open circulatory system is:
    a. the ability to direct the flow of blood to certain tissues
    b. the use of a heart to create pressure gradients necessary for blood flow
    c. the ability to return blood to the heart
    d. the use of a respiratory pigment to transport O_2

13. What prevents the atria and the ventricles from contracting at the same time?
    a. Pacemaker cells located in the atria fire before the pacemaker cells in the ventricles.
    b. It takes time for epinephrine to diffuse from the atria to the ventricles to trigger contraction.
    c. The electrical signal generated in the right atrium is delayed at the AV node before passing to the ventricles.
    d. The Na^+ channels responsible for initiating ventricular contraction are inactivated and need to return to activated configuration to be electrically stimulated.

14. Your blood pressure is reported to you as two values (e.g., 120/80). These represent:
    a. the pressure at the beginning of atrial contraction and the beginning of ventricular contraction
    b. peak pressure during ventricular contraction and lowest pressure at end of ventricular relaxation phase
    c. peak pressure at the end of atrial contraction and lowest pressure at the end of atrial relaxation
    d. the pressure at the end of atrial contraction and at the end of ventricular contraction

CHAPTER 44—ANSWER KEY

Check Your Understanding

(a) $760 \times 0.2095 = 159.2$ mm Hg

$760 \times 0.7889 = 599.6$ mm Hg

$760 \times 0.0003 = 0.228$ mm Hg

The percentages of these gases would remain the same as you climbed up in altitude; however, the partial pressures of these gases would all decrease because the total atmospheric pressure decreases as you increase altitude.

(b) The amount of gas that dissolves in a liquid depends on several factors:
   1. the solubility of the gas in that liquid,
   2. the temperature of the liquid, and
   3. the presence of other solutes.

CO_2 is almost 30 times more soluble in water than O_2 is. As a result, fish rid themselves of CO_2 more easily than they can obtain oxygen.

(c) The rate of diffusion depends on (1) solubility of the gas, (2) temperature, (3) surface area available for diffusion, (4) difference in partial pressures of the gas across the respiratory medium, and (5) thickness of the barrier to diffusion. Of these, (1) is a physical-chemical factor and is unalterable. Factors (3) and (5) represent characteristics of the gas exchange organs, which cannot be altered in the short term. An animal could alter factor (2) if it were a cold-blooded animal, moving to a colder or warmer environment; it could alter factor (4) by increasing ventilation of the gas exchange organs.

(d) In insects, the tracheal system transports air close enough to cells for gas exchange to take place directly. As a result, insects do not require a circulatory system to transport gases from a respiratory structure to the tissues.

Trachea connect to the exterior via an opening called a spiracle that can be closed. Spiracles are interpreted as adaptations to minimize water loss.

(e) The functions of blood include
1. Carries O_2 and CO_2 between mitochondria and lungs
2. Transports nutrients from the digestive tract to other tissues in the body
3. Moves waste products to the kidney and liver for processing
4. Conveys hormones from glands to target tissues
5. Delivers immune system cells to sites of infection
6. Distributes heat from deeper organs to the surface

(f) Decreases in pH and increases in temperature alter hemoglobin's conformation such that it is more likely to release O_2 at all values of P_{O_2}. This is known as the Bohr shift.

(g) **Carbonic anhydrase** in red blood cells catalyzes the formation of carbonic acid from carbon dioxide and water. As a result, CO_2 that diffuses into red blood cells is quickly converted into bicarbonate ions and protons. The same reaction occurs much more slowly in the plasma surrounding the red blood cells.

(h) A cardiac cycle consists of one complete contraction phase, called systole, and one complete relaxation phase, called diastole. Peak pressure is due to the contraction of the ventricle and is called systolic pressure. The lowest pressure occurs at the end of ventricular relaxation and is called diastolic pressure.

(i) The "pacemaker cells" that initiate contraction are located in a region called the SA (sinoatrial) node. The pace at which the SA node initiates electrical signals determines heart rate. The pace, in turn, is regulated by electrical signals from the brain.

Assessing What You've Learned

1. c; 2. a; 3. d; 4. c; 5. a; 6. d; 7. d; 8. a; 9. c; 10. b; 11. c; 12. a; 13. c; 14. b

45

Electrical Signals in Animals

A. KEY CONCEPTS

45.1 Principles of Electrical Signaling

- **Receptor cells** transmit the information they receive via a **sensory neuron**, and this travels to the brain or spinal cord. Together, the brain and spinal cord form the **central nervous system** (CNS).

- The CNS integrates information from many sensory neurons and sends signals to **effector cells** via **motor neurons**. All of the components outside the CNS are part of the **peripheral nervous system** (**PNS; Figure 45.1**).

The Anatomy of a Neuron

- Most neurons have a **cell body**, a highly branched group of short projections called **dendrites**, and one or more long projections called **axons** (**Figure 45.2**).

- A dendrite receives electrical signals from the axons and dendrites of adjacent cells. The axon then sends the signal to the dendrites of other neurons.

⬤ **CHECK YOUR UNDERSTANDING**

(a) Describe the anatomy of a neuron.

_____

_____

_____

_____

_____

An Introduction to Membrane Potentials

- Because of the separation of ions across a membrane, and because ions carry charge, there is a separation of charge across the membrane called a **membrane potential** (**Figure 45.4**).

- If the difference in charges on either side of the membrane is large, so is the membrane voltage.

How Is the Resting Potential Maintained?

- When a neuron is not transmitting an electrical signal but is merely sitting in extracellular fluid at rest, its membrane has a voltage called the **resting potential**.

- Most proteins inside the cell are negatively charged. The major positively charged ion inside neurons is potassium. In the extracellular fluid, however, sodium and chloride predominate.

- In neurons, only K^+ can cross the membrane easily along its concentration gradient. It does so via potassium channels.

- As K^+ moves from the interior of the cell to the exterior via K^+ channels, the inside of the cell becomes more negatively charged relative to the outside. Eventually, the buildup of negative charge inside the cell begins to attract K^+ and counteract the concentration gradient that favors movement of K^+ out. As a result, the membrane reaches a voltage where there is equilibrium between the concentration gradient that moves K^+ out and the electrical gradient that moves K^+ in.

- The voltage where there is no net movement of K^+ is called the **equilibrium potential**.

- The **Nernst equation** converts energy stored in a concentration gradient into the energy stored as an electrical potential.

⬤ **CHECK YOUR UNDERSTANDING**

(b) How is K^+ involved in the maintenance of the resting membrane potential?

_____

_____

_____

- The equilibrium potential given by the Nernst equation is calculated independently for each ion.
- The **Goldman equation** accounts for the fact that the nerve membrane is permeable to three different ions, and so interrelates the potentials of Na^+, K^+ and Cl^- to generate the membrane potential, in volts.

Using Microelectrodes to Measure Membrane Potentials

- Hodgkin and Huxley pioneered the study of electrical signaling in animals. To record membrane voltages in the giant squid axon, they cut off an intact section of the axon, bathed it in seawater or other known solution, and inserted a microelectrode. Their recordings suggested that the giant squid axon had a resting potential of −45 mV, but later work confirmed that most neurons have a resting potential of around −70 mV.

What Is an Action Potential?

- The initial event is a rapid depolarization of the membrane. Current flow causes the inside of the membrane to become less negative and then positive with respect to the outside.

- The membrane then experiences a rapid repolarization. The membrane actually becomes more negative than the resting potential. This is called the undershoot.
- An action potential is all or none. There is no such thing as a partial action potential.
- Action potentials can be triggered by artificially depolarizing the membrane with an injection of electrical current through a microelectrode.

- Action potentials are propagated down the length of the axon.

45.2 Dissecting the Action Potential

Distinct Ion Currents Are Responsible for Depolarization and Repolarization

- The action potential consists of a strong inward flow of Na^+ followed by a strong outward flow of K^+ (**Figure 45.7**).

How Do Voltage-Gated Channels Work?

- **Voltage-gated channels** are proteins that change conformation in response to charges present at the membrane surface. These changes will open or close the channel.
- Cole and colleagues came up with a technique—**voltage clamping**—that holds the membrane potential of a neuron constant and measures the flows of ions (**Figure 45.8**).
- They found that action potentials result from staggered activity of voltage-gated Na^+ and K^+ channels.

CHECK YOUR UNDERSTANDING

(c) What are voltage-gated channels?

_____

_____

_____

_____

Patch Clamping and Studies of Single Channels

- Neher and Sakmann perfected a technique called patch clamping whereby they suctioned a microelectrode to the membrane. They could then document the currents that flowed through individual transmembrane channels.
- The opening of sodium channels exhibits **positive feedback**. This takes place when the occurrence of an event makes the same event more likely to occur.

Using Neurotoxins to Identify Channels and Dissect Currents

- In giant axons from lobsters treated with the tetrodotoxin (found in pufferfish), the outward-directed K^+ current was normal but the inward-directed flow of Na^+ was blocked.

How Is the Action Potential Propagated?

- Positive charges inside the cell are repulsed by the influx of Na^+ and negative charges are attracted. As a result, sections of the membrane close to the site of an action potential are depolarized. Nearby voltage-gated channels pop open in response, positive feed-back occurs, and an action potential results (**Figure 45.11**).

- Na+ channels are **refractory**. Once they have opened and closed, they are less likely to open again for a short period of time.

- **Myelination** acts as insulation. It prevents charge from leaking out as it spreads down an axon. An unmyelinated area of the neuron called a **node of Ranvier** has a dense concentration of voltage-gated channels. Electrical signals jump from node to node much faster than if the entire neuron were unmyelinated.

45.3 The Synapse

- Otto Loewi (1920) showed that the mechanism that

CHECK YOUR UNDERSTANDING

(d) Why does an action potential not go backward?

_____

_____

_____

_____

(e) How does mylenation aid in the propagation of an action potential?

_____

_____

_____

transmits electrical signals from cell to cell involves **neurotransmitters**. He stimulated the vagus nerve and slowed the frog heart. He then took some of the solution that this heart was bathed in and applied it to another heart, and it slowed as well. This result supplied conclusive evidence for the chemical transmission of electrical signals.

- The interface between two neurons is called a **synapse**. Just inside the synapse, the axon contains numerous sac-like structures called synaptic vesicles (**Figure 45.15**). At this site, three steps occur:

1. The action potential arrives and triggers entry of Ca^{2+}.

2. In response to Ca^{2+}, synaptic vesicles fuse with membrane and release neurotransmitter.

3. Ion channels open when neurotransmitter binds change the postsynaptic cell potential.

What Do Neurotransmitters Do?

- Many receptors function as **ligand-gated ion channels**, meaning that a molecule binds to a specific site on the receptor to activate it.

- Some other receptors activate enzymes that lead to the production of a **second messenger**, which may induce changes in enzyme activity, gene transcription, or membrane potential.

CHECK YOUR UNDERSTANDING

(f) What are ligand-gated ion channels?

_____

_____

_____

Postsynaptic Potentials and Summation

- If the receptors at the synapse admit an influx of sodium ions in response to the arrival of the neurotransmitter (NT), then the postsynaptic membrane depolarizes. These are called **excitatory postsynaptic potentials (EPSPs)**.

- In other receptors, binding of the NT leads to the influx of K^+ or Cl^-. These events hyperpolarize the membrane and make action potentials less likely to occur. These are called **inhibitory postsynaptic potentials (IPSPs)**.

CHECK YOUR UNDERSTANDING

(g) Compare and contrast the EPSP and the IPSP.

_____

_____

_____

_____

- EPSPs and IPSPs are not all-or-none events. They are graded and short lived. The size depends on the amount of NT released at the synapse.

- The sodium channels that trigger action potentials in the postsynaptic cell are located near the start of the axon at a site called the **axon hillock**.

45.4 The Vertebrate Nervous System

- The vertebrate nervous system is functional and anatomically divided into the central nervous system (CNS) and the peripheral nervous system (PNS; **Figure 45.18**).

What Does the Peripheral Nervous System Do?

- The PNS consists of two distinct systems:
    1. A somatic system controls the skeletal muscles.
    2. An autonomic system controls internal processes like digestion and heart rate.

- The autonomic system is composed of the parasympathetic nerves, which stimulate digestion and other processes, and the sympathetic nerves, which prepare organs for stressful situations.

Functional Anatomy of the CNS

- Anatomically, the brain has four major lobes (temporal, occipital, parietal, and frontal) and two hemispheres (left and right; (**Figure 45.20**).

- Functional areas of the brain have been mapped by lesion studies.

- Broca (1861) studied an individual who could understand language but could not speak. After this person's death, Broca found a damaged area in the left front lobe of the cerebrum. Broca's claim that functions are localized to specific brain areas has been verified through extensive efforts to map the cerebrum.

- Wilder Penfield worked with several epileptics scheduled to have seizure-prone areas of their brains surgically removed. While patients were awake, Penfield electrically stimulated portions of their cerebrum and recorded the patient's sensations.

- This technique was used to map the general areas of brain function.

How Does Memory Work?

- George Ojemann and Julie Schoenfield-McNeill recorded individual neurons in the temporal lobes of humans.

- Work on the molecular basis of memory is founded on two main ideas:
    1. Learning and memory must involve some type of short- or long-term change in neurons responsible for these processes.
    2. It will be much easier to understand what these changes are if an extremely simple system of neurons can be studied.

- Eric Kandel and colleagues have focused on the sea slug *Aplysia californica*. The sea slugs have a simple reflex that can be modified by learning. Follow-up studies have shown that the neurons involved in learning release the NT **serotonin**, and that serotonin causes an EPSP in the motor neuron in the gill.

- More serotonin is released after learning takes place. Therefore, EPSPs are higher and the motor neuron is more likely to generate action potentials.

CHECK YOUR UNDERSTANDING

(h) How were the functional areas of the brain mapped out?

_____

_____

_____

_____

_____

B. INTEGRATING YOUR KNOWLEDGE

Looking Back—Key Concepts from Earlier Chapters

Osmosis and Diffusion—Chapter 6

Chapter 6 introduced the processes called diffusion and osmosis. Diffusion describes the movement of substances from regions of higher concentration to regions of lower concentration. The movement of water from regions of higher to lower concentration is called osmosis. This chapter also discussed the electrochemical gradient and ion channels.

Looking Forward—Key Concepts in Later Chapters

Senses—Chapter 46

Before we discuss how hormones work, the electrical signals involved in vision, hearing, taste, and movement must be discussed. This is the subject of **Chapter 46**.

Second Messengers—Chapter 47

The second messenger may induce changes in enzyme activity, gene transcription, or membrane potential. The role of second messengers and hormones is explored in detail in **Chapter 47**.

C. MASTERING CHALLENGING TOPICS

The resting and action potentials are some of the first concepts you will encounter in any physiology or neurology course and are, therefore, the foundation of this chapter on electrical signals. The movement of ions generating a resting potential across a membrane is sometimes difficult to visualize as occurring in the human body. Nevertheless, visualizing these ion movements is a great place to start. First, visualize the movement of ions across the membrane in a resting potential and then try to draw this process from memory while explaining it aloud. Once you are comfortable with how the resting potential is maintained, try your skills on the action potential. Concentrate on where it starts, how it is propagated, and where it ends. Visualize the ions moving through channels as the action potential is created, and graph the voltage across the membrane as the action potential proceeds. Lastly, as you consider how this process works in the entire body, integrate the concepts of EPSP and IPSP so that you are sure not to confuse them in a broader sense.

D. ASSESSING WHAT YOU'VE LEARNED

1. Which of the following describes the role of a dendrite on a nerve cell?
   a. receives electrical signals
   b. transmits electrical signals from the neuron
   c. maintains the DNA of the neuron in a nucleus

2. Which of the following is true about the distribution of ions when a cell is at rest?
   a. Potassium and sodium ion concentrations are higher inside the cell than outside.
   b. Potassium and chloride ion concentrations are higher inside the cell than outside.
   c. Sodium and chloride ion concentrations are higher inside the cell than outside.
   d. Potassium ion concentration is higher inside, and sodium ion concentration is higher outside.

3. What would happen if voltage-gated sodium channels on a resting neuron were to open?
   a. Sodium ions would move into the cell, and the membrane potential would become more negative.
   b. Sodium ions would move into the cell, and the membrane potential would become more positive.
   c. Potassium ions would move out of the cell, and the membrane potential would become more negative.
   d. Sodium ions would move out of the cell, and the membrane potential would become more negative.

4. What would happen if voltage-gated potassium channels on a resting neuron were to open?
   a. Potassium ions would move into the cell, and the membrane potential would become more negative.
   b. Sodium ions would move into the cell, and the membrane potential would become more positive.
   c. Potassium ions would move out of the cell, and the membrane potential would become more positive.
   d. Potassium ions would move out of the cell, and the membrane potential would become more negative.

5. What phase of the action potential is associated with the influx of sodium ions that occurs when voltage-gated sodium ion channels open?
   a. repolarization
   b. depolarization
   c. resting potential
   d. undershoot

6. The action potential is produced by a combination of an influx of sodium ions and an efflux of potassium ions. Since these ions carry similar charge and are moving in opposite directions, why don't their movements simply cancel each other out so that there is no change in the membrane potential?
   a. Voltage-gated potassium channels open at the same time as voltage-gated sodium channels.
   b. Voltage-gated potassium channels open at the same voltage as voltage-gated sodium channels.
   c. The opening of voltage-gated potassium channels is delayed.
   d. The opening of voltage-gated sodium channels is delayed.

7. What would happen to the resting membrane potential of a neuron if the extracellular fluid were flooded with a high concentration of potassium ions?
   a. The resting membrane potential would become more negative.
   b. The resting membrane potential would become more positive.
   c. There would be no change in the resting membrane potential.

8. In patients who suffer from multiple sclerosis (MS), the myelin surrounding the axons degenerates. Which of the following describes the effect of this loss of myelination?
   a. Action potentials would spread down the axon more quickly if the myelin is removed.
   b. Action potentials would spread down the axon more slowly if the myelin is removed.

c. Action potentials would spread down the axon more slowly because voltage-gated ion channels would open more slowly.

d. Action potentials would spread down the axon more quickly because voltage-gated ion channels would open more quickly.

9. Otto Loewi determined that the vagus nerve of the frog releases a neurotransmitter that causes the heart to slow down. Which of the following explains the effect that this neurotransmitter is having?
   a. It is causing voltage-gated sodium channels in the heart muscle to open more quickly.
   b. It is causing voltage-gated sodium channels in the heart muscle to become more "sensitive" to changes in voltage.
   c. It is causing an increase in the rate at which action potentials can be generated in the heart muscle.
   d. It is causing a decrease in the rate at which action potentials can be generated in the heart muscle.

10. Eventually, it was discovered that the neurotransmitter being released from the vagus nerve was acetylcholine (ACh). Heart muscle contracts following the depolarization during an action potential. In these cells, the negative resting potential must become less negative and eventually reach what is called a threshold. Once reaching this threshold potential, the action potential will spontaneously generate and the muscle can contract. Which of the following descriptions could explain the effect of acetylcholine on the heart muscle?
    a. The cells become leaky to potassium ions. Potassium ions leak out of the cell, which causes the resting potential to become more negative.
    b. The cells become leaky to sodium ions. Sodium ions leak out of the cell, which causes the resting potential to become more negative.
    c. The cells become leaky to potassium ions. Potassium ions leak into the cell, which causes the resting potential to become more negative.
    d. The cells become leaky to sodium ions. Sodium ions leak into the cell, which causes the resting potential to become more positive.

11. Under which of the following conditions would a ligand-gated ion channel open?
    a. when the membrane potential becomes more positive
    b. when a molecule binds to a specific site on the channel
    c. when the membrane potential becomes more negative

12. The ability of a postsynaptic neuron to generate an action potential (a signal) depends on information that is being received from surrounding presynaptic

neurons. These "connections" can either be inhibitory or excitatory. What action would be associated with an inhibitory presynaptic neuron?
   a. The signal from the presynaptic neuron causes an influx of sodium into the postsynaptic neuron.
   b. The signal from the presynaptic neuron causes an efflux of sodium ions from the postsynaptic neuron.
   c. The signal from the presynaptic neuron causes an influx of potassium ions from the postsynaptic neuron.
   d. The signal from the presynaptic neuron causes an efflux of potassium ions from the postsynaptic neuron.

13. Neurotransmitters released from the vagus nerve cause a decrease in heart rate. What part of the vertebrate nervous system would this nerve be a part of?
    a. parasympathetic nervous system
    b. sympathetic nervous system
    c. somatic system

14. Which of the following two structures make up the central nervous system (CNS)?
    a. the spinal cord and parasympathetic system
    b. the brain and sympathetic nervous system
    c. the brain and spinal cord
    d. the spinal cord and sympathetic nervous system

15. A 42-year-old male patient comes into the emergency room with paralysis on the right side of his body. The attending physician also notices that the patient is having trouble with language. Assuming that there is a head injury, which part of the head did the injury occur on?
    a. the back
    b. the left side
    c. the right side

CHAPTER 45—ANSWER KEY

Check Your Understanding

(a) Most neurons have a cell body, a highly branched group of short projections called dendrites, and one or more long projections called axons. A dendrite receives electrical signals from the axons and dendrites of adjacent cells. The axon then sends the signal to the dendrites of other neurons.

(b) In neurons, only K^+ can cross the membrane easily along its concentration gradient. It does so via potassium channels. As K^+ moves from the interior of the cell to the exterior via K^+ channels, the inside of the cell becomes more negatively charged relative to the outside. Eventually, the buildup of negative charge inside the cell begins to attract K⁺ and counteract the concentration gradient that favors movement of K⁺ out. As a result, the membrane reaches a voltage where there is equilibrium between the concentration gradient that moves K⁺ out and the electrical gradient that moves K⁺ in. The voltage where there is no net movement of K⁺ is called the equilibrium potential.

(c) Voltage-gated channels are proteins that change conformation in response to charges present at the membrane surface. These changes will open or close the channel.

(d) Na⁺ channels are refractory. Once they have opened and closed, they are less likely to open again for a short period of time.

(e) Myelination acts as insulation. It prevents charge from leaking out as it spreads down an axon. An unmyelinated area of the neuron called a node of Ranvier has a dense concentration of voltage-gated channels. Electrical signals jump from node to node much faster than if the entire neuron were unmyelinated.

(f) Many receptors function as ligand-gated ion channels, meaning that a molecule binds to a specific site on the receptor to activate it.

(g) If the receptors at the synapse admit an influx of sodium ions in response to the arrival of the neurotransmitter (NT), then the postsynaptic membrane depolarizes. These are called excitatory postsynaptic potentials (EPSPs). In other receptors, binding of the NT leads to the influx of K⁺ or Cl⁻. These events hyperpolarize the membrane and make action potentials less likely to occur. These are called inhibitory postsynaptic potentials (IPSPs). EPSPs and IPSPs are not all-or-none events. They are graded and short lived. The size depends on the amount of NT released at the synapse.

(h) Wilder Penfield worked with several epileptics scheduled to have seizure-prone areas of their brains surgically removed. While patients were awake, Penfield electrically stimulated portions of their cerebrum and recorded the patient's sensations. This technique was used to map the general areas of brain function.

Assessing What You've Learned

1. a; 2. c; 3. b; 4. d; 5. b; 6. c; 7. a; 8. b; 9. d; 10. a; 11. b; 12. d; 13. a; 14. c; 15. b

46

Animal Sensory Systems and Movement

A. KEY CONCEPTS

46.1 How Do Sensory Organs Convey Information to the Brain?

Sensory Transduction

- Although animals have sensory receptors that detect a remarkable variety of stimuli, they all **transduce** sensory input (light, sounds, touch, and odors) to a change in membrane potential. In this way, different types of information are transduced into a common type of signal.

- The amount of depolarization that occurs in a sound receptor cell is proportional to the loudness of the sound. Louder sounds induce a higher frequency of action potentials than do softer sounds.

Transmitting Information to the Brain

- Receptor cells tend to be highly specific. For example, each receptor in a human ear responds best to certain frequencies of sound. Therefore, the train of action potentials from a cell contains information about the frequency of the sound, its intensity, and how long the stimulus lasts.

- Each type of sensory neuron sends its signal to a specific portion of the brain. Axons from sensory neurons in the human ear project to a particular area in the side of the brain.

46.2 Hearing

- **Hearing** is the ability to sense the changes in pressure called **sound**. A sound consists of **waves** of air or water pressure. The number of pressure waves that occur in 1 second is called the **frequency**.

How Do Sensory Cells Respond to Sound Waves and Other Forms of Pressure?

- In **pressure receptors**, direct physical pressure on a cell membrane or distortion by bending causes ion channels on a cell membrane to open or close.

In response to change in ion flows, the membrane depolarizes or hyperpolarizes; the result is a new pattern of action potentials from the sensory neuron.

- In many pressure-sensing organs, **hair cells** are named for a set of stiff outgrowths, called **stereocilia**, that occur at one end of the cell (**Figure 46.2**). If they bend in one direction in response to pressure, channels open and depolarize the cell. If they bend in the other direction, channels close and the cell hyperpolarizes.

CHECK YOUR UNDERSTANDING

(a) Define a pressure receptor.

_____

_____

_____

The Mammalian Ear

- When sound waves reach the head, they pass through a canal and strike the **tympanic membrane**, which separates the outer ear from the middle ear (**Figure 46.3**). The membrane vibrations are passed to three tiny bones called **ear ossicles**. The last ossicle, called the **stapes**, vibrates against a membrane called the **oval window** that separates the middle ear from the inner ear.

- The oval window vibrates and generates waves in the fluid inside the **cochlea**. These pressure waves are sensed by **hair cells** in the cochlea.

The Middle Ear Amplifies Sounds

- The **size difference** between the tympanic membrane and oval window is important. The amount of vibration induced by sound waves is increased by a factor of 15.

The ossicles act as **levers** and further amplify the sound such that the overall amplification factor is 22.

The Cochlea Detects the Frequency of Sounds

- Georg von Békésy (1920s) performed experiments on cochleas painstakingly dissected from human cadavers. He was able to vibrate the oval window and record how the cochlea's internal membranes moved in response. His key finding was that sounds of different frequencies caused the basilar membrane to vibrate in specific spots along its length.

- Certain portions of the basilar membrane vibrate in response to specific pitches and result in the bending of hair-cell stereocilia. In this way, hair cells in a particular place on the membrane respond to sounds of certain frequency.

CHECK YOUR UNDERSTANDING

(b) How does the ear amplify sound?

_____

_____

_____

_____

_____

Sensory Worlds: What Do Other Animals Hear?

- Elephants use **infrasounds** (sound frequencies too low for humans to hear) to communicate.

- Bats use ultrasonic sound to **echolocate** (navigate by sound). Bats with cotton in their ears or their mouths taped shut are unable to fly without running into objects. Blindfolded bats, in contrast, are fully able to navigate.

46.3 Vision

The Vertebrate Eye

- The outermost layer of the eye is a tough rind of white tissue called the **sclera** (**Figure 46.7**). The front of the sclera is transparent and forms the **cornea**. Just inside the cornea is a colored, round muscle called the **iris**. The hole in the center of the iris is called the **pupil**.

- Together, the cornea and **lens** focus the light onto the retina in the back of the eye. The **retina** contains a layer of **photoreceptors** and several layers of neurons.

What Do Photoreceptor Cells Do?

- **Photoreceptors** in the vertebrate eye are small rod- or cone-shaped cells called rods and cones. **Rods** are very sensitive to dim light but not to color. **Cones** are much less sensitive to faint light but are stimulated by different colors.

How Do Rods and Cones Detect Light?

- Rods and cones are packed with membrane-rich disks containing large quantities of a transmembrane protein called **opsin**. Each opsin molecule is associated with a much smaller molecule called **retinal**. The two-molecule complex is called **rhodopsin** (**Figure 46.9**).

- Retinal changes shape when it absorbs a photon of light, leading to a change in opsin's conformation. This in turn leads to a series of events that culminates in a change in the cell's membrane potential.

Color Vision: The Puzzle of Dalton's Eye

- John Dalton and his brother (1794) could not differentiate between the colors red and green (a condition called red-green color blindness). Thomas Young analyzed cones and found that there were three different cone types, each having a different type of opsin. These proteins are called the blue, green, and red (or S, M, and L for short, medium, and long wavelength) opsins.

- David Hunt (1990) showed that color-blind people lack either functional M or L cones, or both. Hunt analyzed Dalton's DNA and found that he had a perfectly normal *L* allele but lacked a functional *M* allele. As a result, Dalton did not have green-sensitive cones.

CHECK YOUR UNDERSTANDING

(c) Compare and contrast the rods and cones of the eye.

_____

_____

_____

_____

_____

Sensory Worlds: Do Other Animals See Color?

- A marine fish called a coelacanth lives in water 200 meters deep and has two opsins that respond to the blue region of the spectrum. It is likely that these fish perceive several hues of blue that we would consider one color. Birds and insects can also see ultraviolet light, which has shorter wavelengths than the human eye can see. Why would this ability to detect UV light be adaptive?

- Rattlesnakes and other pit vipers can sense **infrared light** via specialized infrared-sensing organs on their snouts. Because endothermic animals radiate heat in the form of infrared radiation, pit vipers can detect and strike at prey in complete darkness.

CHECK YOUR UNDERSTANDING

(d) How are rattlesnakes able to hunt prey at night?

_____

_____

_____

46.4 Taste and Smell

Taste: Detecting Molecules in the Mouth

- A **taste bud** contains about 100 spindle-shaped **taste cells** that synapse to taste neurons (**Figure 46.14**).
- The four basic tastes are salty, sweet, sour, and bitter.
- The sensation of **saltiness** is primarily due to sodium ions dissolved in food. These ions flow into certain taste cells through open channels and depolarize their membranes.
- **Sourness** is due to the presence of protons that flow directly into certain taste cells through channels. In general, the lower the pH of food, the more it depolarizes a taste-cell membrane and the more sour the food tastes.

Why Do Many Different Foods Taste Bitter?

- Arthur Fox (1931) blew some phenylthiocarbamide (PTC) into the air and found that a colleague could taste bitterness in the air while he could not. Follow-up research confirmed that the ability to taste PTC is inherited and polymorphic.
- Independently, two teams identified a family of 40–80 genes that encode transmembrane receptor proteins. Each protein in this family binds to just one particular type of bitter molecule. A taste cell can have many different receptor proteins from this family.

What Is the Molecular Basis of Sweetness and Other Tastes?

- Membrane receptors that respond to sugars have yet to be identified by the same techniques employed in the search for bitterness receptors.

CHECK YOUR UNDERSTANDING

(e) Describe and characterize the four basic tastes.

_____

_____

_____

_____

Olfaction: Detecting Molecules in the Air

- **Smell** allows animals to monitor airborne molecules that convey information. Wolves and domestic dogs can distinguish millions of different odors at extremely small concentrations.
- When odor molecules reach the nose, they diffuse into a **mucus layer** in the roof of the structure. There they activate **olfactory receptor neurons** via membrane-bound receptor proteins. Axons from these neurons project up to the **olfactory bulb** of the brain (**Figure 46.15**).
- A gene family containing hundreds of distinct coding regions encodes receptor proteins on the surface of olfactory receptor neurons.
- Each olfactory neuron has only one type of receptor, and neurons with the same type of receptor project to different regions in the olfactory region of the brain.

46.5 Movement

Skeletons

- Skeletons provide attachment sites for muscles and a support system for the body's soft Tissues. **Exoskeletons** are hard, hollow structures that envelop the body. **Hydrostatic skeletons** use the pressure of internal body fluids to support the body. **Endoskeletons** are hard structures buried inside the body.
- Endoskeletons are composed of cartilage and bone (**Figure 46.17**). **Cartilage** is made up of cells scattered in a gelatinous matrix of polysaccharides and scattered

protein fibers. **Bone** is made up of cells in a hard extra-cellular matrix of calcium phosphate with small amounts of calcium carbonate and protein fibers.

- The ends of skeletal muscle are often attached to two different bones by **tendons**, which are bands of tough, fibrous connective tissue.

How Do Muscles Contract?

- A **muscle fiber** is a long, thin muscle cell. Within each cell there are many small strands called **myofibrils**.

The Sliding-Filament Model

- The **sarcomere**, which is the functional unit of skeletal muscle, appears as light and dark bands on the myofibril.

- Andrew Huxley and Jean Hanson (1954) proposed that the banding patterns in the sarcomere are actually caused by two types of long filaments—thick and thin—and that the filaments slide past one another during contraction. This explanation became known as the **sliding-filament model.**

- Follow-up research showed that the thin filaments are composed of two coiled chains of a globular protein called **actin**. One end of each chain is bound to the sarcomere; the other interacts with the thick filament.

- Thick filaments are composed of multiple strands of a long protein called **myosin**, which is anchored to the middle of the sarcomere.

How Do Actin and Myosin Interact?

- The myosin head can bind actin and catalyze the hydrolysis of ATP into ADP and a phosphate ion. Myosin and actin are locked together when an animal dies, and its muscles enter the stiff state known as **rigor mortis**. This evidence suggested that ATP must be present for myosin to release from actin.

- Ivan Rayment and colleagues proposed the following model:
  1. When a molecule of ATP binds to the myosin head, the myosin releases from actin.
  2. When ATP is subsequently hydrolyzed, the conformation of the protein changes. Specifically, the myosin neck straightens out, the head pivots, and the myosin head then binds to action in a new location.
  3. When inorganic phosphate is released, the myosin neck bends back into its original position. This bending is called the **power stroke** because it moves the entire actin filament.
  4. A new ATP molecule then binds to myosin, and the cycle starts again.

CHECK YOUR UNDERSTANDING

(f) Why does an animal that dies undergo rigor mortis?

_____

_____

_____

_____

_____

How Does Relaxation Occur?

- Sarcomeres contain proteins called **tropomyosin** and **troponin**. They work together to block the myosin binding sites on actin. As a result, actin and myosin cannot slide past each other.

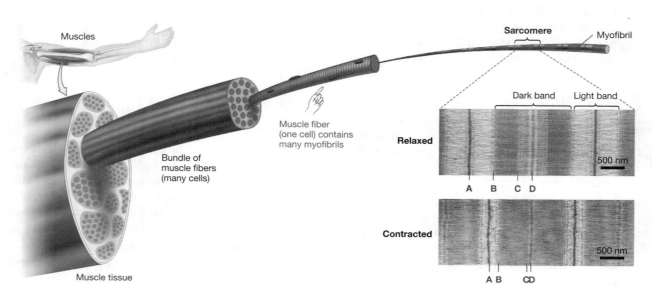

An Overview of Events at the Neuromuscular Junction

1. Action potentials trigger the release of **acetylcholine (ACh)** from the **motor neuron** onto the muscle cell (**Figure 46.23**).

2. Membrane depolarization occurs on the muscle fiber in response to ACh. Action potentials propagate throughout muscle cells via axon-like structures called **T tubules**.

3. T tubules intersect extensive sheets of smooth endoplasmic reticulum called the **sarcoplasmic reticulum (SR)**. As an action potential passes down a T tubule, a protein in the T tubule membrane changes conformation and opens a **calcium channel** in the SR.

4. Calcium ions are released from the SR, causing a conformational change in troponin, which then moves tropomyosin away from the myosin binding sites on actin.

B. INTEGRATING YOUR KNOWLEDGE

Looking Back—Key Concepts from Earlier Chapters

Osmosis and Diffusion—Chapter 6

Chapter 6 introduced the processes called diffusion and osmosis. Diffusion describes the movement of substances from regions of higher concentration to regions of lower concentration. The movement of water from regions of higher to lower concentration is called osmosis. This chapter also discussed the electrochemical gradient and ion channels.

Gene Hunting—Chapter 19

To find the gene responsible for tasting PTC, biologists compared the distribution of genetic markers observed in "tasters" and "nontasters." The effort narrowed its location down to several candidates. **Chapter 19** introduced how this type of gene hunt is done.

Amoebae and Slime Molds—Chapter 29

Actin and myosin are also responsible for the amoeboid movement observed in amoebae and slime molds and the streaming of cytoplasm observed in algae and land plants.

Pollination—Chapter 40

Certain flowers have patterns only apparent in ultraviolet wavelengths that serve as signals for pollinating insects.

Electrical Signals—Chapter 45

Because intercalated discs of cardiac muscle contain many gaps, electrical signals pass directly from one cell to the next. **Chapter 45** explored how electrical signals in animals are initiated and propagated.

Looking Forward— Key Concepts in Later Chapters

Second Messengers—Chapter 47

The second messenger may induce changes in enzyme activity, gene transcription, or membrane potential. The role of second messengers and hormones is explored in detail in **Chapter 47.**

C. MASTERING CHALLENGING TOPICS

You may have had difficulty with the action potential in the last chapter; the difficulty you may have in this chapter relates to the action potential in muscle cells. The sliding-filament model and the electrical and mechanical events that occur in muscle contraction and relaxation are quite difficult to absorb in one sitting. As you prepare for the test, break muscle contraction down into a few discrete sections. First, go over how the action potential gets to the motor neuron and depolarizes the muscle surface. Once you have mastered that part, focus on how the depolarization travels to the actin and myosin filaments and how the release of calcium is involved in initiating muscle contraction. Lastly, go over the molecular action of muscle contraction itself and how the myosin heads ratchet along the actin filaments. Once you have mastered the process in smaller pieces, try to put it all together and write out all the steps as if you were answering an essay question. This should prepare you for an essay question you may see on your test.

D. ASSESSING WHAT YOU'VE LEARNED

1. When sound strikes a sound-receptor cell, the cell depolarizes. Which of the following could explain why this depolarization occurs?
   a. an influx of chloride ions
   b. b an influx of positively charged ions
   c. an efflux of potassium ions
   d. an efflux of sodium ions

2. A light-receptor cell becomes hyperpolarized in response to a flash of light. Which of the following could explain why this hyperpolarization occurs?
   a. The light promotes an influx of sodium ions into the receptor cell.
   b. The light promotes an efflux of chloride ions.
   c. The light promotes the opening of sodium on channels.
   d. The light promotes the closing of sodium on channels.

3. When sound waves enter the ear, they cause a membrane called the tympanic membrane to vibrate. These vibrations are then transmitted across the middle ear to another membrane called the oval window. Why is it beneficial to have the middle ear involved in the delivery of these vibrations to the oval window?
   a. The lever action of the sound waves and the smaller surface area of the oval window help to boost sound energy in the ear so that the ear is more sensitive to sound.
   b. The ear ossicles produce electrical signals that help to amplify the sound waves as they are transmitted to the oval window.
   c. The middle ear is not important, and our hearing would be the same if the ear canal led directly to the oval window.
   d. The middle ear helps to dampen the sound energy to prevent the sound energy from damaging the inner ear.

4. In what part of the human ear are sound waves transmitted in a fluid?
   a. the outer ear
   b. the middle ear
   c. the inner ear

5. Which of the following lists represents the pathway of light through the vertebrate eye?
   a. cornea, iris, pupil, retina, lens
   b. cornea, iris, pupil, lens, retina
   c. cornea, iris, lens, retina, pupil
   d. cornea, lens, pupil, iris, retina

6. What role does the cornea play in the vertebrate eye?
   a. controls the amount of light entering the eye
   b. focuses light on the retina
   c. contains light-sensitive cells that send signals to the brain to process the visual image
   d. the opening through which light passes into the eye

7. John Dalton, an 18th-century chemist, spent time working on the cause of the color blindness that he and his brother both experienced. To them, red and green appear to be the same color. Dalton incorrectly hypothesized that the problem was in the color of the fluid that filled his eye. We now know that he was missing cones that are important for telling the difference between red and green light. How do we know that he had a normal amount of S opsin cones?
   a. S opsin cones reflect red light, and Dalton could see red objects.
   b. M and L opsin cones absorb blue light, and Dalton could see blue objects.
   c. S opsin cones absorb blue light, and his problem was with the colors red and green.
   d. S opsin cones are not involved in sensing color.

8. Taste research has focused on four basic tastes (salty, sour, sweet, and bitter). Which of these tastes is caused by elevating the sodium ion concentration and sensed by a depolarization of taste cells due to an influx of sodium ions?
   a. salty
   b. sour
   c. sweet
   d. bitter

9. Which of the following muscle types is associated with the intestinal tract and arteries?
   a. smooth muscle
   b. skeletal muscle
   c. cardiac muscle

10. Which of the following muscle types is considered voluntary?
    a. smooth muscle
    b. skeletal muscle
    c. cardiac muscle

11. Which of the following muscle types contains multinucleated fibers?
    a. smooth muscle
    b. skeletal muscle
    c. cardiac muscle

12. What component of the sarcomere contains a binding site for ATP?
    a. myosin heads
    b. actin
    c. troponin
    d. tropomyosin

13. Which of the following events produces the "pull" that is associated with myosin and actin sliding past one another?
    a. the binding of ATP to the myosin head
    b. the hydrolysis of ATP
    c. the release of a phosphate from the myosin head after ATP is hydrolyzed
    d. the myosin head with an empty ATP binding site

14. Research has shown that a contraction occurs in response to a release of calcium ions from the sarcoplasmic reticulum found in muscle fibers. What is the role of these calcium ions in the contraction?
    a. The calcium ions are incorporated into ATP, which fuels the contraction process.
    b. The calcium ions cause the thick filaments to shorten. This shortening of the thick filaments is what produces the contraction.
    c. The calcium ions bind to the thin filaments and pull on them to produce the contraction.
    d. The binding of calcium to troponin causes the tropomyosin to rotate out of the way of the binding sites for myosin on the actin molecules. This allows the myosin to bind to the actin to initiate a contraction.

15. In animal joints, at least two muscles are required to flex and extend at the joint. Which of the following explains why a single muscle cannot act as both a flexor and an extensor?
    a. A single muscle would not be strong enough to produce both actions.
    b. The muscle would have to move to a different bone in order to produce both actions.
    c. A muscle can only "pull" at a joint through a contraction. Since muscles cannot push, a different muscle would be required to move the joint in an opposite direction.
    d. The contraction of a muscle that is acting as a flexor is different from the contraction that produces extension.

CHAPTER 46—ANSWER KEY

Check Your Understanding

(a) In pressure receptors, direct physical pressure on a cell membrane or distortion by bending causes ion channels on a cell membrane to open or close. In response to change in ion flows, the membrane depolarizes or hyperpolarizes, and the result is a new pattern of action potentials from the sensory neuron.

(b) The size difference between the tympanic membrane and oval window is important. The amount of vibration induced by sound waves is increased by a factor of 17. The ossicles act as levers and further amplify the sound such that the overall amplification factor is 22.

(c) Photoreceptors in the vertebrate eye are small rod- or cone-shaped cells called rods and cones. Rods are very sensitive to dim light but not to color. Cones are much less sensitive to faint light but are stimulated by different colors.

(d) Rattlesnakes and other pit vipers can sense infrared light using specialized infrared-sending organs on their snouts. Because endothermic animals radiate heat in the form of infrared radiation, pit vipers can detect and strike at prey in complete darkness.

(e) The four basic tastes are salty, sweet, sour, and bitter. The sensation of saltiness is primarily due to sodium ions dissolved in food. Sourness is due to the presence of protons that flow directly into certain taste cells through channels. For bitterness, researchers identified a family of 40–80 genes that encode transmembrane receptor proteins. Each protein in this family binds to just one particular type of bitter molecule. A taste cell can have many different receptor proteins from this family. Membrane receptors that respond to sugars have yet to be identified by the same techniques employed in the search for bitterness receptors.

(f) The myosin head can bind actin and catalyze the hydrolysis of ATP into ADP and a phosphate ion. Myosin and actin are locked together when an animal dies, and its muscles enter the stiff state known as rigor mortis. This evidence suggests that ATP must be present for myosin to release from actin.

Assessing Your Knowledge

1. b; 2. d; 3. a; 3. c; 5. b; 6. b; 7. c; 8. a; 9. a; 10. b; 11. b; 12. a; 13. c; 14. d; 15. c

Chemical Signals in Animals

47

A. KEY CONCEPTS

- The **endocrine system** is responsible for the production and secretion of hormones. A **hormone** is a chemical signal that circulates through the blood or other bodily fluids to affect **target cells**.

47.1 Cell-to-Cell Signaling: An Overview

- Chemical messengers can function in six ways (**Figure 47.1**).

  1. **Autocrine** signals act on the same cell that secreted them.

  2. **Paracrine** signals diffuse locally, acting on neighboring cells.

  3. **Endocrine** signals are produced by hormones, molecules carried to distant cells by blood or other body fluids.

  4. **Neural** signals diffuse extremely short distances between adjacent neurons at synapses.

  5. **Neuroendocrine** signals are hormones that are released from neurons but act remotely on tissues other than synapses.

  6. **Pheromone** signals are released into the environment and act on a different individual.

Hormone Signaling Pathways

- Hormonal signaling pathways take one of three forms (**Figure 47.2**):

  1. Direct hormonal signaling from an endocrine cell

  2. Direct hormonal signaling from the central nervous system

  3. Central nervous system–to–endocrine system signaling

- Signaling pathways are regulated by negative feedback (feedback inhibition), where the product being produced (e.g., hormone) inhibits its own production if high enough levels are present. When levels then fall, production begins again, thus regulating product levels.

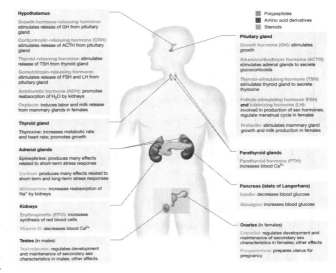

What Are the Components of the Endocrine System?

- Hormone-secreting organs are called **endocrine glands**. They are diverse in size, shape, and location.

Chemical Characteristics of Hormones

- There are three types of chemical messengers in animals (**Figure 47.4**):

  1. Polypeptides

  2. Amino acid derivatives

  3. Steroids

- The key **similarity** in these hormone types is that all are present in extremely small concentrations, yet have large effects. For example, 1000 grams of cow pituitary tissue yields just 0.04 grams of growth hormone.

- The key **difference** in these hormone types is that steroids are lipid soluble while polypeptides and amino acid derivatives are not. As a result, steroids cross cell membranes much more readily than do other types of hormones.

(a) What are the similarities and differences between the three major groups of hormones?

_____

_____

_____

_____

_____

How Do Researchers Identify a Hormone?

- The following steps are usually taken in identifying and purifying hormones:

  1. Remove suspected tissue and study it.

  2. Extract a solution from the tissue in intact animals.

  3. Purify the active ingredient in the raw tissue extract, inject the hormone into animals that cannot produce the molecule, and determine if their symptoms are cured.

(b) What steps are typically taken in the identification and purification of a hormone?

_____

_____

_____

_____

_____

(c) What do you think could go wrong with the process?

_____

_____

_____

_____

47.2 What Do Hormones Do?

- Hormones coordinate the activities of cells in response to three general situations:

  1. Environmental challenges

  2. Growth, development, and reproduction

  3. Homeostasis

How Do Hormones Coordinate Responses to Environmental Change?

- When a person is put into a dangerous situation, hormones regulate both the short-term and long-term response. The short-term reaction is called the **fight-or-flight response** and occurs in conjunction with the activation of the sympathetic nervous system (**Figure 47.8**).

- If you were being chased by a grizzly bear, action potentials from sympathetic nerves would stimulate the adrenal medulla and lead to the release of epinephrine.

- Glucocorticoids prepare an individual for long-term stress by conserving glucose for use by the brain at the expense of other tissues and organs. They also conserve glucose by actively suppressing wound healing and other aspects of immune system function.

How Do Hormones Direct Developmental Processes?

T_3's Role in Amphibian and Insect Metamorphosis

- Thyroid-stimulating hormone in the amphibian brain's pituitary stimulates the production of the thyroid hormone **triiodothyronine**, or T_3, which is responsible for most of the changes observed in amphibian metamorphosis.

- Insect metamorphosis depends upon interactions between **juvenile hormone (JH)** and **20-hydroxyecdysone** (**Figure 47.10**), which stimulates developmental changes from larva to pupa to adult in holometabolous insects (insects that pass through larval stages before reaching adulthood).

Other Examples of How Hormones Regulate Development

- **Primary sex determination** in early embryonic development dictates whether the sex organs of an embryo become male or female. Once testes or ovaries develop, they begin producing male- or female-specific hormones.

- **Testosterone** leads to early development of the male reproductive tract and genitalia and inhibits the formation of breast primordia.

- **Estrogens** induce the initial formation of the female reproductive tract, genitalia, and breast tissue.

- Juvenile-to-adult transition—In boys, surges of sex hormones lead to changes that include enlargement of the penis and testes and growth of facial and body hair.

- In girls, increased concentrations of estrogen called **estradiol** lead to the enlargement of breasts, the onset of menstruation, and other changes.

CHECK YOUR UNDERSTANDING

(d) What are the specific effects of hormone surges in males and females entering puberty?

_____

_____

_____

- Seasonal or cyclical sexual activity—Most long-lived animals reproduce seasonally. In many species, environmental cues trigger the release of sex hormones.

How Are Hormones Involved in Homeostasis?

- Messages from sensory receptors travel to the brain for integration, and the messages often travel from integrators to effectors in the form of hormones.

- **Erythropoietin (EPO)** is released from the kidneys and other tissues when blood oxygen levels are low. EPO acts to restore homeostasis by stimulating the production of red blood cells, which increase the oxygen-carrying capacity of the blood.

- **Calcitonin** and **parathyroid hormone** work in tandem to keep Ca^{2+} levels in the blood close to a set point, much as **insulin** and **glucagon** interact to maintain blood glucose concentrations at preferred levels.

47.3 How Is the Production of Hormones Regulated?

The Hypothalamus and Pituitary Gland

Controlling the Release of Glucocorticoids

- Physiologists showed that rats suffered from a variety of debilitating symptoms when their pituitary gland was removed. In addition, their genitals, thyroid glands, and adrenal cortexes atrophied. Smith proposed the existence of a molecule that affects the adrenal gland.

- In 1943, **adrenocorticotropic hormone (ACTH)** was purified and characterized, confirming that glucocorticoids are secreted from the adrenal cortex when ACTH is released from the pituitary. ACTH is therefore called a **regulatory hormone**.

- The presence of cortisone inhibits the release of ACTH. When the presence of a molecule inhibits its production, **feedback inhibition** is occurring (**Figure 47.14**).

- All of these hormones act as regulators, and they are all involved in feedback inhibition. CRF triggers ACTH production and ACTH triggers glucocorticoid release, but ACTH also inhibits CRF function through a feedback loop, and the glucocorticoids provide feedback by inhibiting ACTH release.

The Hypothalamic-Pituitary Axis—an Overview

- The posterior and anterior sections of the pituitary gland are each influenced by different populations of neurons in the **hypothalamus**. The populations of hypothalamic neurons are called **neurosecretory cells** because they synthesize the release of hormones (**Figure 47.17**).

- The release of hormones by the cells in the hypothalamus is under the control of brain regions responsible for integrating information about the external or internal environment.

- The posterior pituitary is actually an extension of the hypothalamus itself. Neurosecretory cells that project from the hypothalamus produce the hormones **ADH** and **oxytocin**, which are then stored in the posterior pituitary. From there, these hormones are released into the bloodstream.

- The hypothalamus and anterior pituitary are connected indirectly by blood vessels. In response to the arrival of releasing hormones from the hypothalamus, the anterior pituitary secretes hormones that enter the bloodstream and act on target tissues and glands.

CHECK YOUR UNDERSTANDING

(e) Describe the functional relationship between the anterior pituitary, the posterior pituitary, and the hypothalamus.

_____

_____

_____

_____

_____

Control of Epinephrine by Sympathetic Nerves

- Epinephrine functions as a neurotransmitter as well as a hormone, and in some cases, the mode of action of hormones and neurotransmitters is similar.
- By altering synapses during memory functions in the brain, neurotransmitters play a central role in learning and memory. Similarly, many hormones exert their effects by activating particular genes in target cells.

47.4 How Do Hormones Act on Target Cells?

Steroid Hormones and Intracellular Receptors

Identifying the Estrogen Receptor

- In 1964 biologists succeeded in isolating the estradiol receptor in laboratory rats (**Figure 47.18**).
- Follow-up experiments confirmed that the estradiol receptor is located in the nucleus, but not associated with the nuclear envelope. Further, it is found in all target tissues including the uterus, hypothalamus, and mammary glands.
- Subsequently it was found that the gene for the estradiol receptor has a sequence and structure that are similar to the receptors for the glucocorticoids, testosterone, and other steroids.

Documenting Changes in Gene Expression

- A distinctive DNA-binding region occurs in the steroid hormone receptor (**Figure 47.19**). The region codes for the DNA-binding domain called a **zinc finger**. Zinc fingers are found in all of the proteins in the steroid hormone receptor family.
- Investigators also confirmed that steroid hormone-receptor complexes bind to specific sites in DNA. These sites are called **hormone-response elements**. They function as the type of regulatory DNA sequence called an enhancer.

Hormones That Bind to Cell-Surface Receptors

- Epinephrine and the peptide hormones are not lipid soluble, so they have to bind to receptors on the cell surface. The messenger never enters the target cell, so it must activate a receptor on the cell—**signal transduction**.

Identifying the Epinephrine Receptor

- Epinephrine and its agonists produce two distinct patterns of responses because two types of receptor exist—**alpha receptors** and **beta receptors**—each of which has two subtypes.

Signal Transduction and the Role of Second Messengers

- To understand how epinephrine triggers the release of glucose from target cells, biologists focused on an enzyme called **phosphorylase**. It catalyzes the cleavage of glucose molecules off of glycogen. Phosphorylase exists in active and inactive forms, and as the enzyme (**Figure 47.20**).
- Researchers at Earl Sutherland's laboratory found that when they added epinephrine to their cell-free system, large amounts of phosphorylase were activated. By purifying components of the cell-free system, they found **cyclic adenosine monophosphate (cAMP)** as the molecule that activated phosphorylase.
- A **second messenger** is a signaling molecule that increases in concentration inside a cell in response to a molecule that binds at the surface.
- cAMP binds to a protein called **cAMP-dependent protein kinase**. This enzyme responds by phosphorylating an enzyme that then phosphorylates phosphorylase. cAMP is produced from ATP in a reaction that is catalyzed by the enzyme **adenylyl cyclase**.
- The second messenger cAMP plays a primary role in transmitting the signal from the cell surface to the signaling cascade.
- The primary function of the other events in the sequence is to amplify the signal.

B. INTEGRATING YOUR KNOWLEDGE

Looking Back—Key Concepts from Earlier Chapters

Lab Techniques—Chapters 16 and 17

As **Chapter 17** explained, DNA-binding domains are protein regions that make physical contact with DNA. As **Chapter 16** showed, transcription of a gene begins when a regulatory protein like a steroid-receptor complex binds to an enhancer element for that locus.

Homeostasis—Chapter 41

Chapter 41 introduced the concept of homeostasis, or the maintenance of relatively constant physical and chemical conditions inside the body.

Antidiuretic Hormone (ADH)—Chapter 42

Recall from **Chapter 42** that when an individual is dehydrated, ADH is released from the pituitary gland. ADH increases the permeability of the kidney's distal tubules and collecting ducts to water. As a result, water is reabsorbed from the urine and saved. In this way, ADH is instrumental in achieving homeostasis with respect to water balance. If ADH's action is inhibited, homeostasis fails and illness may occur.

Metamorphosis—Chapter 32

Insect metamorphosis is controlled by hormones. If the steroid called juvenile hormone is present at a high concentration, surges of the steroid hormone ecdysone induce growth of a juvenile insect via molting. But if JH levels are low, ecdysone triggers metamorphosis and the transition to adulthood and sexual maturity.

Electrical Signals—Chapter 45

Recall from **Chapter 45** that neurotransmitters have two major effects on target cells. All neurotransmitters trigger a postsynaptic potential, which can make the postsynaptic neuron more or less likely to deliver an action potential. This chapter also illustrated the effect of neurotransmitters on transcription by reviewing experiments on the sea slug *Aplysia*.

Looking Forward—Key Concepts in Later Chapters

Reproduction—Chapter 48

After discovering what is at the forefront of research in chemical signaling, the way has been paved for exploring how hormones regulate reproduction by humans and other animals.

Animal Mating—Chapter 48

Chapter 48 details how during development, a flush of testosterone or estrogen induces the development of seasonal traits like singing in male birds and sexual receptivity in female lizards.

C.   MASTERING CHALLENGING TOPICS

The most complicated system to learn in this chapter is the last one you studied—the second messenger system. As with other physiological cascades, break the system down bit by bit, learn each piece, and then integrate the puzzle. For chemical reactions in the cAMP second messenger system, drawing the reactions in the cascade is of great benefit. This is the best way to answer an essay question. Rather than memorize a lot of key points, practice drawing a diagram and you will find great amounts of information to extract.

D.   ASSESSING WHAT YOU'VE LEARNED

1. All of the following accurately describe the functions of hormones *except*:
   a. Hormones are released into the bloodstream to act on distant targets.
   b. Hormones coordinate the activities of groups of cells.
   c. Hormones respond to changes in the internal, but not the external, environment.
   d. Hormones act only on cells that possess specific receptors for that hormone.

2. All of the following would help in the identification of a suspected hormone *except*:
   a. removal of hormone target tissues
   b. removal of suspected source (endocrine) gland
   c. injection of source gland extract to animal that lacks source gland
   d. purification of source gland extract

3. All of the following statements concerning the chemical classes of hormones are correct *except*:
   a. Steroid hormones are lipid soluble, while polypeptide hormones are water soluble.
   b. Steroid hormones readily cross cell membranes, while polypeptide hormones do not.
   c. Some amino acid derivative hormones are lipid soluble, while most are water soluble.
   d. Steroid hormones are lipid soluble, while polysaccharide hormones are water soluble.

4. Which of the following represents a consequence of the observation that hormones fall into different chemical classes?
   a. Polypeptide hormones bind to receptors located inside target cells.
   b. Steroid hormones bind to receptors located inside target cells.
   c. Steroid hormones bind to receptors located on the surface of target cells.
   d. Most amino acid derivative hormones bind to receptors located inside target cells.

5. Hormones coordinate the activities of cells in response to which of the following?
   a. maintaining homeostasis
   b. growth and development
   c. environmental challenges
   d. all of the above

6. Which of the following situations do you predict might lead to epinephrine release by the adrenal glands, but *not* to cortisol release?
   a. witnessing a car accident
   b. preparing for final exams
   c. skipping meals for two days
   d. training for a marathon

7. Predict the effects of the administration of ACTH on a normal rat.
   a. decrease in cortisol release and an increase in CRH release
   b. increase in cortisol release and a decrease in CRH release
   c. increase in cortisol release and an increase in CRH release
   d. increase in growth hormone release and a decrease in CRH release

8. Predict the effects of the administration of ACTH to a rat that has had its pituitary gland removed.
   a. The adrenal glands will grow in size.
   b. The adrenal glands will shrink (or atrophy).
   c. There will be an increase in CRH levels.
   d. There will be a decrease in cortisol production.

9. All of the following effects would be expected to occur if a normal rat was given cortisol injections *except*:
   a. a decrease in size of the adrenal glands
   b. a decrease in ACTH release
   c. a decrease in CRH release
   d. an increase in cortisol release from the adrenal glands

10. Which of the following provides the best definition of a hormone-response element?
    a. A hormone-response element is a receptor with hormone bound to it.
    b. A hormone-response element is a DNA binding region located on the hormone receptor.
    c. A hormone-response element is a DNA sequence that binds a specific hormone-receptor complex.
    d. A hormone-response element is an RNA transcript.

11. What is the significance of the discovery of different hormone receptor types for the hormone epinephrine?
    a. Two types of epinephrine hormone response elements were identified.
    b. Different tissues responded differently to the presence of epinephrine.
    c. It was discovered that slightly different forms of epinephrine (agonists) were secreted by the adrenal glands.
    d. All target tissues for epinephrine possess all four types of epinephrine receptors.

12. Which of the following molecules serves as a second messenger for epinephrine actions on glucose production?
    a. phosphorylase
    b. cAMP
    c. protein kinase
    d. adenylyl cyclase

13. Which of the following represents the correct sequence of events following the binding of epinephrine to a liver cell?
    a. increase in cAMP levels; phosphorylation of protein kinase; activation of phosphorylase; decrease in glucose levels
    b. increase in cAMP levels; activation of protein kinase; inhibition of phosphorylase; decrease in glucose levels
    c. increase in cAMP levels; activation of protein kinase; phosphorylation of phosphorylase; increase in glucose levels
    d. increase in cAMP levels; phosphorylation of protein kinase; inhibition of phosphorylase; increase in glucose levels

14. The primary functions of a hormonally regulated second messenger system are:
    a. to relay a signal from the cell surface to the interior and to stimulate expression of certain genes
    b. to relay a signal from the cell surface to the interior and to initiate long-term metabolic effects in that cell
    c. to relay and amplify the signal from the cell surface to the interior
    d. to change levels of cAMP in target cells

15. Which of the following best explains the observation that certain hormones, like epinephrine, can have different effects on different cells?
    a. Second messenger molecules, like cAMP, can bind to different proteins within cells.
    b. Second messenger molecules, like cAMP, can result in the transcription of different genes.
    c. Different receptor types can activate the same signal transduction pathway.
    d. Different receptor types can activate different signal transduction pathways.

CHAPTER 47—ANSWER KEY

Check Your Understanding

(a) The key similarity in these hormone types is that all are present in extremely small concentrations, yet have large effects. For example, one kilogram of cow pituitary tissue yields 0.04 grams of growth hormone. The key difference in these hormone types is that steroids are lipid soluble while polypeptides and amino acid derivatives are not. As a result, steroids cross cell membranes much more readily than do other types of hormones.

(b) The following steps are usually taken in identifying and purifying hormones:

1. Remove suspected tissue and study it.

2. Extract a solution from the tissue in intact animals.

3. Purify the active ingredient in the raw tissue extract, inject the hormone into animals that cannot produce the molecule, and determine if their symptoms are cured.

(c) Interference with the process of hormone identification could result from tissue contamination (multiple tissue types in a tissue sample) or from the use of inappropriate solvents for extraction. Test animals may also give ambiguous results unless their current physiological state is fully known.

(d) In boys entering puberty, surges of sex hormones lead to changes that include enlargement of the penis and testes and growth of facial and body hair. In girls, increased concentrations of estrogen (called estradiol) lead to the enlargement of breasts, onset of menstruation, and other changes.

(e) The posterior pituitary is actually an extension of the hypothalamus itself. Neurosecretory cells that project from the hypothalamus produce the hormones ADH and oxytocin, which are then stored in the posterior pituitary. From there, these hormones are released into the bloodstream. The hypothalamus and anterior pituitary are connected indirectly by blood vessels. In response to the arrival of releasing hormones from the hypothalamus, the anterior pituitary secretes hormones that enter the bloodstream and act on target tissues and glands.

Assessing What You've Learned

1. c; 2. a; 3. d; 4. b; 5. d; 6. a; 7. b; 8. a; 9. d; 10. c; 11. b; 12. b; 13. c; 14. c; 15. d

48

Animal Reproduction

A. KEY CONCEPTS

48.1 Asexual and Sexual Reproduction

How Does Asexual Reproduction Occur?

- The sponge generates offspring by **budding**. Budding is completed when a miniature version of the parent breaks free and begins to grow on its own.

- Some groups, like the rotifers, some species of fish, and a few lizards, reproduce exclusively by parthenogenesis.

- In ants, honeybees, and related insects, males are produced by parthenogenesis, but females develop from fertilized eggs produced through sexual reproduction.

Switching Reproductive Modes: A Case History

- *Daphnia* reproduce asexually throughout the spring and summer, with the females producing diploid eggs without sex. The production of offspring via unfertilized eggs is called **parthenogenesis**.

- In late summer or early fall, many females begin producing male offspring by parthenogenesis. Sperm from these males fertilize haploid eggs that females produce via meiosis.

- High population densities and food availability are also factors (in addition to day length) in causing this switch.

CHECK YOUR UNDERSTANDING

(a) What is parthenogenesis, and which animals use it?

_____

_____

_____

Mechanisms of Sexual Reproduction: Gametogenesis

- The mitotic cell divisions, meiotic cell divisions, and developmental events that result in the production of male and female gametes are collectively called **gametogenesis**. This process usually occurs in a sex organ, or **gonad**, in which the male gonads are called **testes** and the female gonads are called **ovaries**.

- In the male and female gonad, diploid cells called spermatogonia and oogonia divide by mitosis to generate cells that undergo meiosis.

48.2 Fertilization and Egg Development

- **Fertilization** is the joining of a sperm and egg to form a diploid zygote. The most basic aspect of diversity in fertilization is where the union of sperm and egg takes place.

External Fertilization

- Most animals that rely on external fertilization live in aquatic environments and tend to produce exceptionally large numbers of gametes.

- Gametogenesis occurs in response to environmental cues like lengthening days and warmer water temperatures.

Internal Fertilization and Sperm Competition

- **Internal fertilization** occurs in the vast majority of terrestrial animals as well as in a significant number of aquatic animals.

- Males deposit sperm directly into the female reproductive tract with an organ called the **penis**, or males may package their sperm into a structure called a **spermatophore**, which is then placed into the female's reproductive tract by the male or female.

- Geoff Parker (1970) published results of an experiment that confirmed the existence of **sperm competition**. Whichever male was last to copulate fathered an average of 85 percent of the offspring produced.

- In species where multiple mating is common, males have extraordinarily large testes for their size and

Spermatogenesis

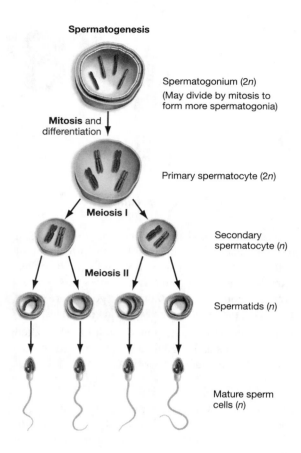

Spermatogonium (2n)
(May divide by mitosis to
form more spermatogonia)

Mitosis and
differentiation

Primary spermatocyte (2n)

Meiosis I

Secondary
spermatocyte (n)

Meiosis II

Spermatids (n)

Mature sperm
cells (n)

Oogenesis

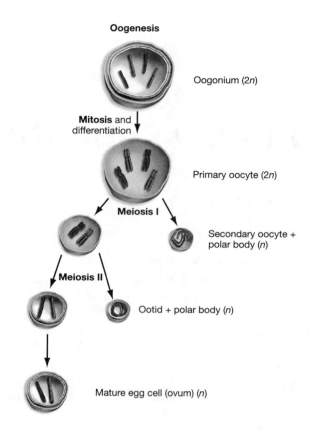

Oogonium (2n)

Mitosis and
differentiation

Primary oocyte (2n)

Meiosis I

Secondary oocyte +
polar body (n)

Meiosis II

Ootid + polar body (n)

Mature egg cell (ovum) (n)

produce proportionately larger numbers of sperm. Although this was first documented in primates, it has also been observed in many other animals, including fruit bats.

✏️ CHECK YOUR UNDERSTANDING

(b) Compare and contrast external and internal fertilization.

_____

_____

_____

_____

Why Do Some Females Lay Eggs while Others Give Birth?

• In **oviparous** animals, the embryo develops in the external environment. In some animals there is no

parental care, but a substantial number of species continue to care for their young after the eggs have emerged from the mother's body.

• In **viviparous** species, development takes place within the mother's body. The embryo attaches to the female's reproductive tract and receives nutrition directly from the mother's circulatory system.

• Egg laying probably represents the ancestral condition in the lizard *Sceloporus*, and viviparity evolved from this, probably due to cold or high-altitude habitats.

✏️ CHECK YOUR UNDERSTANDING

(c) What are the similarities and differnces between oviparity and viviparity?

_____

_____

_____

_____

48.3 Reproductive Structures and Their Functions

The Male Reproductive System

How Does External Anatomy Affect Sperm Competition?

- The external anatomy of the male reproductive system consists of a sac-like **scrotum** and the **penis** (**Figure 48.6**). The scrotum holds the testes; the penis functions as the organ of intromission prior to fertilization.

External and Internal Anatomy of Human Males

- Following are the structural components of the internal reproductive structures in the human male:
  1. *Spermatogenesis and sperm storage*—Sperm are produced in the **testes** and stored in the **epididymis**.
  2. *Production of additional fluids*—Complex solutions that form in the **seminal vesicles, prostate gland**, and **bulbourethral gland** are added to sperm prior to **ejaculation**. The combination of sperm and these **accessory fluids** is called **semen**.
  3. *Transport and delivery*—These functions are the domain of the **vas deferens, urethra**, and **penis**.

CHECK YOUR UNDERSTANDING

(d) Describe the three main functions of the internal anatomy of the male reproductive system.

_____

_____

_____

The Female Reproductive System

The Reproductive Tract of Female Birds

- Birds produce eggs in a single ovary through a series of steps beginning with meiosis and maturation of the follicle, fertilization, addition of egg white and membranes, and finally the formation of a calcium carbonate egg shell prior to laying (**Figure 48.7**).
- Dual oviducts exist, but only one is functional.

External and Internal Anatomy of Human Females

- The external anatomy of the female reproductive system features two folds of skin that cover the **clitoris**, the opening of the **urethra**, and the opening of the **vagina** (**Figure 48.8**). The **clitoris** develops from the same population of embryonic cells that give rise to the penis in males. The **vagina** is where semen is deposited during sexual intercourse and where the fetus is delivered during birth.

- Eggs are produced in the paired **ovaries**. When **ovulation** occurs, a mature egg is expelled from the ovary and enters the **oviduct**, where fertilization may take place. Fertilized eggs are transported to the muscular sac called the **uterus**, where embryonic development continues.

48.4 The Role of Sex Hormones in Mammalian Reproduction

Which Hormones Control Puberty in Mammals?

- Changes that occur during puberty are triggered by increased levels of **testosterone** in boys, and **estradiol** in girls (**Figure 48.11**).
- Researchers isolated a hormone from the hypothalamus called **gonadotropin-releasing hormone (GnRH)**. Investigators also noted that boys and girls who were entering puberty experienced distinctive pulses in the pituitary hormone called **luteinizing hormone (LH)**.

CHECK YOUR UNDERSTANDING

(e) What hormonal changes occur during male and female puberty, and what physiological changes do they induce?

_____

_____

_____

_____

Which Hormones Control the Menstrual Cycle in Mammals?

- Although the length of the **menstrual cycle** varies among women, 28 days is about average. In conjunction with changes in the ovary, the lining of the uterus undergoes a dramatic thickening and regression (**Figure 48.13**). **Menstruation**, by definition, is the expulsion of the uterine lining, which designates day 0 in the menstrual cycle.
- **Follicular phase** (14 days)—a **follicle** matures, and **ovulation** occurs when the follicle is mature; it releases its **oocyte** into the oviduct.

- **Luteal phase** (14 days)—the **corpus luteum** forms from the ruptured follicle and subsequently degenerates.

How Do Pituitary and Ovarian Hormones Change during a Menstrual Cycle?

- Dramatic changes in hormone concentrations occur during the cycle (**Figure 48.14**), including estradiol, LH, FSH, and progesterone.

- LH stays constant except during a spike just prior to ovulation, whereas FSH levels are high during the follicular phase and low during the luteal phase.

- Progesterone is low during the follicular phase and high during the luteal phase.

How Do the Pituitary and Ovarian Hormones Interact?

- Changes in the concentration of estradiol and progesterone affect the release of the pituitary hormones LH and FSH.

- The following steps summarize the interaction of these hormones:

    1. As the uterus is shedding its lining, a follicle is being stimulated to develop under the influence of FSH.

    2. As the follicle grows, its production of estradiol begins to increase. While estradiol is low, it suppresses LH secretion via negative feedback inhibition.

    3. When the follicle produces large quantities of estradiol, it begins to exert positive feedback on LH, producing a spike in both hormones.

    4. The LH surge triggers ovulation and ends the follicular phase.

    5. As the corpus luteum develops from the ruptured follicle, it begins to secrete progesterone in response to LH.

    6. The rise in progesterone converts the thickened uterine lining to an actively secreting tissue with a well-developed blood supply.

    7. If fertilization does *not* occur, the corpus luteum degenerates and progesterone levels fall, causing the uterine lining to degenerate. With progesterone levels down, LH and FSH are released from inhibitory control, their levels rise, and a new cycle starts again.

48.5 Pregnancy and Birth in Mammals

- Once an egg is released from the ovary, it is viable for less than 24 hours. Sperm can survive for 5 days, so sexual intercourse must occur less than 5 days prior to ovulation for pregnancy to result.

CHECK YOUR UNDERSTANDING

(f) Describe the two phases of the menstrual cycle.

_____

_____

_____

_____

_____

- Although an ejaculate may contain hundreds of millions of sperm, only about 100–300 actually reach the oviduct.

Major Events during Pregnancy

- As smooth-muscle contractions in the oviduct gradually move the zygote toward the uterus, the cell begins to divide by mitosis.

- Once the embryo has become implanted in the uterine lining, its cells begin synthesizing and secreting a hormone called **human chorionic gonadotropin (hCG)**, which is a chemical messenger that prevents the corpus luteum from degenerating.

The First Trimester

- The mass movement of cells results in the formation of the embryonic tissues called ectoderm, endoderm, and mesoderm. By eight weeks, these tissues have differentiated into the various organs and systems of the body.

- Another key event in the first trimester is the formation of the **placenta**. This structure starts to form on the uterine wall a few weeks after implantation.

The Second and Third Trimesters

- During this time, the remainder of development focuses mainly on growth. The brain and lungs undergo dramatic growth and development.

How Does the Mother Nourish the Fetus?

- The developing embryo is completely dependent on the mother for oxygen; chemical energy in the form of sugars, amino acids, and other raw materials for growth; and waste removal.

- A mother's blood volume expands by as much as 50 percent during pregnancy, and dramatic increases occur in the stroke volume of the heart.

- The mother's breathing rate and volume also increase to accommodate the fetus's demand for oxygen and its production of carbon dioxide.

- The fetus's blood has a higher affinity for oxygen than the mother's blood does (**Figure 48.17**).

Birth

- The rapid growth of the fetus's brain during the last trimester makes birth challenging. The pituitary hormone **oxytocin** is clearly important in stimulating smooth-muscle cells in the uterine wall to begin contractions (**Figure 48.19**).

- During the last stage of birth, the uterus contracts at a low frequency; then, as labor progresses, the cervix at the base of the uterus begins to dilate. Once the cervix is fully dilated, uterine contractions become more forceful, longer, and more frequent. Eventually, the fetus is expelled through the cervix and into the vagina.

- After the baby is delivered, the medical staff clamps and then cuts the umbilical cord that connects the child and the placenta. By gently tugging the cord, they remove the placenta.

B.   INTEGRATING YOUR KNOWLEDGE

Looking Back—Key Concepts from Earlier Chapters

Fertilization—Chapter 22

Chapter 22 introduced this process by describing how sperm makes contact with an egg and penetrates its membrane. The discussion in that chapter focused on the molecular mechanisms of fertilization using the sea urchin as a model organism. Recall that contact between sperm and egg triggers the acrosomal reaction, in which enzymes released from the head of the sperm chew a path through the material surrounding the egg membrane.

Countercurrent Exchangers—Chapter 44

Maternal and fetal arteries in the placenta are arranged in a countercurrent fashion. As **Chapter 44** explained, countercurrent flows maintain a concentration gradient that makes diffusion efficient.

Hormones—Chapter 47

Chapter 47 introduced the female sex hormone estradiol and the male sex hormone testosterone. Also recall that the human sex hormones play a key role in three events—the development of the male and female reproductive tract in embryos, the maturation of the reproductive tract during the transition from childhood to adulthood, and the regulation of spermatogenesis and oogenesis in adults.

C.   MASTERING CHALLENGING TOPICS

You have studied mitosis and meiosis in previous chapters, so make sure you have reviewed these processes extensively. They were presented as difficult topics in the chapters in which they first appeared. As far as new material is concerned, the menstrual cycle is by far the most complex portion of this chapter. As with everything else you've studied, begin by learning how each major hormone changes during this cycle. Then, study the physical and physiological changes that occur because of the hormonal changes. Finally, try to put everything together by learning how all of the hormones interact by positive and negative feedback. Drawing diagrams and explaining them to classmates will help to reinforce these concepts.

D.   ASSESSING WHAT YOU'VE LEARNED

1. The production of offspring via unfertilized eggs is called:
   a. gametogenesis
   b. parthenogenesis
   c. oogenesis
   d. spermatogenesis

2. All of the following conditions led to a switch from asexual to sexual reproduction in the aquatic crustacean *Daphnia except*:
   a. longer day lengths
   b. low food availability
   c. high population densities
   d. season

3. Many animals, including *Daphnia*, are able to reproduce both sexually and asexually. Which of the following represents the leading hypothesis explaining the advantage associated with the ability to switch to sexual reproduction?
   a. Sexual reproduction is advantageous when energy is readily available.
   b. Sexual reproduction does not require the process of meiosis.
   c. Sexual reproduction produces more genetically diverse offspring capable of surviving stressful environmental conditions.
   d. Sexual reproduction allows for more rapid production of offspring.

4. All of the following represent differences between the processes of spermatogenesis and oogenesis in humans *except*:
   a. The production of oogonia stops before birth, while the production of spermatogonia occurs throughout adult life.
   b. Sperm are haploid, while eggs are diploid.
   c. Oogenesis produces polar bodies; spermatogenesis does not.
   d. Sperm cells are smaller in size than oocytes.

5. All of the following are general characteristics of external fertilization *except*:
   a. production of large numbers of gametes
   b. use of terrestrial environments
   c. synchronization of release of gametes
   d. production of diploid offspring

6. Which of the following experimental results support the hypothesis that, for some species, pheromones may coordinate spawning?
   a. Fish maintained in groups, but not individually, exhibited coordinated spawning activity.
   b. Only 10% of sea cucumbers raised in isolation under natural conditions released their gametes.
   c. Observations of sea stars in the wild have demonstrated the coordination of spawning activity.
   d. Male fruit flies mate with a female only when her eggs have achieved the appropriate level of maturity.

7. Which of the following mechanisms likely play a role in conferring "the second male advantage" during internal fertilization in fruit flies? *Note:* There may be more than one correct choice.
   a. The fluid surrounding the sperm is poisonous to the sperm of other males.
   b. The second male's sperm physically dislodge the first male's sperm from storage areas within the female reproductive tract.
   c. The female reproductive tract produces toxins that eventually harm stored sperm.
   d. The most recently deposited sperm are more vigorous in their swimming movements toward the egg.

8. What characteristics are common to animals for which multiple matings are common?
   a. They tend to live in aquatic habitats.
   b. This characteristic is observed only in insects (like the fruit fly).
   c. The females tend to produce large numbers of offspring.
   d. The males tend to have large testes.

9. Which of the following best explains the difference between oviparity and viviparity?
   a. Oviparous animals utilize external fertilization, while viviparous animals use internal fertilization.
   b. Oviparous animals do not care for their young following fertilization, while viviparous animals do.
   c. Oviparous animals are egg bearing, and embryonic development takes place in the environment; viviparous animals retain their embryos for development within the mother's body.
   d. Oviparous animals receive nutrition directly from the mother's body during development, while the embryos of viviparous animals receive nourishment from yolk sacs.

10. How have studies of the lizard species *Sceloporus* helped scientists understand how natural selection has favored changes between oviparity and viviparity?
    a. Because oviparity evolved independently in two different groups of lizards, it is thought to represent the ancestral condition.
    b. Viviparous lizard species appear to have evolved from oviparous species.
    c. There appears to be a correlation between cold weather and the evolution of oviparity.
    d. Oviparity is favored in high-altitude environments.

11. The seminal vesicles, bulbourethral glands, and prostate gland contribute all of the following components to the semen *except*:
    a. an energy source in the form of sucrose
    b. prostaglandins to stimulate smooth-muscle contractions in the female reproductive tract
    c. alkaline compounds to counteract the acidity of the female reproductive tract
    d. antibiotic substances

12. Puberty in both males and females is triggered by:
    a. testosterone in males and estrogen in females
    b. LH release from the hypothalamus in both males and females
    c. GnRH release from the hypothalamus in both males and females
    d. GnRH release from the anterior pituitary gland in both males and females

13. How does a new menstrual cycle become initiated when fertilization fails to occur?
    a. Progesterone levels drop due to regression of the corpus luteum, releasing the hypothalamus and pituitary from feedback inhibition. FSH levels rise.
    b. hCG produced by the uterine lining signals the corpus luteum to regress, releasing the hypothalamus and pituitary from feedback inhibition. FSH levels rise.
    c. The nervous system detects the absence of implantation, signaling the release of GnRH from the hypothalamus, which triggers the release of FSH from the pituitary.
    d. hCG, produced by the embryo following implantation in the uterus, signals the corpus luteum to regress. Progesterone levels drop and FSH levels rise.

14. Which statement is correct concerning the length of viability of the human egg and sperm?
    a. Both the egg and the sperm are viable for 1 day.
    b. Both the egg and the sperm are viable for 5 days.

c. Sperm are viable for 1 day, while eggs are viable for 5 days.

d. Sperm are viable for 5 days, while eggs are viable for 1 day.

15. Which of the following best describes the function of the placenta?

a. The placenta produces the hormone FSH, needed for embryonic development.

b. The placenta provides the embryo with a protective cushion.

c. The placenta allows for gas and nutrient exchange between the maternal and fetal circulations.

d. The placenta is one of the three main types of embryonic tissues. Eventually, it develops into the fetal heart.

CHAPTER 48—ANSWER KEY

Check Your Understanding

(a) *Daphnia* reproduce asexually throughout the spring and summer; the females produce diploid eggs without sex. The production of offspring via unfertilized eggs is called parthenogenesis. In late summer or early fall, many females begin producing male offspring parthenogenetically. Sperm from these males fertilize haploid eggs that females produce via meiosis.

(b) Most animals that rely on external fertilization live in aquatic environments and tend to produce exceptionally large numbers of gametes. Gametogenesis occurs in response to environmental cues like lengthening days and warmer water temperatures. Internal fertilization occurs in the vast majority of terrestrial animals as well as in a significant number of aquatic animals. Males deposit sperm directly into the female reproductive tract with an organ called the penis; or males may package their sperm into a structure called a spermatophore, which is then placed into the female's reproductive tract by the male or female.

(c) In oviparous animals, the embryo develops in the external environment. In some animals there is no parental care, but a substantial number of species continue to care for their young after the eggs have emerged from the mother's body. In viviparous species, development takes place within the mother's body. The embryo attaches to the female's reproductive tract and receives nutrition directly from the mother's circulatory system.

(d) The structural components of the internal reproductive structures in the human male:

1. *Spermatogenesis and sperm storage.* Sperm are produced in the testes and stored in the epididymis.

2. *Production of additional fluids.* Complex solutions that form in the seminal vesicles, prostate gland, and bulbourethral gland are added to sperm prior to ejaculation. The combination of sperm and these accessory fluids is called semen.

3. *Transport and delivery.* These functions are the domain of the vas deferens, urethra, and penis.

(e) Changes that occur during puberty are triggered by increased levels of testosterone in boys, and estradiol in girls. Researchers isolated a hormone from the hypothalamus called gonadotropin-releasing hormone (GnRH). Investigators also noted that boys and girls who were entering puberty experienced distinctive pulses in the pituitary hormone, called luteinizing hormone (LH).

(f) (1) Follicular phase (14 days)—a follicle matures, and ovulation occurs when the follicle is mature; it releases its oocyte into the oviduct. (2) Luteal phase (14 days)—the corpus luteum forms from the ruptured follicle and subsequently degenerates.

Assessing What You've Learned

1. b; 2. a; 3. c; 4. b; 5. b; 6. b; 7. a and b; 8. d; 9. c; 10. b; 11. a; 12. c; 13. a; 14. d; 15. c

49

The Immune System in Animals

A. KEY CONCEPTS

49.1 Innate Immunity

- **Innate immunity** refers to immune system cells that are ready to respond to foreign invaders at all times.

- An **antigen** is any foreign molecule. Most antigens are proteins or glycoproteins from bacteria or viruses or other invaders, but foreign carbohydrates and lipids also function as antigens.

- The innate immune system responds the same way to all antigens.

Barriers to Entry

- In humans and other animals, the most important barrier to pathogen entry is the **skin**. In addition to a physical barrier, skin cells present a chemical barrier because they secrete **lactic acid** and **fatty acids** that lower pH of the surface and prevent bacterial growth

The Innate Immune Response

- The cells involved in the innate response are alerted to the presence of foreign invaders by the presence of certain molecules (**Figure 49.2**). For example, all bacteria have proteins that begin with a N-formyl-methionyl molecule instead of methionine. When **Pattern-recognition receptors** in the innate immune systems detect specific shapes or patterns present in foreign molecules, **leukocytes** (white blood cells) respond.

- The main steps in the **inflammatory response** (**Figure 49.3**) are as follows:

1. Bacteria and other pathogens enter the wound.

2. At the wound, clotting reduces bleeding.

3. Mast cells secrete factors that mediate **vasodilation** and **vascular constriction**, which bring blood, plasma, and cells to the injured area.

4. **Neutrophils** secrete factors that kill and degrade pathogens.

5. Neutrophils and **macrophages** remove pathogens by **phagocytosis**.

6. Macrophages secrete hormones called **cytokinins** that attract immune system cells to the site and activate cells involved in tissue repair.

7. The inflammatory response continues until the foreign material is eliminated and the wound is repaired.

CHECK YOUR UNDERSTANDING

(a) Outline the main steps in the inflammatory response.

_____

_____

_____

_____

_____

49.2 The Acquired Immune Response: Recognition

An Introduction to Lymphocytes and the Immune System

- An **antibody** is a protein that binds to a specific antigen. The binding of an antibody to an antigen leads to the destruction of the foreign molecule or cell.

- The cells involved in the acquired immune response are called **lymphocytes**. Lymphocytes form and mature in the primary organs of the immune system: the **bone marrow** and **thymus**.

- The **lymph nodes** and **spleen** are important sites where lymphocytes encounter antigens. The fluid and lymphocytes present in the lymph nodes and ducts are called **lymph**.

The Discovery of B Cells and T Cells

- Investigations of the immune system's response to *Salmonella typhimurium* show that antibodies are important in neutralizing the antigen and that the **bursa** is critical for antibody production.

- Mice lacking a thymus developed pronounced defects in their immune system. Follow-up experiments showed that lymphocytes from the bursa and thymus have two different functions.

- The bursa- and thymus-dependent lymphocytes became known as **B cells**, which produce antibodies, and **T cells**, which are involved in an array of functions.

CHECK YOUR UNDERSTANDING

(b) What are the bursae, and how do they relate to the immune system?

_____

_____

_____

Antigen Recognition and Clonal Selection

- The clonal-selection theory consists of three central claims:

  1. Each lymphocyte formed in the bone marrow or thymus has a unique receptor on its surface that recognizes one antigen.

  2. When the receptor on a lymphocyte binds to an antigen, the lymphocyte is activated.

  3. Some of these cloned cells persist after the pathogen is eliminated. As a result, they are able to respond quickly and effectively if the infection recurs in the future.

The Discovery of B-Cell Receptors and T-Cell Receptors

- The **B-cell receptor (BCR)** and antibodies belong to a family of proteins called the gamma globulins and are more specifically called **immunoglobulins** (**Figure 49.6**).

- The BCR has three distinct components:

  1. The **light chain** of about 25,000 daltons

  2. The **heavy chain** of about 50,000 daltons

  3. A **transmembrane domain** that anchors the protein in the B-cell membrane. Antibodies lack this domain and are secreted from the cell.

- The **T-cell receptor (TCR)** binds to antigens only after they have been processed by other immune system cells and presented on their cell membranes. It is composed of a single **alpha chain** and a single **beta chain**. The shape is similar to the arm of the BCR molecule.

- Antibodies and lymphocyte receptors do not bind to the entire antibody molecule, but to a selected region called an **epitope**. An antigen may have many different epitopes where binding by antibodies and lymphocytes actually takes place.

What Is the Molecular Basis of Antigen Specificity and Diversity?

- Researchers have shown that certain amino acids in the light chain are extremely variable among B cells.

- The presence of specific amino acids at certain positions in the light chain is responsible for the ability of the BCR and antibodies to bind to unique epitopes.

- The binding occurs at the tips of the antibody and BCR molecule's arms.

The Discovery of Gene Recombination

- Dryer and Bennett (1965) suggested that a single gene codes for the C region of the light chain and that a separate gene codes for the V region. They hypothesized that early in the life of a lymphocyte, sequences from the *V* gene are inserted into the *C* gene.

- Hozumi and Tonegawa (1976) reasoned that if light-chain diversity was produced by DNA recombination, then the variable and constant regions of mature B cells should be shorter than in the same region in immature cells.

- Two large DNA fragments from the immature cell hybridized with the entire L chain, and the larger of these two fragments also hybridized with the **constant-region mRNA**.

- Biologists proposed that the smaller fragment contained the variable region and that the larger fragment included the constant region. This was strong evidence that DNA recombination brings *V* and *C* genes closer together as B cells mature (**Figure 49.9**).

- The following steps summarize how the light-chain locus changes:

  1. Start with immature DNA containing full complement of V, J, and C segments.

  2. One V segment joins with one J segment and the C segment by recombination.

  3. Transcription occurs.

  4. RNA processing occurs (removal of introns, poly(A) tail etc.).

5. Translation results in a protein with a unique amino acid sequence. Depending on which V and J segments join during recombination, 629 other amino acid sequences are possible.

(c) Outline the steps involved in the light-chain locus changes in gene recombination.

_____

_____

_____

_____

How Does the Immune System Distinguish Self from Nonself?

- Researchers introduced B cells and T cells with anti-self receptors into mice. These experimental lymphocytes were eliminated.

- The great majority of all immature B cells and T cells produced in the bone marrow and thymus are anti-self and are destroyed before they mature via a process termed **negative selection**.

- The mechanism for recognition of self is being intensively studied, since it holds the key to **autoimmunity**, where immune system cells turn on the normal cells of the body and destroy them as if they were nonself.

49.3 The Acquired Immune Response: Activation

Antigen Presentation of MHC Proteins: Activating T Cells

- If bacteria migrate rapidly at a wound site, leukocytes called **dendritic cells** are recruited to the area. They take up some of the antigens present and migrate to the nearest lymph node.

- As CD4+ T cells move through the lymph node in search of an antigen, those with complementary receptors interact with their corresponding epitope on dendritic cells.

- Once a T cell receives signal 1 and 2, it begins to divide and produce a series of daughter cells. This process, called **clonal expansion**, is a crucial step in the acquired immune response because it leads to a large population of lymphocytes capable of responding specifically to the antigen that has entered the body.

- If the original T cell is CD4+, the daughters differentiate into one of two types of **helper T cells**: T_H^1 and T_H^2.

- If the original T cell is CD8+, its daughter cells develop into **cytotoxic T lymphocytes (CTLs)**.

- Helper T cells and CTLs are often referred to as **effector T cells**.

(d) What is clonal expansion?

_____

_____

_____

B-Cell Activation and Antibody Secretion

- Receptors on B cells interact directly with the bacterial or viral antigens that are floating free in lymph or blood. Once a free antigen is bound, B cells **internalize** the molecule and process it via the same Class II pathway used by dendritic cells.

- The second part of the activation process occurs when an activated CD4+ T_H^2 cell with a **complementary receptor** arrives. Helper T cell binds to the **receptor–MHC complex** on the B cell.

- The interaction between the B cell and helper T cell supplies an initial activation signal, which is followed by a co-stimulatory signal 2 analogous to the one that occurs during T-cell activation.

49.4 The Acquired Immune Response: Culmination

How Are Bacteria and Other Foreign Cells Killed?

- During the innate response to bacteria that enter a wound, macrophages at the site **phagocytose** some of the invaders.

- Macrophages also process and present antigens via the **Class II pathway**. As a result, macrophages at the site of infection display epitopes on their surfaces that can be recognized by helper T cells.

- If an activated T_H^1 cell binds to these antigen-laden macrophages, two things happen:

  1. The phagocytic activity of the macrophages is enhanced.

2. The $T_H{}^1$ cells secrete cytokines that kill bacteria and viruses, recruit additional phagocytic cells to the site, and increase the inflammatory response.

How Are Viruses Destroyed?

- The immune system has two major ways to eliminate viruses, the **cell-mediated response** and the **humoral response**.

Cell-Mediated Response

- Once T cells and B cells have been activated, most of the major elements of the acquired immune response are in place.

- Infected cells respond to the arrival of a virus by processing antigens from the invader. Viral antigens are attached to **MHC Class I proteins** and presented on the surface of the infected cell. This sends a message to the immune system to kill the cell.

- In the cell-mediated response, cytotoxic T lymphocytes (CTLs) contain granules that punch holes in cells. CTL makes contact with virus-infected cell and releases granules causing the cell to lyse.

Humoral Response

- In the humoral response, antibodies coat free virus particles. The virus envelope cannot fuse with the host-cell membrane. The antibody-coated virus is recognized and phagocytosed by the macrophage.

(e) Compare and contrast the cell-mediated response and the humoral response.

_____

_____

_____

_____

_____

Why Does the Immune System Reject Foreign Tissues and Organs?

- Organ transplants must be matched to the donor, because the MHC cells of the transplanted tissue will otherwise be recognized as foreign antigens.

- Transplanted human organs are carefully chosen to match the immune system characteristics of the recipients, who are also given immune-suppressing drugs to ensure the success of the transplant operation.

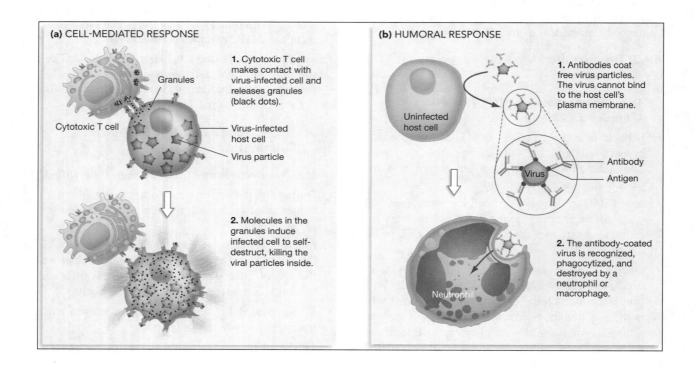

(a) CELL-MEDIATED RESPONSE

Granules

Cytotoxic T cell

1. Cytotoxic T cell makes contact with virus-infected cell and releases granules (black dots).

Virus-infected host cell

Virus particle

2. Molecules in the granules induce infected cell to self-destruct, killing the viral particles inside.

(b) HUMORAL RESPONSE

Uninfected host cell

1. Antibodies coat free virus particles. The virus cannot bind to the host cell's plasma membrane.

Antibody

Virus

Antigen

Neutrophil

2. The antibody-coated virus is recognized, phagocytized, and destroyed by a neutrophil or macrophage.

Responding to Future Infections:
Immunological Memory

- **Memory cells** do not participate in the initial, or primary immune response. Instead, they provide a surveillance service after the original infection has been cleared.

- If the same antigen enters a body a second time, a **secondary acquired immune response** takes place. This response is faster and more efficient than the primary response (**Figure 49.14**).

- The process of **somatic hypermutation** leads to a fine-tuning of the immune response. The B cells that result from somatic hypermutation produce antibodies that bind to the antigen more tightly than their ancestor cells did during the primary response.

- A **vaccine** contains epitopes from a pathogen, but not the pathogen itself. After vaccination occurs, the body mounts a primary immune response that results in the production of memory cells.

CHECK YOUR UNDERSTANDING

(f) How does a vaccination impart immunity to an individual?

_____

_____

_____

_____

B.   INTEGRATING YOUR KNOWLEDGE

Looking Back—Key Concepts
from Earlier Chapters

Viruses—Chapter 35

The only evidence that Dryer and Bennett could offer in support of their gene recombination idea was that viral genes insert themselves into the genomes of host cells.

DNA Recombination—Chapter 19

Children who are born with a genetic defect in the enzyme responsible for DNA recombination in maturing lymphocytes are unable to generate normal T-cell and B-cell receptors. As **Chapter 19** detailed, their acquired immune response is badly impaired as a result.

C.   MASTERING CHALLENGING TOPICS

The topics covered in this chapter are very visual and involve multistep processes, therefore a lot of diagram drawing will be helpful in learning the immune responses. The most confusing aspect of the immune system is the different responses, including the differences in the B-cell and T-cell activation. Inevitably, compare and contrast questions will appear on your exam, so study similarities as well as differences in these systems. As with every chapter, learn the responses and cell types separately until you are comfortable with them. Next, attempt to draw charts outlining similarities and then differences.

D.   ASSESSING WHAT YOU'VE LEARNED

1. The barriers faced by invading pathogens are considered part of the innate division of the immune system. Which of the following statements regarding innate immunity is true?
   a. Components of the innate immune system must first be activated to respond to an antigen.
   b. Components of the innate immune system respond the same way to all antigens.
   c. Components of the innate immune system respond in a specific way to each antigen.
   d. The response of the innate immune system depends on the type of proteins or glycoproteins found on the bacteria, viruses, or other invaders.

2. Which of the following would be associated with innate immunity?
   a. lymphocytes
   b. antibodies
   c. inflammation
   d. B cells

3. A very important step in our understanding of acquired immunity came when scientists discovered the receptors on B- cells and T- cells that bind to antigens. What part of these protein receptor molecules shows the highest variation in amino acid sequence from cell to cell?
   a. the antigen-binding sites
   b. the transmembrane domains
   c. the heavy chain of the B-cell receptor and alpha chain of the T-cell receptor
   d. the light chain of the B-cell receptor and the beta chain of the T-cell receptor

4. In acquired immunity, the epitope is:
   a. the type of bacteria or pathogen that is invading
   b. the molecule produced by B cells for antigen recognition
   c. a molecule involved in promoting inflammation
   d. a selected region of an antigen where antigen recognition occurs

5. When T cells encounter an antigen, they begin the process of clonal expansion. What is meant by clonal expansion?
   a. The T cell divides rapidly to produce a large population of genetically identical cells that are all capable of recognizing the same antigen.
   b. The T cell divides rapidly to produce a large population of genetically diverse cells that are capable of recognizing a variety of antigens.
   c. The T cell secretes chemicals to destroy the antigen.
   d. The T cells bind to B cells to activate the B cells and promote the production of antibodies that are capable of recognizing the antigen.

6. What process leads to the activation of T cells and triggers the clonal expansion of these cells?
   a. T-cell clonal expansion is triggered when T cells encounter a free-floating antigen in blood or lymph.
   b. T cells are always active and producing clones.
   c. T-cell clonal expansion is triggered when T cells encounter an antigen that is presented by another cell.

7. During HIV infection, the number of cells drops as the virus infects these cells and destroys many of them. What makes HIV selective for this specific type of cell?
   a. HIV contains surface proteins that bind to the CD4 receptor.
   b. HIV contains surface proteins that bind to the antigen-binding portion of the heavy chain of an antibody receptor.
   c. HIV contains surface proteins that bind to the CD8 receptor.

8. Which of the following is associated with the humoral response to a viral infection?
   a. activated cells
   b. cytotoxic T cells
   c. MHC I proteins bound to an antigen found at the surface of an antigen-presenting cell
   d. plasma cell production of antibodies

9. Which of the following can explain why the secondary immune response is both stronger and faster than the primary immune response?
   a. The exposure to an antigen is always stronger the second time it is encountered.
   b. cells remain activated after they are exposed to the antigen in the primary response.
   c. Memory cells were produced during the first exposure to the antigen.
   d. Antibodies produced during the first exposure remain in the blood for decades and can take part in the secondary response.

10. During the secondary immune response, some of the memory B cells that respond migrate to a specialized area in the lymph node, where they undergo rapid mutation. This process is referred to as:
    a. clonal expansion
    b. clonal selection
    c. somatic hypermutation
    d. inflammation

11. What role does somatic hypermutation play in the immune response?
    a. This helps the body to prevent other infections while the immune system responds to a specific antigen.
    b. This helps the immune system become more efficient as antibodies are produced that can bind even more tightly to the antigen during the second exposure.
    c. Somatic hypermutation is a defect in the immune system, which produces a secondary immune response that is inefficient.

12. HIV and influenza are viruses that have been difficult to fight with vaccines. Explain why this has been such a problem.
    a. The epitopes on these viruses change frequently as the viruses replicate.
    b. The vaccines would cause serious infections such as HIV infection and the flu.
    c. These viruses do not contain any epitopes on their surface.
    d. These viruses cause infections that have only recently been discovered, and an effective vaccine has not been found yet.

13. During an allergic response, the body produces a large number of IgE antibodies. What do these antibodies bind to en route to producing the well-known symptoms of an allergic response?
    a. T_H^1 cells
    b. plasma cells and memory B cells
    c. T_H^2 cells
    d. mast cells and basophils

14. During the innate response to bacteria that enter a wound, macrophages at the site phagocytize some of the invaders. What are the macrophages doing in this process?
    a. They are engulfing or "eating" the bacteria.
    b. They are producing chemicals that kill the bacteria.
    c. They are producing antibodies that bind to the bacteria.
    d. They are releasing histamine, which causes an inflammatory response around the bacteria.

15. The production and maturation of T cells occurs in two different places in the body. Where do T cells become mature?
    a. bone marrow
    b. thyroid
    c. thymus
    d. pancreas

CHAPTER 49—ANSWER KEY

Check Your Understanding

(a) The main steps in the inflammatory response are as follows:

1. Bacteria and other pathogens enter the wound.
2. At the wound, clotting reduces bleeding.
3. Mast cells secrete factors that mediate vasodilation and vascular constriction, which bring blood, plasma, and cells to the injured area.
4. Neutrophils secrete factors that kill and degrade pathogens.
5. Neutrophils and macrophages remove pathogens by phagocytosis.
6. Macrophages secrete hormones called cytokinins that attract immune system cells to the site and activate cells involved in tissue repair.
7. The inflammatory response continues until the foreign material is eliminated and the wound is repaired.

(b) Antibodies in the immune system's response to *Salmonella typhimurium* are important in neutralizing the antigen. The bursa is critical for antibody production. The bursa- and thymus-dependent lymphocytes became known as B cells, which produce antibodies, and T cells, which are involved in an array of functions.

(c) The following are the summary of steps in how the light-chain locus changes:

1. Start with immature DNA containing full complement of V, J, and C segments.
2. One V segment joins with one J segment and the C segment by recombination.
3. Transcription occurs.
4. RNA processing occurs (removal of introns, poly(A) tail etc.).
5. Translation results in a protein with a unique amino acid sequence. Depending on which V and J segments join during recombination, 629 other amino acid sequences are possible.

(d) Once a T cell receives signal 1 and 2, it begins to divide and produce a series of daughter cells. This is called clonal expansion and is a crucial step in the acquired immune response because it leads to a large population of lymphocytes capable of responding specifically to the antigen that has entered the body.

(e) The immune system has two major ways to eliminate viruses—the cell-mediated response and the humoral response. In the cell-mediated response, cytotoxic T lymphocytes (CTLs) contain granules that punch holes in cells. CTL makes contact with a virus-infected cell and releases granules causing the cell to lyse. In the humoral response, antibodies coat free virus particles. The virus envelope cannot fuse with the host-cell membrane. The antibody-coated virus is recognized and phagocytosed by macrophage.

(f) A vaccine contains epitopes from a pathogen, but not the pathogen itself. After vaccination occurs, the body mounts a primary immune response that results in the production of memory cells.

Assessing Your Knowledge

1. b; 2. c; 3. a; 4. d; 5. a; 6. c; 7. a; 8. d; 9. c; 10. c; 11. b; 12. a; 13. d; 14. a; 15. c

50

An Introduction to Ecology

A. KEY CONCEPTS

50.1 Areas of Ecological Study

Organismal Ecology

- This ecology focuses on how individual organisms interact with their environment.

- Organismal ecologists explore the morphological, physiological, and behavioral adaptations that allow individuals to live successfully in a particular area.

Population Ecology

- A **population** is a group of individuals of the same species that live in the same area at the same time.

- Population ecologists focus on how the numbers of individuals in a population change over time.

Community Ecology

- A biological **community** consists of the species that interact with each other within a particular area.

- Community ecologists ask questions about the nature of the interactions between species and the consequences of those interactions.

Ecosystem Ecology

- An **ecosystem** consists of all of the organisms in a particular region along with nonliving, or **abiotic**, components.

- Biologists analyze how chemical elements that act as nutrients cycle through ecosystems and how energy flows through the organisms in a region.

How Do Ecology and Conservation Efforts Interact?

- **Conservation biology** is the effort to study, preserve, and restore threatened populations, communities, and ecosystems.

CHECK YOUR UNDERSTANDING

(a) What is the difference between population ecology and community ecology?

_____

_____

_____

50.2 Types of Aquatic Ecosystems

What Physical Factors Play a Key Role in Aquatic Ecosystems?

- Light has a major influence on **productivity**: the total amount of carbon fixed by photosynthesis per unit area per year.

- Water flow and depth also have a major influence on aquatic environments.

Freshwater Environments > Lakes and ponds

- Lakes and ponds are distinguished by size, and lakes are larger.

Water Depth

- The **littoral zone** consists of the shallow water along the shore.

- The **limnetic zone** is offshore and comprises water that received enough light to support photosynthesis.

- The **benthic zone** is made up of the substrate.

Water Flow

- Water movement in lakes and ponds is driven by wind and temperature.

- Mixing driven by wind and temperature changes allows well-oxygenated water from the surface to reach the benthic zone and nutrient-rich water from the benthic zone to reach the surface.

Organisms

- Cyanobacteria, algae, and other microscopic organisms are collectively called **plankton** and live in the photic zone.
- Animals that consume **detritus** are common in the benthic zone.

Freshwater Environments > Wetlands

Water Depth

- Wetlands are shallow-water habitats where the soil is saturated with water for at least part of the year and can be distinguished from lakes and ponds by these traits:
  1. They have shallow water.
  2. They have **emergent vegetation** (plants grow above water surface).

Water Flow

- Marshes and swamps have slow but steady flow of water.
- **Bogs** develop in depressions where there is no water flow.
- Bog water is low in oxygen and very acidic.

Organisms

- **Marshes** lack trees and typically feature grasses.
- **Swamps** are dominated by trees and shrubs.
- Because their physical environments are so different, there is little overlap in the types of species found in bogs, marshes, and swamps.

Freshwater Environments > Streams

- **Streams** are bodies of water that move constantly in one direction. Creeks are small streams, and rivers are large.

Water Depth

- Most streams are shallow enough that light reaches the bottom.

Water Flow

- Where it originates at a mountain glacier, lake, or spring, a stream tends to be cold, narrow, and fast. These areas tend to be well oxygenated.
- As it descends, a stream tends to become larger, warmer, and slower. These areas tend to be oxygen poor.

Organisms

- It is rare to find photosynthetic organisms in small, fast-moving streams.

- As streams widen and slow down, conditions become more favorable for growth of algae and plants.

Freshwater/Marine Environments > Estuaries

- An **estuary** is the environment that forms where rivers meet the ocean.
- Salinity has dramatic effects on osmosis and water balance; therefore, the species living in estuaries have adaptations that allow them to cope with variations in salinity.

Marine Environments > The Ocean

Water Depth

- Biologists describe the structure of an ocean by naming six regions:
  1. The **intertidal zone** (littoral zone) is along the shore.
  2. The **neritic zone** extends from the intertidal zone to an ocean depth of about 200 m (the edge of the **continental shelf**).
  3. The **oceanic zone** encompasses the remainder and largest of the ocean environments.
  4. The **benthic zone** is the area at the bottom.
  5. The **photic zone** receives sun to support photosynthesis.
  6. The **aphotic zone** is the area too dark to support photosynthesis.

Water Flow

- Tides and wave action are major influences in the intertidal zone.
- Large-scale currents circulate water in the oceanic zone in response to prevailing winds and the Earth's rotation.

Organisms

- Each zone in the ocean is populated by distinct species.
- In the tropics, shallow portions of the neritic zone may support **coral reefs**—among the most productive environments in the world.

CHECK YOUR UNDERSTANDING

(b) What are the three main factors that influence aquatic ecosystems?

_____

_____

_____

50.3 Types of Terrestrial Ecosystems

- A **biome** is a type of terrestrial ecosystem that is unique to a given region and characterized by a distinct type of vegetation.

- **Climate** is the prevailing long-term weather conditions in a particular region. **Weather** consists of the specific short-term atmospheric conditions of temperature, moisture, sunlight, and wind.

- **Net primary productivity** is the total amount of carbon fixed per year minus the amount of carbon oxidized during cellular respiration. **Aboveground biomass** is the total mass of living plants excluding roots.

Terrestrial Biomes > Tropical wet forests

- **Tropical wet forests**, or rain forests, are found in equatorial regions where temperatures and rainfall are high and variation is low. The plants grow all year round.

- Equatorial forests are also renowned for their **species diversity**. A 10 m × 100 m plot may contain over 200 tree species, and some researchers contend that the rain forests house over 30 million arthropod species.

Terrestrial Biomes > Subtropical deserts

- Subtropical desert bands exist along 30° of **latitude** north and south of the equator.

- Most of the world's great deserts, including the Sahara, Gobi, Kalahari, and Australian Outback, lie at this latitude.

- The striking feature of this biome is the low precipitation.

- Desert species generally cope with the extreme temperatures and aridity in one of two ways:

  1. Growing at a low rate year round

  2. Growing rapidly in response to any rainfall

Terrestrial Biomes > Temperate grasslands

- Precipitation is quite low, and the temperature profiles in temperate grasslands are quite seasonal. Temperature variation is important because it dictates a well-defined growing season. These regions usually exist because conditions are too dry to support tree growth but too cold and seasonal for drought-adapted desert species.

- Although the productivity of temperate grasslands is generally lower than that of forest communities, grassland soils are often highly fertile.

Terrestrial Biomes > Temperate forests

- In temperate areas with high precipitation, grasslands give way to forests (e.g., the region around Chicago). The abundance of moisture allows trees to dominate the landscape.

- Trees experience a dormant period in the winter, and therefore temperate forests are dominated by deciduous species whose leaves fall in the autumn and regrow in the spring.

- Productivity and diversity of temperate forests are moderate.

Terrestrial Biomes > Boreal forests

- The **boreal forest**, or **taiga**, stretches across most of Canada, Alaska, Russia, and northern Europe. This climate is characterized by very cold winters, cool, short summers, and extraordinarily high annual variation in temperature. Precipitation is virtually identical to that in the temperate grasslands.

- The subarctic landscape is dominated by highly cold-tolerant conifers, including spruce, pine, fir, and larch trees. Except for larch, these species are evergreen.

- Although aboveground biomass is high, productivity and diversity in boreal forests are low.

Terrestrial Biomes > Arctic tundra

- The growing season is 6–8 weeks at the most; otherwise, temperature is below freezing. The annual precipitation is also very low.

- The arctic is treeless; however, the soils are saturated year round due to the low evaporation rates.

- Tundra has low species diversity, productivity, and aboveground biomass. Most tundra soils are in the perennially frozen state known as **permafrost**.

CHECK YOUR UNDERSTANDING

(c) What would you predict about the productivity and aboveground biomass in a tropical wet forest that has year-round growing conditions?

_____

_____

_____

_____

50.4 The Role of Climate and the Consequences of Climate Change

Global Patterns in Climate

Why Are the Tropics Warm and the Poles Cold?

- Areas of the world are warm if they receive a large amount of sunlight per unit area; they are cold if they receive a small amount of sunlight per unit area.
- The Earth's shape dictates that the regions at or near the equator receive a great deal of sunlight per unit area relative to regions that are closer to the poles.

Why Are the Tropics Wet?

- The **Hadley cell** is responsible for making the Amazon River basin wet and the Sahara dry.
- Air that is heated by the strong sunlight along the equator expands and rises. Warm air can also hold a great deal of moisture. When the air cools as it rises, it can no longer hold moisture; thus, it rains.
- As the old, cold air is pushed out of the way, it falls and warms over the desert areas. Thus little rain occurs in these areas.

What Causes Seasonality in Weather?

- **Seasons** are regular, annual fluctuations in temperature, precipitation, or both.
- Seasons are the result of Earth's tilt on its axis. Tilting areas toward the Sun causes summer, and tilting areas away from the Sun causes winter.

Mountains and Oceans: Regional Effects on Climate

- The most important regional effects are mountains and oceans.
- The presence of mountain ranges tends to produce extremes in precipitation, while the presence of oceans has a moderating influence on temperature.

How Will Global Warming Affect Ecosystems?

- Recent increases in average global temperatures may represent the second-most rapid period of climate change ever.
- Scientists use three tools to predict how global warming will affect ecosystems:
  1. *Simulation studies* are based on computer models of weather patterns.
  2. *Observational studies* are based on long-term monitoring at fixed sites around the globe.
  3. *Experiments* are designed to simulate changed climate conditions.

Experiments That Manipulate Temperature

- Many of these experiments have been done in the tundra biome because these regions are projected to experience dramatic changes in temperature and precipitation.
- Most experiments are performed in the fashion outlined in the following figure:

Experiment

Question: **How does increasing variation in rainfall—without changing total amount—affect grassland biomes?**

Hypothesis: Increasing variability in rainfall affects soil moisture, net primary productivity (NPP), and species diversity.

Null hypothesis: Increasing variability in rainfall has no effect on grassland biomes.

Experimental setup:

Record soil moisture, aboveground biomass (as an index of NPP), and species diversity, over a 4-year period.

Prediction: Soil moisture, NPP, and/or species diversity will differ between experimental plots and control plots.

Prediction of null hypothesis: Soil moisture, NPP, and species diversity will not differ between experimental plots and control plots.

Results:

Conclusion: **Altering variability in rainfall has dramatic effects on soil moisture (data not shown), NPP, and species diversity.**

- Results support the conclusion that species diversity decreases and arctic tundra environments are giving way to boreal forest.

Experiments That Manipulate Precipitation

Simulation and observational studies have supported the conclusion that increases in average global temperature increases the variability in temperature and precipitation.

CHECK YOUR UNDERSTANDING

(d) What is the difference between climate and weather?

_____

_____

_____

50.5 Biogeography: Why Are Organisms Found Where They Are?

- **Biogeography** is the study of how organisms are distributed in space.
- No one species can survive the full range of environmental conditions present on Earth.

The Role of History

- **Dispersal** refers to the movement of an individual from the place of birth to the location where it lives and breeds as an adult.
- If an **exotic** (one that is not native) organism is introduced to a new area, spreads rapidly, and eliminates native species, it is an **invasive species**.

Biotic Factors

- The distribution of species is often limited by **biotic factors**—interactions with other organisms.

Abiotic Factors

- The distribution of species is also limited by **abiotic factors**—particularly temperature and moisture.

CHECK YOUR UNDERSTANDING

(e) List one additional biotic and abiotic factor not mentioned above that could affect species distribution.

_____

_____

B. INTEGRATING YOUR KNOWLEDGE

Looking Back—Key Concepts from Earlier Chapters

Temperature—Chapter 3

Temperature is critical because the enzymes that make life possible work at optimal efficiency only in a narrow range of temperatures.

Water Balance—Chapter 42

Aquatic organisms are surrounded by water and are exquisitely sensitive to changes in ionic composition.

Desert Plant Adaptations—Chapter 10

Desert plant species have adapted in that they have small leaves, a waxy coating on the leaves, and the CAM pathway for photosynthesis as discussed in **Chapter 10**.

Bacteria and Archaea—Chapter 28

Bacteria and archaea can grow profusely in deep ocean environments, using hydrogen gas, hydrogen sulfide, and other inorganic molecules as a source of energy.

Speciation—Chapter 26

Chapter 26 explored how continental drift has contributed to the current distributions of species; here the focus is on how interactions with the abiotic and biotic environments affect where a particular species is able to live.

Looking Forward—Key Concepts in Later Chapters

Animal Interactions—Chapters 51–53

Chapters 51, 52, and 53 investigate how different species interact with one another and with their physical environment. Here in Chapter 50, we have introduced the topics.

C. MASTERING CHALLENGING TOPICS

The most challenging concept in this chapter, apart from memorizing all the new terms, is the dynamics of the Hadley cell. Remember that the Hadley cell is responsible for maintaining the wet conditions in the Amazon River basin and dry conditions at 30 degrees north and south latitude. The easiest way to learn the dynamics of the Hadley cell is to go over Figure 50.24 and read the accompanying text. The figure allows to you really understand the movements of cold air and warm air. Drawing the movements without the aid of the figure and explaining the concept to a fellow student is a sure way to ensure that you have a grasp of this process.

D. ASSESSING WHAT YOU'VE LEARNED

1. The ecology that focuses on how individuals in a group change over time is called:
   a. organismal ecology
   b. population ecology
   c. community ecology
   d. ecosystem ecology

2. The effort to preserve, study, and restore threatened populations, communities, and ecosystems is called:
   a. community ecology
   b. biogeography
   c. conservation biology
   d. climate biology

3. The overall factor dictating whether an area will be generally warm or cold is:
   a. the amount and intensity of sunlight
   b. the amount of wind
   c. the amount of precipitation
   d. volcanic activity

4. Seasonality in climate is caused directly by:
   a. geomagnetic differences year round
   b. Earth's tilt on its axis
   c. ocean currents
   d. Sun position relative to other planets

5. The most productive biome is:MOST
   a. subtropical deserts
   b. temperate forests
   c. boreal forests
   d. tropical wet forests

6. The biome dominated by deciduous trees is:
   a. subtropical deserts
   b. temperate forests
   c. boreal forests
   d. tropical wet forests

7. The biome with the lowest species diversity is:
   a. subtropical deserts
   b. temperate forests
   c. boreal forests
   d. tundra

8. The wetlands with the slowest water flow and the most acidic water are typically:
   a. marshes
   b. bogs
   c. swamps
   d. streams

9. The ocean zone over the continental shelf is called:
   a. the oceanic zone
   b. the benthic zone
   c. the intertidal zone
   d. the neritic zone

10. If an exotic organism is introduced to a new area and becomes successful, it is called:
    a. an invasive species
    b. a foreign species
    c. a foreign successful species
    d. an alien species

CHAPTER 50—ANSWER KEY

Check Your Understanding

(a) A population is a group of individuals of the same species that live in the same area at the same time. Population ecologists focus on how the numbers of individuals in a population change over time. A biological community consists of the species that interact with each other within a particular area. Community ecologists ask questions about the nature of the interactions between species and the consequences of those interactions.

(b) Light has a major influence on productivity: the total amount of carbon fixed by photosynthesis per unit area per year. Water flow and depth also have a major influence on aquatic environments.

(c) The favorable year-round growing conditions in wet tropical forests produce extensive growth, leading to extremely high productivity and above-ground biomass.

(d) Climate refers to the prevailing long-term weather conditions in a particular region. Weather consists of the specific short-term atmospheric conditions of temperature, moisture, sunlight, and wind.

(e) A biotic factor involves living organisms. A bacterium could cause disease, and that could affect species dispersal. Besides temperature and moisture, altitude is another important abiotic factor.

Assessing What You've Learned

1. b; 2. c; 3. a; 4. b; 5. d; 6. b; 7. d; 8. b; 9. d; 10. a

51

Behavior

A. KEY CONCEPTS

51.1 Types of Behavior: An Overview

- **Behavior** is action—a response to a stimulus.

- **Proximate causes** determine how actions occur.

- **Ultimate causes** determine why actions occur.

- **Learning** is defined as a change in behavior that results in a specific experience in the life of an individual.

Innate Behavior

- Highly inflexible behavior patterns are called **fixed action patterns** or **FAPs**.

- These highly stereotypical behavior patterns have three characteristics:

  1. There is almost no variation in how they are performed.

  2. They are species specific.

  3. Once the sequence of actions begins, it typically continues until completion.

- FAPs are examples of **innate behavior**—types of behavior that are inherited and show little variation based on learning.

- Marla Sokolowski discovered an important behavioral trait controlled by a single gene. She noticed that some fruit-fly larvae tended to move away from food after eating, while others remained in place. By breeding "rovers" and "sitters" that possessed other distinct genetic markers, Sokolowski and colleagues were able to map the gene responsible for the behaviors.

Conditional Strategies and Decision Making

- It is common for an individual's behavior to change in response to learning and to show flexibility.

- **Cost-benefit analysis**—animals appear to weigh the costs and benefits of responding in various ways. The responses are measured in terms of their impact on fitness.

What Decisions Do White-Fronted Bee-Eaters Make When Foraging?

- **Optimal foraging** defines animals maximizing their feeding strategy.

- Some individuals forage a few meters away from their colony, while others forage hundreds of meters away.

- Individuals foraging far from the colony actually bring back a much larger mass of insects on each trip.

How Do Female Barn Swallows Choose Mates?

- Male barn swallows with particularly long tails are more efficient in flight and thus more successful in finding food.

- Long-tailed males mated earliest—meaning that females preferred them. Early nesting pairs have time to complete a second nest later in the season.

Why Do Some Bluehead Wrasses Undergo a Sex Change?

- A group of female wrasses live inside a boundary dominated by one large male fish. When that male dies, the largest female wrasse changes sex and becomes the dominant male.

CHECK YOUR UNDERSTANDING

(a) What are the three characteristics of a fixed action pattern?

_____

_____

_____

51.2 Learning

Simple Types of Learning: Classical Conditioning and Imprinting

- In **classical conditioning**, individuals are trained by experience to give the same response to more than one stimulus—even a stimulus that has nothing to do with the normal response.

- Upon hatching, ducklings and goslings adopt as their mother the first moving thing they see—this is called **imprinting**.

- Imprinting occurs only during a **critical** or **sensitive period**.

More Complex Types of Learning: Birdsong

- Depending on the species involved, song-learning behavior falls at various locations along the learning continuum described in Figure 51.2.

- For example, song learning is innate in certain species (phoebes); in other species (white-crowned sparrows), it has to be learned during a certain critical period.

Can Animals Think?

- **Cognition** can be defined as the recognition and manipulation of facts about the world.

- New Caledonian crows have the ability to make tools and solve complex problems, which suggests that they can think.

What Is the Adaptive Significance of Learning?

- Learning is an adaptation that allows individuals to change their behavior in response to a changing and unpredictable environment.

- Norway rats are exceptionally good at learning to avoid foods that contain poisons. They taste only small quantities of novel foods and avoid foods that induce illness.

- Scrub jays instead are good at remembering where they cached seeds.

CHECK YOUR UNDERSTANDING

(b) What is the difference between classical conditioning and imprinting?

_____

_____

_____

51.3 How Animals Act: Hormonal and Neural Control

Sexual Activity in Anolis Lizards

- *Anolis carolinensis* live in the woodlands of the southeastern United States. Males emerge in January and establish breeding territories, and females become active a month later. By May, females are laying an egg every 10–14 days and by the end of the breeding season will have produced an amount of eggs equaling twice their body mass.

- Testosterone and estradiol produce these behaviors in males and females, respectively. If production of these hormones is the proximate cause, what environmental cues trigger these breeding events?

- Researchers brought a large group of inactive lizards into the laboratory and divided them into five groups:
  1. Single females isolated
  2. Females placed in female-only groups
  3. Females paired with males
  4. Females placed with groups of castrated males
  5. Females placed with groups of uncastrated males

- Females exposed to breeding males began producing eggs much earlier than the females placed in the other treatment groups. Females need to experience springlike light and temperatures and exposure to breeding males.

- To test this idea, researchers repeated the previous experiment, but added a twist. They placed females with males that had intact dewlaps or with males whose dewlaps had been surgically removed. The result? Females grouped with dewlap-less males were slow to produce eggs—just as slow, in fact, as the females in the first experiment that had been grouped with castrated males. These females had not been courted at all. The result suggests that the dewlap is a key visual signal.

Escape Behavior in Noctuid Moths

- Bats use **echolocation** to pursue prey at night.

- Moths can detect the bat sounds and then either avoid the bat if it is at a distance or power dive when it is close.

- Researchers have actually identified and characterized the neurons responsible for this behavior.

51.4 Communication

Modes of Communication

- **Communication** is defined as any process in which a signal from one individual modifies the behavior of a recipient individual.

(c) How did researchers investigate the importance of the dewlap, and what was the result?

_____

_____

_____

_____

- A **signal** is any information-containing behavior.
- Communication can be acoustic, visual, olfactory, or tactile.

A Case History: The Honeybee Dance

- Karl von Frisch suspected that successful food finders communicate the location of food to other individuals.
- He observed bees displaying a "round dance" that contained information about the location of food. Workers got the information about the location of food through the dancer's movements.
- He also observed a "waggle run" that tells the length of distance the workers must fly, as well as the direction.

When Is Communication Honest or Deceitful?

Deceiving Individuals of Another Species

- The anglerfish has an appendage that looks remarkably like a minnow and dangles near its mouth.
- Predatory *Photinus* fireflies can mimic the pattern of flashes by females of several other species.
- Individuals increase their fitness by providing inaccurate or misleading information to members of a different species.

Deceiving Individuals of the Same Species

- In some cases, natural selection has also favored the evolution of lying to the same species.
- A male bluegill will imitate a female and get close to another male while he is courting an actual female. The other male bluegill then thinks he is courting two females; and while he does this, the female mimic secretly fertilizes the real female's eggs but does not have to care for them.

When Does Deception Work?

- Lying works only when it is relatively rare.
- As deceit becomes extremely common, then natural selection will strongly favor individuals that can detect and avoid or punish liars.

(d) Describe what Karl von Frisch found about honeybee dances.

_____

_____

_____

51.5 Migration and Navigation

- A movement that results in a change in position is called **orientation**.
- The simplest type of orientation is called **taxis**, which involves positioning the body toward or away from a stimulus.
- **Phototaxis** is orientation toward light, while **phototaxis** is orientation toward sound.

Why Do Animals Move with a Change of Seasons?

- **Migration** is defined as the long-distance movement of a population associated with a change of seasons.
- Arctic terns nest along the Atlantic coast of North America, fly south along the coast of Africa to wintering grounds off Antarctica, then fly back along the eastern coast of South America.
- Many of the monarch butterflies native to North America spend the winter in the mountains of central Mexico or southwest California.
- Salmon that hatch in rivers along the Pacific coast of North America and northern Asia migrate to the ocean when they are a few months to several years old. They return to the stream where they hatched to mate and then die.

How Do Animals Find Their Way?

Piloting

- **Piloting,** or using familiar landmarks, is used by many species to find their way. Young offspring follow their parents during migrations and appear to memorize the route.

- Homing pigeons can always find their way back home. But if their eyes are covered, they fail to navigate only the final stage, suggesting they may use piloting for the final part of a journey.

Compass Navigation

- To determine where north is, animals appear to use the Sun during the day and stars at night.
- During cloudy conditions, birds appear to use Earth's magnetic field to orient.

CHECK YOUR UNDERSTANDING

(e) What is migration? Give two examples taken from animal behavior.

_____

_____

_____

_____

51.6 The Evolution of Self-Sacrificing Behavior

- **Altruism** is an act that has cost to the actor, in terms of his or her ability to survive and reproduce, and a benefit to the recipient of the altruistic act.

Kin Selection

- The **coefficient of relatedness** is a measure of how closely the actor and beneficiary are related. **Hamilton's rule** states that if the benefits of altruistic behavior are high, if the benefits are dispersed to close relatives, and if the costs are low, then alleles associated with altruistic behavior will be favored by natural selection and will be spread throughout the population.
- **Kin selection** is natural selection that acts through benefits to relatives.
- Individuals can pass their alleles on to the next generation not only by having their own offspring but also by helping close relatives produce more offspring.

Reciprocal Altruism

- **Reciprocal altruism** is an exchange of fitness benefits that are separated in time.
- For example, vervet monkeys are most likely to groom unrelated individuals that have groomed or helped them in the past.

CHECK YOUR UNDERSTANDING

(f) Compare and contrast altruism and kin selection.

_____

_____

_____

_____

_____

B. INTEGRATING YOUR KNOWLEDGE

Looking Back—Key Concepts from Earlier Chapters

Behavioral Traits—Chapter 13

To use the terms introduced in **Chapter 13**, most behavioral traits are polygenic—exhibiting a quantitative variation, with individuals differing by degree and many different phenotypes being observed when the population is considered as a whole.

The *for* Gene—Chapters 19 and 20

Recall from **Chapter 20** that the entire fruit-fly genome has been sequenced. **Chapter 19** introduced the techniques that Sokolowski used to map the *for* gene.

Flexible Behavior—Chapter 45

Chapter 45 introduced how learning works at the proximate level and concluded that learning occurs because individual neurons and connections between neurons are modified in response to experience.

C. MASTERING CHALLENGING TOPICS

The study of animal behavior seems to be a more tangible subject in biology because we can directly observe data with our own eyes as it is occurring. However, there are still a few topics that require extra time and thought. The concepts of **altruism** and **kin selection** are relatively straightforward to define, but even professors have a difficult time applying these terms to the observations they record in nature. It would be wise to locate a current ecology text and learn how to apply altruism or kin selection to examples given in the text. An ecology text will provide you with much more detail and reinforce the knowledge that you have already acquired.

252225222222222

D. ASSESSING WHAT YOU'VE LEARNED

1. A change in behavior that results in a specific experience in the life of an individual is termed:
   a. behavior
   b. proximate cause
   c. ultimate cause
   d. learning

2. An innate behavior is one that _____.
   a. is not inherited
   b. shows some variation
   c. is inherited and shows little variation
   d. is inherited and shows wide variation

3. Individuals are trained by experience to give the same response to more than one stimulus. This is termed:
   a. classical conditioning
   b. imprinting
   c. proximate conditioning
   d. sensitization

4. The recognition and manipulation of facts about the world is termed:
   a. imprinting
   b. learning
   c. cognition
   d. memory

5. Inflexible behavior is adaptive when _____.
   a. mistakes would be costly
   b. there are many safe opportunities to learn
   c. mistakes can be made without dying
   d. parents can teach their young

6. The key signal prompting a hormonal response and breeding in female *Anolis* lizards is:
   a. light
   b. male dewlap display
   c. female dewlap display
   d. male hormones

7. Any process in which the signal from one individual modifies the behavior of a recipient individual is called:
   a. a connecting signal
   b. communication
   c. imprinting
   d. learning

8. Natural selection favoring the deceit of the same species is best described by which of the following examples?
   a. anglerfish
   b. fireflies
   c. monkeys
   d. bluegills

9. A movement that results in a change in position is called:
   a. orientation
   b. taxis
   c. phototaxis
   d. instant taxis

10. An exchange of fitness benefits that are separated in time is termed:
    a. kin selection
    b. altruism
    c. reciprocal altruism
    d. coefficient of relatedness

CHAPTER 51—ANSWER KEY

Check Your Understanding

(a) Fixed action patterns are highly stereotypical behavior patterns that have three characteristics: (1) There is almost no variation in how they are performed; (2) they are species specific; and (3) once the sequence of actions begins, it typically continues until completion.

(b) In classical conditioning, individuals are trained by experience to give the same response to more than one stimulus—even a stimulus that has nothing to do with the normal response. Upon hatching, ducklings and goslings adopt as their mother the first moving thing they see—this is called imprinting. Imprinting occurs only during a critical or sensitive period.

(c) Researchers placed female *Anolis* lizards with males that had intact dewlaps or with those whose dewlaps had been surgically removed. The result? Females grouped with dewlap-less males were slow to produce eggs—just as slow, in fact, as females that had been grouped with castrated males. These females had not been courted at all. The result suggests that the dewlap is a key visual signal.

(d) Von Frisch observed bees displaying a "round dance" that contained information about the location of food. Workers got the information about the location of food through the dancer's movements. He also observed a "waggle run" that tells the length of distance the workers must fly, as well as the direction.

(e) Migration is defined as the long-distance movement of a population associated with a change of seasons. Arctic terns nest along the Atlantic coast of North America, fly south along the coast of Africa to wintering grounds off Antarctica, then fly back along the eastern coast of South America. Many of the monarch butterflies native to North America spend the winter in the mountains of central Mexico or southwestern California. Salmon that hatch in rivers along the Pacific coast of North America and northern Asia migrate to the ocean when they are a few months to several years old. They return to the stream where they hatched to mate and then die.

(f) Altruism is an act that has cost to the actor, in terms of his or her ability to survive and reproduce, and a benefit to the recipient. Kin selection is natural selection that acts through benefits to relatives.

Assessing What You've Learned

1. d; 2. c; 3. a; 4. c; 5. a; 6. b; 7. b; 8. d; 9. a; 10. c

52

Population Ecology

A. KEY CONCEPTS

52.1 Demography

- A **population** is a group of individuals from the same species that live in the same area at the same time.

- **Population ecology** is the study of how and why the number of individuals in a population changes over time.

- **Demography** is the study of factors that determine the size and structure of populations through time. Analyzing birth rates, death rates, **immigration** rates, and **emigration** rates is fundamental to demography.

Life Tables

- Formal demographic analyses of populations are based on a type of data set called a **life table**, which summarizes the probability that an individual will survive and reproduce in any given year over the course of its lifetime.

Survivorship

- **Survivorship** is the proportion of offspring produced that survive, on average, to a particular age (l_x, where x represents the age class being considered). It is calculated by dividing the number of individuals in that age class by the number of individuals in the first age class.

$$I_x = N_x/N_0 \quad \textbf{(Box 52.1, Eq. 52.1)}$$

- Type I curve—high survivorship throughout life
- Type II curve—constant mortality throughout life
- Type III curve—high death rate early in life

Fecundity

- **Fecundity** is the number of female offspring produced by each female in the population.

$$R_0 = \Sigma l_x m_x \quad \textbf{(Box 52.1, Eq. 52.2)}$$

Lifetime reproduction is a function of fecundity at each age (m_x) and survivorship to each age class (l_x). If R_0 is greater than 1, the population is increasing. If R_0 is less than 1, the population is declining.

The Role of Life History

- An organism's **life history** consists of how the organism allocates resources to growth, reproduction, and activities related to survival.

CHECK YOUR UNDERSTANDING

(a) What is fecundity, and how does it relate to population growth? Include the equation in your answer.

_____

_____

_____

_____

_____

52.2 Population Growth

Exponential Growth

- **Exponential growth** occurs when r does not change over time. It does *not* depend on the number of individuals in the population. Therefore, it is **density independent**.

- In reality, it is not possible for growth to continue indefinitely. Eventually, the habitat would not have enough resources for the number of individuals present. When a population stabilizes at the maximum number that can be supported by the resources available, that population has reached the habitat's **carrying capacity**.

Logistic Growth

- Biologists recognize two types of logistic growth: *discrete* growth and *continuous* growth.

- In calculating discrete growth, biologists use N to symbolize population size. N_0 is the population size at time zero, and N_1 is the population size one breeding interval later.

$$N_1/N_0 = \lambda \quad \text{(Box 52.2, Eq. 52.3)}$$

where λ is the **finite rate of increase**.

$$N_t = N_0\lambda^t \quad \text{(Box 52.2, Eq. 52.5)}$$

This equation summarizes how populations grow when breeding takes place seasonally.

- In calculating continuous growth, a population's **per capita increase** is symbolized r and is defined as the per capita birth rate minus the per capita death rate.

$$\lambda = e^r \quad \text{(Box 52.2, Eq. 52.6)}$$

where e is the natural logarithm, or about 2.72.

$$N_t = N_0 e^{rt} \quad \text{(Box 52.2, Eq. 52.7)}$$

This equation summarizes how populations grow when they breed continuously. Because r represents the growth rate at any given time and because r and λ are so closely related, biologists routinely calculate r, even for species that breed seasonally.

What Limits Growth Rates and Population Sizes?

- **Density-independent factors** change birth rates and death rates irrespective of the number of individuals in a population.
- **Density-dependent factors** are usually biotic in nature and change in intensity as a function of population size.
- Researchers studied the bridled goby, a coral-reef fish. They stocked artificial reefs with varying densities of adult gobies; then, after 2.5 months, they captured all the gobies and computed the growth rate of individuals, the survival rate, and the immigration rate.
- Adult gobies survive better when population density is *low*. More juvenile gobies immigrate successfully when population density is low. Higher rates of predation and disease might occur in dense populations.

CHECK YOUR UNDERSTANDING

(b) Describe the characteristics of exponential growth.

_____

_____

_____

52.3 Population Dynamics

How Do Metapopulations Change through Time?

- Studying population dynamics means studying changes in populations through time.
- If individuals from a species occupy many small patches of habitat, they may form many independent populations, or **metapopulations**.
- Ilkka Hanski and colleagues have shown that metapopulations exist in nature. They performed a mark-and-recapture study on the Glanville fritillary butterfly. They found a migration rate of 9 percent, which was high enough to recolonize extinct populations. They also confirmed that some populations had gone extinct while others had been created.
- Therefore, the dynamics of the metapopulation depend on the births and deaths of the smaller populations.

Why Do Some Populations Cycle?

- **Population cycle**s are regular fluctuations in size that some populations exhibit.
- Researchers have concluded that hare populations are limited by availability of food as well as by predation and that both of these factors interact. (See graph on next page.)

How Does Age Structure Affect Population Growth?

- A population's **age structure**—meaning the proportion of individuals that are at each possible age—dramatically influences the population's growth over time.

Age Structure in a Woodland Herb

- The common primrose has a complex population dynamic:
  1. Populations dominated by juveniles experience rapid growth and then are limited by the shading of larger trees.
  2. The long-term trajectory of the primrose population seems dependent on the frequency and severity of windstorms.
  3. Primroses have metapopulation structure.

Age Structure in Human Populations

- In nations where industrial and technological development is advanced and average incomes are relatively high, the **age pyramid** of the population tends to be even.
- The predicted population structure in 2050 highlights a major policy concern in developed countries: how to care for an increasingly aged population.

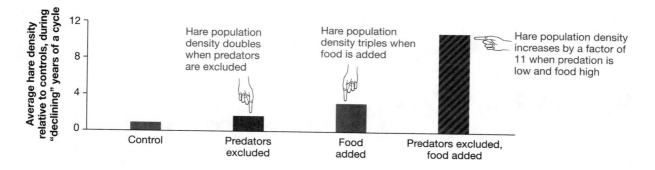

In the graph: Average hare density relative to controls, during "declining" years of a cycle (y-axis: 0, 4, 8, 12).

- Control
- Predators excluded — "Hare population density doubles when predators are excluded"
- Food added — "Hare population density triples when food is added"
- Predators excluded, food added — "Hare population density increases by a factor of 11 when predation is low and food high"

- In contrast, the age distribution is "bottom-heavy" in the less-developed countries.

- The projected age distribution in 2050 illustrates the major public policy concerns in these countries: providing education and jobs for an enormous influx of young people who will want to be starting families.

Analyzing Change in the Growth Rate of Human Populations

- The growth rate for humans has increased over time since about 1750, leading to a very steeply rising curve over the past few centuries. The highest values occurred between 1965 and 1970, when population growth averaged 2.04 percent per year.

Will Human Population Size Peak in Your Lifetime?

- The population growth has slowed over the past few years, and now it is 1.2 percent per year. Humans are ending a period of rapid growth that lasted well over 500 years. How quickly growth rates decline and the maximum population size ultimately reached will be decided by changes in fertility rates and the course of the AIDS epidemic.

- Extrapolating the world's population to the year 2050 is based on three different fertility rates:

    1. High fertility = nearly 11 billion
    2. Medium fertility = about 9 billion
    3. Low fertility = 7.4 billion

CHECK YOUR UNDERSTANDING

(c) What is a metapopulation, and what are some of its characteristics?

_____

_____

_____

52.4 How Can Population Ecology Help Endangered Species?

Preserving Metapopulations

- Over time, each population within the larger metapopulation is likely to be wiped out. Migration from nearby populations can reestablish populations in these empty habitats, thereby stabilizing the overall population.

- Areas being protected need to be substantial enough in area to maintain large populations.

- If that is not possible, smaller tracts of land should be connected by corridors of habitat so that migration is possible.

- Land where populations have emigrated should be preserved so that immigration is possible.

Using Life-Table Data to Make Population Projections

- Population projections allow biologists to alter values for survivorship and fecundity at particular ages and then assess the consequences.

- Studied performed on this aspect support some general conclusions:

    1. Whooping cranes, sea turtles, spotted owls, and many other endangered species have high juvenile mortality, low adult mortality, and low fecundity.

    2. In humans and species with high survivorship in most age classes, rates of population growth are extremely sensitive to changes in age-specific fecundity.

Population Viability Analysis

- **Population viability analysis (PVA)** is a model used to estimate the likelihood that a population will avoid extinction for a given time period. PVAs attempt to combine basic demographic models for the species in question with data on geographic structure and the rate and severity of habitat disturbance.

- Researchers constructed PVAs for an endangered marsupial, Leadbeater's possum. Their analysis allowed them to assess the impact of migration and to simulate the effects of logging, fires, storms, and other types of disturbances.

CHECK YOUR UNDERSTANDING

(d) What is the difference between demography and a life table?

_____

_____

_____

_____

B. INTEGRATING YOUR KNOWLEDGE

*Looking Back—Key Concepts
from Earlier Chapters*

Animal Interaction—Chapter 51

Chapter 51 explored how organisms respond to environmental stimuli and how they interact with the environment. Here in Chapter 52, the focus is on population size.

*Looking Forward—Key Concepts
in Later Chapters*

Animal Interaction—Chapters 53–55

Chapters 53–55 investigate how different species interact with one another and with their physical environment. Here in Chapter 52, our focus is on population size.

C. MASTERING CHALLENGING TOPICS

The difficult aspect of this chapter is undoubtedly the mathematical equations and quantitative nature of this chapter. For those of us who have trouble with math and quantitative analysis, this chapter will present challenges. The amount of time you will spend on working through equations and solving problems will ultimately depend on how much your instructor focuses on these issues. If you are required to solve equations and word problems associated with population size, mortality and fecundity, then you will need to invest a significant amount of time in learning these concepts and solving problems. As with any quantitative subject, you need to learn the concepts and the origins of the equations before you can solve word problems off the top of your head. Study qualitatively, learning all the definitions of terms

and meanings of the equations, and then start plugging in numbers and solving equations. Do as many problems as you can; then start making up your own problems for your classmates and ask them to do the same. In problem solving, practice makes perfect!

D. ASSESSING WHAT YOU'VE LEARNED

1. The study of how and why the number of individuals in a group of animals changes over time is called:
   a. demography
   b. population ecology
   c. population
   d. fecundity

2. The proportion of offspring produced that survive, on average, to a particular age is called:
   a. survivorship
   b. life table
   c. demography
   d. fecundity

3. An organism's life history _____.
   a. is the number of female offspring produced by each female
   b. is the probability an organism will survive and reproduce
   c. consists of how an organism allocates resources to growth, reproduction, and activities related to survival
   d. consists of how an organism interacts with other organisms

4. Factors that are usually biotic in nature and change in intensity as a function of population size are termed:
   a. density-dependent factors
   b. density-independent factors
   c. density-neutral factors
   d. exponential growth factors

5. Some populations exhibit regular fluctuations in size called:
   a. age structure
   b. population crashes
   c. population tides
   d. population cycles

6. In nations where industrial and technological development are advanced, _____.
   a. age distribution is top-heavy
   b. age distribution is bottom-heavy
   c. age distribution is even
   d. age distribution is bimodal

7. Adult gobies survive better when population density is:
   a. low
   b. high
   c. intermediate
   d. dynamic

8. A huge population of animals consisting of smaller isolated populations is termed a:
   a. megapopulation
   b. metapopulation
   c. metropolis
   d. cohort

9. The model that estimates the likelihood that a population will avoid extinction for a given time period is termed:
   a. mortality
   b. age structure
   c. population viability analysis
   d. population fecundity

10. A small, isolated population _____.
    a. will do extremely well if protected
    b. is unlikely to survive, even within a reserve
    c. will grow, but extremely slowly
    d. will explode given the right conditions

CHAPTER 52—ANSWER KEY

Check Your Understanding

(a) Fecundity is the number of female offspring produced by each female in the population: $R_0 = \Sigma\, l_x m_x$.

Lifetime reproduction is a function of fecundity at each age (m_x) and survivorship to each age class (l_x). If R_0 is greater than 1, the population is increasing. If R_0 is less than 1, the population is declining.

(b) Exponential growth occurs when r does not change over time. It does *not* depend on the number of individuals in the population. Therefore, it is density independent.

(c) A metapopulation is made up of many small populations isolated in fragments of habitat. Over time, each population within the larger metapopulation is likely to be wiped out. Migration from nearby populations can reestablish populations in these empty habitats, thereby stabilizing the overall population.

(d) Demography is the study of factors that determine the size and structure of populations through time. Formal demographic analyses of populations are based on a type of data set called a life table, which summarizes the probability that an individual will survive and reproduce in any given year over the course of its lifetime.

Assessing What You've Learned

1. b; 2. a; 3. c; 4. a; 5. d; 6. c; 7. a; 8. b; 9. c; 10. b

53

Community Ecology

A. KEY CONCEPTS

53.1 Species Interactions

- A biological **community** consists of interacting species, usually within a defined area.

Competition

- **Competition** is a – / – interaction that occurs when individuals use the same resources and when those resources are limited.

- Competition occurring between the same species is called **intraspecific competition**, while **interspecific competition** occurs between individuals from different species.

Using the Niche Concept to Analyze Competition

- A niche is the range of resources that a species is able to use or the range of conditions it can tolerate.

What Happens When One Species Is a Better Competitor?

- The principle of **competitive exclusion** states that it is not possible for species within the same niche to coexist. However, if the niches do not overlap completely, then the weaker species should be able to retreat into an area of non-overlap—a mechanism called **niche differentiation**.

- Under **asymmetric competition**, one species suffers a greater fitness decline whereas under **symmetric competition**, both species have an equal fitness decline.

- The **fundamental niche** is the combination of conditions that the species will occupy in the absence of competitors, and its **realized niche** is that portion of resources used when competition occurs.

Experimental Studies of Competition

- Joseph Connell noticed two species of barnacles with interesting distributions on an intertidal rocky shore of Scotland.

- Connell's study is one of the first experimental examples of how competition can modify a distribution of a species. In this case, the poorer competitor (*Chthamalus*) is restricted to the upper intertidal zone, which is a more severe habitat.

Experiment

Question: Why is the distribution of adult *Chthamalus* restricted to the upper intertidal zone?

Hypothesis: Adult *Chthamalus* are competitively excluded from the lower intertidal zone.

Alternative hypothesis: Adult *Chthamalus* do not thrive in the physical conditions of the lower intertidal zone.

Experimental setup:

1. Transplant rocks containing young *Chthamalus* to lower intertidal zone.

2. Let *Semibalanus* colonize the rocks.

3. Remove *Semibalanus* from half of each rock. Monitor survival of *Chthamalus* on both sides.

Prediction: *Chthamalus* will survive better in the absence of *Semibalanus*.

Prediction of alternative hypothesis: *Chthamalus* survival will be low and the same in the presence or absence of *Semibalanus*.

Results:

Chthamalus survival is higher when *Semibalanus* is absent

Conclusion: *Semibalanus* is competitively excluding *Chthamalus* from the lower intertidal zone.

*Mechanisms of Coexistence: Fitness
Trade-offs and Niche Differentiation*

- Change in resource use is called **niche differentiation**.

- The change in species' traits is called **character displacement**.

Consumption

- Consumption occurs when one organism eats another:
  1. **Herbivory**—when **herbivores** consume plant tissues.
  2. **Parasitism**—When a **parasite** consumes small amounts of tissues from another organism, or **host**.
  3. **Predation**—When a **predator** kills and consumes most or all of another individual, or **prey**.

How Do Prey Defend Themselves?

- Some species run or hide in response to predators; others sequester or spray toxins or employ weaponry for defense (**standing** or **constitutive defenses**).

- **Mimicry** occurs when one species closely resemble another species.

- **Müllerian mimicry** occurs when harmful species resemble each other. For example, brightly colored wasps resemble each other and by this, predators tend to avoid these colors.

- **Batesian mimicry** occurs when a harmless species resembles a harmful one.

- Researchers hypothesized that if blue mussels possess **inducible defenses** against crabs, they should occur in the low-flow area where predation pressure is higher. They found that mussels in the high predation area had thicker shells.

*Are Animal Predators Efficient Enough
to Reduce Prey Populations?*

- Species interactions have a strong impact on the evolution of predator and prey populations.

- A wolf control program in Alaska during the 1970s decreased predator abundance to 55–80 percent below pre-control density. Concurrently, the population of moose tripled. This observation suggests that predators reduced this moose population far below the maximum number that the habitat can support.

- In the arctic, tests have shown that hare population is limited by the availability of food as well as by predation, and that food availability and predation intensity interact—the combined effect of

food and predation is much larger than their impact in isolation.

*Why Don't Herbivores Eat Everything—
Why Is the World Green?*

- Biologists routinely consider three possible answers to the above question:
  1. **Top-down** controlover herbivores occurs when predation or disease limits herbivore populations.
  2. In the **poor nutrition hypothesis**, the quality of the resources available at the base of the food chain limits herbivore density. More specifically, growth and reproduction of herbivores could be limited by the availability of nitrogen.
  3. In the **plant defense hypothesis**, plants produce toxins to repel herbivores.

Adaptation and Arms Races

- A **coevolutionary arms race** between parasites and hosts begins when a parasitic species develops a trait that allows it to survive and reproduce in a host. In response, natural selection favors host individuals that are able to defend themselves against the parasite.

- *Plasmodium* are unicellular protists that cause malaria. They have a complex life cycle that includes the mosquito as the main vector.

- Researchers have found a strong association between a certain allele and protection against malaria. HLA-B53 proteins bind to a particular protein found in *Plasmodium*.

- Researchers have also shown that *Plasmodium* populations in West Africa now have a variety of alleles for the protein recognized by HLA-B53. This suggests that natural selection has favored the evolution of *Plasmodium* strains with weapons to counter HLA-B53.

Can Parasites Manipulate Their Hosts?

- Snails infected with flukes become attracted to light and are therefore more easily eaten by birds, thus continuing the life cycle of the fluke.

Mutualism

- **Mutualisms** are + / + interactions that involve a wide variety of organisms and rewards. For example, bees visit flowers to harvest nectar and pollen, and the flowers benefit by being fertilized.

- Examples are the association between fungi and roots, bacteria and certain species of plants, and rancher ants that protect aphids in exchange for sugar.

(a) Compare and contrast competition and mutualism.

_____

_____

_____

_____

53.2 Community Structure

How Predictable Are Communities?

- Clements promoted a view that biological communities are stable, integrated and orderly entities with a highly predictable composition. He argued that communities develop by passing through a series of predictable stages and culminates in a **climax community**.

- Gleason contended that the community found in a particular area is neither stable nor predictable. He claimed that plant and animal communities are associations of species that just *happen to share climatic requirements*.

Mapping Current and Past Species' Distributions

- Historical data on plant communities showed that groups of species did not change their distributions in close association. Instead, species tended to change their ranges independently of one another.

Experimental Tests

- Researchers constructed 12 identical ponds, sterilized them, and then inoculated them with various species of zooplankton. They found a total of 61 species in all the ponds but only 31 to 39 species in individual ponds.

- Researchers contended that some species are particularly good at dispersing and are likely to colonize all or most of the available habitats. Other species disperse more slowly and tend to reach only one or a few of the available habitats. As a result, the specific details of community assembly and composition are contingent and unpredictable. As Gleason maintained, communities are a product of chance and history.

How Do Keystone Species Structure Communities?

- A **keystone species** has a much greater impact on the surrounding species than its abundance would suggest.

- Wolves, sea stars, and sea otters are examples of species that vary in abundance over space and time, which significantly affects their communities.

(b) Briefly state Clements's view on community dynamics.

_____

_____

_____

53.3 Community Dynamics

Disturbance and Change in Ecological Communities

- **Disturbance** is an event that removes some individuals or biomass from a community. The important feature or disturbance is that it alters light levels, nutrients, unoccupied space, or some other aspect of resource availability.

- Each community experiences a characteristic type, frequency, and severity of disturbance, known as a **disturbance regime**. For example, forest fires kill all or most of the trees in a boreal forest every 100 to 300 years, whereas disturbances in temperate and tropical forests are more frequent.

How Do Researchers Determine a Community's Disturbance Regime?

- Ecologists use two strategies to determine the natural pattern of disturbance:

  1. They infer long-term patterns from data obtained in short-term analysis. Unless sampling is extensive, it is difficult to avoid errors caused by extrapolating from particularly disturbance-prone or disturbance-free years or areas.

  2. By reconstructing the history of a particular site, flood frequency (for example) can be estimated. Forest fires are most often studied by using historical techniques.

Why Is It Important to Understand Disturbance Regimes?

- Researchers studied the fire history of giant sequoia groves in California. They found that in most of the groves, 10 to 53 fires had occurred each century for the past 1530 to 2000 years. They also established that fires are extremely frequent and of low severity in this community.

- Biologists are now able to better manage these forests by allowing, monitoring, and controlling burns in this area.

Succession: The Development of Communities after Disturbance

- Severe disturbances remove all or most of the organisms from an area. The recovery that follows is **succession**. **Primary succession** occurs when the disturbance removes the soil and its organisms as well as the organisms above the surface. **Secondary succession** occurs when a disturbance removes some or all of the organisms from an area but leaves the soil intact.
- The specific sequence of species that appears over time is called a **successional pathway**.

Theoretical Considerations

- Biologists focus on three factors to predict the outcome of succession in a community:

  1. Particular traits of the species involved
  2. How species interact
  3. Historical and environmental circumstances such as the size of the area involved and short-term weather conditions

- **Facilitation** takes place when the presence of an earlier-arriving species makes conditions more favorable for the arrival of certain later species. **Inhibition** is the opposite of facilitation.
- **Tolerance** means existing species do not affect the probability that subsequent species will become established.

A Case History: Glacier Bay, Alaska

- An extraordinarily rapid and extensive glacial recession is occurring at Glacier Bay. In just 200 years, glaciers have retreated about 100 km, thus uncovering an important site for studying succession.

CHECK YOUR UNDERSTANDING

(c) Hypothesize what would happen if California prevented small wildfires from occurring.

_____

_____

_____

53.4 Species Richness in Ecological Communities

Predicting Species Richness: The Theory of Island Biogeography

- **Species richness** is a simple count of how many species are present in a given community, whereas **species diversity** is a weighted measure that incorporates abundance and presence.
- Islands in the ocean have smaller numbers of species than do areas of the same size on continents.
- Immigration rates should be higher on large islands closer to the mainland, and extinction rates should be highest on small islands that are far from shore.

Global Patterns in Species Richness

- The communities of the tropics contain more species than temperate or subarctic environments do. As latitude increases, species diversity decreases.
- *High productivity in the tropics promotes high diversity.* Increased biomass supports more herbivores, thus more predators.
- In addition, speciation rates should increase when niche differentiation occurs within populations of herbivores, predators, parasites, and scavengers. This hypothesis is still in the experimental stage.

CHECK YOUR UNDERSTANDING

(d) Compare and contrast species richness and species diversity.

_____

_____

_____

_____

_____

B. INTEGRATING YOUR KNOWLEDGE

Looking Back—Key Concepts from Earlier Chapters

Animal Interaction—Chapter 51

Chapter 51 explored how organisms respond to environmental stimuli and how they interact with the environment.

1

Here in Chapter 53, however, the focus is on large numbers of species within a community.

Population Dynamics—Chapter 52

Chapter 52 explored the dynamics of populations—how and why they grow or decline, and how they change over time and space. Although that chapter considered populations in isolation, the reality is that individuals of different species constantly interact within a community. As a result, the fate of a community may be tightly linked to the other species that share its habitat.

Growth Rings—Chapter 36

Researchers estimate the frequency and impact of storms by finding wind-killed trees and determining the date of their death. This can be done by comparing patterns in the growth rings of the dead trees with those of living individuals nearby.

C. MASTERING CHALLENGING TOPICS

The development of communities after disturbance is probably the most complex event described in this chapter. Because communities respond differently after disturbances, you cannot simply memorize one type of succession. Instead, learning various examples will help you understand how succession proceeds in different communities and after different types of disturbances. Also, by understanding the theoretical considerations (traits of species, species interaction, and environment), you will gain an appreciation for the difficulty in determining the exact events of succession in various areas.

D. ASSESSING WHAT YOU'VE LEARNED

1. It is not possible for species within the same niche to coexist. This statement describes:
   a. competition
   b. niche theory
   c. competitive exclusion
   d. mimicry

2. The combination of conditions that a species will occupy in the absence of competitors is:
   a. the fundamental niche
   b. the realized niche
   c. competitive exclusion
   d. mimicry

3. Batesian mimicry can be best described as:
   a. when one species closely resembles another
   b. when a harmless species resembles a harmful one
   c. when harmful species resemble each other
   d. when harmless species resemble each other

4. When predation or disease limits herbivore populations, this is:
   a. top-down control
   b. the poor nutrition hypothesis
   c. the plant defense hypothesis
   d. the disease hypothesis

5. The + / + interactions that involve a wide variety of organisms and rewards are termed:
   a. predation
   b. competition
   c. parasitisms
   d. mutualisms

6. An event that removes some individuals or biomass from a community is called:
   a. competition
   b. predation
   c. disturbance
   d. resistance

7. The recovery that follows a severe disturbance is:
   a. succession
   b. growth
   c. colonization
   d. seeding

8. The primary difference between primary and secondary succession is:
   a. the complete removal of species during the disturbance
   b. the complete removal of soil during the disturbance
   c. time
   d. nothing

9. Which one of the following statements is *not* representative of a global pattern in species diversity?
   a. High productivity in the tropics promotes high diversity.
   b. Habitats with simple physical structures have more niches.
   c. Tropical regions have had more time for speciation to occur.
   d. Tropical areas experience fewer disturbance events.

10. Net primary productivity is:
    a. the total amount of photosynthesis per unit area
    b. the total amount of plant mass per unit area
    c. the amount of primary productivity that ends up as biomass
    d. the amount of primary productivity above the ground

CHAPTER 53—ANSWER KEY

Check Your Understanding

(a) Competition and mutualism are both types of species interactions. Competition is a – / – interaction that occurs when individuals use the same resources and when those resources are limited. Mutualisms are + / + interactions that involve a wide variety of organisms and rewards. For example, bees visit flowers to harvest nectar and pollen, and the flowers benefit by being fertilized.

(b) Clements promoted a view that biological communities are stable, integrated, and orderly entities with a highly predictable composition. He argued that communities develop by passing through a series of predictable stages and culminates in a climax community.

(c) One could hypothesize that by preventing small fires from occurring in giant sequoia groves, the underbrush would continue to grow instead of remaining low. This would continue and not be a physical threat (although biologically, the species diversity could be affected); if a fire were to occur, it would burn longer, higher, and with more intensity such that the sequoia groves would likely be harmed. In smaller, more regular wildfires, the sequoias remain unharmed.

(d) Species richness and species diversity are two different measures of the number of species in a community. Species richness is a simple count of how many species are present in a given community, whereas species diversity is a weighted measure that incorporates abundance and presence.

Assessing What You've Learned

1. c; 2. a; 3. b; 4. a; 5. d; 6. c; 7. a; 8. b; 9. b; 10. c

54

Ecosystems

A. KEY CONCEPTS

54.1 Energy Flow and Trophic Structure

- An **ecosystem** consists of the organisms that live in an area along with certain nonbiological components. The relevant abiotic components include the nutrients used by growing organisms and the energy supplied by the Sun.

- Energy is the currency of an ecosystem, and every ecosystem has four components that are linked by energy flow.

  1. The **abiotic environment** includes the soil, climate, atmosphere, and the particulates and solutes in water.

  2. **Primary producers** use solar energy or chemical energy contained in reduced inorganic compounds to manufacture their own food (autotroph).

  3. **Consumers** eat other organisms. **Herbivores** are consumers that eat plants; **carnivores** are consumers that eat animals.

  4. **Decomposers** obtain energy by feeding on the dead remains of other organisms or waste products.

Global Patterns in Productivity

- **Tropical rain forests** have the highest productivity, whereas desert and arctic regions have the lowest. Marine productivity is the highest along coasts.

What Limits Productivity?

- The overall productivity of terrestrial ecosystems is limited by a combination of temperature and the availability of water and sunlight.

- The overall productivity of marine ecosystems is limited by the availability of nutrients.

How Does Energy Flow Through an Ecosystem?

- Net primary productivity results in **biomass**—organic material that non-photosynthetic organisms can eat.

- A **primary consumer** is an herbivore.

- **Detritus** is consumed by a variety of **primary decomposers**: usually bacteria, archaea, and fungi.

Trophic Levels, Food Chains, and Food Webs

- Organisms that obtain their energy from the same type of source occupy the same **trophic level**. Productivity declines from one trophic level to the next.

Trophic level	Feeding strategy	Decomposer food chain	Grazing food chain
5	Quaternary consumer		Cooper's hawk
4	Tertiary consumer	Robin	Cooper's hawk
3	Secondary consumer	Earthworm	Robin
2	Primary decomposer or consumer	Bacteria, archaea	Cricket
1	Primary producer	Dead maple leaves	Maple tree leaves

- A **food chain** describes the species occupying each trophic level in a particular ecosystem.

- A **grazing food chain** is made up of the network of organisms that eat plants, along with the organisms that eat the herbivores.

- Consumers that eat herbivores are called **secondary consumers**.

- The **decomposer food chain** is comprised of species that eat the dead remains of organisms. Plant litter and dead animals, or detritus, are consumed by bacteria, archaea, fungi, protozoa, and millipedes.

- Food chains are often embedded in more complex **food webs** because multiple species are present at several trophic levels.

Why Is Energy Lost at Each Trophic Level?

- Productivity at the second trophic level (primary consumers) must be less than productivity at the first trophic level (primary producers); this produces a pyramid of productivity. Productivity is always highest at the lowest trophic level.

What Limits the Length of Food Chains?

- *Hypothesis 1: Energy Transfer*—Food chain length is limited by productivity, so there should be more trophic levels in ecosystems with higher productivity or higher energy transfer. To date there is no support for this hypothesis.

- *Hypothesis 2: Stability*—Long food chains are easily disrupted by environmental perturbations and thus tend to be eliminated.

- Pimm demonstrated that longer food chains took longer to return to their previous state following a disturbance. He also predicted that food chains should be longer in more stable environments. There have been challenges to this hypothesis, and so far support remains tentative.

- *Hypothesis 3: Environmental Complexity*—Food chain length is a function of an ecosystem's physical structure.

- Researchers predicted that three-dimensional ecosystems should have longer food chains than two-dimensional sites. The data support their predictions, and they concluded that dimensionality does affect food web structure. The mechanism for the pattern remains to be determined, however.

Analyzing Energy Flow: A Case History

- **Gross primary productivity** is the total amount of photosynthesis in a given area and time period.

- **Gross photosynthetic efficiency** is the ratio of gross photosynthesis to solar radiation in kcal/m2.

How Does Energy Flow through the Hubbard Brook Ecosystem?

- Researchers extensively measured how energy flows through a forest ecosystem. They found that:

- 0.8 percent of the light was captured by photosynthesis.

- Plants used 55 percent of this energy for growth and 45 percent for maintenance (lost as respiration)

- In chipmunks, 80.7 percent of energy intake goes to maintenance, 17.7 percent is excreted, and only 1.6 percent goes into the production of new tissue (**secondary production**).

CHECK YOUR UNDERSTANDING

(a) Define an ecosystem.

_____

_____

CHECK YOUR UNDERSTANDING

(b) Compare and contrast a food chain and a food web.

_____

_____

_____

_____

54.2 Biogeochemical Cycles

- The path that an element takes as it moves from abiotic systems through living organisms and back again is referred to as its **biogeochemical cycle**.

- There are three aspects studied in biogeochemical cycling:

1. The nature and size of the **reservoirs** where elements are stored for a period of time

2. The rate of movement between reservoirs and the factors that influence these rates

3. How different biogeochemical cycles interact

Biogeochemical Cycles in Ecosystems

- The combination of breakdown products and microscopic decomposers forms the soil organic matter. **Soil organic matter** is a complex mixture of partially decomposed detritus rich in a family of carbon-containing molecules called humic acids. (Soil organic matter is also called **humus**).

- A key feature of nutrient cycling is that nutrients are reused. Not all of the nutrients are reused, however. Nutrients leave ecosystems when animals leave or when flowing wind or water removes nutrients.

What Factors Control the Rate of Nutrient Cycling in Ecosystems?

- Decomposition of detritus most often limits the rate at which nutrients move through an ecosystem. Decomposition rate is influenced by:

  1. Abiotic conditions—temperature, precipitation, etc.

  2. The quality of detritus as a nutrient source for fungi, bacteria, and archaea that accomplish the decomposition

What Factors Influence the Rate of Nutrient Loss?

- Nutrient loss is an important characteristic of an ecosystem. Farming, logging, and soil erosion accelerate nutrient loss.

- Vegetation removal has a huge impact on nutrient export. In the absence of plants to recycle nutrients, they are quickly washed out of the soil and lost to the ecosystem.

Global Biogeochemical Cycles

The Global Water Cycle

- Water typically evaporates from the ocean and then precipitates back into the ocean. The key here is that evaporation exceeds precipitation—a net gain of water into the atmosphere.

- This vapor moves over land and precipitates. The cycle is completed by water that moves from land to the oceans via streams and **groundwater**.

- The **water table** is the upper limit below the surface that the ground is saturated with stored water.

The Global Carbon Cycle

- The **global carbon cycle** involves a relatively rapid movement of carbon between terrestrial ecosystems, the oceans, and the atmosphere.

- Photosynthesis is responsible for taking carbon from the atmosphere and incorporating it into tissue. Respiration releases carbon that has been incorporated into living organisms to the atmosphere.

- The geological reservoir of carbon-rich sediments (fossil fuels) is estimated to contain 5000–10,000 gigatons of carbon. Burning fossil fuels moves carbon from an inactive geological reservoir to an active pool—the atmosphere.

- Carbon dioxide is called a **greenhouse gas** because it traps heat that has been radiated from Earth and keeps it from being lost to space—in the same way that the glass of a greenhouse traps heat underneath it.

The Global Nitrogen Cycle

- The key aspect of the **global nitrogen cycle** is that plants are able to use nitrogen only in the form of ammonium or nitrate ions (NH_4^+ or NO_3^-).

- Nitrogen fixation (converted from N_2 to NH_4^+ or NO_3^-) results from lightning-driven reactions in the atmosphere or enzyme-catalyzed reactions in bacteria that live in the soil and oceans.

- Three sources of human-fixed nitrogen are as follows:

  1. Industrially produced fertilizers

  2. The cultivation of crops such as soybeans and peas that harbor nitrogen-fixing bacteria

  3. Release of nitric oxide during the combustion of fossil fuels

CHECK YOUR UNDERSTANDING

(c) How are photosynthesis and respiration involved in the global carbon cycle?

_____

_____

_____

_____

_____

54.3 Human Impacts on Ecosystems

Global Warming

- In 1998, researchers summarized the studies and data available on **global warming** and concluded that current evidence suggests a "discernible human influence on climate."

- In 2001 the new report stated that "there is new and stronger evidence that most of the warming observed over the last 50 years is attributable to human activities."

- In its most recent report, issued in 2007, the Intergovernmental Panel on Climate Change (IPCC) declared that the evidence for global warming is unequivocal and that it is "very likely" due to human-induced changes in greenhouse gases.

- A few documented impacts on organisms:

- There is strong evidence that most of the 122 frog species that have gone extinct recently succumbed to a parasitic infection that increased in frequency due to warming.

- Increases in water temperature cause reef-building coral to expel their photosynthetic algae. Corals begin to die of starvation.

- Malarial parasite and copepod distributions are changing.
- Timing of events is changing in many seasonal environments, including early flowering by *Baptisia leucantha*.
- Changing temperatures are causing allele frequencies to change, such as increased fitness to hot habitats in *Drosophila*.

Productivity Changes

- If grassland habitats are fertilized with nitrogen, productivity increases but species richness declines over time.
- Nitrate ions derived from fertilizers are washing into the Gulf of Mexico. This has stimulated the growth of planktonic organisms, whose subsequent decomposition has used up available oxygen and triggered the formation of anaerobic "dead zones."
- Increased nitrate runoff from fertilizers has also cause toxic algal blooms.
- Massive phosphate runoff in the 1960s and 1970s resulted in explosive growth in cyanobacteria, large die-off, and eventual lake eutrophication—the conversion of a lake to a highly productive ecosystem with rapid decomposition, low oxygen levels, and rapid filling with decomposing organic matter.

CHECK YOUR UNDERSTANDING

(d) Provide an example of how global warming is directly affecting a species.

_____

_____

_____

B. INTEGRATING YOUR KNOWLEDGE

Looking Back—Key Concepts
from Earlier Chapters

Chemosynthesis—Chapter 2

In some ecosystems, energy flow begins not with photosynthesis but with chemosynthesis. For example, bacteria that live near the deep-sea vents introduced in **Chapter 2** derive energy from hydrogen sulfide rather than sunlight.

Producers—Chapter 10

Producers form the basis of ecosystems by transforming the energy in sunlight or reduced inorganic compounds into the chemical energy stored in sugars. As **Chapter 10** pointed out, producers are called autotrophs or self-feeders.

Nitrogen Fixation—Chapter 38

Nitrogen fixation results from lightning-driven reactions in the atmosphere or enzyme-catalyzed reactions in bacteria that live in the soil and oceans. **Chapter 38** explains how bacteria fix nitrogen.

Animal Interaction—Chapter 51

Chapter 51 explored how organisms respond to environmental stimuli and how they interact with the environment. Here in Chapter 54, the focus is on large numbers of species within an ecosystem.

Population Dynamics—Chapter 52

Chapter 52 explored the dynamics of populations—how and why they grow or decline, and how they change over time and space. Although that chapter considered populations in isolation, the reality is that individuals of different species constantly interact within a community, and communities within an ecosystem. As a result, the fate of an ecosystem may be tightly linked to the other species that share its habitat.

C. MASTERING CHALLENGING TOPICS

The most complex information you will process in this chapter relates to the biogeochemical cycles. It is not difficult to memorize the facts and specifics about each cycle; however, you may find it difficult to understand concepts when these types of questions are asked on a test. It is not sufficient to simply memorize the carbon or nitrogen cycle; you must learn the concepts and trends as well. For example, what happens to the carbon cycle when fossil fuels are burned? Not only do you need to know that atmospheric levels of CO_2 would increase, but you also need to understand how this will affect the rest of the cycle.

D. ASSESSING WHAT YOU'VE LEARNED

1. These use solar energy or chemical energy contained in reduced inorganic compounds to manufacture their own food.
   a. consumers
   b. primary producers
   c. secondary producers
   d. decomposers

2. The areas that typically have the highest productivity are:
   a. tropical rain forests
   b. grasslands
   c. arctic regions
   d. marine ecosystems

3. The overall productivity in terrestrial ecosystems is limited by:
   a. temperature
   b. water
   c. sunlight
   d. a combination of the above

4. Productivity at the second trophic level is always _____.
   a. greater than the productivity at the first trophic level
   b. less than the productivity at the first trophic level
   c. equal to the productivity at the first trophic level
   d. extremely variable compared to the productivity at the first trophic level

5. Which hypothesis regarding what limits food chain length has the most support?
   a. energy transfer hypothesis
   b. stability hypothesis
   c. environmental complexity hypothesis
   d. all of the above have equal support

6. The path that an element takes as it moves from abiotic systems through living organisms and back again is referred to as its _____.
   a. biogeochemical cycle
   b. biological cycle
   c. nutrient cycle
   d. geochemical cycle

7. Which of the following is *not* one of the three aspects studied in biogeochemical cycling?
   a. the nature and size of element reservoirs
   b. the rate of movement between reservoirs
   c. how different biogeochemical cycles interact
   d. how new species create their own biogeochemical cycles

8. Which of the following *most often* limits the rate at which nutrients move through an ecosystem?
   a. species composition
   b. decomposition rate
   c. primary production
   d. none of the above

9. Perturbation of which of the following cycles contributes most to global warming?
   a. the global carbon cycle
   b. the global water cycle
   c. the global nitrogen cycle
   d. all of the above cycles contribute equally

10. The most recent evidence on global warming concludes _____.
    a. nothing about human activity causing global warming
    b. possible links to human activity linked to global warming
    c. strong evidence of human activity linked to global warming
    d. humans are responsible for every catastrophe for the past 1000 years

CHAPTER 54—ANSWER KEY

Check Your Understanding

(a) An ecosystem consists of the organisms that live in an area along with certain nonbiological components. The relevant abiotic components include the nutrients used by growing organisms and the energy supplied by the Sun.

(b) Food chains and food webs both describe the relationships between species at different trophic levels. A food chain describes the species occupying each trophic level in a particular ecosystem. Food chains are often embedded in more complex food webs because multiple species are present at several trophic levels.

(c) Photosynthesis is responsible for taking carbon from the atmosphere and incorporating it into tissue. Respiration releases carbon that has been incorporated into living organisms to the atmosphere.

(d) Any one of the following examples were in the text:
   • There is strong evidence that most of the 122 frog species that have gone extinct recently succumbed to a parasitic infection that increased in frequency due to warming.
   • Increases in water temperature cause reef-building coral to expel their photosynthetic algae. Corals begin to die of starvation.
   • Malarial parasite and copepod distributions are changing.
   • Timing of events is changing in many seasonal environments, including early flowering by *Baptisia leucantha*.
   • Changing temperatures are causing allele frequencies to change, such as increased fitness to hot habitats in *Drosophila*.

Assessing What You've Learned

1. b; 2. a; 3. d; 4. b; 5. c; 6. a; 7. d; 8. b; 9. a; 10. c

55

Biodiversity and Conservation Biology

A. KEY CONCEPTS

55.1 What Is Biodiversity?

Biodiversity Can Be Measured and Analyzed at Several Levels

- Biodiversity defines all the distinctive populations and species living today. There are three levels of biodiversity:

  1. **Genetic diversity**—the total genetic information contained within all individuals of a species.

  2. **Species diversity**—the variety of life-forms on Earth

  3. **Ecosystem diversity**—the variety of biotic communities in a region along with abiotic components.

How Many Species Are Living Today?

Taxon-Specific Surveys

- Terry Erwin and J. C. Scott used insecticidal fog to knock down species from the top of a rain-forest tree called *Luehea seemannii*. They identified over 900 different species of beetles among the individuals that fell.

- If each of the 50,000 tropical tree species harbors the same number of arthropod specialists, then the world total of arthropod species would exceed 30 million. Based on these types of studies, biologists estimate that there are over 100 million species.

All-Taxa Surveys

- To obtain a more direct estimate of total species numbers, the first effort to find and catalog *all* forms of life present at a single site is now under way. The location is the Great Smoky Mountains National Park in the southeastern United States. A consortium of biologists and research organizations initiated this all-taxon survey in 1999. When it is complete, in 2015, biologists will have a much better database to use in estimating the extent of biodiversity.

(a) What are the three levels of biodiversity?

_____

_____

_____

55.2 Where Is Biodiversity Highest?

- Biologists use the term **hotspot** to denote areas in the tropics with exceptionally high species richness.

- In terms of bird species richness, the Andes Mountains, the Amazon River basin, portions of East Africa, and southwest China are important hotspots.

- Researchers are also interested in understanding which regions have a high proportion of **endemic species**—meaning species that are found in an area and nowhere else.

- In 2000, a team set out to identify regions of the world meeting two criteria:

  1. They contain at least 1500 endemic plant species.

  2. At least 70 percent of their traditional or primary vegetation has been lost.

- Although these areas cover only 1.4 percent of Earth's land area, they contain 44 percent of all known plant species and 35 percent of all known vertebrate species.

CHECK YOUR UNDERSTANDING

(b) Identify some regions considered hotspots because of their species richness.

_____

_____

_____

55.3 Threats to Biodiversity

Humans Have Affected Biodiversity throughout History

- Fossil evidence on islands in the South Pacific suggests that about 2000 bird species were wiped out as people colonized this area between about A.D. 400 and 1600.

- In 1722, when European settlers arrived on Easter Island, 1000 people lived there. The island was treeless. Soil samples analyzed confirmed that the island was once a lush forest.

Current Threats to Biodiversity

- Most extinctions that have occurred over the past 1000 years took place on islands and were caused by overhunting or the introduction of **exotic species**—nonnative competitors, diseases, or predators.

- **Endangered species,** which are almost certain to go extinct unless effective conservation programs are put in place, are now more likely on continents than on islands.

- A recent analysis on 488 endangered species in Canada shows:

  – Virtually all the endangered species are affected by more than one factor.

  – Habitat loss is the single most important factor in the decline of these species.

  – Overharvesting is the dominant problem for marine species, while pollution plays a large role for freshwater species.

  – Factors beyond human control can be important—background extinctions will continue to occur.

Habitat Destruction

- We cause **habitat destruction** by logging and burning forests, damming rivers, dredging and trawling the oceans, plowing prairies, grazing livestock, filling in wetlands, and excavating and extracting minerals to build housing developments, golf courses, shopping centers, office complexes, airports, and roads.

Habitat Fragmentation

- This activity turns large, contiguous areas of natural habitats into small isolated fragments. **Habitat fragmentation** concerns biologists for these reasons:

  1. Habitats can be too small to support some species.

  2. It reduces the ability for some animals to disperse.

  3. It creates a large amount of vulnerable edge habitat.

- When habitats are fragmented, the quality *and* quantity of habitat decline drastically.

An Introduction to Conservation Genetics (Box 55.2)

- When populations are fragmented into a metapopulation structure, gene flow between isolated groups is reduced or eliminated.

- Genetic drift is much more pronounced in small populations.

- Small populations become inbred, and inbreeding often leads to lowered fitness—the phenomenon known as inbreeding depression.

How Can Biologists Predict Future Extinction Rates?

Estimates Based on Direct Counts

- The best information comes from fossil birds.

- Background extinctions occur at a rate of about 1 extinction per million species alive per year.

- Over the past 1000 years, data suggest that birds have been going extinct at about 100 times the background rate.

- This trend appears to be similar for mammals, reptiles, amphibians, and conifers.

Species–Area Relationships

- **Species–area relationships** measure rates of habitat destruction to projected rates of species loss.

- Researchers compared satellite images of the Amazon Basin from 1978 and 1988. Their analysis showed that 5000 km^2 was deforested in the Amazon each year during this decade.

- Researchers have used species–area curves to analyze the number of bird species found on islands in the Bismarck Archipelago near New Guinea (**Figure 55.13**).

- Biologists want to understand how many species are currently threatened so they can estimate how many will go extinct in the near future.

S = 18.9 A^{0.15}

Slope of line

$S = 18.9\,A^{0.15}$

CHECK YOUR UNDERSTANDING

(c) Define habitat fragmentation, and explain why biologists are concerned about it.

_____

_____

_____

_____

55.4 Preserving Biodiversity

Why Is Biodiversity Important?

- **Ecosystem services** are processes that increase the quality of the abiotic environment.

- Economists are attempting to quantify the dollar value of ecosystem services as a way of justifying the cost of preserving natural areas, and biologists continue to document how the loss of biodiversity is affecting the quality of the abiotic environment.

Biodiversity Increases Productivity

- Many experiments have shown that species richness has a positive impact on **net primary productivity (NPP)**.

- Several mechanisms may be at work:
  - _Resource use efficiency._ Diverse assemblages of plant species make more efficient use of the sunlight.
  - _Facilitation._ Certain species facilitate the growth of other species by providing them with nutrients, partial shade, or other benefits.

- _Sampling effects._ In many habitats, it is common to observe that one or two species are extremely productive. Therefore, if the number of species in a study is low, the likelihood of missing the big producer is high.

Economic Benefits of Biodiversity (**Box 55.3**)

- Research programs collectively known as bioprospecting focus on assessing bacteria, archaea, plants, and fungi as novel sources of drugs or ingredients in consumer products.

- Agricultural scientists are preserving diverse strains of crop plants in seed banks and continue to use wild relatives of domesticated species in breeding programs aimed at improving crop traits.

- The production of almonds, apples, cherries, chocolate, alfalfa, and an array of other crops depends on the presence of wild pollinators.

- Strategies for cleaning up oil spills, abandoned mines, and contaminated industrial sites are incorporating bioremediation—the use of bacteria, archaea, and plants to metabolize pollutants and render them harmless.

- Recreation based on visiting wild places, or ecotourism, is a major industry internationally and is growing rapidly.

- High ecosystem diversity dramatically reduces flood damage and the danger posed by mudslides.

Does Biodiversity Lead to Stability?

- Stability of a community is the ability to:
  1. Withstand a disturbance without changing
  2. Recover to former levels of productivity or species richness after a disturbance
  3. Maintain productivity and other aspects of ecosystem function as conditions change over time

- **Resistance** is a measure of how much a community is affected by a disturbance. **Resilience** is a measure of how quickly a community recovers following a disturbance.

- Species richness has a strongly positive effect on how ecosystems function.

Designing Effective Protected Areas

- A Gap Analysis Program (GAP) analysis tries to identify gaps between geographic areas that are particularly rich in biodiversity and areas that are actually managed for conservation.

- The analysis revealed that many species' ranges occur completely outside any protected areas.

- It also revealed that the 11.5 percent of Earth's surface area that is now being managed for biodiversity will not be enough to conserve many species.
- The leading hypothesis in reserve design is to make sure that strips of undeveloped habitat, called **wildlife corridors**, connect populations that would otherwise be isolated.
- The goals of the corridors are as follows:
  1. Allow areas to be recolonized if a species was lost.
  2. Introduce new alleles that would counteract the deleterious effects of genetic drift and inbreeding.

CHECK YOUR UNDERSTANDING

(d) List at least two economic benefits of preserving biodiversity.

_____

_____

_____

_____

_____

B. INTEGRATING YOUR KNOWLEDGE

Looking Back—Key Concepts from Earlier Chapters

Extinction Rates—Chapter 27

Chapter 27 analyzed evidence that extinction rates are now accelerating to between 10 to 100 times the normal, or "background" extinction rates, indicating that a mass extinction event is under way.

Inbreeding—Chapter 25

Genetic factors play a role in the fate of populations. You may recall from **Chapter 25** that small populations tend to become inbred, and that inbreeding is usually detrimental to the fitness of organisms.

Animal Interaction—Chapter 51

Chapter 51 explored how organisms respond to environmental stimuli and how they interact with the environment; but here in Chapter 55, the focus is on large numbers of species, their diversity, and their conservation.

Population Dynamics—Chapter 52

Chapter 52 explored the dynamics of populations and metapopulations—how and why they grow or decline, and

how they change over time and space. Although that chapter considered populations in isolation, the reality is that individuals of different species constantly interact within a community, and communities within an ecosystem. As a result, the fate of an ecosystem may be tightly linked to the other species that share its habitat.

C. MASTERING CHALLENGING TOPICS

The most challenging aspect of this chapter is the concept of species–area relationships. Curves that are drawn from these relationships indicate how many species may be lost by the destruction of a particular size of habitat. Theoretical and conservation biologists derive these curves from complex data sets so those amateur biologists can calculate the effect of habitat loss on a particular species. The tricky part of these curves is that they are based on a logarithmic scale. If you are not familiar with logarithms, it may be a good idea to review them. If you are familiar with logarithms, this should help you in drawing and analyzing the species–area curves.

D. ASSESSING WHAT YOU'VE LEARNED

1. Research programs focusing on assessing bacteria, archaea, plants, and fungi as novel sources of drugs or ingredients in consumer products.
   a. bioremediation
   b. bioprospecting
   c. bioprocessing
   d. ecosystem services

2. Why are economists trying to quantify the dollar value of ecosystem services?
   a. because this allows them to sell abiotic components
   b. because this allows them to manage the area
   c. because this allows them to justify the cost of preservation
   d. because this allows them to prevent habitat loss

3. What is genetic diversity?
   a. the total genetic information contained within all individuals of a species
   b. the total phenotypic information contained within all individuals of a species
   c. the variety of life-forms on Earth
   d. the variety of biotic communities in a region along with abiotic components

4. An area with an exceptionally high species richness is called:
   a. an endemic area
   b. a protected area

c. a threatened area

d. a hotspot

5. What defines how much a community is affected by a disturbance?
   a. hardiness
   b. toughness
   c. resistance
   d. resilience

6. An exotic species that is introduced to a new area, spreads rapidly, and eliminates native species is called:
   a. an exotic species
   b. an immigrant species
   c. an invasive species
   d. a destructive species

7. Which of the following does not constitute habitat destruction?
   a. fishing
   b. logging
   c. burning forests
   d. filling in wetlands

8. Which of the following is not a concern regarding habitat fragmentation?
   a. Habitats can change type.
   b. Habitats can be too small to support some species.
   c. It reduces dispersal ability of some animals.
   d. It creates a large amount of vulnerable edge habitat.

9. Impact on other species from the loss of a different species is called:
   a. a linked effect
   b. a dependent effect
   c. a domino effect
   d. an edge effect

10. Which species descriptor implies that the species will almost certainly go extinct unless swift and decisive action is taken to protect it?
    a. threatened species
    b. red flag species
    c. extinct species
    d. endangered species

CHAPTER 55—ANSWER KEY

Check Your Understanding

(a) Genetic diversity—the total genetic information contained within all individuals of a species. Species diversity—the variety of life-forms on Earth. Ecosystem diversity—the variety of biotic communities in a region along with abiotic components.

(b) In terms of bird species richness, the Andes Mountains, the Amazon River basin, portions of East Africa, and southwest China are important hotspots.

(c) Habitat fragmentation is the transformation of large, continuous blocks of habitat into isolated parcels surrounded by pasture, cropland, or urban development. Habitat fragmentation concerns biologists because
1. Habitats can be too small to support some species.
2. It reduces the ability for some animals to disperse.
3. It creates a large amount of vulnerable edge habitat.

(d) Research programs collectively known as bioprospecting focus on assessing bacteria, archaea, plants, and fungi as novel sources of drugs or ingredients in consumer products. Agricultural scientists are preserving diverse strains of crop plants in seed banks and continue to use wild relatives of domesticated species in breeding programs aimed at improving crop traits. The production of almonds, apples, cherries, chocolate, alfalfa, and an array of other crops depends on the presence of wild pollinators. Strategies for cleaning up oil spills, abandoned mines, and contaminated industrial sites are incorporating bioremediation—the use of bacteria, archaea, and plants to metabolize pollutants and render them harmless. Recreation based on visiting wild places, or ecotourism, is a major industry internationally and is growing rapidly. High ecosystem diversity dramatically reduces flood damage and the danger posed by mudslides.

Assessing What You've Learned

1. b; 2. c; 3. a; 4. d; 5. c; 6. c; 7. a; 8. a; 9. c; 10. d

Notes

Notes

Notes

Notes

Notes

Notes

Notes

Notes

Notes

Notes

Notes

Notes